S0-CDW-981

Application of Biological Markers to Carcinogen Testing

ENVIRONMENTAL SCIENCE RESEARCH

Recent Volumes in this Series

A Continuation Order Plan is available for this series. A continuation order will bring delivery of each new volume immediately upon publication. Volumes are billed only upon actual shipment. For further information please contact the publisher.

Application of Biological Markers to Carcinogen Testing

Edited by

Harry A. Milman

United States Environmental Protection Agency
Washington, D.C.

and

Stewart Sell

University of Texas Health Science Center
Houston, Texas

With

Doris Balinsky, Claudio Basilico, T. Ming Chu,
Alexander Hollaender, K. Robert McIntire, Ralph Reisfeld,
Arthur C. Upton, and Elizabeth K. Weisburger

Technical Editor: Claire M. Wilson

PLENUM PRESS • NEW YORK AND LONDON

Library of Congress Cataloging in Publication Data

Symposium on the Application of Biological Markers to Carcinogen Testing (1982: Bethesda, Md.)
Application of biological markers to carcinogen testing.

(Environmental science research; v. 29)
"Proceedings of the Symposium on the Application of Biological Markers to Carcinogen Testing, held November 15-19, 1982, in Bethesda, Maryland."
Includes bibliographical references and index.
1. Carcinogenicity testing—Congresses. 2. Biological chemistry—Congresses. 3. Biological assay—Congresses. I. Milman, Harry A. II. Sell, Stewart. III. Balinsky, Doris. IV. Title. V. Series.
RC268.65.S96 1982 616.99′4071 83-19097
ISBN 0-306-41490-2

Proceedings of the Symposium on the Application of Biological Markers to Carcinogen Testing held November 15-19, 1982, in Bethesda, Maryland

This report has been reviewed in accordance with the U.S. Environmental Protection Agency's peer and administrative review policies and approved for presentation and publication. Approval does not signify that the contents necessarily reflect the views and policies of the U.S. Environmental Protection Agency, nor does mention of trade names or commercial products constitute endorsement or recommendation for use.

A Division of Plenum Publishing Corporation
233 Spring Street, New York, N.Y. 10013

Printed in the United States of America

ACKNOWLEDGEMENTS

On behalf of the Office of Toxic Substances, United States Environmental Protection Agency, I would like to extend my sincere thanks to the Scientific Organizing Committee composed of Dr. Stewart Sell (Chairman), Dr. Doris Balinsky, Dr. T. Ming Chu, Dr. K. Robert McIntire, Dr. Ralph Reisfeld, Dr. Arthur C. Upton, and Dr. Elizabeth K. Weisburger, and to all the participants for the success of the symposium on the "Application of Biological Markers to Carcinogen Testing" which was held in Bethesda, Maryland on November 15-19, 1982.

The excellent administrative support by Dr. Alexander Hollaender, Dr. Raymond Tice, and Ms. Claire Wilson of The Council for Research Planning in Biological Sciences at the Associated Universities, and Mr. Richard J. Burk, Jr. and his assistant, Ms. Diane Taub, of the Society & Association Services Corporation is greatly appreciated.

The financial support for the Symposium by the U.S. Environmental Protection Agency and the National Cancer Institute is gratefully acknowledged.

Harry A. Milman

CONTENTS

ENZYMES

ONCOFETAL ANTIGENS AND HORMONES

MONOCLONAL ANTIBODIES

CARCINOGEN-INDUCED MODIFICATION IN DNA

VIRUSES

ROUNDTABLE DISCUSSION

POSTERS

WELCOMING REMARKS

John A. Todhunter

Pesticides and Toxic Substances
U.S. Environmental Protection Agency
Washington, D.C. 20460

Welcome to the Office of Toxic Substances' Conference on the "Application of Biological Markers to Carcinogen Testing." This meeting marks the signal development in the evolution of the application of science at the Office of Toxic Substances of the United States Environmental Protection Agency (U.S. EPA). The tone of the conference can perhaps best be expressed by the words of Henry Pitot in the second edition of his Fundamentals of Oncology, "Although the production of neoplasia in animals at a statistically higher level than in controls has been considered indicative of carcinogenicity of the agent under study, modern concepts of the natural history of neoplastic development require that this simplistic evaluation of the data be reconsidered." In other words, it is time that we look beyond the "black box" approach to carcinogen testing and evaluation.

Undoubtedly the rodent model will remain at the core of bioassay programs for the detection of chemicals which have carcinogenic potential, but we must begin to use intelligently the scientific tools that are available to understand and unravel the "how" and the "why" of tumorgenic responses. Along this line, Weisburger and Williams have proposed a five-subcategory scheme for the classification of carcinogens, based largely on putative mechanisms whereby these carcinogenic agents may act. Certainly at this time there is not sufficient evidence on the mechanisms of carcinogenic action of chemicals for one to construct and employ definitive categories such as these, but it helps to point out the strong suggestive evidence that there are, in fact, different types of carcinogens and different modalities of tumorgenic induction. If we fail to use the scientific tools available to us, we will remain in what Phillipe Shubick has very aptly termed "the pre-history of carcinogenesis."

We at EPA remain committed to a preventative stance with respect to the evaluation of potential carcinogenic risks for the human population. We are eager to identify carcinogenic agents before they become recognized human problems. To do the best job possible we must commit ourselves to the intelligent use of the scientific tools that are available and help stimulate the development of tools that may become available in the future. In this way we will make better use of public resources to detect, to assess the risks, and, where necessary, to control the risks of carcinogenic substances.

High on that list of significant developments for the near term and future is the work on biological markers. We can expect that advances in the area of markers for carcinogenesis will help in the development of short-term screening procedures which will allow us to more rapidly, effectively, and efficiently select compounds for more detailed examination and long-term cancer bioassays. These investigations will probe the developmental sequences of carcinogenesis and shed light on potential mechanisms that underlie the development of neoplasms. Undoubtedly, biological markers will also be used more extensively in clinical medicine for a number of different applications.

I believe that science owes the public no less than to bring all of its available methodologies and insights to bear on the pressing, long-standing problem of assessment of tumorgenic responses and the risks which these responses may indicate for humans who are exposed to these agents.

I welcome you to the Conference. The program is bursting with excellent papers and presentations. Your participation, reflective consideration, and dialogue will bring this Conference to a fruitful and productive conclusion.

AN OVERVIEW OF CURRENT RESEARCH EFFORTS ON THE APPLICATION OF BIOLOGICAL MARKERS TO CARCINOGEN TESTING

Harry A. Milman

Health and Environmental Review Division
Office of Toxic Substances
U.S. Environmental Protection Agency
Washington, DC 20460

The Toxic Substances Control Act (TSCA) was enacted in 1976 on the assumption that: 1) there is substantial exposure of humans and the environment to chemicals; 2) that some of these exposures may present an unreasonable risk of injury; and 3) that existing laws do not fully protect against such injury. The U.S. Environmental Protection Agency (EPA) has a broad range of authorities under this law including screening new chemicals in commerce, requiring industry to test chemicals of concern, and assessing chemicals for control actions.

In carrying out its responsibilities under TSCA, EPA must assess the carcinogenic potential of certain chemicals. For example, examination of a new chemical under Section 5 of TSCA begins with a comprehensive evaluation of any available data on the chemical itself and related analogs. After such an evaluation it may be concluded that the new chemical may have the potential for carcinogenicity but that additional short-term or long-term testing is needed to verify this assumption.

The long-term animal bioassay in rodents is the best available method for detecting carcinogens, however, it is expensive, time consuming, and often provides ambiguous results. Efforts, therefore, are being expended to identify short-term assays with potential application for the identification of chemical carcinogens.

The etiology of cancer is believed to be multifactorial. In a simplified form, it can be envisioned that through the action of chemicals, radiation, viruses, hormones, or other initiating agents, normal cells may be altered. Such "initiated" cells may lie dormant or undergo expression and replication, and form tumors. Nutritional factors, stress, hormonal imbalance, aging, and the immune system all have been suggested to play an important role in the expression phase of carcinogenesis.

At the biochemical level, we can see that carcinogens, x-rays or alkylators can alter DNA to form DNA-adducts which, if the altered DNA is not repaired, will produce transformed genotypes. These, in turn, will be passed on to daughter cells during cellular replication. With continued cellular proliferation, tumors will be formed.

Biological changes which can be correlated with the carcinogenic process may be used as early indicators or markers of chemical carcinogenesis. For example, four areas of investigation with potential application to short-term testing for carcinogenicity have been identified: 1) tests based on correlating biochemical and immunological changes with the carcinogenic process (tumor markers); 2) tests based on correlating carcinogenicity with the ability of chemicals to cause mutations or to inhibit DNA repair; 3) tests based on the ability of chemicals to transform cells; and 4) tests based on the ability of chemicals to induce benign tumors which correlate with carcinogenicity (limited bioassays).

The search for a blood constituent which is useful as an early indicator of the onset of cancer has attracted the attention of biochemists for some time. This search has not been completely successful because of poor sensitivity of the methods employed and lack of specificity for neoplastic cells. A comprehensive review of the literature on biochemical and immunological markers of carcinogenicity was recently completed. This review identified over 120 markers including alphafetoprotein, CEA, pancreatic glycoprotein, and others. The markers under consideration fell into one of the following ten categories: 1) hepatic and renal enzymes; 2) enzymes of nucleic acid metabolism; 3) carbohydrate metabolizing enzymes; 4) glycotransferases; 5) glycosidases and blood carbohydrates; 6) modified nucleosides of ribonucleic acid; 7) glycoproteins and glycolipids; 8) immunological markers; 9) hormones; and 10) others. The report is now being reviewed by scientists at the Environmental Protection Agency (EPA) and elsewhere for the potential application of tumor markers to carcinogen testing. It is envisioned that following this review, and in conjunction with this symposium, recommendations will be made on the further validation of markers which appear to have potential utility in the identification of chemical carcinogens.

In the area of mutagenesis, the ability of selected mutagenicity and related assay systems to correlate with carcinogenic activity of chemicals is systematically being evaluated. The assays which have been selected for evaluation are based on bacterial and mammalian gene mutation, primary DNA damage, and chromosomal effects. These evaluations are being conducted by scientists representing academia, industry, and government working through a series of committees to examine the feasibility of using these methods in a pre-chronic testing battery for carcinogenicity.

In the area of cell transformation tests, several systems are currently under consideration. These fall into three basic types: 1) cell strain, those cells with a limited lifespan; 2) cell lines, those cells with an unlimited lifespan; and 3) oncogenic viral-chemical interactions involving cells. These tests are being examined by the appropriate committees of the GENE-TOX program of the EPA for potential use in a screening battery.

In the area of limited bioassays, a comprehensive review of the published literature on limited bioassays yielded over 20 different methods for consideration. Of these, five were selected for additional validation. These are: 1) the Sencar mouse skin tumorigenesis assay; 2) pulmonary tumor induction in strain A mice assay; 3) pulmonary tumor induction in newborn mice assay; 4) mammary tumor induction in female Sprague-Dawley rats assay; and 5) the induction of iron-resistant liver foci assay. When these methods were evaluated for overall accuracy for detecting chemicals which have been shown by the National Cancer Institute, the International Agency for Research on Cancer, or the EPA's Carcinogen Assessment Group to be animal carcinogens, all but the pulmonary tumor induction in strain A mice assay showed potential utility as a pre-screen for the long-term animal bioassay (>86% accuracy in detecting proven carcinogens). Increasing the number of chemicals being examined to include all chemicals judged to be carcinogenic by any investigator reduced the level of accuracy slightly, but again, only the strain A mouse bioassay was found not to be useful.

The utility of these methods in a short-term testing battery for carcinogenicity is still being investigated. Further validation of the methods is necessary before any definitive conclusion could be made.

In summary, biological markers of carcinogenicity may be applied to the investigation of four areas of short-term testing methodologies for carcinogens, namely, tumor markers, mutagenesis, cell transformation, and limited bioassays. I would like to propose that the goals for this symposium on biological markers of carcinogenicity be: 1) to review what is known in this area; 2) to identify areas of research and short-term assays which have potential application to carcinogen testing; and 3) to recommend future directions in research.

CHAIRMAN'S OVERVIEW ON *IN VIVO* TESTS

Elizabeth K. Weisburger

National Cancer Institute

Bethesda, Maryland 20205

For a number of years, there have been attempts to delineate biological indicators of the progression of the process of carcinogenesis, other than purely morphologic or histopathologic characteristics. Such indicators or "markers" would facilitate following the course of model experiments, might facilitate earlier intervention and treatment in clinical situations, and might be followed as an indicator that there was favorable or unfavorable response to treatment. Within the past decade, the increased emphasis in the area of markers has led to reports of antigenic and enzymic indicators of the presence of tumors. Although further research has not always supported these preliminary results, in other cases it has strengthened the case for the validity of the markers.

The purpose of this symposium is to provide information on advances in the identification and application of useful tumor markers. Coupled with this was the concept that consideration should be given to model tumor systems of relevance to humans, including intestinal and pancreatic cancer. In cancers of these organs, it often is the case that clinical indications of a neoplastic state often are not apparent until the tumor is too far advanced for surgical intervention. Thus, a need certainly exists for identifying markers which will provide a forewarning of the development and growth of a tumor.

The first session will begin with a presentation by Dr. Thomas Hamm, Chemical Industry Institute of Toxicology. Dr. Hamm has also been associated with the National Toxicology Program where he participated in administration of a bioassay program. Dr. Hamm will discuss the lack of specificity of tumor markers in a bioassay program and why histopathologic examination thus far remains the best diagnostic tool for carcinogenicity studies in animals.

One of the best studied model tumor systems in animals has been the induction of hepatocellular carcinoma in rats. Dr. Emmanuel Farber and associates from the University of Toronto will review the developments in this area and explain why no specific marker for neoplasia has yet been identified in the rat liver system, as well as the basis for their conclusions.

Dr. Daniel S. Longnecker, Dartmouth Medical School, will discuss the experimental rat pancreas model as well as morphologic and biochemical indicators of the neoplastic process in this system.

Human colon cancer represents a continuing problem to the clinician. Within the past 10 or 15 years, reliable and reproducible methods for inducing large bowel neoplasia in rodents have facilitated the identification of the stages of tumor development. Dr. Martin Lipkin, Memorial Sloan-Kettering Cancer Center, has been active in the study of experimentally induced large bowel cancer. In addition, he has applied basic research to the clinical situation, especially in humans with a hereditary predisposition to colorectal cancer. In the same field, that of intestinal cancer, Dr. Bandaru S. Reddy, American Health Foundation, will describe how the results from epidemiologic investigations led to the design and conduct of animal experiments which tend to substantiate the epidemiologic studies. Furthermore, the results of his experiments may point toward feasible dietary modifications for our population.

Skin cancer, although one of the most readily treated and curable forms of cancer, is also one of the most numerous types of tumors. Dr. Margaret L. Kripke, Frederick Cancer Research Facility, has developed a very relevant animal model, namely ultraviolet radiation-induced skin cancers in mice. She will present data on a UV radiation-associated antigen which may lead to interesting advances in this area.

THE OCCURRENCE OF NEOPLASMS IN LONG-TERM *IN VIVO* STUDIES

Thomas E. Hamm, Jr.

Toxicology Department
Chemical Industry Institute of Toxicology
Research Triangle Park, North Carolina 22077

INTRODUCTION

Long term *in vivo* carcinogenicity bioassays are usually designed to study the tumors which occur after administration of a chemical for at least two years to two species, usually rats and mice. The measurements made include body weights, clinical signs, hematology, serum chemistries, organ weights, and histopathology. These studies take from 3 to 5 years to complete and cost from ½ to 1½ million dollars at today's rates. The use of markers which could shorten the length of these studies, reduce the cost, or simplify the diagnosis of chemically induced neoplasms would be extremely useful. Any marker will have to be validated by comparing its use to the results of long-term *in vivo* studies.

It is my purpose to discuss the occurrence of tumors in two-year carcinogenicity bioassays to serve as a baseline of data for the chapters that follow. The National Cancer Institute (NCI) bioassays reported between 1977-1980 have been selected since they have several advantages. First, the B6C3F1 mouse was used in all the studies that included mice. This mouse was produced by commercial animal suppliers using parental lines from the National Institutes of Health (NIH) colonies. Rats were usually either F-344/Ns or Osborne Mendels (OM), also produced by NCI contractors. This minimized the effect of genetic differences between groups of animals in different bioassays. Second, both sexes were used so differences between the sexes can be evaluated. Third, husbandry procedures were standardized to minimize environmental variability within and between laboratories (1,2). Fourth, necropsy at a standardized time helped minimize variable tumor incidence caused by accelerated tumor formation in older animals. Finally, and most importantly, a large number of these similar bioassays are available for evaluation and comparison.

More information on the evaluation and design of these bioassays can be found in References 1 and 2, respectively. Animals of both sexes from each species were divided into three groups: Control, low dose, and high dose. Usually, each group consisted of 50 animals, although some earlier bioassays used only 20 control animals. The major route of chemical administration was oral. The high dose was selected to be the maximum tolerated dose (MTD) based on the results of short-term tests including a ninety day study. The MTD was defined as the highest dose that could be given that would not cause a significant decrease in survival from effects other than carcinogenicity. The ideal MTD was expected to cause no more than a 10% reduction in body weight gain and did not exceed 5% of the diet to minimize affecting the nutritive balance of the diet. The test period was 104-110 weeks after which all survivors were necropsied and examined for lesions.

This review was conducted using data in the original reports[1] and the paper in the International Agency for Research in Cancer (IARC) scientific publication #27 by Griesemer and Cueto (3). The paper by Griesemer and Cueto should be consulted for additional information about the interpretation and classification of these bioassay experiments.

One hundred ninety-two bioassays were available for review. In approximately one half (94 bioassays) there was no evidence (66 bioassays) or only equivocal evidence (28 bioassays) of any type of chemically induced tumor in either species. In 98 studies there was evidence of carcinogenicity in at least one sex of one species. These chemicals are called positive throughout the remainder of this chapter. Three of these positive studies were eliminated from consideration since rats only were used. Ten more studies were not considered since they were conducted, using B6C3F1 mice and Sprague-Dawley rats, by intraperitoneal injection of the test chemical and did not provide a large enough group of similar experiments to make useful comparisons. Thus, 85 bioassays were used for this review. The positive chemicals listed by tumor type or site are shown in Table 1.

One potential problem with the use of markers for <u>in vivo</u> studies is the high background incidence of tumors in control animals. Table 2 lists the incidence of those neoplasms occurring in more than 1% of control animals in NCI bioassays at sites where

---

[1]Reports may be obtained from the National Technical Information Service, U.S. Department of Labor, 5285 Port Royal Road, Springfield, Virginia 22161, U.S.A., (703) 487-4650.

Table 1. Site or type of tumor induced by "positive" NCI bioassay chemicals.

Site or Type	Number of	B6C3F1 Mice			F-344 Rats			OM Rats		
of Tumor	Positive Chemicals[a]	M[b]	F[c]	Both	M[b]	F[c]	Both	M[b]	F[c]	Both
1. Liver	54	31	43	24	14	12	11	2	2	1
2. Bladder	10	2	2	2	6	9	5			
3. Hemangiosaroma	10	7	5	5	2	2	0	2	0	0
4. Mammary Gland	9	0	2	0	0	6	0	0	2	0
5. Thyroid Gland	8	4	3	3	5	5	4			
6. Stomach	8	4	4	4	3	3	3	3	2	2
7. Lung	8	3	6	3	1	2	1	0	0	0
8. Kidney	7	2	0	0	4	2	2	2	1	1
9. Uterus	7		2			5			1	
10. Zymbal's Gland	7	0	1	0	6	7	6			
11. Skin and Skin Glands	5	0	0	0	5	2	2			
12. Lymphoma or Leukemia	5	1	1	0	3	1	1			
13. Subcutaneous Fibroma	4	0	0	0	2	1	0	1	0	0
14. Spleen	4	0	0	0	4	3	3			
15. Clitoral Gland	3		0			3				
16. Adrenal	3	1	1	1	2	0	0			
17. Colon	2	0	0	0	2	1	1			
18. Nasal Cavity	2				1	1	1	1	1	1
19. Pancreas	2	0	0	0	1	0	0	0	1	0
20. Sarcomas-Multiple Sites	2	0	0	0	2	2	2			
21. Harderian Gland	1	0	1	0	0	0	0			
22. Duodenum	1	1	0	0	0	0	0			
23. Mesothelioma	1	0	0	0	1	0	0			
24. Seminal Vesicle	1	1			0					
25. Ovary	1		1			0				

[a] A statistically increased incidence of neoplasms in at least one sex of one species.

[b] M = Males

[c] F = Females

neoplasms were induced by chemicals. Data on all neoplastic lesions found in control animals from these bioassays have been published for B6C3F1 mice (4), F-344 rats (5), and OM rats (6).

Table 2. Neoplasms occurring in more than 1% of NCI bioassay control animals at sites where neoplasms were induced by chemicals.

Site	Type	B6C3F1 Mice[a]		F-344 Rats[b]		OM Rats[c]	
		Males (2543)[d]	Females (2522)[d]	Males (1794)[d]	Females (1754)[d]	Males (975)[d]	Females (970)[d]
Liver	Hepatocellular Adenoma	7.9%	1.6	----	----	---	---
	Hepatocellular Carcinoma	13.7	2.3	----	----	---	---
	Neoplastic Nodules	-----	---	1.3	2.7		1.6
Circulatory System	Hemangiosarcoma	1.6	1.1	---	---	2.6	1.5
	Hemangioma	1.2	---	---	---	---	---
Mammary Gland	Adenocarcinoma	---	---	---	1.4	1.7	2.9
	Fibroadenoma	---	---	1.2	15.8	1.3	22.0
	Adenoma	---	---	---	1.3	---	---
Thyroid Gland	C-Cell Adenoma	---	---	3.8	4.0	1.8	3.3
	C-Cell Carcinoma	---	---	1.7	1.7	---	1.6
	Follicular Cell Adenoma	---	1.2	---	---	4.3	1.9
	Follicular Cell Carcinoma	---	---	---	---	2.8	1.5
Kidney	Lipomatous Tumor	---	---	---	---	1.9	2.4
Lung	Alveolar/Bronchiolar Adenoma	---	---	1.9	1.4	---	---
Uterus	Endometrial Stromal Polyp	NA[e]	---	NA	12.7	NA	4.5
	Carcinoma	NA	---	NA	1.5	NA	---
Skin	Fibroma	---	---	---	---	2.4	2.5
	Fibrosarcoma	---	---	---	---	1.9	---
Hematopoietic System	Leukemia/Lymphoma	8.3	16.3	12.3	9.9	2.1	1.9
Subcutis	Fibroma	---	---	2.6	---	1.7	3.4
Adrenal Gland	Cortical Adenoma	---	---	---	1.3	2.2	---
	Pheochromocytoma	---	---	8.8	3.1	---	---
Pancreas	Islet Cell Adenoma	---	---	3.5	1.0	1.9	1.4
Body Cavities	Mesothelioma	---	---	2.3	---	---	---

[a]Adapted from Reference 6, Ward *et al.*, JNCI, 1979

[b]Adapted from Reference 7, Goodman *et al.*, TAP, 1979

[c]Adapted from Reference 8, Goodman *et al.*, TAP, 1979

[d]Total number of control animals

[e]Not applicable; this tissue not present in males

Table 3 contains a list of the 20 chemicals which caused an increased incidence of heptocellular carcinomas in at least one sex of B6C3F1 mice or OM rats. Both sexes of mice had an increased incidence of hepatocellular carcinomas when exposed to 14 of 20 chemicals. Males only were positive in two bioassays (aldrin and tetrachlorvinphos) and females only were positive in two bioassays (chloramben and trifluralin). Neither sex of mice was positive in one bioassay (1,2-dibromoethane). Both sexes of OM rats were positive in one bioassay (chlordecone) and females only were positive in one bioassay (1,2-dibromoethane).

Table 4 contains a list of the chemicals which induced an increased incidence of hepatocellular carcinomas in at least one sex

Table 3. NCI bioassay chemicals which induced hepatocellular carcinomas in B6C3F1 mice or OM rats.

	Chemical	B6C3F1 Mice		OM Rats		NCI Carcinogenesis Technical Report Number
		M[a]	F[b]	M[a]	F[b]	
1.	Aldrin	+	-	-	-	21
2.	Chloramben	-	+	-	-	25
3.	Chlordane	+	+	-	-	8
4.	Chlordecone	+	+	+	+	None
5.	Chlorobenzilate	+	+	-	-	75
6.	Chloroform	+	+	-	-	None
7.	p,p'-DDE	+	+	-	-	131
8.	1,2-Dibromoethane	-	-	-	+	86
9.	Dicofol	+	-	-	-	90
10.	1,4-Dioxane	+	+	-	+[d]	80
11.	Heptachlor	+	+	-	-	9
12.	Hexachloroethane	+	+	-	-	68
13.	Nitrofen	+	+	0[c]	-	26
14.	1,1,2,2,-Tetrachloroethane	+	+	-	-	27
15.	Tetrachloroethylene	+	+	-	-	13
16.	Tetrachlorvinphos	+	-	-	-	33
17.	Toxaphene	+	+	-	-	37
18.	1,1,2-Trichloroethane	+	+	-	-	74
19.	Trichloroethylene	+	+	-	-	2
20.	Trifluralin	-	+	-	-	34

[a]M = Males
[b]F = Females
[c]0 = Inadequate study caused by poor survival
[d] = Adenomas significant, Adenocarcinomas not significant

of B6C3F1 mice or F-344 rats. Male mice had an increased incidence of that tumor in 10 of 26 bioassays, females in 20 of 26, and both sexes in 8 of 26. In one bioassay (1-amino-2-methylanthraquinone), female mice had an increased combined incidence of hepatocellular carcinomas and neoplastic nodules. Male rats had an increased incidence of hepatocellular carcinomas in 9 of 26 bioassays, females in 10 of 26, and both sexes in 8 of 26. F-344 rats had an increased incidence of a combination of neoplastic nodules and hepatocellular carcinomas in 4 of these 26 bioassays. Both sexes showed this result in two (2,4-diaminotoluene and 2,4,5-trimethylaniline) and males only in two (2-aminoanthraquinone and p-nitrosodiphenylamine). In one bioassay (p-cresidine) the females had an increased incidence

Table 4. NCI bioassay chemicals which induced hepatocellular carcinomas in B6C3F1 mice or F-344 rats.

	Chemical	B6C3F1 Mice		F-344 Rats		NCI Carcinogenesis Technical Report Number
		M[a]	F[b]	M[a]	F[b]	
1.	2-Aminoanthraquinone	+	+	+[c]	-	144
2.	3-Amino-9-ethylcarbazole HCl	+	+	+	+	93
3.	1-Amino-2-methylanthraquinone	-	+[c]	+	+	111
4.	4-Chloro-m-phenylenediamine	-	+	-	-	85
5.	4-Chloro-o-phenylenediamine	+	+	-	-	63
6.	5-Chloro-o-toluidine	+	+	-	-	187
7.	p-Cresidine	-	+	+[e]	+	142
8.	Cupferron	-	+	+	+	100
9.	2,4-Diaminotoluene	-	+	+[c]	+[c]	162
10.	Direct black 38	0	0[d]	+	+	108
11.	Direct blue 6	0	0	+	+	108
12.	Direct brown 95	0	0	-	+	108
13.	Hydrazobenzene	-	+	+	+	92
14.	Michler's ketone	-	+	+	+	181
15.	1,5-Naphthalenediamine	-	+	-	-	143
16.	5-Nitroacenaphthene	-	+	-	-	118
17.	5-Nitro-o-anisidine	-	+	-	-	127
18.	6-Nitrobenzimidazole	+	+	-	-	117
19.	Nitrofen	+	+	-	-	184
20.	p-Nitrosodiphenylamine	+	-	+[c]	-	190
21.	5-Nitro-o-toluidine	+	+	-	-	107
22.	Piperonyl sulfoxide	+	-	-	-	124
23.	Selenium sulfide	-	+	+	+	194
24.	4,4'-Thiodianiline	+	+	+	-	47
25.	2,4,5-Trimethylaniline	-	+	+[c]	+[c]	160
26.	Tris(2,3-dibromopropyl)-phosphate	-	+	-	-	76

[a]M = Males
[b]F = Females
[c] = A combination of neoplastic nodules and hepatocellular carcinomas was significant. Hepatocellular carcinomas alone were not statistically significant.
[d] = Inadequate since lasted only 13 weeks
[e] = A combination of neoplastic nodules, hepatocellular carcinomas and mixed carcinomas

of hepatocellular carcinomas and the males had an increased combined incidence of neoplastic nodules, hepatocellular carcinomas, and mixed hepatocellular/cholangiocarcinomas. Table 5 contains the list of the 8 chemicals which did not induce an increased incidence of hepatocellular carcinomas but did induce an increased incidence of the combination of hepatocellular adenomas and carcinomas.

Liver tumors in the B6C3F1 mouse were the most frequently diagnosed tumor (50 bioassays were positive in at least one sex of mice out of 179 total or 85 positive bioassays). They occurred in both sexes in 24 bioassays; in the male, only in 7; and in the female, only in 19. There is a high "background" incidence of liver tumors in this mouse cross. When data from the control B6C3F1 mice from 104 week or longer NCI bioassays is pooled, the incidence of liver tumors in the control animals is 31.3% (range of 17.8 to 46.9%) in 2,333 males and 6.4% (range of 2.4 to 8%) in 2,499 females (7). Tarone et al. (8), using data from the same bioassays, have shown

Table 5. NCI bioassay chemicals which induced an increased incidence of the combination of hepatocellular adenomas and carcinomas.

Chemical	B6C3F1 Mice M^a	B6C3F1 Mice F^b	F344 Rats M^a	F344 Rats F^b	NCI Carcinogenesis Technical Report Number
1. Cinnamyl anthranilate	+	+	-	-	196
2. 4,4'-Methylene-bis-(N,N-dimethyl)-benzeneamine	-	+	-	-	186
3. Nithiazide	+	-	-	-	146
4. 3'-Nitro-p-aceto-phenetide	+	-	-	-	133
5. 2-Nitro-p-phenylene-diamine	-	+	-	-	169
6. Phenazopyridine·HCl	-	+	-	-	99
7. o-Toluidine·HCl	-	+	-	-	153
8. 2,4,6-Trichlorophenol	+	+	-	-	155

[a]M = Males

[b]F = Females

significant inter—laboratory variation in the incidence of liver tumors in control B6C3F1 mice. Laboratory 1 had a mean of 40.1% (range 24-58%) for males and 9.7% (2-21%) for females, while the corresponding values for the other laboratories were: Laboratory 2, 31.3% (17-39%); 4.6% (2-10%); Laboratory 3, 25% (16-39%), 7.3% (0-13%); Laboratory 4, 32.2% (15-55%), 5.1% (0-21%); and Laboratory 5, 27.4% (7-45%), 4.8% (0-17%). For a more complete discussion of the B6C3F1 mouse liver tumor, see Reference 9.

Liver tumors were found in at least one sex of OM rats in only two bioassays. Liver tumors in F-344 rats were the second most common site occurring in 15 bioassays.

Bladder tumors and hemangiosarcomas were the next most frequently induced tumors (10 positives out of 179 total or 85 positive bioassays). Table 6 is a list of the 10 chemicals which induced bladder tumors in B6C3F1 mice or F-344 rats. All 10 chemicals induced bladder tumors in F-344 rats. Only two induced bladder tumors in mice. All chemicals listed induced transitional cell carcinomas. In two bioassays (p-cresidine and p-quinone dioxime), the incidence of tumors was significantly elevated only when all types of induced bladder carcinomas were combined.

Table 7 is the list of NCI bioassay chemicals which induced hemangiosarcomas in B6C3F1 mice, OM rats, or F-344 rats. Hemangiosarcomas were induced in both mice and rats in only one bioassay

Table 6. NCI bioassay chemicals which induced transitional cell carcinomas of the urinary bladder.

Chemical	B6C3F1 Mice M[a]	B6C3F1 Mice F[b]	F-344 Rats M[a]	F-344 Rats F[b]	NCI Carcinogenesis Technical Report Number
1. 4-Amino-2-nitrophenol	-	-	+	-	94
2. o-Anisidine·HCl	+	+	+	+	89
3. 4-Chloro-o-phenylenediamine	-	-	+	+	63
4. m-Cresidine	0[c]	-	+	+	105
5. p-Cresidine	+[d]	+[d]	+[d]	+[d]	142
6. Nitrilotriacetic acid, trisodium salt	-	-	-	+	6
7. Nitrilotriacetic acid	-	-	-	+	6
8. N-Nitrosodiphenylamine	-	-	+	+	164
9. p-Quinone dioxime			-	+[d]	179
10. o-Toluidine·HCl	-	-	-	+	153

[a]M = Males

[b]F = Females

[c]No data, inadequate study caused by poor survival of male mice only.

[d]A combination of all bladder carcinomas.

(cupferron). Rats were positive in 3 bioassays when mice were negative, and mice were positive in 7 bioassays when rats were negative.

Table 8 is a list of the 9 chemicals which induced mammary tumors. In four bioassays (dibromochloropropane, 1,2-dichloroethane, hydrazobenzene, and 5-nitroacenaphthene) adenocarcinomas were induced. In one bioassay (2,4-dinitrotoluene) fibroadenomas occurred. In one bioassay (2,4-diaminotoluene) a combination of carcinomas and adenomas were induced. In two bioassays (nithiazide and o-toluidine-HCl), a combination of different types of adenomas occurred, and in one bioassay (reserpine), a combination of malignant tumors occurred.

Table 9 is a list of the eight chemicals which induced follicular cell neoplasms in B6C3F1 mice or F-344 rats. In addition, 2,4-diaminoanisole sulfate induced C-cell neoplasms in male F-344 rats and 1,5-naphthalenediamine induced C-cell neoplasms in female B6C3F1 mice.

Table 10 is a list of the eight chemicals which induced tumors of the forestomach. Two of the chemicals [(3-(chloromethyl)pyridine-HCl and pivalolactone] were administered by gavage and caused

Table 7. NCI bioassay chemicals which induced hemangiosarcomas.

	Chemical	B6C3F1 Mice		OM Rats		F-344 Rats		NCI Carcinogenesis Technical Report Number
		M[a]	F[b]	M[a]	F[b]	M[a]	F[b]	
1.	Aniline·HCl	-	-			+	+	130
2.	1,2-Dibromoethane	-	-	+	-			86
3.	1,2-Dichloroethane	-	-	+	-			55
4.	5-Chloro-o-toluidine	+	+			-	-	187
5.	4-Chloro-o-toluidine·HCl	+	+			-	-	165
6.	Cupferron	+	+			+	+	100
7.	Michler's ketone	+	-			-	-	181
8.	Nitrofen	+	+	0[c]	-			26
9.	5-Nitro-o-toluidine	+	+			-	-	107
10.	o-Toluidine·HCl	+	-			-	-	153

[a]M = Males

[b]F = Females

[c]Inadequate study for males since there was poor survival.

Table 8. NCI bioassay chemicals which induced mammary tumors.

	Chemical	B6C3F1 Mice		OM Rats		F-344 Rats		NCI Carcinogenesis Technical Report Number
		M[a]	F[b]	M[a]	F[b]	M[a]	F[b]	
1.	2,4-Diaminotoluene	-	-			-	+	162
2.	Dibromochloropropane	-	-	-	+			28
3.	1,2-Dichloroethane	-	+	-	+			55
4.	2,4-Dinitrotoluene	-	-			-	+	54
5.	Hydrazobenzene	-	-			-	+	92
6.	Nithiazide	-	-			-	+	146
7.	5-Nitroacenaphthene	-	-			-	+	118
8.	Reserpine	-	+			-	-	193
9.	o-Toluidine·HCl	-	-			-	+	153

[a]M = Males

[b]F = Females

Table 9. NCI bioassay chemicals which induced thyroid tumors.

	Chemical	B6C3F1 Mice		OM Rats		F-344 Rats		NCI Carcinogenesis Technical Report Number
		M[a]	F[b]	M[a]	F[b]	M[a]	F[b]	
1.	3-Amino-4-ethoxy-acetanilide	+	-			-	-	112
2.	o-Anisidine·HCl	-	-			+	-	89
3.	2,4-Diaminoanisole sulfate	+	+			+	+	84
4.	N,N'-Diethylthiourea	-	-			+	+	149
5.	4,4'-Methylene-bis(N,N-dimethyl)-benzenamine	-	-			+	+	186
6.	1,5-Naphthalenediamine	+	+			-	-	143
7.	4,4'-Thiodianiline	+	+			+	+	47
8.	Trimethylthiourea	-	-			-	+	129

[a]M = Males

[b]F = Females

Table 10. NCI bioassay chemicals which induced tumors of the fore-stomach.

	Chemical	B6C3F1 Mice		OM Rats		F-344 Rats		NCI Carcinogenesis Technical Report Number
		M[a]	F[b]	M[a]	F[b]	M[a]	F[b]	
1.	3-(Chloromethyl)pyridine·HCl	+	+			±	-	95
2.	4-Chloro-o-phenylenediamine	-	-			+	+	63
3.	Cupferron	-	-			+	+	100
4.	Dibromochloropropane	+	+	+	+			28
5.	1,2-Dibromoethane	+	+	+	+			86
6.	1,2-Dichloroethane	-	-	+	-			55
7.	Pivalolactone	-	-			+	+	140
8.	Tris(2,3-dibromo-propyl)phosphate	+	+			-	-	76

[a]M = Males

[b]F = Females

Table 11. NCI bioassay chemicals which induced lung tumors.

	Chemical	B6C3F1 Mice M[a]	B6C3F1 Mice F[b]	OM Rats M[a]	OM Rats F[b]	F-344 Rats M[a]	F-344 Rats F[b]	NCI Carcinogenesis Technical Report Number
1.	1,2-Dibromoethane	+	+	-	-			86
2.	1,2-Dichloroethane	+	+	-	-			55
3.	1,5-Naphthalenediamine	-	+			-	-	143
4.	5-Nitroacenaphthene	-	-			+	+	118
5.	Selenium sulfide	-	+			-	-	194
6.	2,4,5-Trimethylaniline	-	-			-	+	160
7.	Trifluralin	-	+	-	-			34
8.	Tris(2,3-dibromopropyl)-phosphate	+	+			-	-	76

[a]M = Males

[b]F = Females

only tumors of the forestomach. Three (dibromochloropropane; 1,2-dibromoethane; and 1,2-dichloroethane) were administered by gavage and caused tumors of the forestomach and various other sites. The remaining three [4-chloro-o-phenylenediamine, cupferron, and tris(2,3-dibromopropyl) phosphate] were administered in feed and caused tumors of the forestomach and various other sites. Only two chemicals (dibromochloropropane and 1,2-dibromoethane) caused tumors of the forestomach in both rats and mice.

Table 11 is a list of the eight chemicals which induced lung tumors. Four chemicals (1,2-dibromoethane; 1,2-dichloroethane; 1,5-naphthalenediamine; and trifluralin) induced only adenomas, two chemicals (5-nitroacenaphthene and 2,4,5-trimethylaniline) induced carcinomas, and two chemicals [tris(2,3-dibromopropyl)phosphate and selenium sulfide] induced a combination of adenomas and carcinomas. None of these chemicals induced lung tumors in both rats and mice.

All other tumor sites or types occurred in fewer than 5% of the total bioassays (7 or fewer positives in 166 bioassays). These chemicals are listed in Table 12.

In summary, the following points should be emphasized:

1) When testing these chemicals at the maximum tolerated dose, approximately one half of the chemicals (94 of 192) did not cause an increase in tumors at any site in either sex of two rodent species.

Table 12. NCI bioassay chemicals which induced tumors in less than 5% of the total bioassays listed by site or type of tumor.

Chemical	B6C3F1 Mice M[a]	B6C3F1 Mice F[b]	OM Rats M[a]	OM Rats F[b]	F-344 Rats M[a]	F-344 Rats F[b]	NCI Carcinogenesis Technical Report Number
A. Kidney Tumors							
1. 1-Amino-2-methyl-anthraquinone	-	-			+	-	111
2. Chloroform	-	-	+	-			None
3. Chlorothalonil	-	-	+	+			41
4. Cinnamyl anthranilate	-	-			+	-	196
5. Nitrilotriacetic acid trisodium salt	-	-			+	+	6
6. Nitrilotriacetic acid	+	-			-	-	6
7. Tris(2,3-dibromopropyl)-phosphate	+	-			+	+	76
B. Uterine Tumors							
1. 3 Amino-9-ethylcarbazole·HCl		-			0^c	+	93
2. Diaminozide	0^c	+			0^c	+	83
3. 1,2-Dichloroethane	0^c	+	0^c	+			55
4. 3,3'-Dimethoxy-benzidine-4,4'-diisocyanate	0^c	-			0^c	+	128
5. 1,5-Naphthalenediamine	0^c	-			0^c	+	143
6. 4,4'-Thiodianiline	0^c	-			0^c	+	47
7. Trimethyl phosphate	0^c	+			0^c	-	81
C. Zymbal's Gland Tumors							
1. 3-Amino-9-ethyl-carbazole·HCl	-	-			+	+	93
2. Cupferron	-	+			-	+	100
3. 3,3'-Dimethoxy-benzidine-4,4'-diisocyanate	-	-			+	+	128
4. Hydrazobenzene	-	-			+	+	92
5. 5-Nitroacenaphthene	-	-			+	+	118
6. 5-Nitro-o-anisidine	-	-			+	+	127
7. 4,4'-Thiodianiline	-	-			+	+	47

(Table 12. continued)

Chemical	B6C3F1 Mice		OM Rats		F-344 Rats		NCI Carcinogenesis Technical Report Number
	M[a]	F[b]	M[a]	F[b]	M[a]	F[b]	
D. Tumors of the Skin & Skin Glands							
1. 3-Amino-9-ethylcarbazole·HCl	-	-			+	-	93
2. 2,4-Diaminoanisole sulfate	-	-			+	+	
3. 3,3'-Dimethoxybenzidine-4,4'-diisocyanate	-	-			+	-	128
4. 2,4-Dinitrotoluene	-	-			+	-	54
5. 5-Nitro-o-anisidine					+	+	127
E. Lymphoma or Leukemia							
1. 2-Aminoanthraquinone	-	+			-	-	144
2. 2-Amino-5-nitrothiazole	-	-			+	-	53
3. C. I. Vat Yellow	+	-			-	-	134
4. 3,3'-Dimethoxybenzidine-4,4'-diisocyanate	-	-			+	+	128
5. 2,4,6-Trichlorophenol	-	-			+	-	155
F. Subcutaneous Fibroma							
1. 1,2-Dichloroethane	-	-	+	-			55
2. 1,2-Diaminotoluene	-	-			-	+	162
3. o-Toluidine·HCl	-	-			+	-	153
4. Trimethyl phosphate	-	-			+	-	81
G. Spleen (except hemangiosarcomas)							
1. Aniline·HCl	-	-			+	+	130
2. Azobenzene	-	-			+	+	154
3. Dapsone	-	-			+	-	20
4. o-Toluidine·HCl	-	-			+	+	153
H. Clitoral Gland/Preputial Gland							
1. 1,5-Naphthalene-diamine	-	-			-	+	143
2. 5-Nitroacenaphthene	-	-			-	+	118
3. 5-Nitro-o-anisidine	-	-			-	+	127
I. Adrenal Medulla							
1. 4-Chloro-m-phenylene-diamine	-	-			+	-	85
2. Reserpine	-	-			+	-	193
3. 1,1,2-Trichloroethane	+	+	-	-			74
J. Colon							
1. Phenazopyridine·HCl	-	-			+	+	99
2. 4,4'-Thiodianiline	-	-			+	-	47

(Table 12, continued)

Chemical	B6C3F1 Mice M[a]	B6C3F1 Mice F[b]	OM Rats M[a]	OM Rats F[b]	F-344 Rats M[a]	F-344 Rats F[b]	NCI Carcinogenesis Technical Report Number
K. Nasal Cavity							
1. 1,4-Dioxane	-	-	+	+			80
2. p-Cresidine	-	-			+	+	142
L. Pancreatic Tumors							
1. Nitrofen	-	-	I[d]	+			26
2. Cinnamyl anthranilate	-	-			+	-	196
M. Sarcomas, Multiple Sites							
1. Aniline·HCl	-	-			+	+	130
2. Azobenzene	-	-			+	+	154
N. Harderian Gland							
1. Cupferron	-	+			-	-	100
O. Duodenum							
1. Captan	+	-	-	-			15
P. Mesothelioma							
1. o-Toluidine·HCl	-	-			+	-	153
Q. Seminal Vesicle							
1. Reserpine	+				-	I[d]	193
R. Ovary							
1. 5-Nitroacenaphthene		+				-	118

[a]M = Males

[b]F = Females

[c]O = Tissue not present in that sex

[d]I = Inadequate study in that sex since poor survival

2) Of the 85 chemicals that were carcinogenic in at least one sex of one species, almost 60% (50) caused liver tumors in the B6C3F1 mouse. This site was positive in 29% (50 of 179) of the total bioassays reviewed.

3) The B6C3F1 mouse has a high incidence of liver tumors in control animals, especially in the male. This incidence varies significantly between control groups in different laboratories.

4) No other tumor site or type occurred in more than 6% (10 of 179) of total bioassays.

REFERENCES

1. Chu, K.C., C. Cueto, Jr., and J.M. Ward (1981) Factors in the evaluation of 200 National Cancer Institute Bioassays. J. Tox. & Env. Hlth. 8:251-280.
2. Sontag, J.M., N.P. Page, and U. Saffioti (1976) Guidelines for carcinogen bioassay in small rodents, NCI-CG-TR-1 U.S. DHEW, Stock No. 17-042-00118-8. Supt. of D.C., U.S. Govt. Printing Office, Washington, D.C.
3. Griesemer, R.A., and C. Cueto (1980) Toward a classification scheme for degrees of experimental evidence for the carcinogenicity of chemicals for animals. In Molecular and Cellular Aspects of Carcinogen Screening Tests, R. Montesano, H. Bartsch, and L. Tomatis, eds. IARC Scientific Publication #27, Lyon.
4. Ward, J.M., D.G. Goodman, R.A. Squire, K.C. Chu, and M.S. Linhart (1979) Neoplastic and non-neoplastic lesions in aging C57BL/6XC3H/HeN/F_1 (B6C3F1) mice. JNCI 63:849-853.
5. Goodman, D.G., J.M. Ward, R.A. Squire, K.C. Chu, and M.S. Linhart (1979) Neoplastic and non-neooplastic lesions in aging F-344 rats. Tox. & Appl. Pharm. 48:237-248.
6. Goodman, D.G., J.M. Ward, R.A. Squire, M.B. Paxton, W.D. Reichardt, K.C. Chu, and M.S. Linhart (1980) Neoplastic and non-neoplastic lesions in aging Osborne-Mendel rats. Tox. & Appl. Pharm. 55:433-447.
7. Lee, J. (1980) Pers. comm.
8. Tarone, R.E., K.C. Chu, and J.M. Ward (1981) Variability in the rates of some commonly occurring tumors in Fischer-344 rats and (C57BL/6NXC3H/HeN)F1 (B6C3F1) mice. JNCI 66:1175-1181.
9. Hamm, Jr., T.E. Occurrence of mouse liver neoplasms in the National Cancer Institute Bioassays. In Current Perspectives in Mouse Liver Neoplasia, J.A. Popp, ed. Hemisphere Publishing Corporation, New York (in press).

MARKERS OF LIVER NEOPLASIA - REAL OR FICTIONAL?

M. Tatematsu, T. Kaku, A. Medline, L. Eriksson,
W. Roomi, R.N. Sharma, R.K. Murray, and E. Farber

Departments of Pathology and Biochemistry
University of Toronto
Toronto, Ontario, Canada M5G 1L5

INTRODUCTION

The development of hepatocellular carcinoma in rats by chemicals is a long-term process in which discrete focal populations of new hepatocytes and other liver cells appear regularly and quite reproducibly. Included among the cellular changes are "oval cell proliferation" (1,2), foci or islands of altered hepatocytes (3-6), hepatocyte nodules ("hyperplastic" or "neoplastic" nodules or "adenomas") (7,8), and, later, nodules (persistent nodules) that appear to be closely associated with the precancerous steps of hepatocellular carcinoma (9-12). As with any cells showing new biological behavior, these new hepatocytes and other liver cells are associated with altered phenotypic expression. These are recognized as altered "phenotypic markers" during the carcinogenic process.

What is their significance and utility in the analysis of carcinogenesis and in the development of reliable short term indices of the potential carcinogenicity of environmental chemicals? I would like to discuss these from three points of view: (a) the use of phenotypic markers to identify and characterize new populations at different steps or stages in carcinogenesis; (b) the use of phenotypic markers to quantitate new populations with different properties; and (c) the use of phenotypic markers in the attempt to identify cells of origin of cancer, i.e., in the study of possible precursor-product relationships of cells.

PHENOTYPIC MARKERS TO IDENTIFY NEW POPULATIONS

An increasing number of markers have been described for some of the new hepatocyte populations that appear during carcinogenesis (4-6, 9-11, 13) (Tables 1-5). The most widely used are the histochemical enzyme methods, especially γ-glutamyl transferase (γ-GT), glucose-6-phosphatase (G-6-Pase), and nucleotide polyphosphatase ("ATPase"). The γ-GT is "positive" (13) (much stronger in the new population), while the other two are "negative" (much weaker in the new population).

So far, no phenotypic enzyme marker can be considered as being specific for liver neoplasia or for the new hepatocyte populations that are seen during the carcinogenic process. For example, γ-GT, G-6-Pase, and ATPase have received the most attention. In the resistant hepatocyte model (6,13,14), 90-95% of hepatocyte nodules are positive for γ-GT as well as for another enzyme, DT-diaphorase (DTD). The majority of the nodules also show a loss of G-6-Pase and ATPase. Early prenodular foci are positive for γ-GT and DTD to the same degree. However, the loss of G-6-Pase and ATPase at earlier time periods is much less consistent (9). A similar dissociation, in principle, between the staining for γ-GT, G-6-Pase, and ATPase is also seen in two other models of liver carcinogenesis (15,16).

It is well known that hepatocytes lose G-6-Pase and ATPase as they undergo irreversible cell damage leading to cell death with many hepatotoxins (17). It may be difficult, therefore, to distinguish between reversible or irreversible hepatocyte injury and foci of new hepatocytes undergoing expansion during the early phases of liver carcinogenesis.

From our own experience over the past 5 years, it is clearly evident that γ-GT is "turned on" in hepatocytes, especially in

Table 1. Architectural and vascular "markers" during liver carcinogenesis.

1. Hepatocytes in hepatocyte nodules ("hyperplastic" or "neoplastic" nodules or "adenomas") arranged in two or more cell thick plates and in glandular formations (10,11,22).

2. Hepatocyte nodules and hepatocellular carcinomas show a decrease in portal blood supply with a relative increase in arterial blood supply (see ref. 24).

Table 2. Positive biochemical-enzymatic markers during liver carcinogenesis.

1. α-Fetoprotein (25-27).
2. γ-Glutamyl transferase (28,29).
3. DT-Diaphorase (30,31).
4. Preneoplastic antigen (32,33).
5. Epoxide hydrolase (33,34).
6. Esterase (35,36).
7. β-Subunit-chorionic gonadotropin (HCG) (37).
8. Glutathione (see ref. 23).
9. Glutathione-S-transferase (see ref. 23).
10. UDP-Glucuronyl transferase (see ref. 23).
11. Aldehyde dehydrogenase isozyme (see ref. 48).

zone 1, under a variety of circumstances apparently unrelated to hepatocarcinogenesis (Table 6). For example, exposure "briefly" to doses of ethanol, methionine, adenine, or other compounds, or the intraportal injection of saline or hepatocytes from normal donors are associated with an intense staining reaction for γ-GT in the hepatocytes in zone 1 and sometimes extending into zone 2. This staining reaction may last for several weeks. However, it is not accompanied by the appearance of foci of altered hepatocytes that are frequently seen after initiation of liver carcinogenesis and that are thought to be relevant to some aspect of carcinogenesis (4-6, 9, 14-16, 18). Thus, considerable care has to be exercised in the use of enzyme histochemistry in the identification and study of foci of altered hepatocytes during carcinogenesis.

It should also be emphasized that in only one model, the resistant hepatocyte model for carcinogenesis, has it been shown that the foci of altered hepatocytes expand into hepatocyte nodules (9,19). In this model, the expansion of foci into nodules appears to be essential for cancer development and no cancer without the earlier occurrence of nodules has been found (19,20). Although the opinion has been expressed that foci of altered hepatocytes may conceivably act as direct precursors for liver cancer without their expansion to nodules (21), no objective data in support of this possibility has been presented.

The most widely used "marker" for liver neoplasia is alphafetoprotein (AFP). This is present in a significant number of liver

Table 3. Negative biochemical-enzymatic markers during liver carcinogenesis.

1. Glucose-6-phosphatase (see ref. 56).
2. Nucleotide polyphosphatase ("ATPase") (see refs.5,6).
3. β-Glucuronidase (38).
4. Serine dehydratase (38).
5. Glycogen phosphorylase (39).
6. Ribonucleases and deoxyribonucleases (40,41).
7. Loss of concentration ability for iron (21,42).
8. Retention of glycogen on fasting (39,43).
9. Cytochromes P450 (see ref. 23).
10. Aminopyrine N-demethylase, aryl hydrocarbon hydroxylase, other microsomal drug metabolizing enzymes (see ref. 23).

cancers in experimental animals and in humans (25-27). Since the serum levels of AFP increase very much in some patients with liver cancer, it is most useful as a guide to diagnosis and recurrence.

However, as discussed below, there are serious problems in the use of AFP as a marker for early changes in carcinogenesis. As with almost every other biochemical "marker", the "appearance" of AFP by histochemical or quantitative methods is by no means specific for cellular or phenotypic changes in liver carcinogenesis (2, 25-27).

Serum levels of useful phenotypes have not been exploited sufficiently, except with very few methods. A potentially useful serum "marker" for liver carcinogenesis is γ-GT. The activity of this enzyme is markedly elevated during liver carcinogenesis, including late in the process when cancer is already present (64).

So far, very few phenotypic markers appear to be characteristic of hepatocyte foci or nodules and of hepatocellular carcinoma. The most specific to date has been the architectural arrangement of the hepatocytes in these lesions (Table 1). Instead of being arranged as single cell plates, hepatocytes in nodules are arranged in plates two or more cells thick or in glandular patterns. Commonly associated with these changes are hypertrophied cytoplasm of the hepatocytes related to proliferation of the endoplasmic reticulum and an obvious "loosening" of the chromatin structure of the hepatocyte nuclei (Table 4).

The hepatocytes in the majority of hepatocellular carcinomas also show new patterns of organization including trabeculae several

Table 4. Cytologic "markers" during liver carcinogenesis.

1. Nuclei: Hepatocyte nuclei in hepatocyte nodules show loose chromatin organization with apparent decrease in condensed chromatin (19,44).
2. Cytoplasm: Abundant smooth endoplasmic reticulum with altered enzyme pattern (7,9,10,13).
3. Plasma membrane: Cell separation with irregular bile canaliculi and long microvilli (10,13).
4. Peroxisomes: Progressive change in inducibility of peroxisomes by clofibrate (49).

cells thick and a special vascular pattern. These have been described many times and need not be discussed further (see 22 for references).

Another highly characteristic pattern of phenotypic change is the appearance of what we call the "resistance phenotype" (23) (Tables 2,3, and 5). This is a special set of biochemical changes in the microsomes and in the cytosol that can be easily related to the acquisition by nodule hepatocytes of resistance to some cytoxic effects of hepatotoxins. Using nodules induced by several different carcinogens and promoted by several procedures, all the nodules show a large decrease in phase 1 enzymes, i.e., enzymes in the microsomes that activate carcinogens and other xenobiotic agents, a large increase in glutathione, in a glutathione-S-transferase, in γ-GT, in a microsomal glucuronyl transferase, in DT-diaphorase (cytosolic), in epoxide hydrolase (microsomal), and a large decrease in the production of DNA adducts with 2-AAF and with dimethylnitrosamine. All of these changes are compatible with a large overall decrease in the cytotoxic effects of xenobiotics on hepatocyte nodules as compared to normal or surrounding liver (7,13,14,19,45). In our opinion, the acquisition of resistance occurs at the time of or shortly after initiation of liver carcinogenesis and is a biochemical-physiological pattern or complex that is used to promote or select the initiated hepatocytes to generate the first crop of hepatocyte nodules (19,23,45). If this view is valid, the "resistance phenotype" becomes the key set of markers that is critical for the further evolution to cancer, i.e., it is the major marker for the first phase of carcinogenesis that generates the hepatocyte nodules (46).

Recently, we have been studying the pattern of polypeptides in the cytosol of hepatocyte nodules induced by one of several models

Table 5. Physiological or functional "markers" during liver carcinogenesis.

1. Resistance Phenotype - microsomal and cytosolic components (see ref. 23).

2. Resistance to inhibitory effect of dietary 2-AAF and other carcinogens (19,45,47).

of liver carcinogenesis. The polypeptides of the cytosols of hepatocyte nodules produced according to the resistant hepatocyte model (RH) were analyzed using SDS-PAGE. Approximately 40 polypeptides were separated in the gel system used. The pattern of the nodules (N = 8) showed a series of differences from that of normal rat liver: a marked decrease in the amounts of polypeptides of 12, 26, and 44 Kd, an increase of polypeptides of 22, 27, and 90 Kd, and a major increase of a polypeptide of 21 Kd. Cytosols from nodules generated by the Peraino model [2-AAF and phenobarbital (PB) (n = 2)], the Pitot model [diethylnitrosamine (DEN) and PB] (n = 3), and the Lombardi-Shinozuka model [DEN and choline-methionine deficient (CD) diet] (n = 5) showed the same pattern of alterations. Liver cytosols from adult rats treated with DEN plus partial hepatectomy (PH) without 2-AAF (n = 2), 2-AAF, PB and CD (n = 2), PB alone or 3-methylcholanthrene (n = 4), from fetal and neonatal rats (n = 16) and from rats subjected to PH (n = 10) were quite different. We conclude that nodular cells, independent of their mode of production, may show a pattern of cytosolic polypeptides useful as an early indicator of liver carcinogenesis.

PHENOTYPIC MARKERS TO QUANTITATE FOCI AND NODULES OF ALTERED HEPATOCYTES

The availability of histochemical enzyme methods has allowed the quantitation of early (microscopic) foci or islands of altered hepatocytes as well as larger grossly visible nodules. This approach is being used increasingly to develop rapid assay procedures for the potential carcinogenicity of chemicals of diverse structure (i.e., 5,6,15,16,18,21).

There are important technical considerations that must be used in the accurate measurement of small focal changes in any two-dimensional representation of a three-dimensional process. These have been considered in great detail by Scherer et al. (50) and Pitot and co-workers (51). These calculations are particularly important when

Table 6. Increase in intensity of staining for γ-glutamyl transferase in cell populations after various treatments.

Treatment[a]	Site at which γ-GT is-"turned on"		
	Oval Cell Proliferation	Zone 1 Hepatocytes	Random Foci of Hepatocytes
Carcinogen plus selection or promotion	+	±	+
Ethanol (3 x daily)	-	+	-
Adenine (3 x daily)	-	+	-
Methionine (3 x daily)	-	+	-
Selection with 2-AAF + PH without initiation	+	+	-
Diethylstilbesterol (1x)	-	+	-
Tomoxefen (1x)	-	+	-
α-Hexachlorocyclohexane (2 x daily)	-	+	-
Cyproterone acetate (2 x daily)	-	+	-
Cortisone (6 x daily)	-	+	-
Intraportal injection of normal hepatocytes	-	+	-

[a] In most instances, administered as a single parenteral dose daily for the number of days indicated.

the foci or focal lesions induced are widely different in diameters under the different conditions being compared and are minimized when these differences are small.

Although focal collections of altered hepatocytes are commonly seen with many carcinogens (18,52), there is no positive evidence that these lesions are early precursors for the ultimate development of liver cancer except for one model (19). Also, as already indicated, no evidence is available concerning the precursor role of islands or foci for cancer in the absence of hepatocyte nodules as essential intermediaries.

PHENOTYPIC MARKERS IN THE STUDY OF POSSIBLE PRECURSOR CELLS FOR CANCER

Fundamental to our knowledge of any malignant neoplasm is the identification of the cell or cells of origin and the formulation of the possible sequence(s) of biological, biochemical, and structural changes that initiated cells undergo during the pathogenesis of cancer.

In the case of hepatocellular carcinoma in the rat initiated with diethylnitrosamine (DEN) and promoted with dietary 2-acetylaminofluorene (2-AAF) plus partial hepatectomy (PH), one overall sequence has been established by observation of a direct cellular continuity - carcinogen-induced resistant hepatocyte, hyperplastic liver cell nodules, a small subset of these nodules (persistent nodules) (12), and cancer (19). Findings consistent with this conclusion have been reported with aramite (53), ethionine (39,44), 3'-methyl-4-dimethylaminoazobenzene (3'-Me-DAB) (35), and 2-AAF (16).

A second possible sequence, much less well defined, has been suggested to arise from a subset of oval cells (2,54-62). Proliferation of oval cells is commonly seen early in hepatocarcinogenesis with many carcinogens (see 2,14,22 for references). With one group of carcinogens, the azo dyes, oval cells can be seen to undergo an apparent morphologic transition to hepatocytes (1,54,57,63,65,66) and such a transition has been proposed as a possible early link in a sequence of changes during the development of hepatocellular carcinoma. However, no distinctive or characteristic lesions between these early changes and the ultimate appearance of liver cell cancer other than the hyperplastic liver cell nodules (2,14,22) have been proposed by any investigator. The major evidence in support of the oval cell hypothesis of the origin of liver cancer, in addition to the early transition of oval cells to hepatocytes in some azo dye models, is the finding of several phenotypic markers in common between oval cells or subsets of oval cells and some hepatocellular carcinoma. Alpha-fetoprotein (AFP) (2,27,56,59,60,67-69), γ-glutamyltransferase (γ-GT) (9,15,16,19,29,58,70), and fetal aldolase isozymes (67) are among the many markers studied.

As is evident from the distribution of so-called markers for cell identification in the liver (Table 7), phenotypic markers are obviously unreliable as indices for studies of the conversion of one cell type to another, especially during complex alterations in cellular metabolism that accompany many disease processes. This may also be the case during normal development, differentiation, aging, etc. Specifically, in liver carcinogenesis in the rat (Table 7), the random selection of phenotypic markers can support many different interpretations of histogenesis. Thus, the use of many histochemical markers in rat liver during hepatocarcinogenesis highlights the general unreliability of phenotypic markers in the study of the possible origin of new cell populations during normal development or in disease, including hepatocellular carcinoma. The challenge clearly becomes the search for much more reliable genotypic or chromosomal alterations that might be less subject to metabolic modulation.

GENERAL CONSIDERATIONS

The past two decades have seen the identification of many

Table 7. Phenotypic "markers" for different cell types in normal liver and during hepatocarcinogenesis.

Marker	Normal Hepatocytes	Normal Bile Duct Epithelium	Carcinogenesis Oval Cells	Carcinogenesis Hyperplastic Nodules	Carcinogenesis Hepatocellular Carcinoma	Reference
Keratin	-	+	+	-	-	
γ-GT	-	+	+	+	+	9,15,29,58, 70
Glucose-6-phosphatase	+	+	+	-	-	3,26,38,71, 73,74
Epoxide Hydrolase	+	-	-	+	+	34,75-78
DT-diaphorase	+	-	-	+	+	9,31
Albumin	+	-	+	+	+ or -	27,62,79
ATP-ase	+	-	-	-	-	3,9,71,72, 74
β-Glucuronidase	+	-	-	-		38
Serine Dehydratase	+	-	-	-		38
Aldolase B*	+	-	+	-	+	67
Aldolase A or C*	-	-	+	-	+	67
Deoxyribonuclease	+	-	-	-		40,41
α-Fetoprotein	-	-	+	-	+ or -	27,56,59, 60,62,67, 68,79
Esterase	-	-	-	+		35,36
Retention of Glycogen on fasting	-	-	-	+		39,43,81
Concentrating ability for iron	+	-	-	-		21,42

* In normal liver, B in hepatocytes and in sinusoidal lining cells. During carcinogenesis with 3'-methyl-4-dimethylaminoazobenzene, A and C were found in small "hepatocytes" but not in regular ones.

phenotypic changes in the cells of the focal lesions considered to be highly relevant to the development of liver cancer. Many of these have proven to be useful in the quantitation of some of the changes. However, it is becoming evident that their use must be tempered by a knowledge and understanding of the origin and meaning of the marker at hand.

Some phenotypic markers, especially those that form a coherent package or pattern are the most interesting from a mechanistic point of view. For example, the architectural pattern and the biochemical programming of nodules appears to be highly characteristic and probably diagnostic for this lesion. Other markers are much less well understood and even though valuable, must be used with due caution. Based upon the results to date, there is every indication that many more phenotypic markers are there to be found.

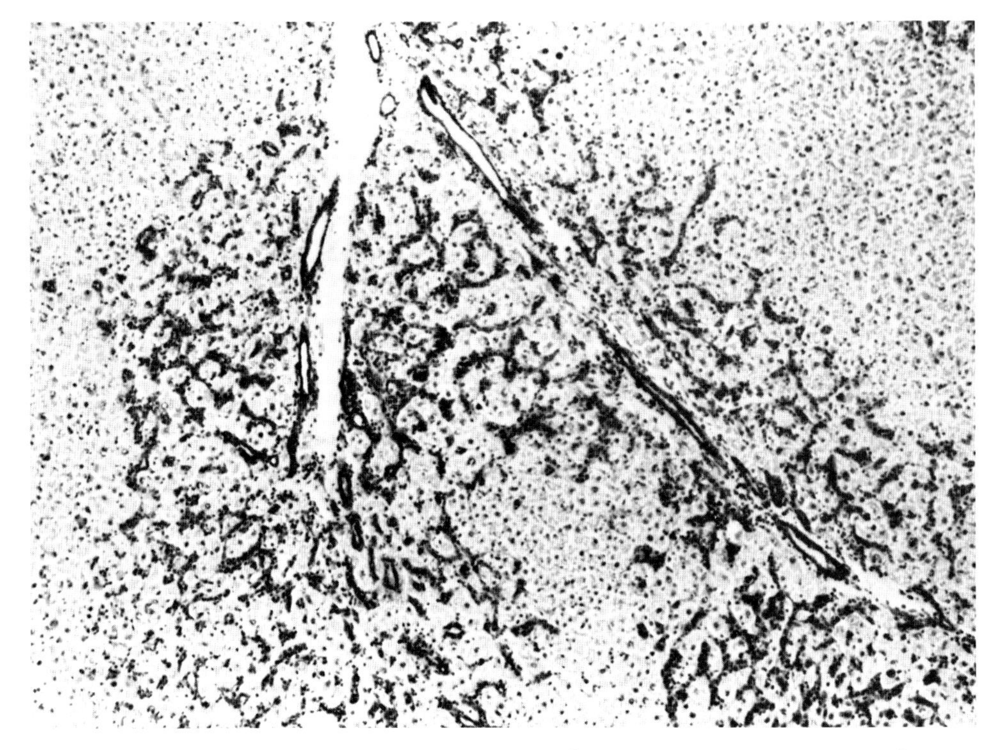

Fig. 1. Section of liver stained for keratin by immunoperoxidase methods. The animal was treated with dietary 2-acetylaminofluorene plus partial hepatectomy to induce a vigorous "oval cell proliferation". Note the strong labelling of the proliferating ductular and oval cells and the negative staining of the liver parenchymal cell (X40).

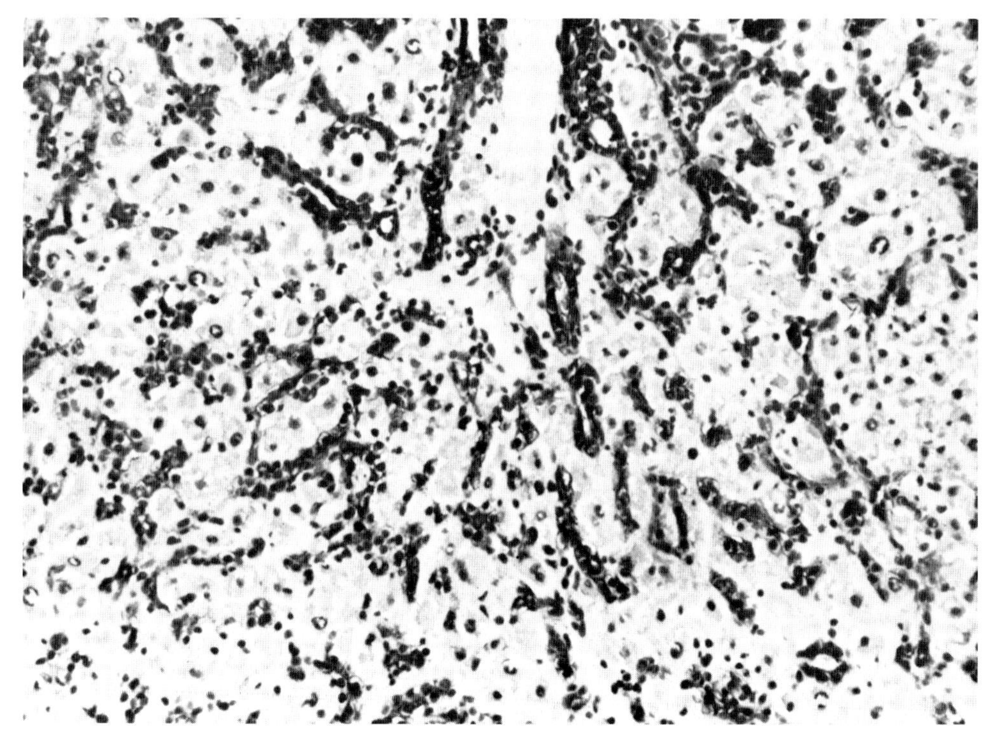

Fig. 2. Higher magnification of liver section in Fig. 1. Note selective labelling for keratin in oval and ductular cells (X100).

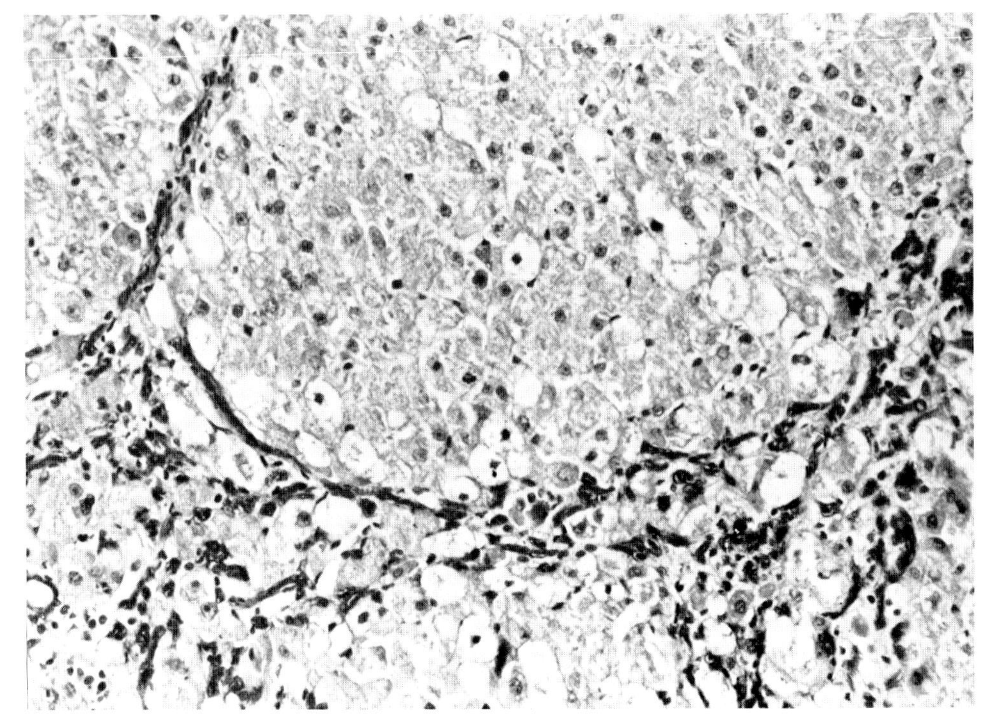

Fig. 3. Section of liver containing a large hepatocyte nodule (upper) surrounded by a rim of oval cells and hepatocytes. Note the positive staining of the oval cells and the negative staining in the hepatocytes in the nodule.

The use of phenotypic markers for studies on histogenesis of liver cancer is the most uncertain. Given the plasticity of many phenotypic expressions of hepatocytes and other cells as a function of their environment, only if the majority if not all the markers point in the same direction, can they be useful in the study of the origin of new cell populations during liver carcinogenesis. In this area, there is an obvious need for chromosomal or other markers, such as altered chromosomes, that would be reliable indices of the origin and fate of a cell or cell population. The successful search for such markers would have a major application in studies of several aspects of liver cancer development.

ACKNOWLEDGEMENT

The authors' research was supported by grants from the National Cancer Institute of Canada, Medical Research Council of Canada (MT-5594), and the National Cancer Institute, NIH, USA (CA-21157). We would like to acknowledge the valuable assistance given by Helene Robitaille in the preparation of this manuscript.

REFERENCES

1. Farber, E. (1956) Similarities in the sequence of early histological changes induced in the liver of the rat by ethionine, 2-acetylaminofluorene and 3'-methyl-4-dimethylaminoazobenzene. Cancer Res. 16:142-148.
2. Sell, S., and H.L. Leffert (1982) An evaluation of cellular lineages in the pathogenesis of experimental hepatocellular carcinoma. Hepatology 2:77-86.
3. Schauer, A., and E. Kunze (1968) Enzymehistochemische und Autoradiographische Untersuchungen der Cancerisierung der Rattenleber mit Diathylnitrosamin. Z. Krebsforsch., 70:252-266.
4. Farber, E., and M.B. Sporn (1976) Early lesions and the development of epithelial cancer - A symposium. Cancer Res. 36:2475-2706.
5. Emmelot, P., and E. Scherer (1980) The first relevant cell stage in rat liver carcinogenesis. A quantitative approach. Biochim. Biophys. Acta 605:247-304.
6. Farber, E. (1980) The sequential analysis of cancer induction. Biochim. Biophys. Acta 605:247-304.
7. Farber, E. (1973) Hyperplastic nodules. Methods Cancer Res. 7:345-375.
8. Squire, R.A., and M.H. Levitt (1975) Report of a workshop on classification of specific hepatocellular lesions in rats. Cancer Res. 35:3214-3223.
9. Ogawa, K., D.B. Solt, and E. Farber (1980) Phenotypic diversity as an early property of putative preneoplastic hepatocyte populations in liver carcinogenesis. Cancer Res. 40:725-733.
10. Ogawa, K., A. Medline, and E. Farber (1979) Sequential analysis of hepatic carcinogenesis. A comparative study of the ultrastructure of preneoplastic, malignant, prenatal, postnatal and regenerating liver. Lab. Invest. 41:22-35.
11. Ogawa, K., A. Medline, and E. Farber (1979) Sequential analysis of hepatic carcinogenesis: The comparative architecture of preneoplastic, malignant, prenatal, postnatal and regenerating liver. Br. J. Cancer 40:782-790.
12. Enomoto, K., and E. Farber (1982) Kinetics of phenotypic maturation of remodelling of hyperplastic nodules during liver carcinogenesis. Cancer Res. 42:2330-2335.
13. Farber, E., R.G. Cameron, B. Laishes, J.-C. Lin, A. Medline, K. Ogawa, and D.B. Solt (1979) Physiological and molecular markers during carcinogenesis. In Carcinogens: Identification and Mechanisms of Action. A.C. Griffin and C.R. Shaw, eds., Raven Press, New York, pp. 319-335.
14. Farber, E., and R. Cameron (1980) The sequential analysis of cancer development. Adv. Cancer Res. 31:125-226.
15. Pitot, H.C., L. Barsness, T. Goldsworthy, and T. Kitagawa (1978) Biochemical characterization of stages of hepatocarcinogenesis after a single dose of diethylnitrosamine. Nature 271:456-458.
16. Pugh, T., and S. Goldfarb (1978) Quantitative histochemical and

autoradiographic studies of hepatocarcinogenesis in rats fed 2-acetylaminofluorene. Cancer Res. 38:4450-4457.

17. Reynolds, E.S., and M.T. Moslen (1980) Environmental liver injury: Halogenated hydrocarbons. In Toxic Injury of the Liver. E. Farber and M.M. Fisher, eds. Marcel Dekker, New York, pp. 541-596.
18. Tsuda, H., G. Lee, and E. Farber (1980) The induction of resistant hepatocytes as a new principle for a possible short term in vivo test for carcinogens. Cancer Res. 40:1157-1164.
19. Solt, D., A. Medline, and E. Farber (1977) Rapid emergence of carcinogen-induced hyperplastic lesions in a new model for the sequential analysis of liver carcinogenesis. Am. J. Pathol. 88:595-618.
20. Solt, D.B., E. Cayama, H. Tsuda, K. Enomoto, G. Lee, and E. Farber (1983) The promotion of liver cancer development by brief exposure to dietary 2-acetylaminofluorene plus partial hepatectomy or carbon tetrachloride. Cancer Res. 43:188-191.
21. Williams, G.M. (1980) The pathogenesis of rat liver cancer caused by chemical carcinogens. Biochim. Biophys. Acta 605:167-189.
22. Farber, E. (1976) The pathology of experimental liver cancer. In Liver Cell Cancer, H.M. Cameron, D.A. Linsell, and G.P. Warwick, eds., Elsevier, Amsterdam, pp. 243-277.
23. Eriksson, L., M. Ahluwalia, J. Spiewak, G. Lee, D.S.R. Sarma, M.W. Roomi, and E. Farber (1983) Distinctive biochemical pattern associated with resistance of hepatocytes in hepatocyte nodules during liver carcinogenesis. Environ. Hlth. Persp. Vol. 49:171-174.
24. Solt, D.B., J.B. Hay, and E. Farber (1977) Comparison of the blood supply to diethylnitrosamine-induced hyperplastic nodules and hepatomas and to the surrounding liver. Cancer Res. 27:1686-1691.
25. Sell, S., and H.T. Wepsic (1975) Alpha-fetoprotein. In The Liver: Normal and Abnormal Function. F.F. Becker, ed., Part B. Marcel Dekker, New York, pp. 773-820.
26. Uriel, J. (1975) Fetal characteristics of cancer. In: Cancer: A Comprehensive Treatise. F.F. Becker, ed., Vol. 3, Plenum Publ. Co., New York, pp. 21-55.
27. Sell, S. (1978) Distribution of α-fetoprotein and albumin-containing cells in the liver of Fischer rats fed four cycles of N-2-fluorenylacetamide. Cancer Res. 38:3107-3113.
28. Fiala, S., A.E. Fiala, and B. Dixon (1972) Gamma-glutamyl transpeptidase in transplantable chemically induced rat hepatomas and "spontaneous" mouse hepatomas. J. Natl. Cancer Inst. 48:1393-1401.
29. Cameron, R., J. Kellen, A. Kolin, A. Malkin, and E. Farber (1978) γ-Glutamyl transferase in putative premalignant liver cell populations during hepatocarcinogenesis. Cancer Res. 38:823-829.
30. Schor, N.A., and H.P. Morris (1977) The activity of DT-diaphorase in experimental hepatomas. Cancer Biochem. Biophys. 2:5-9.

31. Schor, N., K. Ogawa, G. Lee, and E. Farber (1978) The use of DT-diaphorase for the detection of foci of early neoplastic transformation in rat liver. Cancer Lett. 5:167-171.
32. Okita, K., L.H. Kligman, and E. Farber (1975) A new common marker for premalignant and malignant hepatocytes induced in the rat by chemical carcinogens. J. Natl. Cancer Inst. 54:199-201.
33. Lin, J.-C., L. Kligman, and E. Farber (1980) Purification and partial characterization of a preneoplastic antigen in liver carcinogenesis. Cancer Res. 40:3755-3762.
34. Levin, W., A.Y.H. Lee, P.E. Thomas, D. Ryan, K. Kizer, and M.J. Griffin (1978) Identification of epoxide hydrase as the preneoplastic antigen in rat liver hyperplastic nodules. Proc. Natl. Acad. Sci., USA 75:3240-3243.
35. Goldfarb, S.A. (1973) Morphological and histochemical study of carcinogenesis of the liver in rats fed 3'-methyl-4-dimethylaminoazobenzene. Cancer Res. 33:1119-1128.
36. Mori, M., T. Kaku, K. Dempo, M. Satoh, A. Kaneko, and T. Onoe (1980) Histochemical investigation of precancerous lesions induced by 3'-methyl-4-dimethylaminoazobenzene. Gann Monograph. Cancer Res. 35:103-114.
37. Malkin, A., J.A. Kellen, A. Kolin, R. Cameron, and E. Farber (1978) Scand. J. Immunol. 8, Suppl. 8, pp. 603-607.
38. Kitagawa, T., and H.C. Pitot (1975) The regulation of serine dehydratase and glucose-6-phosphatase in hyperplastic nodules of rat liver during diethylnitrosamine and N-2-fluorenylacetamide feeding. Cancer Res. 35:1075-1084.
39. Epstein, S.M., N. Ito, L. Merkow, and E. Farber (1967) Cellular analysis of liver carcinogenesis: The induction of large hyperplastic nodules in the liver with 2-fluorenylacetamide or ethionine and some aspects of their morphology and glycogen metabolism. Cancer Res. 27:1702-1711.
40. Daoust, R. (1963) Cellular populations and nucleic acid metabolism in rat liver parenchyma during azo dye carcinogenesis. Can. Cancer Conf. 5:225-239.
41. Fontaniere, B., and R. Daoust (1973) Histochemical studies on nuclease activity and neoplastic transformation in rat liver during diethylnitrosamine carcinogenesis. Cancer Res. 33:3108-3111.
42. Williams, G.M., and R.S. Yamamoto (1972) Absence of stainable iron from preneoplastic and neoplastic lesions in rat liver with 8-hydroxyquinoline-induced siderosis. J. Natl. Cancer Inst. 49:685-692.
43. Bannasch, P. (1968) The cytoplasm of hepatocytes during carcinogenesis: Electron and light microscopical investigations of the nitrosomorpholine-intoxicated rat liver. Recent Results Cancer Res. Vol. 19: Springer-Verlag, Berlin.
44. Farber, E. (1963) Ethionine carcinogenesis. Adv. Cancer Res. 7:383-474.
45. Solt, D.B., and E. Farber (1976) A new principle for the sequential analysis of chemical carcinogenesis. Nature 263:701-703.

46. Farber, E. (1982) Sequential events in chemical carcinogenesis. In Cancer: A Comprehensive Treatise. F.F. Becker, ed., Vol. 1, 2nd edition, Plenum Publ. Corp., New York, pp. 485-506.
47. Ito, N., M. Tatematsu, K. Nakanishi, R. Hasegawa, T. Takano, K. Imaida, and T. Ogiso (1980) The effect of various chemicals in the development of hyperplastic liver nodules in hepatectomized rats treated with N-nitrosoethylamine or N-2-fluorenylacetamide. Gann 71:832-842.
48. Lindahl, R., S. Evces, and W.-L. Sheng (1982) Expression of the tumor aldehyde dehydrogenase phenotype during 2-acetyl-amino-fluorene-induced rat hepatocarcinogenesis. Cancer Res. 42:577-582.
49. Sawada, N., K. Furukawa, M. Gotoh, Y. Mochizuki, and H. Tsukada (1982) Primary culture of preneoplastic hepatocytes from rats treated with 2-acetylaminofluorene and clofibrate: Relationship between resistance to dimethylnitrosamine and responsiveness of peroxisomes. Gann 73:1-6.
50. Scherer, E., M. Hoffman, P. Emmelot, and H. Friedrich-Freksa (1972) Quantitative study on foci of altered liver cells induced in the rat by a single dose of diethylnitrosamine and partial hepatectomy. J. Natl. Cancer Inst. 49:93-106.
51. Campbell, H.A., H.C. Pitot, V.R. Potter, and B.A. Laishes (1982) Application of quantitative stereology to the evaluation of enzyme-altered foci in rat liver. Cancer Res. 42:465-472.
52. Scherer, E., and P. Emmelot (1976) Kinetics of induction and growth of enzyme-deficient islands involved in hepatocarcinogenesis. Cancer Res. 36:2544-2554.
53. Popper, H., S.S. Sternberg, B.L. Oser, and M. Oser (1960) The carcinogenic effect of aramite in rats. A study of hepatic nodules. Cancer 13:1035-1046.
54. Inaoka, Y. (1969) Significance of the so-called oval cell proliferation during azo-dye hepatocarcinogenesis. Gann 58:355-366.
55. Iwasaki, T., K. Dempo, A. Kaneko, and T. Onoe (1972) Fluctuation of various cell populations and their characteristics during azo-dye carcinogenesis. Gann 63:21-30.
56. Dempo, K., N. Chisaka, Y. Yoshida, A. Kaneko, and T. Onoe (1975) Immunofluorescent study on a α-fetoprotein-producing cells in the early stage of 3'-methyl-4-dimethylaminozaobenzene carcinogenesis. Cancer Res. 35:1282-1287.
57. Onoe, T., A. Kaneko, K. Dempo, K. Ogawa, and T. Minase (1975) α-Fetoprotein and early histological changes of hepatic tissues in DAB-hepatocarcinogenesis. Ann. N.Y. Acad. Sci., 259:168-180.
58. Kalengayi, M.M., and V.J. Desmet (1976) Histochemical patterns of two oncofetal markers, gamma-glutamyltranspeptidase and alpha-1-fetoprotein during aflatoxin hepatocarcinogenesis in the rat. In Liver Cancer, K. Shanmugaratum, R. Nambiar, K.K. Tan, and L.K. Chan, eds. Singapore, Stanford Press, pp. 198-200.
59. Onda, H. (1976) Immunohistological studies of α_1-fetoprotein

and α_1-acid glycoprotein during azo-dye hepatocarcinogenesis in rats. Gann 67:253-262.
60. Kuhlmann, W.D. (1978) Localization of alpha-fetoprotein and DNA synthesis in liver cell populations during experimental hepatocarcinogenesis in rats. Int. J. Cancer 21:368-380.
61. Shinozuka, H., M.A. Sells, S.L. Katyal, S. Sell, and B. Lombardi (1979) Effects of a choline-devoid diet on the emergence of γ-glutamyl-transpeptidase-positive foci in the liver of carcinogen treated rats. Cancer Res. 39:2515-2521.
62. Sell, S., K. Osborn, and L. Leffert (1981) Autoradiography of "oval cells" appearing rapidly in the livers of rats fed N-2-fluorenylacetamide in a chlorine devoid diet. Carcinogenesis 2:7-14.
63. Desmet, V.J. (1963) Histochemical study of the experimental liver carcinogenesis (1963) Brussels Presses Academiques Europeennes, S.C.
64. Ahluwalia, M.B., J. Rotstein, M. Tatematsu, M.W. Roomi, and E. Farber (1983) Failure of glutathione to prevent liver cancer development in rats initiated with diethylnitrosamine in the resistant hepatocyte model. Carcinogenesis 4:119-121.
65. Kinosita, R. (1937) Studies on the carcinogenic chemical substances. Trans. Jap. Pathol. Soc. 27:665-727.
66. Price, J.M., J.W. Harman, E.C. Miller, and J.A. Miller (1952) Progressive microscopic alterations in the livers of rats fed the heptic carcinogens 3'-methyl-4-dimethylaminoazobenzene and 4'-fluoro-4-dimethylaminoazobenzene. Cancer Res. 12:192-300.
67. Guillouzo, A., A. Weber, E. Le Provost, M. Rissel, and F. Schapira (1981) Cell types involved in the expression of fetal aldolases during rat azo-dye hepatocarcinogenesis. J. Cell Sci. 49:249-260.
68. Kitagawa, T., T. Yokochi, and H. Sugano (1972) α-Fetoprotein and hepatocarcinogenesis in rats fed 3'-methyl-4-dimethylaminoazobenzene or N-2-fluorenylacetamide. Int. J. Cancer 10:368-381.
69. Tchipysheva, T.A., V.I. Guelstein, and G.A. Bannikov (1977) α-Fetoprotein-containing cells in the early stages of liver carcinogenesis induced by 3'-methyl-4-dimethylaminoazobenzene and 2-acetylaminofluorene. Int. J. Cancer 20:388-393.
70. Harada, M., K. Okabe, K. Shibata, H. Masuda, K. Miyata, and M. Enomoto (1976) Histochemical demonstration of increased activity of γ-glutamyl transpeptidase in rat liver during hepatocarcinogenesis. Acta Histochem. Cytochem. 9:168-179.
71. Benner, V., H.J. Hacker, and P. Bannasch (1979) Electron microscopical demonstration of a glucose-6-phosphatase in native cryostat sections fixed with glutaraldehyde through semipermeable membranes. Histochemistry 65:41-47.
72. Kitagawa, T. (1971) Histochemical analysis of hyperplastic lesions and hepatomas of the liver of rats fed 2-fluorenylacetamide. Gann 62:207-216.
73. Ogawa, K., T. Minase, and T. Onoe (1974) Demonstration of glucose-6-phosphatase activity in the oval cells of rat liver

and the significance of the oval cells in azo dye carcinogenesis. Cancer Res. 34:3379-3386.

74. Tatematsu, M., T. Shirai, H. Tsuda, Y. Miyata, Y. Shinohara, and N. Ito (1977) Rapid production of hyperplastic liver nodules in rats treated with carcinogenic chemicals: A new approach for an *in vivo* short-term screening of hepatocarcinogens. Gann 68:499-507.
75. Enomoto, K., T.S. Ying, and E. Farber (1981) Immunohistochemical study of epoxide hydrolase during experimental liver carcinogenesis. Cancer Res. 41:3281-8277.
76. Griffin, M.J., and D.E. Kizer (1978) Purification and quantitation of preneoplastic antigen from hyperplastic nodules and normal liver. Cancer Res. 38:1136-1141.
77. Novikoff, A.B., P.M. Novikoff, R.J. Stockert, F.F. Becker, A. Yam, M.S. Poruchynsky, W. Levin, and P.E. Thomas (1979) Immunocytochemical localization of epoxide hydrase in hyperplastic nodules induced in rat liver by 2-acetylaminofluorene. Proc. Natl. Acad. Sci., USA 76:5207-5211.
78. Sharma, R.N., H.L. Gurtoo, E. Farber, R.K. Murray, and R.G. Cameron (1981) Effects of hepatocarcinogens and hepatocarcinogenesis on the activity of rat liver microsomal epoxide hydrolase and observations on the electrophoretic behavior of this enzyme. Cancer Res. 41:3311-3319.
79. Shinozuka, H., B. Lombardi, S. Sell, and R.M. Iammarino (1978) Early histological and functional alterations of ethionine liver carcinogenesis in rats fed a choline-deficient diet. Cancer Res. 38:1092-1098.
80. Daoust, R., and A. Cantero (1958) The distribution of deoxyribonuclease in normal, cirrhotic and neoplastic rat livers. J. Histochem. Cytochem. 7:139-143.
81. Taper, H.S., and P. Bannasch (1976) Histochemical correlation between glycogen, nucleic acids and nuclease in pre-neoplastic and neoplastic lesions of rat liver after short-term administration of N-nitrosomorpholine. Z. Krebsforsch. 87:53065.

EARLY MORPHOLOGIC MARKERS OF CARCINOGENICITY IN RAT PANCREAS

Daniel S. Longnecker

Department of Pathology
Dartmouth Medical School
Hanover, New Hampshire 03755

INTRODUCTION

Carcinomas have been experimentally induced in the pancreas of rats by several chemicals since Hayashi and Hasegawa first reported induction of hyperplastic nodules and acinar cell adenomas in pancreas of rats treated with 4-hydroxyaminoquinoline-1-oxide (4-HAQO) (1). Induction of exocrine pancreatic carcinomas by azaserine, 7,12-dimethylbenz(a)anthracene, and 4-HAQO has been been studied to an extent that each can be regarded as providing a model in rats. Of these three, the azaserine-rat model has been best characterized and can be regarded as a prototype for carcinogenesis by chemicals that appear primarily to affect acinar cells and lead to their transformation. This model will be described in detail and compared with other models that will be less completely described.

AZASERINE-RAT MODEL

Initiation

Azaserine, O-diazoacetyl-L-serine, is an antibiotic made by a Streptomyces species. Azaserine is a bacterial mutagen (2) and damages DNA in rat pancreas, liver, and kidney (3). N-7 Carboxymethylation of guanine in rat pancreas after treatment with azaserine has been reported (4) and there is evidence of alkylation of macromolecules by azaserine in several tissues (5). Although azaserine is a potent direct acting mutagen in the Ames system, it appears that a pyridoxal-dependent enzyme system is involved in its metabolic activation in rodents (6).

Azaserine is carcinogenic when administered by IP (intraperitoneal) injection or by gavage in aqueous solution with total doses ranging from 30 to 300 mg/kg (7). Single injections are most effective when given during a period of rapid acinar cell division, i.e., at age 2 weeks (8) or following partial pancreatectomy (9). Single doses in rats should not exceed 30-50 mg/kg because cytotoxicity in pancreas yielded fewer surviving "initiated" cells when higher doses, i.e., 60 mg/kg and 80 mg/kg, were used (10). Lewis strain rats were more sensitive and uniformly responsive to azaserine than random bred Wistar rats in regard to induction of early focal pancreatic lesions while F-344 rats were much less responsive (10).

Early Azaserine-Induced Lesions in Pancreas

Foci of phenotypically altered acinar cells appear in the pancreas as early as two months after the injection of azaserine (Fig. 1). Foci are initially small, consisting of 10 or fewer cells in a planar transection, but some rapidly increase in size to achieve diameters as great as 3mm by 4 months. Such lesions were originally described in rats killed 6 months after initial exposure to azaserine. Many were several millimeters in diameter and they were designated as hyperplastic nodules (11) or atypical acinar cell nodules (AACN) (2). We and others have subsequently suggested that small lesions (<1 mm diameter) be designated as foci, whereas larger and grossly visible lesions be called nodules (7,12,13). In our early reports, the spectrum of lesions called AACN included both foci and nodules. The average diameter of AACN two months after a single injection of azaserine has been reported to be 0.15 mm (14).

Phenotypic changes in the cells of AACN that distinguish them from acinar cells of the surrounding normal pancreas include: increased mitotic rate, altered cell size, altered RNA content, altered zymogen content, reduced gamma glutamyl transpeptidase activity, and altered nuclear size and/or configuration. One or several of these features may appear in a single AACN. The histologic appearance of cells in a focus or nodule tends to be homogeneous, suggesting clonal origin. AACN are generally spherical and circumscribed but do not have a capsule. The histologic appearance of nodules is as diverse as is their growth potential. Mitotic rate and sizes of AACN vary greatly in rats killed 4 months or longer after treatment with azaserine (Fig. 2). We estimate that fewer than 1 percent of foci progress to become neoplasms (15).

Neoplasms

Adenomas. Rapidly growing AACN compress adjacent pancreas and sometimes become encapsulated while retaining a high degree of acinar cell differentiation. At some rather arbitrary point in this progression it seems appropriate to designate such lesions as adenomas since they seem to have neoplastic growth potential but do not

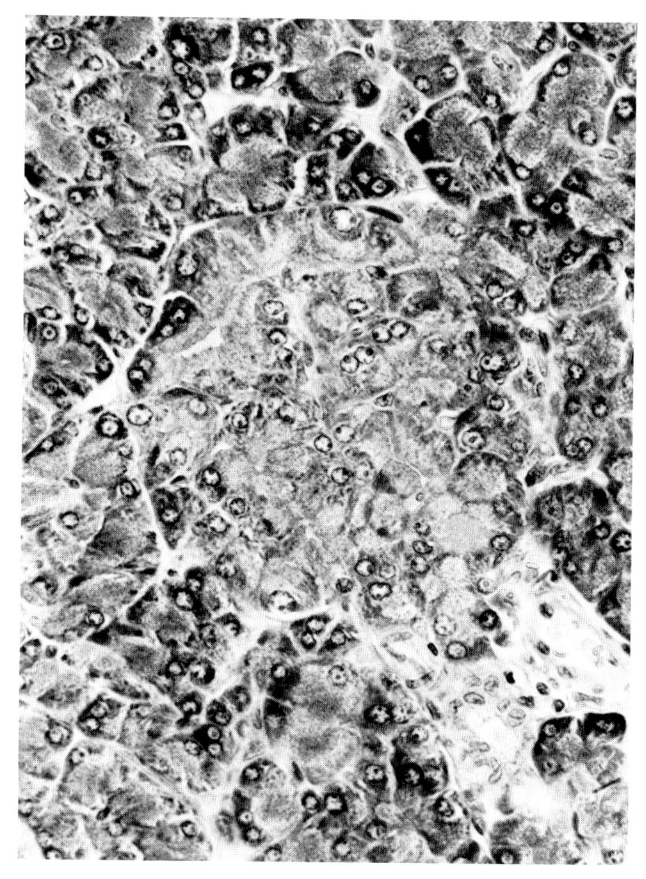

Fig. 1. A focus of atypical acinar cells occupies the center of this field. The pancreas is from a rat injected with azaserine 2 months before necropsy. H and E x 310.

have anaplastic cytologic features that denote malignancy. We have seen such lesions as early as 6 months after treatment with azaserine, and they persist as apparently benign lesions in experiments of 1-2 years duration. Adenomas are usually 2-7 mm in diameter.

Adenomas are composed of well differentiated acinar cells that may exhibit some of the phenotypic changes seen in AACN. Scattered ductules may be included. Vessels may be more prominent than in normal acinar tissue. Islets and interlobular ducts are characteristically absent although they may be enveloped at the periphery of large adenomas.

For some lesions in the size range of 2-5 mm that lack a fibrous capsule, classification as large nodules vs. small adenomas

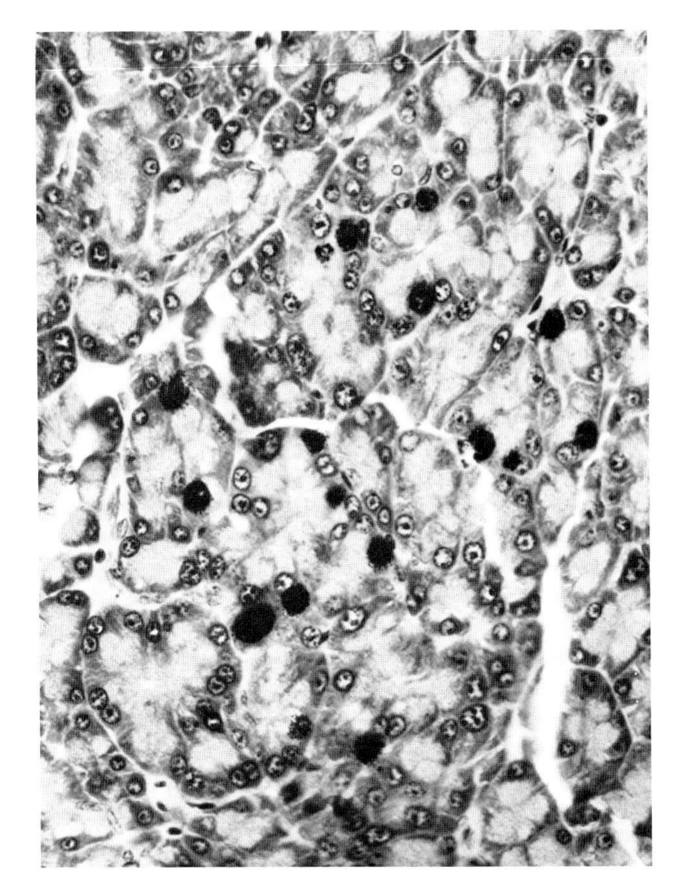

Fig. 2. Autoradiographic evidence of a high labeling index in the cells of an AACN from the pancreas of an azaserine-treated rat that was given a single 1 uci/g injection of ^{3}H-TdR one hour before necropsy. Although AACN typically have a higher labeling index than the normal pancreas, the AACN illustrated contains an unusually high number of labeled nuclei. H and E x 200.

is arbitrary. A solution may be to adopt a diameter, e.g., 3 mm, as the major criterion for such classification when histologic study shows compression or displacement of surrounding pancreas but no capsule.

Finally, it appears that some of the neoplasms classed as adenomas have the potential for progression to invasion and metastasis, i.e., to become well differentiated acinar cell carcinomas.

Carcinomas. Malignant potential of azaserine-induced neoplasms is exhibited by metastasis to regional lymph nodes, liver, and

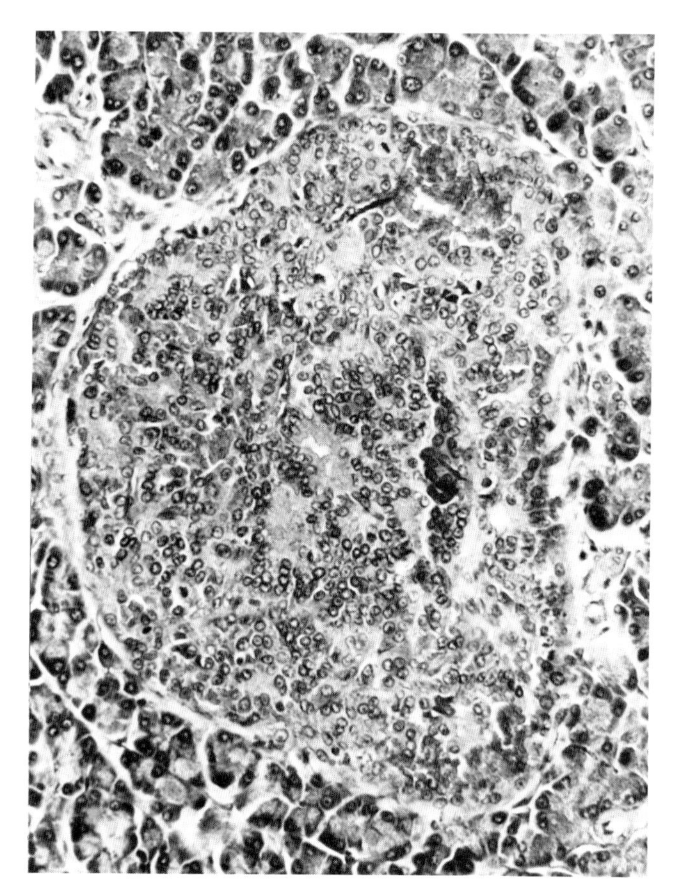

Fig. 3. This focus of atypical cells shows a high degree of anaplasia, vascularity, and desmoplastic stromal response and the lesion is regarded as an early carcinoma of the pancreas (CIS). The rat was treated for 6 weeks with azaserine beginning 6 months before necropsy. H and E x 200.

rarely to lung (2). Several such carcinomas have been transplanted (15, 16). The histologic appearance of carcinomas varies greatly and ranges from well differentiated acinar cell to anaplastic types (7). Most carcinomas show evidence of acinar cell differentiation, but may include cystic, ductlike, scirrhous, and anaplastic areas. Non-scirrhous carcinomas may be highly vascular. Carcinomas sometimes develop thick fibrous capsules.

Metastasis is found 1-2 years after initial exposure to azaserine, but neoplasms with significant degrees of anaplasia have been seen as early as 6-9 months after treatment with azaserine (Fig. 3). Such neoplasms are histologically similar to metastasizing carcinomas and we regard them as an early, localized stage of carcinoma.

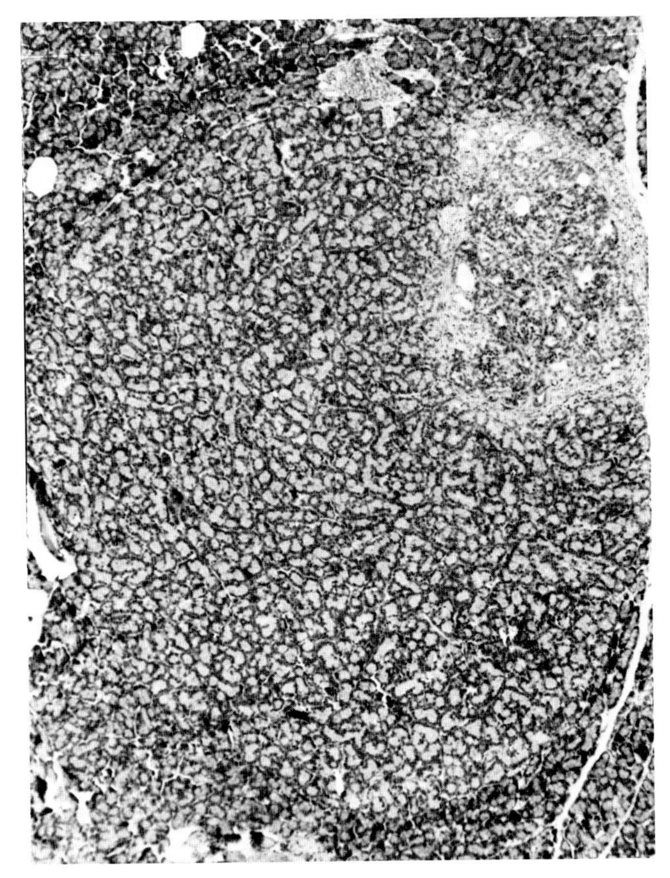

Fig. 4. This large AACN is composed of well differentiated acinar cells except in one area at the right where there is focal anaplasia and fibrosis. H and E x 50.

We have called them localized carcinoma or carcinoma-*in-situ* (CIS). The CIS stage is easily recognized when a neoplasm has a poorly differentiated acinar cell type, ductlike metaplasia with anaplasia, or a completely anaplastic cell type. The presence of a scirrhous stromal response and/or increased vascularity is helpful but inconstant.

Borderlines. There are three problems in the classification of lesions into the CIS group. First, it seems likely that a localized, well differentiated acinar cell carcinoma might be classed as an adenoma since we depend heavily on loss of differentiation for recognition of the CIS stage. It might be useful to set an arbitrary upper limit of size for classification of well differentiated lesions as adenomas. Second, foci of anaplastic change can be seen within large atypical acinar cell nodules (Figs. 4, 5). Thus, the line between AACN and CIS becomes arbitrary at some point. Third, the cells of rare foci that measure less than 1 mm in diameter show degrees of anaplasia similar to those seen in carcinomas (Fig. 3). Thus, it may be desirable to set an arbitrary lower limit of size for classification as CIS. The smallest lesions that we have called CIS have been about 0.5 mm in diameter.

Comments on the Azaserine-Rat Model

Azaserine induces many focal cellular lesions in the rat pancreas. The number per pancreas has exceeded 100 and varies with the dose of azaserine (10,17). The earliest lesions consist of phenotypically altered acinar cells. We have not seen atypical or hyperplastic changes in duct epithelium. Many of the foci grow to become

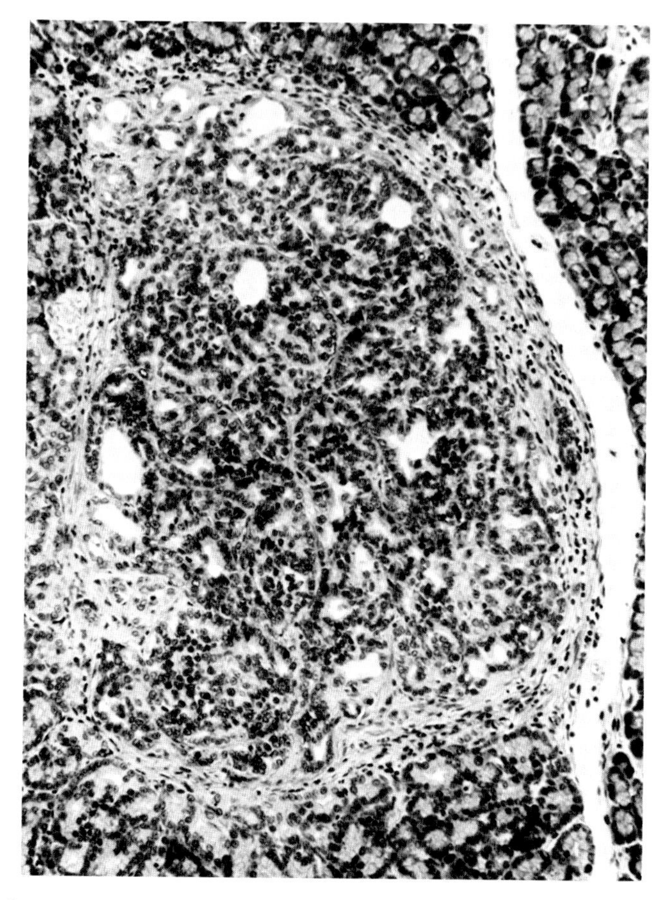

Fig. 5. This focus of anaplastic acinar cells appeared to arise within the large AACN shown in Fig. 4. The rat was treated with 6 weekly injections of azaserine beginning 1 year before necropsy (18). H and E x 125.

nodules and fewer progress to become adenomas or carcinomas. The rate of progression has been increased by feeding diets high in unsaturated fat (14,18) and decreased by feeding retinoid-supplemented diets (7,19).

We have used the number and size of AACN in the pancreas at 4 or 6 months after treatment as a parameter to compare factors, i.e., rat strain, age, sex, and diet (7,8,10), which influence initiation and progression in the azaserine model. There has been complete correlation of the results of such short term *in vivo* studies and 1-2 year studies when data from both are available, i.e., the number and size of AACN seem to correlate positively with the incidence of carcinoma.

OTHER RAT MODELS FOR PANCREATIC CARCINOGENESIS

Chemicals that have caused pancreatic carcinomas in rats are listed in Table 1. Two types of early carcinogen-induced lesions have been described in these studies, i.e., 1) AACN, and 2) ductal complexes. They will be discussed further in connection with specific chemicals.

4-Hydroxyaminoquinoline-1-oxide (4-HAQO)

4-HAQO is a derivative of 4-nitroquinoline-1-oxide that is mutagenic in bacteria and damages DNA in rat pancreas (20). Single IV injections of 4-HAQO initiate pancreatic carcinomas which are preceded by foci and nodules of atypical acinar cells (1,13) that are generally similar to those induced by azaserine. Two classes of foci have been described in the pancreases of 4-HAQO-treated rats (13). These are acidophilic foci and basophilic foci. Acidophilic foci have a higher mitotic index, and grow to larger sizes than basophilic foci (Fig. 6). The most consistent difference between cells in the two types of foci is the relative amount of zymogen (higher in acidophilic nodules) and RNA. Basophilic foci tend to remain small, and it appears that nodules, adenomas, and carcinomas arise from acidophilic foci (13).

Adenomas and carcinomas induced by 4-HAQO have typically maintained a high degree of acinar cell differentiation. When adult Wistar rats have been treated, only adenomas have ensued. Carcinomas have been induced by treating with 4-HAQO when the pancreas is regenerating following partial pancreatectomy or ethionine-induced atrophy (21,22). The 4-HAQO model appears quite similar to the azaserine model in many respects, although the former is less completely characterized.

7,12-Dimethylbenz(a)anthracene (DMBA)

DMBA is a polycyclic hydrocarbon which is best known as a breast carcinogen. It requires metabolic activation, presumably by

Table 1. Chemicals reported to cause carcinomas, and/or atypical acinar cell nodules, in the rat pancreas.

A. Chemicals reported to induce exocrine carcinomas

Azaserine

Clofibrate, ethylchlorophenoxyisobutyrate

7,12-Dimethylbenz(a)anthracene (DMBA)[a]

4-Hydroxyaminoquinoline-1-oxide (4-HAQO)

N^{δ}-(N-Methyl-N-nitrosocarbamoyl)-L-ornithine (MNCO)

Nafenopin, [2-methyl-2-(p-1,2,3,4-tetrahydro-1-naphthyl)-phenoxy]propionic acid

Nitrofen, 2,4-dichloro-1(4-nitrophenoxy)benzene

N-Nitrosobis(2-hydroxypropyl)amine (BHP)

B. Chemicals reported to induce atypical acinar cell nodules

Azaserine

4-HAQO

MNCO

N^{δ}-(N-Methyl-N-nitrosocarbamoyl)-L-diaminobutyric acid (MNDABA)

N^{ϵ}-(N-Methyl-N-nitrosocarbamoyl)-L-lysine (MNCL)

O-(Methyl-N-nitrosocarbamoyl)-L-serine

N-Nitrosobis(2-oxopropyl)amine (BOP)

BHP and raw soya flour [b]

N-Nitroso(2-hydroxypropyl)(2-oxopropyl)amine (HPOP)

[a] Abbreviations are included in parentheses.

[b] Many AACN were observed in rats injected with BHP and fed raw soya flour. The incidence was also high in rats fed raw soya flour alone.

oxidative steps (23). DMBA crystals have induced duct-like adenocarcinomas and anaplastic carcinomas following implantation into the pancreas of Sprague-Dawley rats (24). Carcinomas arise at the site of implantation, and the sequence of morphologic events in this region has been studied in detail by Bockman et al. (25). Inflammation, fibrosis, and atrophy of exocrine elements occur adjacent to the pellet. Some lobules of acinar tissue are replaced by aggregates of duct-like structures which have been called tubular complexes (Fig. 7). Bockman postulates that these complexes arise from acinar tissue by progressive loss from acinar cells of differentiated features such as zymogen production and rough endoplasmic reticulum (25). Tubular complexes have developed by one month after DMBA implantation. These complexes are regarded as a likely origin for the carcinomas. The pancreas remote from the site of the implanted DMBA is apparently free of significant carcinogen-induced lesions. The extent to which tubular complexes represent a specific phenotypic response to the carcinogen vs. non-specific, focal

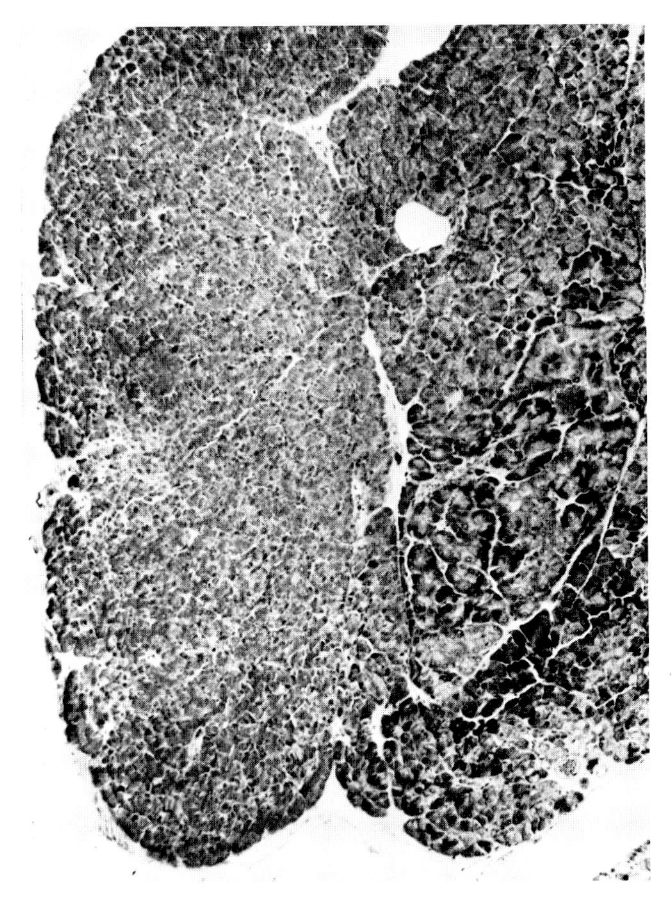

Fig. 6. This lobule of pancreas contains a basophilic focus (right center) and an acidophilic nodule with abundant zymogen (left center). Mitoses were evident in the larger, but not in the smaller lesion. The rat was given a single injection of 4-HAQO one year before necropsy. H and E x 80.

pancreatitis due to cytotoxicity of DMBA is not clear. A low incidence of such lesions was seen around sutures and an implanted dextrose pellet (25).

Other Chemicals

All other carcinogens which have been reported to cause exocrine pancreatic carcinoma in animals have been studied less completely in regard to pancreatic carcinogenesis in the rat than azaserine, 4-HAQO and DMBA. Clofibrate and nafenopin are both reported to have induced a single acinar cell carcinoma in F-344

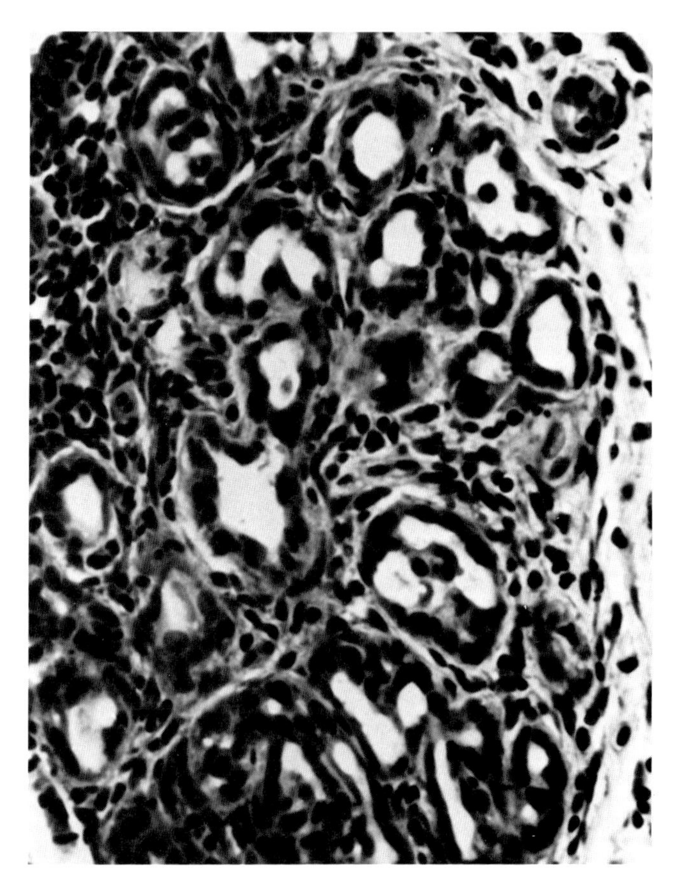

Fig. 7. Tubular complex from a rat pancreas that was implanted with crystals of DMBA 4 months before necropsy. No zymogen containing cells remain. Lymphocytic infiltration and slight stromal fibrosis are evident. The photomicrograph was provided by Dr. Dale Bockman and is reproduced with permission of the American Journal of Pathology. H and E x 300.

rats (26,27). Clofibrate also induced several tubular adenomas which were similar in appearance to the tubular complexes described above. Nitrofen induced a low incidence of acinar cell carcinomas in Osborne-Mendel rats (28). There is only a single report for each of these three carcinogens and little is known about the histogenesis of the carcinomas. Because the carcinomas have been of acinar cell type, it seems probable that early morphologic changes might be similar to those described in the azaserine model.

Table 1 also lists carcinogens which have induced AACN but not carcinomas in rat pancreas. It should be noted that BOP and HPOP are both effective pancreatic carcinogens in the hamster and that they are closely related to BHP, also effective in hamster, which has induced pancreatic carcinomas in rats fed raw soya flour, a promoter of pancreatic carcinogenesis in rats (29). Thus, it seems likely that BOP and HPOP would also cause pancreatic carcinomas if the agents were given to young rats which were then maintained on a diet that promotes pancreatic carcinogenesis (30).

MNDABA and MNCL are nitrosocarbamoyl amino acids which are carcinogens for skin, breast, and kidney (31). They have induced large numbers of AACN in the pancreas in a six month study, but they have not been evaluated as pancreatic carcinogens in long-term studies. They are closely related to MNCO, which has caused acinar cell carcinoma in the pancreas (32). All three agents also induce ductal complexes of the cystic variety (Fig. 8). Cystic ductal complexes replace portions of pancreatic lobules and consist of circumscribed aggregates of small cystic glands or ductules lined predominantly by a single layer of flattened, nondescript epithelium. Zymogen containing cells are occasionally present in the lining layer. Lumens are usually less than 1 mm in diameter and sometimes contain eosinophilic precipitate. Variable amounts of fibrous tissue surround the cystic spaces. Cystic ductal complexes vary greatly in size and measure from less than 1 mm to several mm in diameter. Their number is comparable to that of AACN, i.e., many per pancreas. These lesions appear to have low growth potential. Mitoses are rare in the lining epithelium and we have not seen adenomas or carcinomas of similar histologic type.

Several other carcinogens and nitrosoamino acids have been evaluated in regard to their ability to induce AACN, ductal complexes, intraductal epithelial hyperplasia and/or neoplasms in the pancreas of rats. Chemicals which have failed to induce such lesions in studies of 4-6 months duration include D-azaserine (33), ethyl diazoacetate (34), 6-diazo-5-oxo-L-norleucine (34), N^e-(N-methyl-N-nitroso-β-alanyl)-L-lysine, O-(N-methyl-N-nitroso-β-alanyl)-L-serine, dl-alanosine, L-ethionine, neutral red, and N^e(N-nitrososarcosyl)-L-lysine (31). Thus, the induction of AACN is not a ubiquitous and non-specific response to xenobiotics.

FINAL COMMENTS

It is of interest that no pancreatic carcinogen has been described which induces hyperplasia of the epithelium of the main or interlobular pancreatic ducts in rats. Acinar cells appear to be the most responsive of the exocrine cell types in rats to the effects of carcinogens. A few small cystic ductal complexes have been found in pancreases from azaserine- and 4-HAQO-treated rats but neoplasms of a corresponding type were not found. We also have

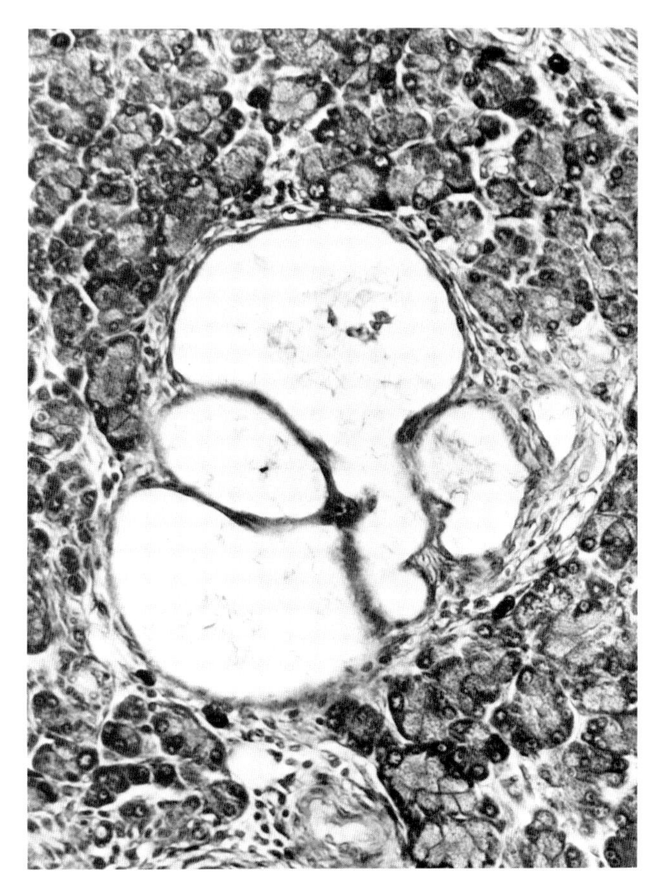

Fig. 8. Cystic ductal complex from a rat which was injected with MNCO 6 months before necropsy. The cystic spaces are lined by flattened epithelium. H and E x 200.

observed a low incidence of lesions in the pancreases of azaserine-treated rats that are similar in appearance to the tubular complexes described by Bockman. There is usually evidence of chronic inflammation and fibrosis in such lesions (Fig. 10), and we have regarded them as representing focal chronic pancreatitis. We do not know if they are carcinogen-induced.

The description by Rao et al. (13) of subgroups of atypical acinar cell foci with differing proliferative potential is noted with interest. Acidophilic AACN and basophilic foci can also be identified in pancreases from azaserine-treated rats and appear to have growth potential generally similar to that described in the 4-HAQO model (35).

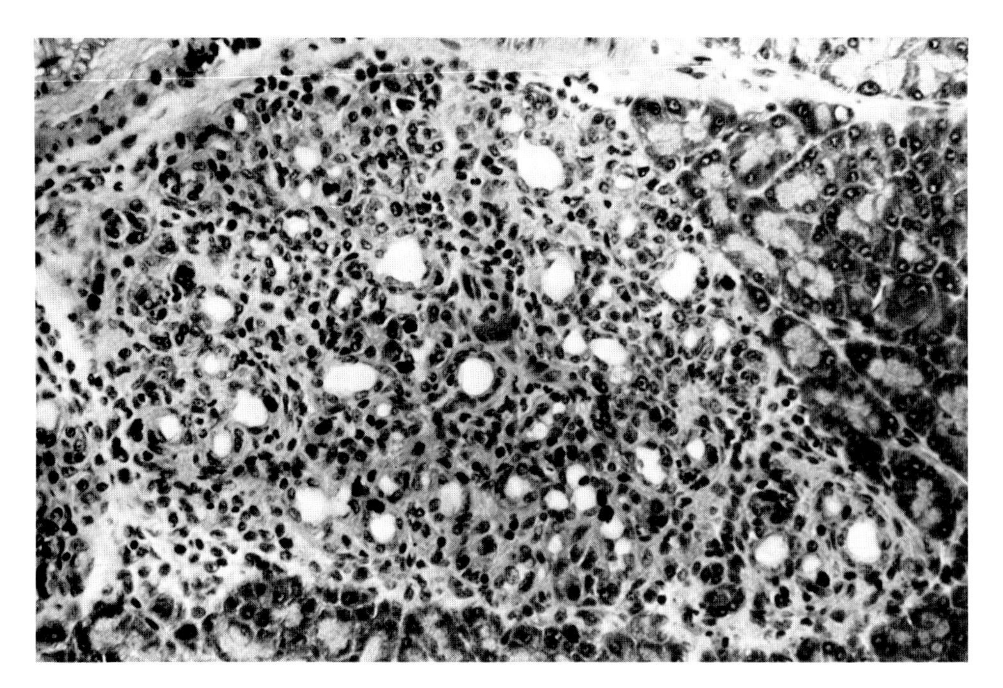

Fig. 9. This circumscribed lesion contains duct-like tubules separated by fibrous tissue infiltrated by lymphocytes. The autoradiogram shows a high labeling index in the epithelium of the tubules. The pancreas is from the same azaserine-treated rat as that shown in Fig. 2. The pancreas contained many AACN and only a single lesion like that shown here. H and E x 200.

The spectrum of adenomas and carcinomas which have been experimentally induced in rats seems to match the type encountered in low incidence among untreated laboratory rats (28,36). "Spontaneous" acinar cell adenomas are more common than carcinomas, and the latter usually have shown evidence of acinar cell differentiation, although undifferentiated carcinomas were described.

Atypical acinar cell foci and nodules also occur spontaneously and have been referred to as focal hyperplasia, hyperplastic nodules, and hyperplastic lobules. Their incidence increases with age. We have usually seen AACN in about 25% of control animals in studies of 4-6 months duration. In a 2 year study in Wistar rats, we found such lesions in 73% of control rats (7). Spontaneous lesions occasionally grow to a diameter of several mm, but most appear to have low-growth potential and remain small. The striking difference between induced and spontaneous AACN is the greater number per pancreas in carcinogen-treated rats. In chow fed control rats the average number per pancreas is usually in the range of 0-2 in studies of 4-6 months duration and the maximum is in the range of 5-10 in 1-2 year studies. These counts are made in completely embedded and sectioned pancreases.

Thus, experimental chemical carcinogenesis in rats appears to be relevant to "spontaneous" neoplasia in this species. The relevance of the rat models to carcinogenesis in humans remains to be established. Our conclusions regarding the histogenesis of pancreatic carcinomas in rats are summarized in Fig. 10. We agree with Bockman that duct-like structures can be derived from acinar tissue and accept the conclusion that tubular ductal complexes are derived from acinar tissue. We are less certain of the origin of cystic ductal complexes. Their origin could be similar to that proposed for tubular complexes or they could arise by proliferation of centroacinar cells. They appear to be of little or no importance in the histogenesis of carcinomas.

SUMMARY

Pancreatic carcinoma has been induced experimentally in rats with several different chemicals. Early carcinogen-induced lesions in rat pancreas fall into two general categories: 1) atypical acinar cell nodules (AACN) and 2) ductal complexes. AACN have been induced by several carcinogens and studied most in the model for pancreatic carcinogenesis in azaserine-treated rats. AACN are focal lesions composed of acinar cells which are phenotypically different from the surrounding normal pancreas. Cellular changes include alteration of nuclear size and/or configuration, increased mitotic rate, altered cell size, altered zymogen content, altered RNA content, and deficient gamma glutamyltranspeptidase activity. A nodule may exhibit one or several of these changes. A small fraction of the AACN appear to progress to become adenomas and/or carcinomas. The distinction between a large AACN and an adenoma is sometimes

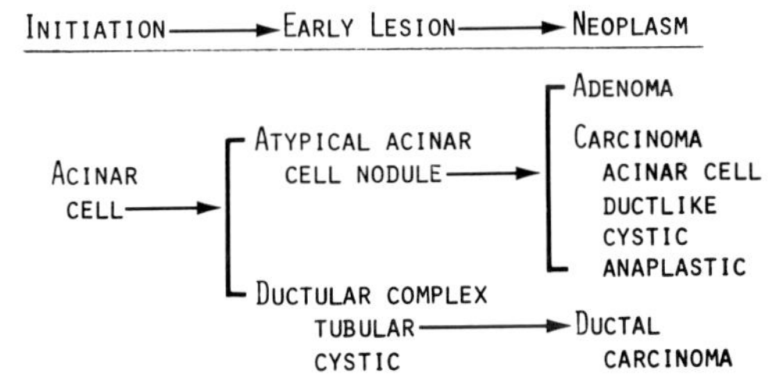

Fig. 10. Proposed scheme for histogenesis of carcinomas in the pancreas of rats. As discussed in the text, origin of

arbitrary, as is the distinction between adenomas and well-differentiated carcinomas. Recent experience indicates that some of the lesions that we have classed as acinar cell adenomas were an early localized stage of well differentiated acinar cell carcinomas. Most carcinomas arising in azaserine-treated rats show evidence of acinar cell differentiation; however, the spectrum of lesions includes carcinomas with duct-like areas, cystic carcinomas, and undifferentiated carcinomas.

The ductal or tubular complex is a focal lesion in which lobules of acinar tissue seem to be replaced by aggregates of ductules. Ductal complexes have been most carefully studied in rats treated by implantation of a pellet of 7,12-dimethyl-benz(a)anthracene, and in this model they appear to arise from acini in which the cells progressively lose cytoplasmic markers of acinar cell differentiation. The resulting ductal complexes are considered to be a possible step in the development of the duct-like carcinomas which are characteristic of this model.

REFERENCES

1. Hayashi, Y., and T. Hasegawa (1971) Experimental pancreatic tumor in rats after intravenous injection of 4-hydroxyamino-quinoline-1-oxide. Gann 62:329-330.
2. Longnecker, D.S., and T.J. Curphey (1975) Adenocarcinoma of the pancreas in azaserine-treated rats. Cancer Res. 35:2249-2257.
3. Lilja, H.S., E. Hyde, D.S. Longnecker, and J.D. Yager, Jr. (1977) DNA damage and repair in rat tissues following administration of azaserine. Cancer Res. 37:3925-3931.
4. Zurlo, J., T.J. Curphey, R. Hiley, and D.S. Longnecker (1982) Identification of 7-carboxymethylguanine in DNA from pancreatic acinar cells exposed to azaserine. Cancer Res. 42:1286-1288.
5. Zurlo, J., C.I. Coon, D.S. Longnecker, and T.J. Curphey (1982) Binding of ^{14}C-azaserine to DNA and protein in the rat and hamster. Cancer Lett. 16:65-70.
6. Zurlo, J., B.D. Roebuck, J.V. Rutkowski, T.J. Curphey, and D.S. Longnecker (1982) Effect of pyridoxal deficiency on pancreatic DNA damage and nodule induction by azaserine. Proc. Amer. Assoc. Cancer Res. 23:68.
7. Longnecker, D.S., B.D. Roebuck, J.D. Yager, Jr., H.S. Lilja, and B. Siegmund (1981) Pancreatic carcinoma in azaserine-treated rats: Induction, classification, and dietary modulation of incidence. Cancer 47:1562-1572.
8. Longnecker, D.S., J. French, E. Hyde, H.S. Lilja, and J.D. Yager, Jr. (1977) Effect of age on nodule induction by azaserine and DNA synthesis in rat pancreas. J. Natl. Cancer Inst. 58:1769-1775.

9. Konishi, Y., A. Denda, H. Maruyama, H. Yoshimura, J. Nobuoka, and M. Sunagawa (1980) Pancreatic tumors induced by a single intraperitoneal injection of azaserine in partial pancreatectomized rats. Cancer Lett. 9:43-46.
10. Roebuck, B.D., and D.S. Longnecker (1977) Species and rat strain variation in pancreatic nodule induction by azaserine. J. Natl. Cancer Inst. 59:1273-1277.
11. Longnecker, D.S., and B.G. Crawford (1974) Hyperplastic nodules and adenomas of exocrine pancreas in azaserine-treated rats. J. Natl. Cancer Inst. 53:573-577.
12. Longnecker, D.S. (1982) Editorial: Experimental pancreatic carcinogenesis. Lab. Invest. 46:543-544.
13. Rao, M.S., M.P. Upton, V. Subbarao, and D.G. Scarpelli (1982) Two populations of cells with differing proliferative capacities in atypical acinar cell foci induced by 4-hydroxyaminoquinoline-1-oxide in the rat pancreas. Lab. Invest. 46:527-534.
14. Roebuck, B.D., and D.S. Longnecker. Dietary lipid promotion of azaserine-induced pancreatic tumors in the rat. In Diet and Cancer: From Basic Research to Policy Implications, D. Roe, ed. Alan R. Liss (in press).
15. Longnecker, D.S., H.S. Lilja, J.I. French, E. Kuhlmann, and W.W. Noll (1979) Transplantation of azaserine-induced carcinomas of pancreas in rats. Cancer Lett. 7:197-202.
16. Rao, K.N., S. Takahashi, and H. Shinozuka (1980) Acinar cell carcinoma of the rat pancreas grown in cell culture and in nude mice. Cancer Res. 40:592-597.
17. Yager, J.D., Jr., B.D. Roebuck, J. Zurlo, D.S. Longnecker, E.O. Weselcouch, and S.A. Wilpone (1981) A single-dose protocol for azaserine initiation of pancreatic carcinogenesis in the rat. Int. J. Cancer 28:601-606.
18. Roebuck, B.D., J.D. Yager, Jr., D.S. Longnecker, and S.A. Wilpone (1981) Promotion by unsaturated fat of azaserine-induced pancreatic carcinogenesis in the rat. Cancer Res. 41:3961-3966.
19. Longnecker, D.S., T.J. Curphey, E.T. Kuhlmann, and B.D. Roebuck (1982) Inhibition of pancreatic carcinogenesis by retinoids in azaserine-treated rats. Cancer Res. 42:19-24.
20. Schaeffer, B.K. (Pers. Comm.)
21. Konishi, Y., A. Denda, S. Inui, S. Takahashi, and H. Kondo (1976) Pancreatic carcinoma induced by 4-hydroxyaminoquinoline 1-oxide after partial pancreatectomy and splenectomy in rats. Gann 67:919-920.
22. Konishi, Y., A. Denda, Y. Miyata, and H. Kawabata (1976) Enhancement of pancreatic tumorigenesis of 4-hydroxyaminoquinoline 1-oxide by ethionine in rats. Gann 67:91-95.
23. Gould, M.N. (1982) Mammary gland cell-mediated mutagenesis of mammalian cells by organ-specific carcinogens. Cancer Res. 40:1836-1841.

24. Dissin, J., L.R. Mills, D.L. Mains, O. Black, and P.D. Webster (1975) Experimental induction of pancreatic adenocarcinoma in rats. J. Natl. Cancer Inst. 55:857-864.
25. Bockman, D.E., O. Black, L.R. Mills, and P.D. Webster (1978) Origin of tubular complexes developing during induction of pancreatic adenocarcinoma by 7,12-dimethylbenz(a)anthracene. Am. J. Pathol. 90:645-658.
26. Svoboda, D.J., D.L. Azarnoff (1979) Tumors in male rats fed ethyl chlorophenoxyisobutyrate, a hypolipidemic drug. Cancer Res. 39:3419-3428.
27. Reddy, J.K., and M.S. Rao (1977) Transplantable pancreatic carcinoma of the rat. Science 198:78-80.
28. Milman, H.A., J.M. Ward, and K.C. Chu (1978) Pancreatic carcinogenesis and naturally occurring pancreatic neoplasms of rats and mice in the NCI carcinogenesis testing program. J. Environ. Pathol. Toxicol. 1:829-840.
29. Levison, D.A., R.G.H. Morgan, J.S. Brimacombe, D. Hopwood, G. Coghill, and K.G. Wormsley (1979) Carcinogenic effects of di(2-hydroxypropyl)nitrosamine (DHPN) in male Wistar rats: Promotion of pancreatic cancer by a raw soya flour diet. Scand. J. Gastroenterol. 14:217-224.
30. Longnecker, D.S., J. Zurlo, T.J. Curphey, and W.E. Adams (1982) Induction of pancreatic DNA damage and nodules in rats treated with N-nitrosobis(2-oxopropyl)amine and N-nitroso(2-hydroxypropyl)(2-oxopropyl)amine. Carcinogenesis 3:715-717.
31. Longnecker, D.S., and T.J. Curphey (Pers. Comm.)
32. Longnecker, D.S., T.J. Curphey, H.S. Lilja, J.I. French, and D.S. Daniel (1980) Carcinogenicity in rats of the nitrosourea amino acid Nδ-(N-methyl-N-nitrosocarbamoyl)-L-ornithine. J. Environ. Pathol. Toxicol. 4:117-129.
33. Zurlo, J., D.S. Longnecker, D.A. Cooney, E.T. Kuhlmann, and T.J. Curphey. Studies of pancreatic nodule induction and DNA damage by D-azaserine. Cancer Lett. 12:75-80.
34. Lilja, H.S., D.S. Longnecker, T.J. Curphey, D.S. Daniel, and W.E. Adams (1981) Studies of DNA damage in rat pancreas and liver by DON, ethyl diazoacetate, and azaserine. Cancer Lett. 12:139-146.
35. Roebuck, B.D., and K.J. Baumgartner. Characterization of two populations of pancreatic atypical acinar cell foci induced by azaserine in the rat (in preparation).
36. Rowlatt, U., and F.J.C. Roe (1967) Epithelial tumors of the rat pancreas. J. Natl. Cancer Inst. 39:18-32.

RODENT AND HUMAN MODELS FOR THE ANALYSIS OF INTESTINAL TUMOR DEVELOPMENT

Martin Lipkin, Eileen Friedman, Paul Higgins,
and Leonard Augenlicht

Memorial Sloan-Kettering Cancer Center
New York, New York 10021

INTRODUCTION

In recent years, new approaches to the detection of early lesions associated with large bowel neoplasia have been explored. These approaches have led to the identification of a series of progressive stages of abnormal cell development, both in rodents exposed to carcinogens and in humans with hereditary predisposition to colorectal cancer.

Currently, measurements of abnormal cell proliferation have provided a basis for analyzing the growth characteristics of gastro-intestinal cells during neoplastic transformation. Some of the changes observed during the development of neoplasms in rodents, following exposure to chemical carcinogens, are similar to those that occur in man. In epithelial cells of the colon, cells gain an increased ability to proliferate, leading to an expansion of the proliferative compartment within the colonic crypts; the cells then develop additional properties which enable them to accumulate in the mucosa as is characteristic of the hyperplastic state.

In this chapter, both proliferative and biochemical abnormalities observed during neoplastic transformation in colonic epithelial cells will be described. Newer techniques that can be applied to analyze the effects of carcinogens and tumor promoters on colonic epithelial cells both *in vivo* and *in vitro* will be reviewed. They include the development of tissue and organ culture methods, improving the growth of normal and neoplastic colonic epithelial cells, measurements of abnormal growth responses of premalignant epithelial cells, and the appearance of abnormal antigens and nucleic acid sequences within the cells.

PROLIFERATION OF NORMAL CELLS IN THE LARGE INTESTINE

The proliferative zone in the mucosa of the large intestine of rodents and humans covers approximately the basal three-quarters of the crypts (illustrated in Figures 1-4 below). Epithelial cells migrate toward the gut lumen and are extruded from the mucosal surface between crypts. These large intestinal crypts are more closely spaced than in either the stomach or small intestine, and the surface of the mucosa is flat. The surface area of the colonic mucosa between crypts is larger than the surface area of the crypt columns, supporting the assumption that the major function of the surface lining cells is absorption, while the major activity of the crypt cells is cell replacement.

All of the colonic epithelial cells, i.e., the columnar, mucous, and enteroendocrine cells, most likely originate in the crypt base (1). Migration takes from 3 to 8 days in man and from 2 to 3 days in rodents. Argentaffin (enteroendocrine) cells of the human rectal mucosa undergo slow renewal in 35 to 100 days (2).

Previous studies with the colons of mice showed the generation time of colonic epithelial cells was 16 hr, DNA synthesis 6.5 hr, G_2 and mitosis 1.5 hr, and G_1 8 hr. Rapid proliferation of some of the cells adjacent to already mature cells took place in the middle one-third of the colonic crypts; in contrast, in the jejunum the separation between the proliferative and nonproliferative regions was more distinct. ^{3}HTdR-labeled epithelial cells then stayed in the colonic mucosa for a longer time than in the jejunum (3-5).

In humans, approximately 10% to 20% of all cells undergo DNA synthesis simultaneously in the lower three-fourths of the crypt columns. In the upper crypt, cell proliferation is less evident as more cells differentiate, and cells cease proliferative activity as they near the surface of the colonic crypts (6-13). In the sigmoid colon, slower re-entry of cells in the cell cycle has been observed compared to other areas; therefore faster recycling of cells occurred in the jejunum (3,4,9).

MODIFICATIONS IN COLONIC EPITHELIAL CELLS INDUCED BY CARCINOGENS

Proliferative and pathological abnormalities in colonic epithelial cells similar to those observed in humans can be induced in rodent colonic epithelial cells. Rodent strains have different susceptibilities to the induction of colon cancer, as occurs among humans. In rodents, colonic cell transformations can be induced by 1,2-dimethylhydrazine (DMH), methylazoxymethanol (MAM), *N*-methyl-*N*'-nitro-*N*-nitrosoguanine (MNNG), and *N*-methyl-N-nitrosourea (MNU) (14-16).

Tumor nodules develop in mice following exposure to DMH. The

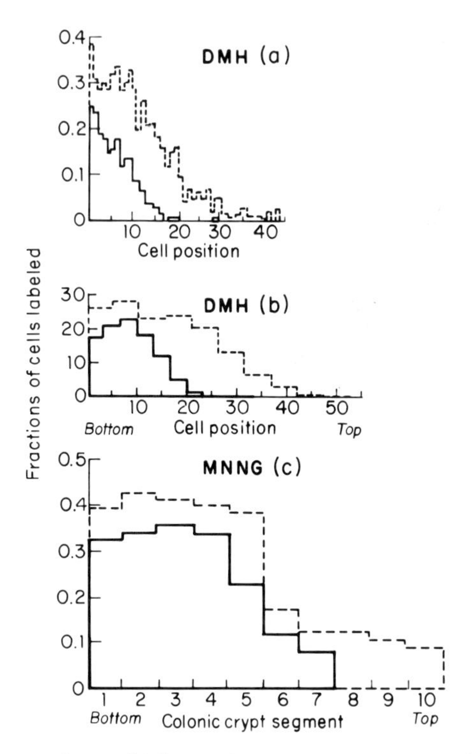

Fig. 1. Extension of proliferative compartment of colonic crypts toward lumenal surface of crypts, after repeated exposure to chemical carcinogens. Control groups: solid lines. Carcinogenic groups: dashed lines.

(a) Fraction of cells labeled at various cell positions within colonic crypts, one hour after injection of ^{3}H-thymidine and 87 days after start of weekly DMH injections into mice (20 μg/g) (redrawn from Thurnherr, et al., ref. 14).

(b) Fractions of cells labeled at various cell positions within colonic crypts, 125 days after beginning of DMH treatment in mice (from Chang, ref. 16).

(c) Fraction of cells labeled in segments of colonic crypt after rectal installation of MNNG in rat (from Kikkawa, ref. 15).

main site of activity is the distal colon, the same distribution noted to occur in humans. In mice, multifocal tumors ranging from adenomatous polyps to metaplasias and carcinomas grow from the mucosa and then protrude into the lumen of the rectosigmoid colon. Early focal atypias and hyperplasias, adenomatous polyps and carcinomas, located mainly on the folds, appear to be part of the progressive pathologic changes in mice and rats that develop following administration of DMH.

Pozharisski (17) has recently demonstrated a marked similarity between these pathologic changes and those found in man. The proliferative changes in adenomas of man also are unusual, and studies are being carried out at present in order to evaluate further correlations that might be present in lesions of differing premalignant potential. In villous adenomas of a subject with familial polyposis, it was recently shown that cells proliferated at the surface of the crypts, and were able to migrate deep into the crypt column instead of being extruded from the surface (18).

In rodents, DMH and MNNG both induce an extension of the proliferative zone of the flat mucosa toward the surfaces of the colonic crypts as seen in Fig. 1, similar to the findings observed in man. Thus, some of the colonic epithelial cells of rodents exposed to repeated injections of DMH also develop the capacity to continue to synthesize DNA throughout most of their life-span. Thymidine-labeled cells show an increase in both their position up from the crypt base and in their total number; these cells also show a shift into expanding adenomas and carcinomas. The tumors that develop manifest a proliferation of neoplastic epithelial cells near the surface of the lesion, with a continued expansion into the colonic lumen.

PROLIFERATION OF DISEASED CELLS IN THE LARGE INTESTINE OF MAN

In individuals with hereditary diseases that predispose to colorectal cancer, an early abnormal characteristic of colonic epithelial cells is modified epithelial cell proliferative activity, identifiable in normal-appearing mucosa of asymptomatic individuals. During progressive stages of abnormal development, epithelial cells gain an increased ability to proliferate and to accumulate in the mucosa (5,10-13). The identification of these changes has aided our understanding of early events that occur during neoplastic transformation of cells in colorectal cancer. *In vitro* and *in vivo* assays that measure the location of cells incorporating ^{3}HTdR within colonic crypts enable these changes to be identified.

As noted above, in the normal human colon cell proliferation occurs in the lower and mid regions of the colonic crypts (Figure 2). About 10 percent to 20 percent of the proliferating cells are in DNA synthesis simultaneously. During migration of these cells, the number that continue to proliferate decreases as they progress to the lumenal region of the crypts. They undergo terminal differentiation within hours, and proliferation ceases as they migrate to the crypt surfaces to be extruded. Our earlier studies in humans showed that the colonic mucosa is replaced by new cells in 4-8 days (6,7).

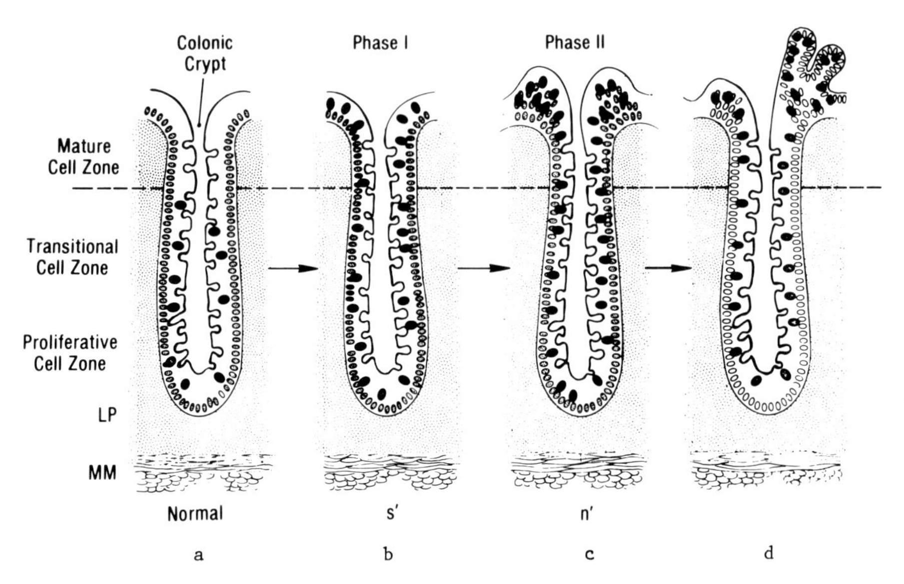

Fig. 2. A sequence of events to account for the location of abnormally proliferating colonic epithelial cells before and during the formation of polypoid neoplasms in humans.

(a) Shows location of proliferating and differentiating epithelial cells in normal colonic crypt. Dark cells illustrate thymidine labeling in cells that are synthesizing DNA and preparing to undergo cell division. As cells pass from the proliferative zone through the transitional zone, DNA synthesis and mitosis are repressed, and migrating epithelial cells leave the proliferative cell cycle to undergo normal maturation before they reach the surface of the mucosa.

(b) Shows the development of a Phase I proliferative lesion in the colonic epithelial cells as they fail to repress the incorporation of ^{3}HTdR into DNA and begin to develop an enhanced ability to proliferate. The mucosa is flat, and the number of new cells born equals the number extruded without excess cell accumulation in the mucosa.

(c) The development of a Phase II proliferative lesion in colonic epithelial cells. The cells incorporate ^{3}H-thymidine into DNA and also have developed additional properties that enable them to accumulate in the mucosa as neoplasms begin to form.

(d) Shows further differentiation of abnormally retained proliferating epithelial cells into pathologically defined neoplastic lesions, including adenomatous polyps and villous papillomas (5).

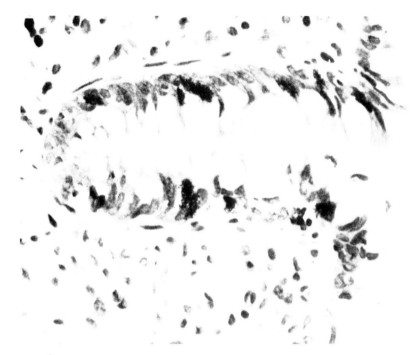

Fig. 3. ^{3}H-Thymidine incorporation into epithelial cells near the crypt surface of flat colonic mucosa in a biopsy specimen from a high risk subject (48).

However, in subjects with familial polyposis, patches of flat mucosa have been found having colonic epithelial cells that fail to repress DNA synthesis during migration to the surface of the mucosa (5,11-13). This is illustrated in Figures 2 and 3; it occurs in normal-appearing colonic epithelial cells before as well as after the cells develop adenomatous changes and begin to accumulate as polyps. Observed in over 80 percent of random biopsy specimens (19,20), a failure of cells to repress DNA synthesis has now been shown to occur with higher frequency in patients with familial polyposis than in population groups at low risk for colorectal cancer (Table 1). More recently, a significant increase in abnormal proliferative activity has also been noted in the colonic mucosa of strongly colon cancer-prone families without polyposis (Table 1).

Our current work is focusing on these findings in other high, average, and low risk populations, and on the development of discriminatory indices in subjects at risk in colon cancer-prone families. For this purpose, these and related measurements are being automated in order to study both genetic and environmental factors contributing to this abnormal cell development. The data from specific population groups are being analyzed using computer-assisted methods to evaluate differences between population groups

Table 1. Labeling distribution (percent of ^{3}HTdR labeled cells) in upper third of colonic crypts (LD μ 1/3) in flat normal appearing mucosa of high- and low-risk populations.

Population Group	No. of Individuals	Ratio of Mean Value of (LD μ 1/3) in Test Group to Mean Value in Control Group(C)
Familial polyposis	18	2.49**
Symptomatic (FP_s)*	18	2.49**
Asymptomatic (FP_a)‡	11	2.68**
Polyp-free branch	10	1.22***
Familial colon cancer		
Symptomatic (FCC_s)*	8	3.21**
Asymptomatic (FCC_a)‡	23	3.16**
Cancer-free branches	16	1.72***
Non-familial colon cancer	13	1.41***
Normal subjects	15	1.00***

*Affected individuals
‡Possible carriers (about 50% risk of developing disease)
**Percentage of labeled cells significantly greater than normal subjects, non-familial colon cancer cases, or polyp-free and cancer-free branches of families ($p < 0.05$).
***Percentage of labeled cells not significantly greater than normal subjects.

(From reference 49)

of interest. An example of the type of computer-generated output that can now be derived from such _in vitro_ organ culture assays is shown in Figure 4.

In the disease familial polyposis, the further modifications of proliferative activity also believed to occur prior to the development of malignancy are shown in Fig. 1. Following the failure of cells that have inherited the germinal mutation to repress DNA synthesis during migration in the colonic crypts, additional events take place, giving rise to new clones from the original cell population. An early event leads to the development of the well-known adenomatous cells that proliferate and accumulate near the surface of the mucosa (18). The further unusual ability of these adenomatous cells to migrate or "intrude" deep into the crypt column, instead of being extruded from the surface, is shown in Figure 5. Further modifications then occur in the cells as they develop into carcinomas in situ, and as they acquire the ability to penetrate through the basal membrane with invasive malignancy. The observed multistage development of these neoplasms suggests a contribution of endogenous or exogenous carcinogenic or promoter elements, which likely interact with the epithelial cells that have a genetic defect predisposing them to neoplasia. It also allows for the introduction of preventive measures to inhibit the steps leading to malignant transformation of the cells.

In the familial polyposis model of human cancer development, as epithelial cells develop an increased capacity to proliferate and to accumulate in the mucosa, areas of microscopic hyperplasia and adenomatous foci develop. As these adenomatous foci become recognizable lesions, they are seen to consist of an overgrowth of hyperplastic, mucus-secreting intestinal epithelium together with a stroma of loose connective tissue and blood vessels. As an

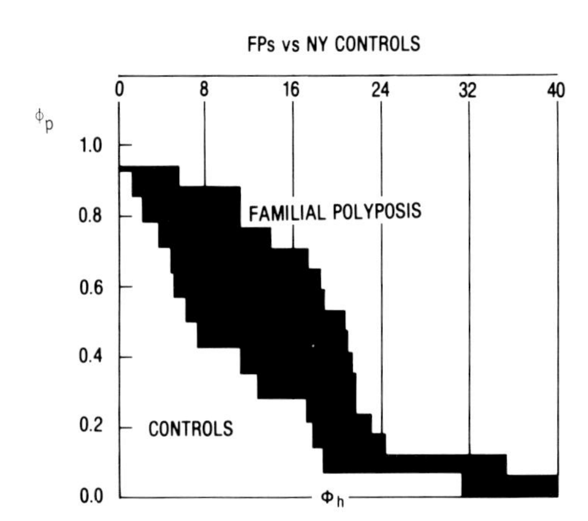

Fig. 4. Diagram of computer-generated curves comparing crypt column occupancy fractions of tritiated thymidine-labeled epithelial cells in a high risk and in a low risk group. Abscissa record ϕ_h, the fraction of labeled cells found within the upper 40 percent of assayed crypt columns adjacent to and including the lumenal surface. The ordinate ϕ_p records the fraction of all individuals in a given group whose measured ϕ_h values equal or exceed the abscissa value. Significant differences are present between high and low risk groups for the labeled cell distributions over all the crypt height compartments ($p < .005$) (48).

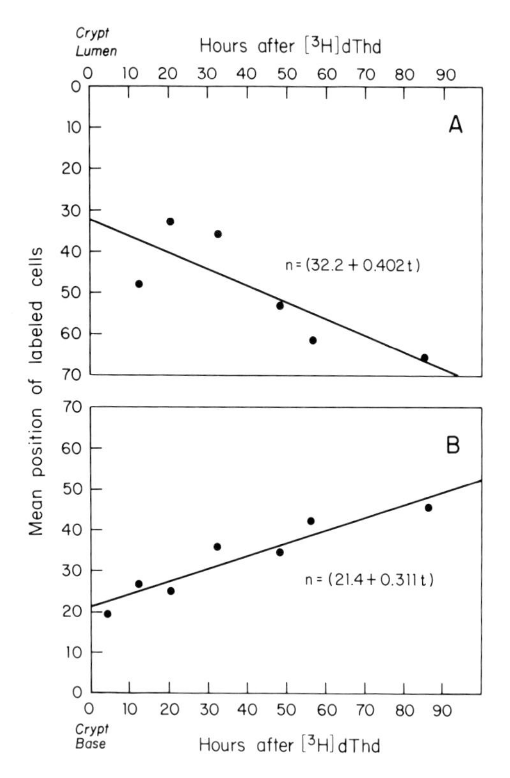

Fig. 5. Mean position of labeled cells versus time of assay for adenomas and flat mucosa, in rectal biopsies removed from a patient with familial polyposis (18).

indication of enhanced cell proliferation, the epithelial cells exhibit hyperchromatism, stratification, and decreased mucus production together with increased mitotic activity. Studies have shown that only a single crypt of Lieberkuhn, or part of its surface, can be affected (21). Occasionally, the epithelium buds like a small diverticulum from the main glandular spaces lined by one layer of normal mucus-secreting cells. The transition from normal to abnormal epithelium can be an abrupt process halfway up to the crypts of Lieberkuhn. When the tubules have branched sufficiently and have expanded to about 3 mm in diameter with 3 mm above the mucosal surface, the lesion can be recognized as an adenoma that will differentiate further into one of the three characteristic types - tubular, villous, or villotubular.

PROLIFERATIVE ABNORMALITIES IN EPITHELIAL CELLS OF OTHER RENEWING TISSUES

A failure of colonic epithelial cells to repress DNA synthesis

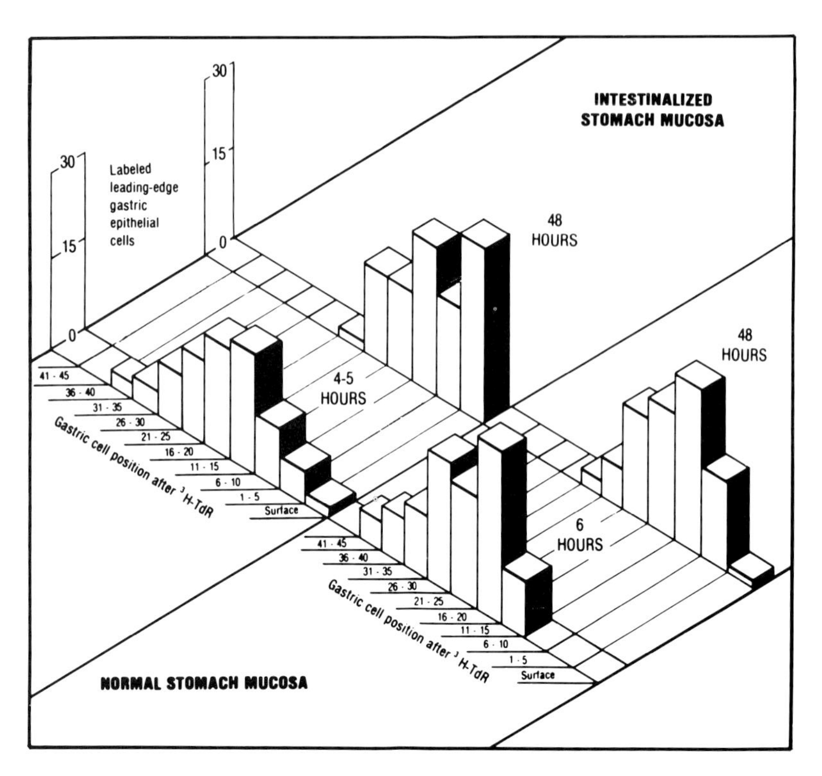

Fig. 6. Cell position below surface of gastric pits occupied by labeled leading-edge gastric epithelial cells at early and late periods after *in vivo* pulse injection of tritiated thymidine into two human subjects. In intestinalized stomach mucosa, the zone of labeled cells extends to the surface of the gastric pits after pulse injection, in contrast to normal stomach mucosa (48).

also occurs during epithelial cell renewal in other human diseases. Among these are ulcerative colitis of the colon (22). In atrophic gastritis, a condition associated with the development of gastric malignancy, epithelial cells also fail to repress DNA synthesis and undergo abnormal maturation as they migrate through the gastric mucosa (23,24) (Figure 6). Analogous changes are found in the stomachs of rodents after exposure to a chemical carcinogen (25). A similar event occurs in precancerous disease of the human cervical epithelium (Figure 7) and in the cervix of rodents (Figure 8) after introduction of a chemical carcinogen (26). Thus, during the development of neoplasms in organs other than the colon that contain renewing epithelium, persistent DNA synthesis occurs in cells that normally would be terminal or end cells. As occurs in the colon in familial polyposis, similar pathological changes accompany this development, and also lead to atypias, dysplasias, and malignancy.

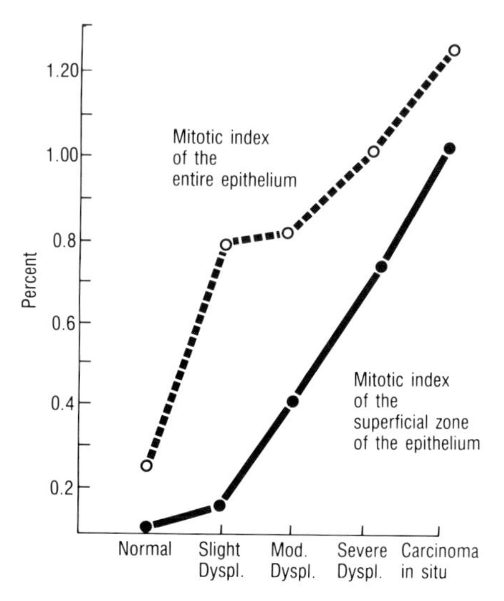

Fig. 7. Relationship of percent of mitotic human cervical epithelial cells to increasing degrees of cervical dysplasia and carcinoma _in situ_ (50).

NEWER METHODS DEVELOPED TO STUDY THE STAGES OF NEOPLASTIC TRANSFORMATION IN CELLS OF RODENTS AND HUMAN COLON, BEFORE AND AFTER THE ADMINISTRATION OF CARCINOGENS AND TUMOR PROMOTERS

Transplantable Technique

Other methods have now been developed to study the characteristics of colonic epithelial cells during the stages of neoplastic transformation induced by carcinogens. Many studies have been carried out indicating that nude mice accept transplants of human malignant tumors. These include tumors of colon and rectum, kidney, breast, pancreas and pharynx, melanoma, sarcoma, Burkitt's lymphoma, hepatoma, and epidermoid carcinoma (27-36). Inoculates of tissue culture cell lines of cancer of colon and breast, melanoma, teratoma, osteosarcoma, and Burkitt's lymphoma have also been used.

In a recent study (37), benign neoplasms of the human colon, i.e., adenomatous polyps, also were successfully transplanted under the kidney capsule of athymic mice, in addition to colonic adenocarcinomas and normal colonic mucosa. The benign human adenomatous cells survived for periods of up to 28 days, normal rodent colonic epithelial cells for 45 days, and colonic carcinoma cells for 43 days (Figure 9-11 and Table 2). This was established by morphologic criteria, and by the incorporation of tritiated thymidine into DNA of the epithelial cells. Thus, the technique of transplantation can supplement _in vitro_ methods for the maintenance of adenomatous tissue derived from human colonic mucosa, in order to facilitate studies of growth characteristics and transformation of the cells.

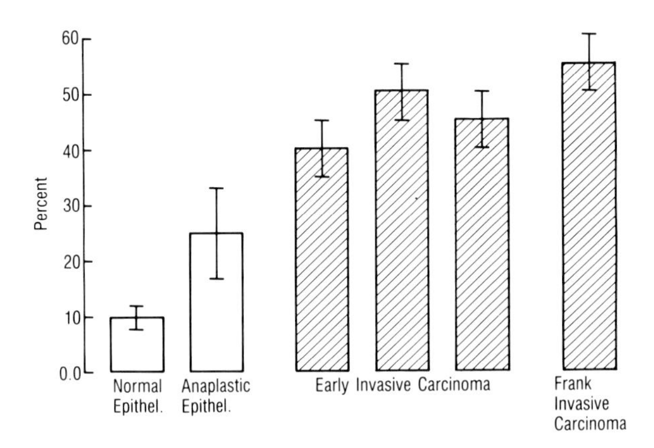

Fig. 8. Relationship of percent of ^{3}HTdR-labeled cervical epithelial cells in rodents to increasing degrees of dysplasia and to carcinoma (26).

Organ Culture

Difficulty in maintaining explants of colonic mucosa in a viable state for more than several days has regularly been encountered, and thus the ability to culture specimens of human gastrointestinal mucosa *in vitro* has been limited. Despite this limitation, short-term organ cultures of specimens of human gastrointestinal mucosa have been valuable in defining the spatial distributions within the mucosa of proliferating and differentiating cells, and abnormalities in this distribution that develop in diseases of the colon and stomach. Earlier studies have illustrated the unique value of explant culture, where cells are maintained in a three-dimensional structure characteristic of the whole tissue, in contrast to the disaggregation of cells which occurs in tissue culture.

In recent years, clinical endoscopic methods have been sufficiently improved to make available for explant culture a wide variety of human gastrointestinal mucosal biopsy specimens. This means that further improvements in explant culture methodology, for the maintenance of both short- and long-term viability *in vitro*, might find application to analyze the effects of carcinogens and tumor promoters and the development of disease states. Studies of morphologic and metabolic characteristics of colonic epithelial cells (38-40) have utilized these advances in explant culture techniques.

In a recent study (40), biopsy specimens of human colonic mucosa taken from the rectosigmoid region of normal subjects were maintained in explant culture for 4 days. Histological, microautoradiographic, and chemical measurements were carried out to evaluate cell replication, the effect of deoxycholic acid, and the

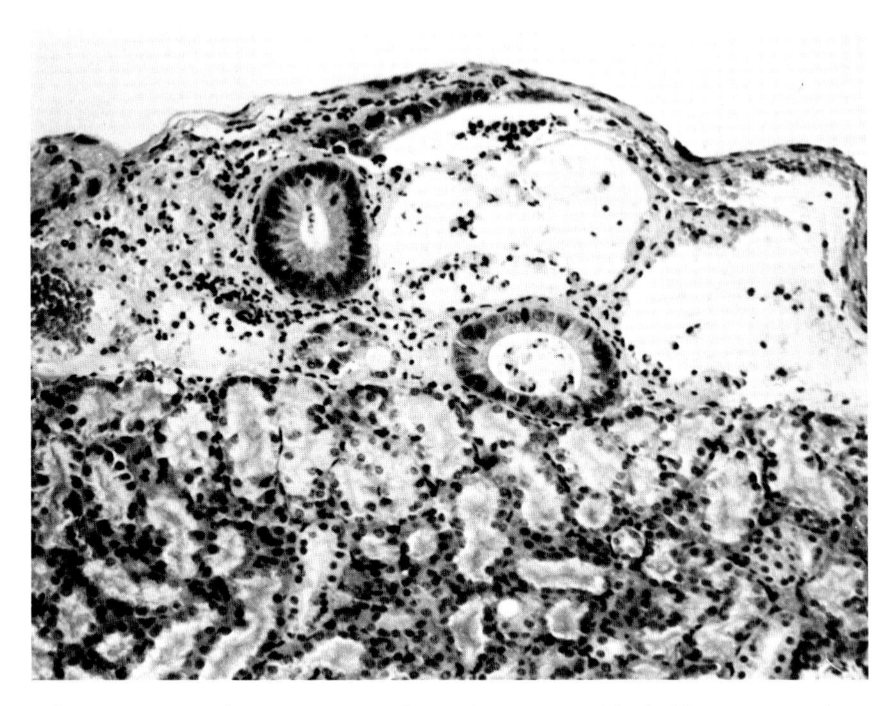

Fig. 9. Human adenomatous polyp showing well-differentiated glandular structure implanted under the kidney capsule of a nude mouse. Specimen survived 15 days. HE x 80 (37).

incorporation of uridine and leucine into RNA and protein. Active cell replication was shown over a period of one day (Figure 12), as well as an increase in the number of cells that synthesized DNA when deoxycholic acid was added to the culture medium. However, at later time periods, despite morphological evidence of tissue viability, the synthesis of DNA, RNA, and protein within colonic epithelial cells decreased, and the total numbers of cells in the crypt columns declined. The findings thus indicated good maintenance of metabolic activities of colonic epithelial cells in short-term explant culture, and the potential to extend studies of this type to analyze the effects of carcinogens and tumor promoters on human cells.

Tissue Culture

Tissue culture has now been successfully applied to the *in vitro* study of colonic adenomas. In a recent study (41) (Figure 13), we described methods to grow reproducibly human premalignant colonic epithelial cells from benign tumors in tissue culture. The cells grew as tightly-packed colonies from small explants free of fibroblasts and remained viable for up to 8 weeks. Cells were identified as epithelial by the presence of characteristic epithelial structures: gap junctions, tight junctions, many desmosomes, and a brush border at the apical surface. The living cells transported

Table 2. Duration of survival of adenoma tissue and microscopic appearance.

Type of Tissues	Duration of Survival When Removed from Kidney Capsule		Microscopic Appearance
	Series	Number of Days	
Adenomatous polyp	1	5	Well-differentiated glands with dense connective tissue stroma
	2	10	Well-differentiated glands with a few inflammatory cells
	3	13	Portions of glands with necrosis and inflammation
	4	15	Well-differentiated glands with a few inflammatory cells
	5	24	Well-differentiated glands and few inflammatory cells
	6	25	Well-differentiated glands with necrotic cells in lumen
	7	28	Acellular mucin-filled spaces with portions of glands and inflammatory cells
Adenocarcinoma	1	10	Well-differentiated glands with fibrous stroma and inflammatory cells
	2	43	Moderately well-differentiated glandular structure with mucinous stroma growing in kidney parenchyma

(From reference 37)

water and salts, forming "domes" or hemicysts in culture, namely areas where the sheet of cells had detached from the plate because of the pressure of the water pumped through the cell from the brush border to the ventral surface.

Cells were cultured with approximately equal frequency from each class of benign tumor (adenoma): 20/23 (87%) tubular (low malignant potential); 15/16 (94%) villotubular (intermediate malignant potential); and 7/8 (88%) villous (high malignant potential). Cultures from these tumors were designated as early-, middle-, and

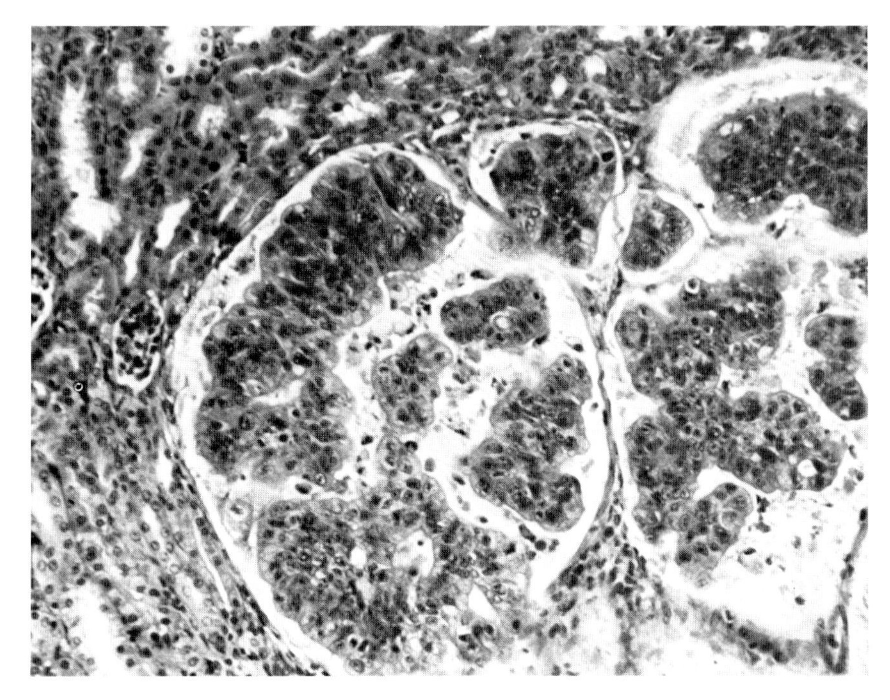

Fig. 10. Human colon carcinoma implanted under kidney capsule of nude mouse, showing moderately well differentiated glandular structure, mucinous stroma, and invasion of carcinoma cells into kidney tissue. Specimen survived 43 days. HE x 300 (37).

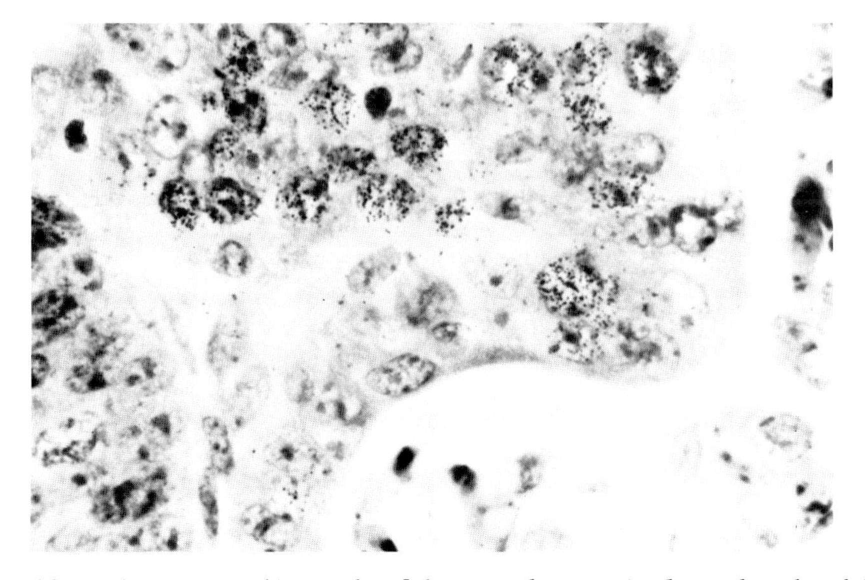

Fig. 11. Microautoradiograph of human adenoma implanted under kidney capsule of nude mouse, showing ^{3}HTdR-labeled epithelial cells x 1,250 (37).

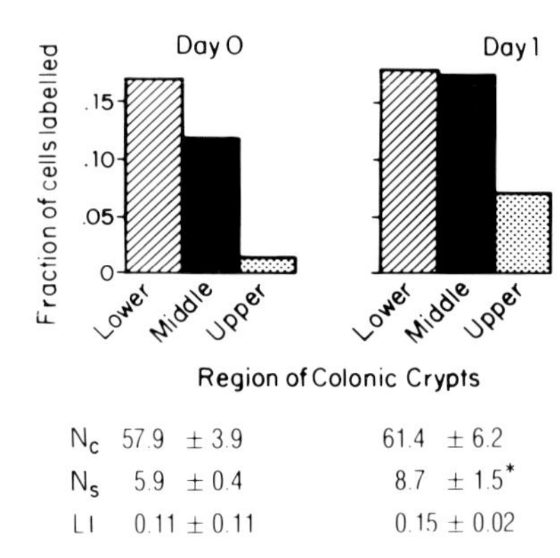

Fig. 12. Fraction of cells labeled with ^{3}HTdR in lower, middle, and upper thirds of colonic crypts in organ culture, after pulse labeling for one hour on Day 0, and chasing with non-radioactive ^{3}HTdR for 1-4 days (40). N_c, number of epithelial cells per crypt column; N_s, number of epithelial cells labeled with ^{3}HTdR; LI, labeling index, (*) $p < .05$ comparing day 0 and day 1.

late-stage premalignant because of the clinical prognosis of the tumors themselves.

A double-blind experiment was performed in which the response of cultured cells from 10 different benign tumors to exogenous epidermal growth factor (EGF) was measured by assaying the fraction of cells incorporating ^{3}H-thymidine in a continuous labeling experiment. EGF increased the fraction of replicating cells cultured from three tubular adenomas an average of 82% while it had no effect on cells cultured from six villous adenomas or the one villotubular adenoma tested (Table 3).

These experiments demonstrated that the early-stage premalignant (tubular adenoma) cells could indeed be grouped into one category because they exhibited a common response to a hormone. Late-stage premalignant (villous adenoma) cultures fell into another category, that of hormone nonresponsive cells. EGF was present in the fetal calf serum used in culture medium, so we might have been seeing only a decreased response to EGF, not necessarily a complete absence of response to EGF. These results are consistent with a loss of the ability to respond to growth modulation by EGF of premalignant colonic epithelial cells as they progress from one stage, the tubular, to stages with more malignant potential, the villotubular and the villous. The EGF has a strong structural homology to human urogastrone, a trophic hormone in the intestine, suggesting that the _in vitro_ results may mirror the _in vivo_ growth modulation of benign tumors.

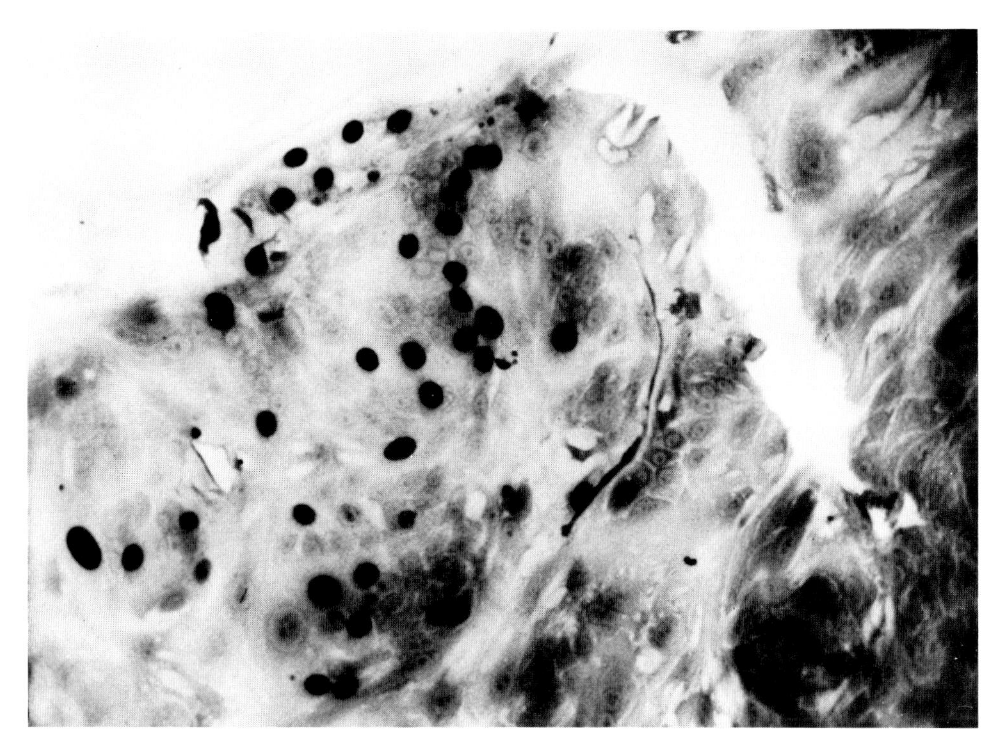

Fig. 13. Microautoradiograph of adenomatous cells growing in tissue culture 1 day after addition of ^{3}HTdR. Blackened cells are labeled with ^{3}HTdR at the periphery of expanding colony (41).

In a further study (42), the concept was strengthened that benign tumor classes could be ordered into a series reflecting cellular progression through stages of premalignancy. An antisera raised against second trimester human fetal tissue and absorbed with pooled adult tissues was found to react with some cultured adenomatous cells. A double-blind study was performed with cells cultured from 28 adenomas of different histopathologic classes and colon carcinomas, both in primary culture and as established cell lines. Fetal-associated antigen was found most frequently in cells cultured from the carcinomas, and the frequency decreased in parallel with a decrease in malignant potential of the adenomas (see below). Antigen expression could thus be graded as carcinoma (71%) > villous adenoma (70%) > villotubular adenoma (43%) > tubular adenoma (18%). Many tumors synthesize proteins whose expression is otherwise limited to the time of fetal development. Benign tumor cells, therefore, were found to express this property with increasing frequency as they progressed toward malignancy.

Table 3. Effect of EGF on epithelial cell proliferation.

Classification of Tumors	Percentage of ^{3}H-Thymidine Labeled Cells		Stimulation by EGF (%)
	Control	EGF Added	
Tubular A[a]	30.8 ± 4.2[b]	48.8 ± 3.5	58
Tubular B	15.4 ± 2.5	26.0 ± 2.9	69
Tubular C	22.2 ± 5.7	48.8 ± 5.2	69
Villotubular A	4.1 ± 0.3	5.2 ± 0.4	Marginal
Villous A	10.7 ± 1.5	11.8 ± 5.7	0
Villous B	32.3 ± 3.4	25.3 ± 4.6	0
Villous C	24.7 ± 3.8	22.9 ± 4.6	0
Villous D	36.1 ± 3.9	37.7 ± 3.9	0
Villous E	19.0 ± 0.1	20.4 ± 5.0	0
Villous F	30.4 ± 4.3	35.9 ± 2.2	0

[a]Cultures derived from villous tumors A, B, and C and villotubular tumor A were pulsed for 2 hr with 5 Ci/ml ^{3}H-thymidine. All other cultures were labeled for 64 hr. After development, nuclei were stained with alkaline hematoxylin. An average of 1,100 cells from five separate colonies were counted for each value listed. All experiments were performed using the same medium preparation, including the same bottle of fetal bovine serum and the same EGF stocks (41).
[b]Mean ± standard deviation.

The effects of two agents, TPA (12-O-tetradecanoylphorbol-13-acetate) and DOC (deoxycholic acid, a secondary bile acid found in the colon), which act as tumor promoters in the gastrointestinal tract of experimental animals, also were compared, using primary cultures of human premalignant colonic epithelial cells at different stages in tumor progression (43) (Figure 14). DOC affected only the early-stage premalignant cells. Of the three mitogenic compounds tested on early-stage cells (DOC, TPA, EGF), DOC had the greatest stimulatory effect on DNA replication. Its major role in tumor progression *in vivo* may be to enlarge significantly the proliferative cell population of normal colonic cells and early-stage premalignant cells, but not late premalignant or tumor cells.

TPA stimulated the growth of these cells to a lesser degree. However, in contrast to DOC, its most pronounced effect was on intermediate and late-stage premalignant cell cultures, where it induced cell clustering, multi-layering, and concomitant release of a protease with many properties similar to a plasminogen activator (Figure 14). Thus, the concept of a differential action of tumor-promoting agents during different stages of normal and abnormal cell growth is presented in Figure 15. In the colon, epithelial cells are arranged in a single layer. Premalignant cells sometimes are pseudo-stratified and show morphological distortion *in vivo*. However, carcinoma *in situ*, that is, without rupture of the basement membrane and invasion of the muscular layer, can appear in the colon as a multilayering of cytologically aberrant cells within a benign tumor or adenoma. Perhaps, *in vivo*, this transition is influenced by release of a protease such as a plasminogen activator from the premalignant late-stage epithelial cells in response to an endogenous promoting agent.

There has been much speculation about the possible interaction between TPA and the hormone EGF. This phorbol ester did not block the binding of EGF in colon cells, and presumably has another receptor. The premalignant epithelial cells which responded to TPA by secreting protease and clustering were predominantly (83%) restricted to the villous and villotubular classes. These cells were not stimulated to divide by EGF added to growth medium (41). Proliferative response to EGF under our culture conditions was thus limited to the early-stage premalignant (tubular-derived) cells. These data also suggest that a premalignant epithelial cell loses responsiveness to EGF as it gains the ability to respond to TPA by multilayering.

The relationship between malignancy and release of plasminogen activator has remained unclear. Many normal cells such as macrophages and trophoblasts secrete this protease. It has also been known for years that some human tumors secrete a fibrinolytic activity. We observed that premalignant cells secreted much more protease in response to TPA than either colonic adenocarcinomas under similar conditions of primary culture or three human colon carcinoma cell lines also exposed to TPA. TPA induced morphological changes in malignant cells in primary culture but not in established colon carcinoma cell lines. Perhaps the capacity to undergo cell clustering due to protease secretion is needed for, or is concomitant with, the transition to malignancy but is not required once malignant status is reached.

Biological Markers

Serological and biochemical markers used to define abnormal stages of colonic cell maturation. Considerable effort has been expended toward the identification of specific immunological and biochemical markers of transformation. The use of syngeneic animal systems has facilitated identification of several serologically demonstrable classes of antigens on murine tumors, including conventional and differentiation alloantigens, murine leukemia virus and mammary tumor virus-related antigens, embryonic or fetal antigens, and anomalous histo-compatibility antigens (44). In man, autologous typing (44) has provided evidence for the existence of cell surface

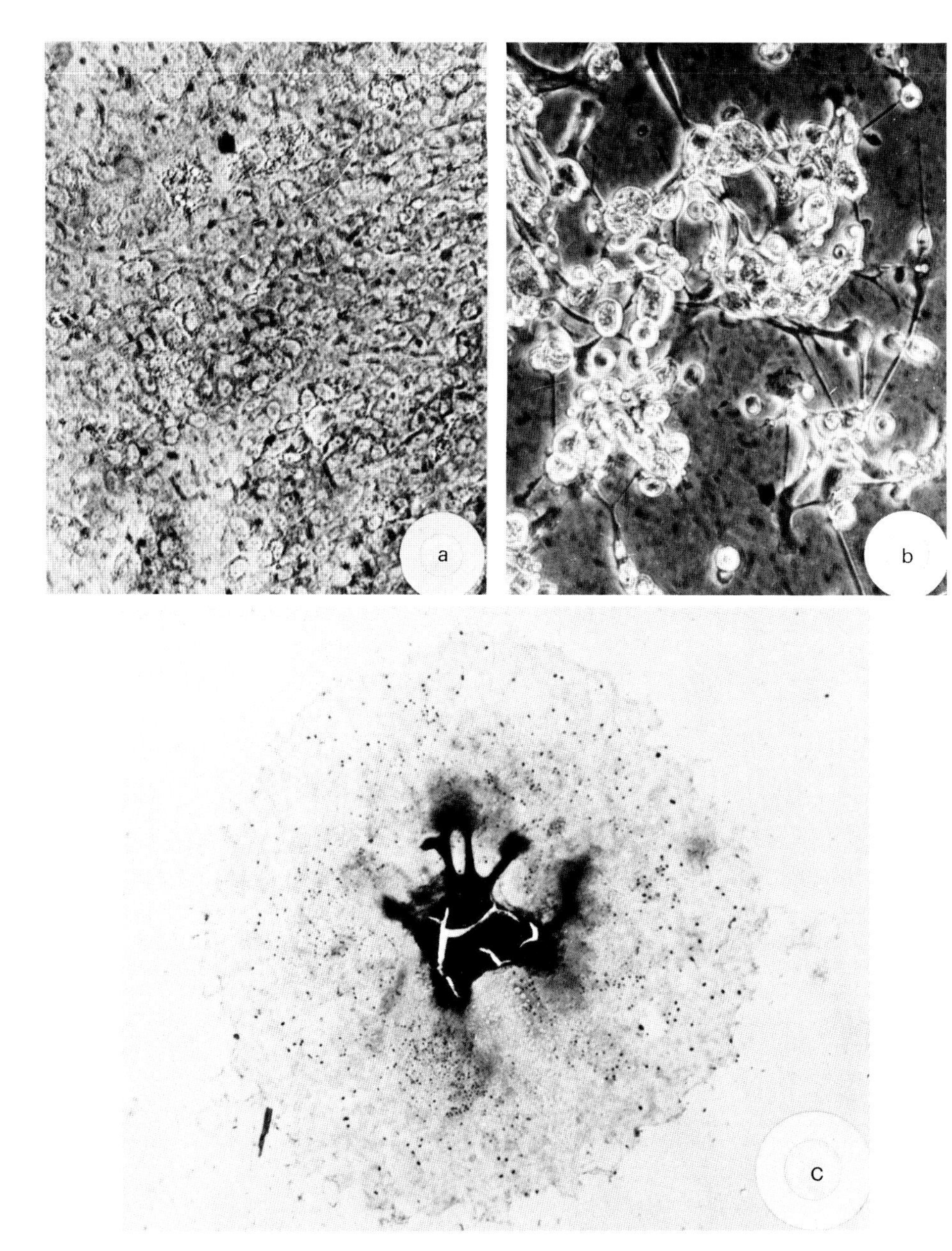

antigens restricted to autologous tumor cells (Class 1), present on autologous and some allogeneic tumor cells (Class 2), or widely distributed on normal and malignant cells (Class 3).

Certain antigens associated with human tumors have been observed to have a relatively restricted distribution among normal tissues and are broadly classified as organ-, tissue-, or site-specific. In some cancers, particularly those involving the gastro-

←Fig. 14. (a) Phase micrograph (X265) of living late-stage premalignant cells, derived from a villous adenoma, incubated for 24 hrs in 10^{-7} M DOC. There was no observable difference in colony morphology compared with untreated parallel cultures (not shown).

(b) Phase micrograph (X265) of a parallel culture to that shown in (a) of late-stage premalignant cells cultured 24 hrs in 100 ng TPA per ml. Cells have formed dense aggregates, and some have detached from the Petri dish.

(c) Lower power magnification (X80) of (a) after labeling for 64 hrs in 5 µCi [^{3}H] thymidine per ml and 10^{-7} M DOC, followed by methanol fixation, autoradiography, and staining with hematoxylin and rhodamine B. The monolayer is intact, with close-packed cells. Nuclei labeled with ^{3}H-thymidine are found throughout the monolayer (43).

intestinal tract, antigens have been described which have a more or less restricted site specificity. Some, such as the intestinal mucosal-specific glycoprotein (IMG), colonic mucoprotein antigen (CMA), colon-specific antigen(s) (CSA), sulfated glycopeptidic antigen (SGA), goblet cell antigen (GOA), and colon-specific antigen-p (CSAp), appear to be cell-specific; none of these antigens, however, is cancer-specific (reviewed in ref. 45).

As detection methods become more sensitive, previous putative "tumor-specific" antigens become reclassified as "tumor-associated" due to quantitative differences in the level of their expression in

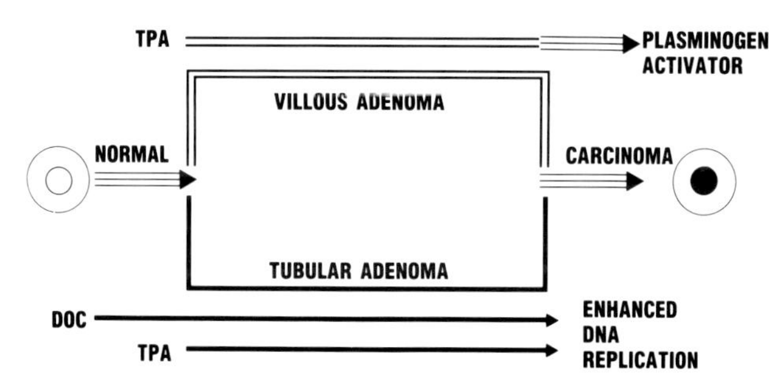

Fig. 15. Diagram illustrating differential action of tumor-promoting agents during stages of normal and abnormal growth of colonic epithelial cells. Deoxycholic acid (DOC) stimulated cell proliferation in normal and early stage premalignant (tubular) adenoma cells, but not in late stage (villous) adenoma cells. TPA stimulated cell proliferation in tubular adenomas, and stimulated production of plasminogen activator in villous adenomas and in carcinomas but not in tubular adenomas.

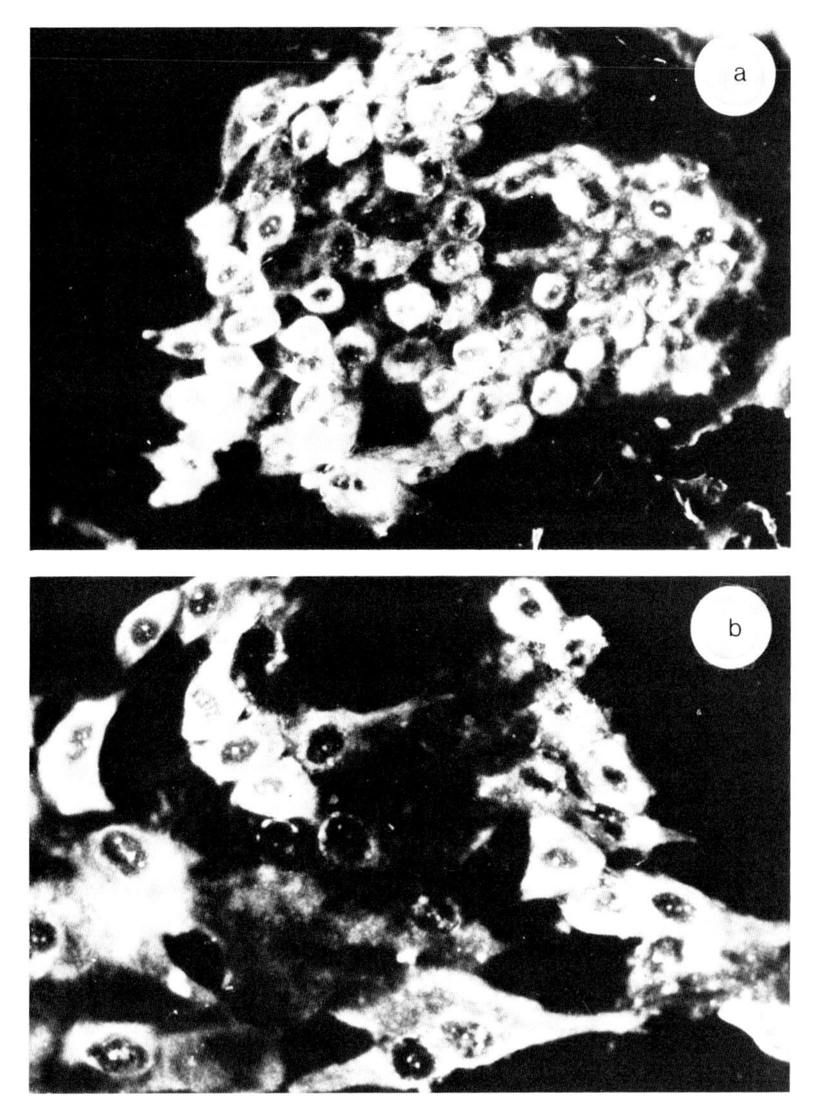

Fig. 16. Distribution of second trimester human fetal antigen-positive cells in cultures derived from colonic villous adenomas. The two colonies shown are comprised of predominantly antigen-positive cells although negative cells are clearly evident on both fields. Indirect immunofluorescence microscopy: a (X240), b (X330) (42).

normal and transformed tissues. Nevertheless, the use of such antigens, particularly those which can be monitored in blood (such as alpha-fetoprotein, carcino-embryonic antigen, galactosyltransferase-isoenzyme, pancreatic oncofetal antigen, basic fetoprotein,

TennaGen, and CSAp) (45), can be clinically relevant as adjunct parameters of prognostic evaluations and for assessments of tumor burden during the course of disease and in response to therapy.

Recently, the inappropriate or ectopic expression of particular gastrointestinal antigens (at anatomic sites in the adult GI tract at which they are not usually expressed) has provided an additional measurement for tumor classification (46). The mucus-associated antigens M1, M2, and M3 occur in three major groups of mucus-secreting cells in the human gastrointestinal tract: M1 is associated with columnar cells of the gastric epithelium, M2 with the mucous cells of gastric Brunner's glands, and M3 with the intestinal goblet cells (47). Of interest is the observation that the gastric M1 antigen occurs in approximately 29% of colonic adenocarcinomas

Table 4. Distribution of fetal large bowel-associated and adult stomach-related antigen(s) among established or primary cultures of human cells as ascertained by indirect immunofluorescence test.

Target Cells		Number Positive/Number Tested
Normal colon cells (primary cultures)		0/4
Colon carcinoma (established cell lines)		3/3
Colon carcinoma (explant cultures)		2/4
Bladder epithelial cells (HCV-29)	(established cultures)	0/4
Breast carcinoma (BT-20)		
Cervical carcinoma (HeLa)		
Hepatic carcinoma (Sk-Hep)		
Colonic tubular adenoma (primary cultures)		2/11
Colonic villotubular adenoma (primary cultures)		3/7
Colonic villous adenoma (primary cultures)		7/10

[a]Indirect immunofluorescence test of methanol-fixed cell cultures using adult tissue-absorbed rabbit antiserum to second trimester human fetal tissue (42) as the 1° antibody (anti-SFTa). Additional absorption experiments indicated that saline extracts of adult brain, heart, lung, liver, kidney, spleen, colon, bone marrow, and skin and extracts of second trimester fetal heart, lung, spleen, kidney, liver, skeletal muscle, brain, thyroid, spinal column, and adrenals were incapable of removing antibody activity to HT-29 colon carcinoma cells. Absorption of anti-STFa serum with extracts of adult stomach, second trimester fetal large bowel, and HT-29 colon carcinoma cells, however, removed all immunofluorescence reactivity to HT-29 target cells (42).

and in certain benign villous adenomas but is not present in normal adult colon (47). While the mechanism involved in the expression of M1 in colonic disease is not known, goblet cells of the second trimester human fetus have been shown to contain both M1 (gastric) and M3 (colonic) antigens (47). One might speculate that the "re-expression" of M1 determinants in the adult colon may reflect either retro-differentiation within specific cellular populations (46) or expansion of the colonic stem cell population (45) with a concomitant arrest in differentiation.

The successful short-term culture of epithelial cells from several classes of benign colonic adenomas referred to above (41) has facilitated additional investigation into the antigenic composition of specific colonic cell subpopulations. Such studies (42) have demonstrated an increased incidence (progressing through the tubular, villotubular, and villous classes adenomatous polyps) with which cultures of epithelial cells derived from benign and malignant colonic lesions exhibited a pattern of expression of second trimester fetal colon-associated and adult gastric-related antigen(s) reminiscent of that observed for the M1 antigen (Table 4, Figure 16). Epithelial cells isolated from adenomas of greatest transformation incidence (villous) tended to express antigens not characteristic of the normal adult colon at a higher frequency compared to cells cultured from adenomas of lower malignant potential (tubular and villotubular). Thus, these data (41,42), and previous observations referred to above, suggest that immunological criteria may provide additional information for pathologic classification of malignant and premalignant lesions of the gastrointestinal tract, and for analysis of the effects of agents on colonic cell development.

<u>Nucleic acid sequences.</u> It is now possible to identify and clone nucleic acid sequences whose pattern of expression is unique to premalignant and malignant colon cells and which may be useful in following the progression of cells to malignancy.

Thus far, we have investigated gene expression in a transplantable, dimethylhydrazine-induced mouse colon tumor (51). Approximately 400 sequences cloned from the messenger population of this tumor were screened for level of expression in the tumor and in normal mouse colon, liver, and kidney. Two general observations were made. First, we could readily detect individual sequences whose level of expression differed, sometimes markedly, in the normal colon as compared to the tumor (Table 5). Second, the sequences could be organized into groups based on their level of expression in each of the tissues, and patterns of expression emerged which characterized each tissue and defined a specific phenotype.

In continuing the screening of each of the sequences in a variety of tissues and cell lines, two sequences have been found to

Table 5. Abundance of each cloned sequence in normal mouse colon and in a mouse colon tumor.

Normal → Tumor	Major	Minor	Total
No Change	-	-	323 (85%)
Increase	7 (2%)	25 (6%)	32 (8%)
Decrease	1 (1%)	22 (6%)	23 (7%)

Each of 378 cloned sequences was screened for level of expression in the normal mouse colon as compared to DMH-induced mouse colon tumor #36 (from ref. 51)

be of particular interest. First, sequence number 432, a 670 base pair insert, is expressed at a very high level in the colon tumor as assayed by colony filter hybridization, dot blot analysis, and Northern blots, but is expressed at a low level in normal colonic mucosa and in most other cells and tissues. Its expression at a modest level in three mouse leukemia cell lines is linked to proliferation and induction of differentiation. Sequence 432 consists of a unique fragment which hybridizes to a transcript of 20S in the tumor, linked to a highly repetitive sequence which hybridizes to several transcripts, several of which may be increased in expression in the tumor. The second sequence of interest is number 541, a sequence which is shut off in both the DMH-induced colon tumor and in an ethyl nitrosourea-induced lung tumor as compared to their respective normal tissues (Augenlicht and Kobrin, unpublished observations).

Our most immediate goal is to investigate the expression of these sequences by _in situ_ hybridization in order to identify the specific proliferating and differentiating cells in the colon in which these sequences are expressed, and further to determine how early during carcinogenesis this genetic reprogramming of individual cells takes place. We are also expanding this work to human colon cancer. Each sequence in an ordered array of 4,000 cloned cDNA sequences from a human colon carcinoma cell line is being assayed for level of expression in biopsy material of normal human colonic mucosa and in benign and malignant colonic neoplasms. The extensive data will provide a profile of expression of sequences which may uniquely define malignant and premalignant states in various population groups at risk for development of colon cancer, and may provide specific nucleic acid probes which are useful in identifying malignant and premalignant cells in human colon biopsy material.

ACKNOWLEDGEMENT

This work was supported by Am. Cancer Soc. grant SIG-7; U.S.PHS grants CA25285, CA28822, CA08748, CA22367, and Contract N01-CP-01050.

REFERENCES

1. Chang, W.W.L., and C.P. Leblond (1971) Renewal of the epithelium in the descending colon of the mouse. 1. Presence of three cell populations: Vacuolated-columnar, mucous and argentaffin. Am. J. Anat. 131:73-100.
2. Deschner, E.E., and M. Lipkin (1966) An autoradiographic study of the renewal of argentaffin cells in human rectal mucosa. Exp. Cell Res. 43:661-665.
3. Lipkin, M., and H. Quastler (1962) Cell retention and incidence of carcinoma in several portions of the gastrointestinal tract. Nature 194:1198-1199.
4. Lipkin, M., and H. Quastler (1966) Cell population kinetics in the colon of the mouse. J. Clin. Invest. 41:141:146.
5. Lipkin, M. (1974) Phase I and Phase II proliferative lesions of colonic epithelial cells in diseases leading to colon cancer. Cancer 34:878-888.
6. Lipkin, M., B. Bell, and P. Sherlock (1963) Cell proliferation kinetics in the gastrointestinal tract of man. 1. Cell renewal in colon and rectum. J. Clin. Invest. 42:767-776.
7. Lipkin, M., B. Bell, G. Stalder, and F. Troncale (1970) The development of abnormalities of growth in colonic epithelial cells of man. In Carcinoma of the Colon and Antecedent Epithelium, H. Burdette, ed. Charles C. Thomas, Springfield, Ill., pp. 213-221.
8. Shorter, R.G., C.G. Moertel, J.L. Titus, and R.J. Reitemeier (1964) Cell kinetics in the jejunum and rectum of man. Am. J. Dig. Dis. 9:760-763.
9. Lipkin, M., and E.E. Deschner (1968) Comparative analysis of cell proliferation in the gastrointestinal tract of newborn hamster. Exp. Cell Res. 49:1-12.
10. Eastwood, G.L., and J.S. Trier (1974) Epithelial cell proliferation during organogenesis of rat colon. Anat. Rec. 179:303-310.
11. Deschner, E.E., C.M. Lewis, and M. Lipkin (1963) In vitro study of human rectal epithelial cells. 1. Atypical zone of H^3 thymidine incorporation in mucosa of multiple polyposis. J. Clin. Invest. 42:1922-1928.
12. Deschner, E.E., and M. Lipkin (1970) Study of human rectal epithelial cells in vitro. III. RNA, protein and DNA synthesis in polyps and adjacent mucosa. J. Natl. Cancer Inst. 44:175-185.
13. Deschner, E.E., M. Lipkin, and C. Solomon (1966) In vitro study of human epithelial cells. II. H^3 thymidine incorporation into

polyps and adjacent mucosa. J. Natl. Cancer Inst. 36:849-857.
14. Thurnherr, N., E.E. Deschner, E. Stonehill, and M. Lipkin (1973) Induction of adenocarcinomas of the colon in mice by weekly injections of 1,2-dimethylhydrazine. Cancer Res. 33:940.
15. Kikkawa, N. (1974) Experimental studies on polypogenesis of the large intestine. Med. J. Osaka Univ. 24:293.
16. Chang, W.W.L. (1978) Histogenesis of sym.1-2 DMH-induced neoplasms of the colon in the mouse. J. Natl. Cancer Inst. 60:1405.
17. Pozharisski, K., and O. Chepeck (1979) The oncological characteristics of colonic polyposis in humans in view of morphogenesis of experimental intestinal tumors. Tumor Research (Sapporo, Japan) 13:40-55.
18. Lightdale, C., M. Lipkin, and E. E. Deschner (1982) In vivo measurements in familial polyposis: Kinetics and location of proliferating cells in colonic adenomas. Cancer Res. 42:4280-4283.
19. Lipkin, M. (1977) Growth kinetics of normal and premalignant gastrointestinal epithelium, B. Drewinko and R.M. Humphreys, eds. 29th Annual Symposium on fundamental cancer research. In Growth Kinetics and Biological Regulation of Normal and Malignant Cells. Williams & Wilkins Co., Baltimore, pp. 562-589.
20. Lipkin, M. (1978) Susceptibility of human population groups to colon cancer. Adv. Cancer Res. 27:281.
21. Bussey, H.J.R. (1975) Familial Polyposis Coli. The Johns Hopkins University Press, Baltimore, p. 104.
22. Eastwood, G.L., and J.S. Trier (1973) Epithelial cell renewal in cultured rectal biopsies in ulcerative colitis. Gastroenterology 64:383.
23. Winawer, S., and M. Lipkin (1969) Cell proliferation kinetics in the gastrointestinal tract of man. IV. Cell renewal in intestinalized gastric mucosa. J. Natl. Cancer Inst. 42:9.
24. Deschner, E.E., S. Winawer, and M. Lipkin (1972) Patterns of nucleic acid and protein synthesis in normal human gastric mucosa and atrophic gastritis. J. Natl. Cancer Inst. 48:1567.
25. Deschner, E.E., K. Tamura, and S.P. Bralow (1979) Early proliferative changes in rat pyloric mucosa induced with N-methyl-N'-nitro-N-nitrosoguanidine. Frontiers Gastrointest. Res. 4:25-31.
26. Hasegawa, I., Y. Matsumira, and S. Tojo (1976) Cellular kinetics and histological changes in experimental cancer of the uterine cervix. Cancer Res. 36:359.
27. Ryggard, J., and C. Povlsen (1969) Heterotransplantation of a human malignant tumor to nude mice. Acta Pathol. Microbiol. Scand., A. Pathol. 77:758-760.
28. Povlsen, C., and J. Ryggard (1971) Heterotransplantation of human adenocarcinoma of the colon and rectum to the mouse mutant nude. A study of nine consecutive transplantations. Acta Pathol. Microbiol. Scand., A. Pathol. 79:159-169.
29. Sordet, B., R. Fritsche, J.P. Mach, et al. (1974) Morphologic

and function evaluation of human solid tumors serially transplanted in nude mouse. Abstr. 1st Int. Workshop on Nude Mice. Copenhagen 1973. Fischer, Stuttgart, p. 15 .

30. Giovanella, B.C., J. Stehlin, and L.J. Williams (1974) Heterotransplantation of human malignant tumors in nude thymus-less mice. J. Natl. Cancer Inst. 52:921-930.
31. Ryggard, J., and C. Povlsen (1969) Heterotransplantation of a human malignant tumor to nude mice. Acta Pathol. Microbiol. Scand., A. Pathol. 77:758-760.
32. Povlsen, C., D.W. Fialk, E. Klein, et al. (1973) Growth and antigenic properties of a biopsy derived Burkitt's lymphoma in nude mice. Int. J. Cancer 11:30-36.
33. Klein, J., J. Gaiser, and J. Hasper (1972) Growth capacity of human tumors in vitro and in nude mice: Rep. Annual Report, 1972, pp. 54-56 (Organization for Health TNO, Rijswijk 1972).
34. Povlsen, C., G. Jacobsen, and J. Ryggard (1972) The mouse mutant nude as a model for the testing of anticancer agents. 5th Symp. Int. Committee on Lab Animals, Hanover 1972, pp. 63-72 (Fischer, Stuttgart, 1973.)
35. Giovanella, B., S. Yim, J. Stehlin, et al. (1972) Development of invasive tumors in the nude mouse after injection of cultured human melanoma cells. J. Natl. Cancer Inst. 48:1531-1533.
36. Giovanella, B., A. Morgan, and J. Stehlin (1973) Development of invasive tumors in the nude mouse injected with human cells cultured from Burkitt's lymphoma. Proc. Am. Assoc. Cancer Res. 14:20.
37. Bhargava, D.K., and M. Lipkin (1981) Transplantation of adenomatous polyps, normal colonic mucosa and adenocarcinoma of colon into athymic mice. Digestion 21:225-231.
38. Autrup, H., L.A. Barrett, F.E. Jackson, M.L. Jesudason, G. Stoner, P.A.B. Phelps, B.F. Trump, and C.C. Harris (1978) Explant culture of human colon. Gastroenterology 74:1248-1257.
39. Reiss, B., and G.M. Williams (1979) Conditions affecting prolonged maintenance of mouse and rat control in organ culture. In Vitro 15:877-890.
40. Usugane, M., M. Fujita, M. Lipkin, R. Palmer, E. Friedman, and L. Augenlicht (1982) Cell replication in explant cultures of human colon. Digestion 24:225-233.
41. Friedman, E.A., P.J. Higgins, M. Lipkin, H. Shinya, and A.M. Gelb (1981) Tissue culture of human epithelial cells from benign colonic tumors. In Vitro 17:632-644.
42. Higgins, P.J., E. Friedman, M. Lipkin, R. Hertz, F. Attiyeh, and E.H. Stonehill (1983) Expression of gastric-associated antigens by human premalignant and malignant colonic epithelial cells. Oncology 40:26-30.
43. Friedman, E. (1981) Differential response of premalignant epithelial cell classes to phorbial ester tumor promoters and to deoxycholic acid. Cancer Res. 41:4588-4599.
44. Old, L.J. (1981) Cancer immunology: The search for specificity. Cancer Res. 41:361-375.

45. Goldenberg, D.M. (1981) Immunobiology and biochemical markers of gastrointestinal cancer. In Gastrointestinal Cancer, J.R. Stroehlein and M.M. Romsdale, eds. Raven Press, N.Y., pp. 65-81.
46. Bara, J., L. Hamelin, E. Martin, and P. Burtin (1981) Intestinal M3 antigen, a marker for the intestinal-type differentiation of gastric carcinomas. Int. J. Cancer 28:711-719.
47. Bara, J., F. Loisillier, and P. Burtin (1980) Antigens of gastric and intestinal mucous cells in human colonic tumors. Br. J. Cancer 41:209-221.
48. Lipkin, M., S. Winawer, and P. Sherlock (1981) Early identification of individuals at increased risk for cancer of the large intestine. Clinical Bulletin 11:13-21 (Part 1), and 66-74 (Part 2).
49. Lipkin, M., W. Blattner, J.F. Fraumeni, Jr., H. Lynch, E. Deschnar, and S. Winawer (1983) Tritiated thymidine (ϕ_p, ϕ_h) labeling distribution as a marker for hereditary predisposition to colon cancer. Cancer Res.
50. Chi, C.H., C.A. Rubio, and B. Lagerlof (1977) The frequency and distribution of mitotic figures in dysplasia and carcinoma in situ Cancer 39(3):1218-1223.
51. Augenlicht, L.H., and D. Kobrin (1982) Cloning and screening of sequences expressed in a mouse colon tumor. Cancer Res. 42:1088-1092.

EFFECT OF DIET ON INTESTINAL TUMOR PRODUCTION

Bandaru S. Reddy

Division of Nutrition
Naylor Dana Institute for Disease Prevention
American Health Foundation
Valhalla, New York 10595

INTRODUCTION

Considerable progress has been made in the basic concepts concerning the role of dietary factors in the development of certain types of human cancer (1). It has been recognized that colon cancer does not stem from intentional or even inadvertant chemical contaminants in the environment, but rather is due to lifestyles and dietary factors. It has been recognized also that the promoting effects operating in colon carcinogenesis are an important component in the overall process because whether or not overt invasive disease is seen depends to a considerable extent on promoting factors (2). Dietary factors appear to modulate carcinogenesis through elevation of agents that act as promoters of tumor development.

During the past two decades, epidemiologic studies have investigated the influence of environmental factors on the development of cancer of the colon. Reliable data, obtained from human epidemiologic and animal model studies and supporting laboratory studies on the underlying mechanisms in humans and in animals, suggest that diets particularly high in total fat and low in certain fibers, vegetables, and micronutrients are generally associated with an increased incidence of colon cancer in man (2-8). Dietary fat may be a risk factor in the absence of factors that are protective, such as the use of high fibrous food and fiber (6,8,9).

This brief review evaluates not only current experimental research on the relationship between dietary factors and colon cancer, but also various initiators and promoters involved in colon carcinogenesis. Finally, it presents an evaluation of the mechanism whereby dietary factors modulate the development of this important cancer.

ETIOLOGY OF COLON CANCER

Cancer of the colon has been the subject of several epidemiologic reviews (5,10). The major differences in intra-country and inter-country distribution, as well as differences in incidence of this cancer in religious groups, suggest that nutritional factors and lifestyles are important in the development of the disease (8,11,12).

In general, with few exceptions, the more economically developed a society, the greater the incidence of colon cancer. The highest incidence rates are found in North America, New Zealand, and Western Europe, with the exception of Finland, where the incidence is one of the lowest in the developed countries. Intermediate rates are found in eastern Europe and the Balkans, and the lowest incidences are found in Africa, Asia, and Latin America (the exception being Uruguay and Argentina, where mortality rates are similar to those found in North America).

Further evidence for the importance of environmental dietary factors in colon cancer is provided by studies of migrants to the United States and Australia (13,14). Colon cancer incidence is higher in the first and second generation Japanese immigrants to the United States and in Polish immigrants to Australia than in native Japanese in Japan and in the native Poles from Poland, respectively.

Further support for dietary influence in colon cancer is derived from the time-trend in Japan showing that colon cancer seems to be increasing in Japan itself, a finding consistent with the increasing Westernization of the Japanese diet (15).

Additional support for considering lifestyle in the risk from colon cancer is derived from studies of special religious groups within a small geographical area. The incidence of colon cancer in Mormons (members of the Church of Jesus Christ of Latter-Day Saints) is lower than other U.S. white populations with the exception of Seventh-Day Adventists (SDA) (12). Traditionally, Mormons eat more whole grain breads, fruits, and vegetables (8). Studies by Phillips et al. (11) indicate that the SDA, who consume less meat and adhere to a lacto-ovo-vegetarian diet, are reported to have about 60% of the colon cancer death rates for the total recorded in California. Jussawalla and Jain (16) found that the Parsis in Bombay exhibited a hgher risk for colon cancer than the other religious groups such as Hindus and Moslems. Since the Parsis lean more toward a Western diet, the dietary habits of these two Indian populations may account for differences in colon cancer rates.

The possible relationship between dietary factors and colon cancer has been investigated by correlational, case-control, and cohort studies. Wynder (17) and Wynder and Reddy (18) proposed that incidence of colon cancer is mainly associated with total dietary fat. A worldwide correlation between the incidence of colon cancer and the consumption of total fat has been established (19). Wynder et al. (20) conducted a large-scale retrospective study in large-bowel cancer patients in Japan, which suggested a correlation between the Westernization of the Japanese diet and colon cancer. Haenszel et al. (13) demonstrated an association between colon cancer and dietary beef in Hawaiian Japanese cases and controls. A case-control study of cancer of the colon in Canada indicated an elevated risk for those with an increased intake of calories, total fat, and saturated fat (21). These results support a role for total dietary fat in the incidence of colon cancer.

Burkitt (22) recognized the rarity of colon cancer in most African populations and suggested that countries consuming a diet rich in fiber have a low incidence of colon cancer, whereas those eating refined carbohydrates with little fiber have a higher incidence of the disease. A recent study comparing populations in Kuopio, Finland (low-risk), Copenhagen, Denmark (high-risk), and New York (high-risk) indicated that one of the factors contributing to the low-risk of colon cancer in Kuopio appears to be high dietary fiber intake (9,23).

In case-control studies in Israel and in San Francisco area Blacks, it was found that among a large variety of dietary constituents investigated, those that were lowest in the diets of patients with colon cancer as compared with controls were foods containing fiber (24,25). Bjelke (26) found less frequent use of vegetables among colo-rectal cancer patients. Another study on a series of cases and controls from Roswell Park Memorial Institute in Buffalo, New York, showed a lower risk of colon cancer for individuals ingesting vegetables such as cabbage, broccoli, and brussel sprouts (27).

The studies cited led us to accept diet as a major etiologic factor in colon cancer. Diets high in total fat, and low in fiber are associated with an increased incidence of colon cancer in man. High dietary fiber acts as a protective factor in a population consuming a high amount of total fat.

CONCEPT OF DIETARY FACTORS AND COLON CANCER

Aries et al. (28) have suggested that: (a) the amount of dietary fat determines the levels of intestinal bile acids and cholesterol metabolites as well as the composition of the gut microflora; and (b) the gut microflora metabolize these acid and neutral sterols to carcinogens which are active in the colon.

Investigators have examined the potential carcinogenic activity of certain bile acids because, presumably, they may be converted to potential carcinogens, and several bile acids induced sarcomas at the site of injection in experimental animals (29). Such reactions are less likely to yield carcinogenic compounds from bile salts, but are much more likely to yield colon tumor promoters or co-carcinogens rather than complete carcinogens. There is no evidence that bile acids or cholesterol per se can initiate a carcinoma in man or animals. Reddy et al. (2) suggest that the dietary fat increases the excretion of bile acids into the gut, and also modifies the activity of gut microflora which enhances the formation of secondary bile acids in the colon. These secondary bile acids act as tumor promoters in the colon (2).

Studies were carried out in our laboratory and elsewhere on the excretion of bile acids in high- and low-risk populations for colon cancer development. This subject has been reviewed recently (30, 31). Briefly, the population with a high risk has an increased amount of colonic secondary bile acids, namely deoxycholic acid and lithocholic acid. In addition, people on a high-fat diet appear to have a higher level of fecal secondary bile acids compared to those on a low-fat diet.

The possible mechanism of the protective effect of dietary fiber against colon cancer has been the subject of a recent workshop (32). Dietary fiber is that part of ingested plant material that is resistant to digestion by the secretions of the gastrointestinal tract. It comprises a heterogeneous group of carbohydrates, including cellulose, hemicellulose, and pectin, and a noncarbohydrate substance, lignin.

The protective effect of dietary fiber may be due to adsorption, dilution, or metabolism of co-carcinogens, promoters, and yet-to-be-identified carcinogens by the components of the fiber (9,33,34). Different types of non-nutritive fibers possess specific binding properties. Dietary fiber could also affect the enterohepatic recirculation of bile salts (35). Fiber not only influences bile acid metabolism, thereby reducing the formation of tumor promoters in the colon, but also exerts a solvent-like effect in that it dilutes potential carcinogens and co-carcinogens by its bulking effect and is able to bind bile acids and certain carcinogenic compounds (9,36).

The effect of dietary fiber on fecal bile acids has been studied in many laboratories. Reddy et al. (9) studied the fecal bile acid excretion in healthy controls in Kuopio (Finland), a low-risk population for colon cancer development. The diet histories indicate that the total fat and protein consumption in Kuopio is quite similar to the New York population, but the consumption of fiber, mainly cereal type, is 3-fold higher in Kuopio compared to the New York population. The daily output of feces is three times higher in Kuopio than that of healthy individuals in the United States. The concentration of fecal secondary bile acids, deoxycholic acid, and lithocholic acid, is decreased in Kuopio due to high fecal bulk, but the daily output remains the same in the two groups because the dietary intake of fat is the same in Kuopio and New York. Cummings et al. (37) and Kay and Truswell (38) report that an increase in wheat bran intake increases the fecal weight and dilutes the fecal bile acids. This suggests that increased fecal bulk dilutes suspected carcinogens and promoters that may be in direct contact with large-bowel mucosa.

Until recently, there were no concepts on the nature of the genotoxic carcinogens associated with the etiology of colon cancer. Based on the discovery of Sugimura et al. (39) that there were powerful mutagens on the surface of fried meat, Weisburger et al. (40) proposed that these mutagens may be responsible for colon cancer. Weisburger et al. (40) also demonstrated that the fat content of food may play a role in mutagen formation. Several of these fried meat mutagens are similar to the aromatic amines related to 3,2'-dimethyl-4-aminobiphenyl (DMAB), an experimental colon carcinogen in animal models. Feeding mice on diets containing fried meat mutagens such as Trp-p-1 or Trp-p-2 demonstrated the hepatocarcinogenicity of these compounds (39). However, the carcinogenic activity of fried meat mutagens for the colon remains to be determined.

The search for mutagenic activity in the feces has been stimulated by the need to understand the nature of genotoxic compunds, if any, relevant to colon cancer. These studies demonstrate that the populations who are at high risk for colon cancer and consuming either a high-fat and/or non-vegetarian diet excrete an increased amount of fecal mutagens compared to the low-risk Seventh Day Adventist and Kuopio populations (31,41-44). Studies are in progress in many laboratories to isolate and identify these fecal mutagens and to determine their carcinogenic activity.

It has been reported that certain compounds are noncarcinogenic or mutagenic alone, but enhance the tumorigenic or mutagenic properties of carcinogens and are termed as co-mutagens (46,47). We have studied two groups of volunteers, namely SDA and non-SDA, who did not show any fecal mutagenic activity in our earlier study (48). Samples collected from non-SDA subjects exhibited a significantly higher co-mutagenic activity in both TA100 and TA98 with S9 activation than those from SDA individuals. This is the first demonstration that the fecal samples contain co-mutagenic activity and that this activity differs in high- and low-risk populations for the development of colon cancer.

Table 1. Modifying factors in colon cancer.

Dietary Fat[a]	Dietary Fibers[a,b]	Micronutrients (include vitamins minerals, antioxidants, etc.)
1. Increases bile acid secretion into gut	1. Certain fibers increase fecal bulk and dilute carcinogens and promoters	1. Modify carcinogenesis at activation and detoxification level
2. Increases metabolic activity of gut bacteria	2. Modify metabolic activity of gut bacteria	2. Act also at promotional phase of carcinogenesis
3. Increases secondary bile acids in colon that act as tumor promoters	3. Modify the metabolism of carcinogens and/or promoters	
4. Alters immune system		
5. Stimulation of mixed function oxidase system		

[a] Dietary factors, particularly high total dietary fat and a relative lack of certain dietary fibers and vegetables have a role.

[b] High dietary fiber or fibrous foods may be a protective factor even in the high dietary fat intake.

Based on the above information, it has been suggested that: (a) the extent of the carcinogenic stress from the exogenous source is probably rather weak; (b) high fat diet alters the concentration of bile acids and the activity of gut microflora which may, in turn, produce tumor-promoting substances from bile acids in the lumen of the colon; (c) certain dietary fibers not only enhance binding of tumorigenic compounds in the gut, but also dilute them so that their effect on the colonic mucosa is minimal; and (d) dietary fat, fiber, and certain vegetables modify the intestinal mucosa, as well as hepatic enzyme inhibitors or inducers that alter the capacity of the animal to metabolize the tumorigenic compounds (2,49-51) (Table 1).

EXPERIMENTAL STUDIES IN ANIMAL MODELS

Development of Animal Model

Studies on the mechanisms of colon carcinogenesis have been advanced by the discovery of several animal models that show the type of lesions observed in man. These models are: (a) induction of colon cancer in rats by aromatic amines such as 3,2'-dimethyl-4-aminobiphenyl (DMAB) or 3-methyl-2-naphthylamine; (b) derivatives and analogs of cycasin such as methylazoxymethanol (MAM), azoxymethane (AOM), and 1,2-dimethylhydrazine (DMH) in rats and mice of selected strains; and (c) intrarectal administration of direct-acting carcinogens, such as methylnitrosourea (MNU) or N-methyl-N'nitro-N-nitroso-guanidine (MNNG), which lead to cancer of the descending colon in every species tested so far.

On the basis of quantitative analysis of colon tumors in DMH-treated rats, it has been suggested that this carcinogen is capable of producing intestinal tumors by a 2-stage mechanism (52). Each DMH injection first induces a stable and transmissible change in a number of colon epithelial cells. In rats receiving small doses or 1 single injection of the carcinogen, the number of such pretransformed cells will be limited and so, too, the risk of further transformation. Frequent injections of DMH would result in the accumulation of pretransformed cells with a rapidly increasing risk of tumor development. Using CF_1 mice and DMH, it has also been shown that: (a) with increased doses of DMH, there was an increased tumor yield and decreased latency period; (b) with repeated doses, there was a rapidly cumulative tumor yield; and (c) new tumors continued to accumulate in the colon even at long intervals after treatment with DMH (53). A recent study by Maskens (54) also indicates that DMH carcinogenesis in the rat colon is a 2-step process. The first change is caused by DMH, the effect of which is additive when similar doses are repeated. This change is transmissible within the renewing mucosa for prolonged periods, and the affected cells have no phenotypic expression in terms of proliferative advantage. The second stage does not require exposure to a carcinogen, although continued presence of the carcinogen can probably contribute to it. This approach to 2-stage carcinogenesis differs from classical initiation-promotion experiments in which the first step is probably mutational and the second can be promoted by a variety of agents which are not necessarily carcinogens _per se_ (3).

Dietary Fat and Colon Carcinogenesis

The effect of dietary fat appears to be exerted during the promotional phase of carcinogenesis rather than during the initiation phase (2).

Nigro et al. (55) induced intestinal tumors in male Sprague-Dawley rats by subcutaneous administration of AOM and compared animals fed Purina chow with 35% beef fat to those fed Purina chow containing 5% fat. Animals fed the high fat developed more intestinal tumors and more metastases into the abdominal cavity, lungs, and liver than the rats fed the low-fat diet. In another study, W/Fu rats fed a 30% lard diet had an increased number of DMH-induced large-bowel tumors compared to animals fed the standard diet (56). Rogers et al. (57) found that a diet marginally deficient in lipotropes, but high in fat, enhanced DMH-induced colon carcinogenesis in Sprague-Dawley rats. These results suggest that total dietary fat may have a function in the pathogenesis of colon cancer.

Reddy et al. (58) studied the effect of a particular type and amount of dietary fat for two generations before animals were exposed to treatment with a carcinogen. Animals fed 20% lard or 20% corn oil were more susceptible to the induction of colon tumor by DMH than those fed 5% lard or 5% corn oil. The type of fat appears to be immaterial at the 20% level, although at the 5% fat level, there is a suggestion that unsaturated fat (corn oil) predisposes to more DMH-induced colon tumors than saturated fat (lard). Broitman et al. (59) showed that rats fed a 20% safflower oil diet had more DMH-induced large-bowel tumors than the animals fed either the 5% or 20% coconut oil diets. However, these data indicate that at low dietary fat levels, diets rich in polyunsaturated fats are more effective tumor promoters than diets rich in saturated fats.

Investigations also were carried out to test the effect of high dietary fat on the induction of colon tumors by a variety of carcinogens, DMH, MAM acetate, DMAB, or MNU, which not only differ in metabolic activation but also represent a broad spectrum of exogenous carcinogens (60,61). Male F344 weanling rats were fed semisynthetic diets containing 20% or 5% beef fat. At seven weeks of age, animals were given DMH (subcutaneous, 150 mg/kg body wt., one dose), MAM acetate (intraperitoneal, 35 mg/kg body wt., one dose), DMAB (subcutaneous, 50 mg/kg body wt., weekly for 20 weeks), or MNU (intrarectal, 2.5 mg/rat weekly for 2 weeks), killed and necropsied 30-35 weeks later. Irrespective of the colon carcinogen, animals fed a diet containing 20% beef fat had a greater incidence of colon tumors than did rats fed a diet containing 5% beef fat (Table 2).

Dietary Fat as a Promoting Agent

The suggestion that promotion may be involved in intestinal cancer has been supported by the observation that the response to a variety of intestinal carcinogens is enhanced by dietary fat which in itself is not carcinogenic. Recent studies indicate that enhanced tumorigenesis in animals fed a high-fat diet is due to promotional effects (62). Ingestion of a high-fat diet increased intestinal tumor incidence when fed after the administration of AOM (carcinogen), but not during or before treatment with AOM. The carcinogenic process in the human may have similar characteristics since there is a good correlation between the results in a variety of studies in animals and those in humans (63).

Dietary Fiber and Colon Carcinogenesis

The relation between the consumption of dietary fiber and colon cancer has been studied in various experiments. Wilson et al. (64) found that Sprague-Dawley rats fed a diet containing 20% corn oil or beef fat and 20% wheat bran had fewer colon tumors induced by DMH than rats fed a control diet containing 20% fat and no bran. Freeman et al. (65) compared the incidence of colon tumors induced by DMH in Sprague-Dawley rats fed either a fiber-free diet or a diet

Table 2. Colon tumor incidence in rats fed diets high in fat and treated with colon carcinogens.

% of dietary fat	% of protein as casein	Carcinogen	% of rats with colon tumors
Lard			
5	25	DMH[a]	17
20	25	DMH	67
Corn Oil			
5	25	DMH[a]	36
25	25	DMH	64
Beef Fat			
5	22	DMH[b]	27
20	22	DMH	60
5	22	MNU[c]	33
20	22	MNU	73
5	22	MAM acetate[d]	45
20	22	MAM acetate	80
5	20	DMAB[e]	26
20	20	DMAB	74

a Female F344 rats, at 7 weeks of age, were given DMH s.c. at a weekly dose rate of 10 mg per kg body weight for 20 weeks,and autopsied 10 weeks later.

b Male F344 rats, at 7 weeks of age, were given a single s.c. dose of DMH, 150 mg per kg body weight and autopsied 30 weeks later.

c Male F344 rats, at 7 weeks of age, were given MNU i.r., 2.5 mg per rat, twice a week for 2 weeks and autopsied 30 weeks later.

d Male F344 rats, at 7 weeks of age, were given a single i.p. dose of MAM acetate, 35 mg per kg body weight, and autopsied 30 weeks later.

e Male F344 rats, at 7 weeks of age, were given DMAB s.c. at a weekly dose rate of 50 mg per kg body weight for 20 weeks, and autopsied 20 weeks later.

containing 4.5% purified cellulose. Among the animals ingesting cellulose, fewer had colonic neoplasms, and the total number of colon tumors in this group was lower.

The effect of a diet containing 15% alfalfa, pectin, or wheat bran on colon carcinogenesis by AOM in F344 rats was studied by Watanabe et al. (66). The frequency of colon tumors induced by AOM in rats fed a diet containing pectin or wheat bran was lower than in rats fed a control diet or the alfalfa diet. The effect of alfalfa, wheat bran, and cellulose on the incidence of intestinal tumors induced by AOM was further studied in Sprague-Dawley rats fed diets containing 10% fiber and 30% beef fat, 20% fiber and 6% beef fat, or 30% fiber and 6% beef fat (44). The presence of 20% bran or cellulose, or 30% of any fiber in a diet containing 6% fat, significantly reduced the frequency of intestinal tumors. All the groups except those with a diet containing 20% alfalfa had a lower frequency of tumors in the proximal half of the large bowel than did the groups not ingesting fiber. The concentration, but not the total daily excretion of fecal steroids, was significantly lower in the groups with a lower frequency of tumors.

The effect of dietary wheat bran (15%) and dehydrated citrus fiber (15%) on AOM or DMAB-induced intestinal tumors was studied in male F344 rats (67). Animals fed the wheat bran and citrus fiber diets and treated with AOM had a lower incidence of colon- and small-intestinal tumors than did those fed the control diet and treated with AOM (Table 3). On the other hand, the animals fed the wheat bran and treated with DMAB developed fewer colon- and small-intestinal tumors than did the rats fed the control diet; animals fed the diet containing citrus fiber and treated with DMAB had a lower incidence of small-intestinal tumors. This study thus indicates that diets containing wheat bran reduce the risk of chemically-induced colon cancer and that the protection against colon cancer depends on the type of fiber.

Bile Acids and Colon Tumor Production

Cholecystectomy, a procedure that increases the concentration of secondary bile acids in feces, predisposes to the development of cancer of the right-sided colon in humans (68). One of the changes associated with cholecystectomy is an increased proportion of secondary bile acids such as deoxycholic acid and lithocholic acid in the total bile acid pool, this change being due to the increased enterohepatic recycling and the exposure of bile acids to gut bacteria.

Table 3. Colon tumor incidence in F344 male rats fed diets containing wheat bran or citrus fiber and treated with azoxymethane.

Diet	Animals with Colon Tumors						Colon Tumors per Tumor Bearing Rat		
	Total[a]		Adenoma		Adeno-carcinoma		Total	Adenoma	Adenocarcinoma
	No.	%	No.	%	No.	%			
Control (96)[b]	86	90	83	86	60	63	3.45±0.16[c]	2.37±0.16	1.08±0.18
Wheat bran (51)	36	71[d]	24	47[d]	20	39[d]	1.55±0.12[e]	0.94±0.13[e]	0.61±0.11[e]
Citrus pulp (51)	32	63[d]	21	41[d]	20	39[d]	1.78±0.18[e]	0.90±0.14[e]	0.88±0.16

a Total represents animals with adenomas and/or adenocarcinomas.

b Effective number of animals in each group is shown in parenthesis.

c Mean ± SEM.

d Significantly different from the group fed the control diet by X^2 test ($P<0.05$ or better).

e Significantly different from the group fed the control diet by Student's t-test ($P<0.05$ or better).

Taking a lead from human epidemiologic and metabolic-epidemiologic information, a test system was devised in animal models to demonstrate the effect of various bile metabolites in colon carcinogenesis.

The role of bile acids in the promotion of colon tumors has received support from studies in animals models. Chomchai et al. (69) demonstrated that the carcinogenic effect of AOM in rats was increased by surgically diverting bile to the middle of the small intestine, a procedure that also raised the fecal excretion of bile salts. Further evidence on the importance of bile acids as promoters of colon tumors came from our studies (70-74) (Table 4). The development of colon tumors increased significantly among conventional rats initiated with limited amounts of intrarectal MNNG to give a definite low yield of colon cancer, and groups intrarectally administered doses of deoxycholic acid, lithocholic acid, or taurodeoxycholic acid as promoters, compared with the groups that were given only the carcinogen. The bile acids themselves did not produce any tumors. A recent study also indicates that the primary bile acids, cholic acid and chenodeoxycholic acid, given intrarectally to conventional rats, increased MNNG-induced colon tumors; these primary bile acids are converted to deoxycholic acid and lithocholic acid, respectively. Cohen et al. (75) reported that cholic acid in the diet increased MNU-induced colon carcinogenesis in rats. Total fecal bile acids, particularly deoxycholic acid output, were elevated in animals fed cholic acid as compared with controls. This increase in fecal deoxycholic acid was due to bacterial 7α-dehydroxylation of cholic acid in the colonic contents. These studies demonstrate that the secondary bile acids have a promoting effect in colon carcinogenesis.

The mechanism of action of bile acids in colon carcinogenesis has not been elucidated. Bile acids affect cell kinetics in the intestinal epithelium (76). The cell renewal system is dynamic and may be influenced by changes in a number of factors. Recently, Cohen et al. (75) reported an enhanced colonic cell proliferation in rats fed cholic acid, as well as in animals treated with intrarectal MNU. Lipkin (77) demonstrated that, during neoplastic transformation of colonic cells, a similar sequence of changes leading to uncontrolled proliferative activity develops in colon cancer in humans and in rodents given a colon carcinogen. Recent studies of Takano et al. (78) suggest that the induction of colonic epithelial orinthine decarboxylase and S-adenosyl-L-methionine decarboxylase activities by the bile acids may play a role in these mechanisms. This study indicates that the colonic epithelial responses of the polyamine biosynthetic enzymes to applications of bile acids are among the earliest changes to occur in this tissue in response to promoting agents.

Sterols and Colon Carcinogenesis

Recent studies have indicated that lowering of serum

Table 4. Colon tumor incidence in germ free and conventional rats treated with intrarectal MNNG and/or bile acids.

	Germfree[b]		Conventional[b]	
	Rats with Tumors(%)	Tumors/Rat	Rats with Tumors(%)	Tumors/Rat
CA (10-12)[a]	0	0	0	0
CDC (10-12)	0	0	0	0
LC (10-12)	0	0	0	0
MNNG (22-30)	27	0.27	37	0.55
MNNG + CA (24-30)	50	0.63	67[c]	0.87
MNNG + CDC (24-30)	54	1.08	70[c]	1.23
MNNG + LC (24)	71[c]	1.04	83[c]	1.83

[a] CA, cholic acid; CDC, chenodeoxycholic acid; LC, lithocholic acid. Number of rats are shown in parentheses.

[b] CA, CDC, or LC group received intrarectally 20 mg of sodium salt of respective bile acid three times weekly for 48 weeks; MNNG group received intrarectally 2 mg of MNNG twice a week for two weeks followed by vehicle for 46 weeks; MNNG+CA, MNNG+LC, or MNNG+CDC group received intrarectally MNNG for two weeks and bile acid thereafter for 46 weeks.

[c] Significantly different from rats given MNNG alone by X^2 test, $P<0.05$.

cholesterol levels is associated with increased colon cancer risk in humans (79). Whether low serum cholesterol levels in these patients precede or follow colon cancer is not completely determined. The effect might not be due to low serum cholesterol, but due to delivery of bile salts to the intestinal tract, resulting from the techniques used to lower serum cholesterol. Cholesterol 3β,5α,6β-triol, a principal metabolite of cholesterol epoxide, has been found in increased levels in feces of patients with colon cancer or ulcerative colitis (72,80).

Cruse et al. (81) proposed that prolonged exposure to dietary cholesterol is co-carcinogenic for human colon cancer, in that it facilitates the development, growth, and spread of the disease because dietary fats promote the action of several experimental carcinogens. Broitman et al. (59) studied the effect of polyunsaturated fat and cholesterol on DMH-induced colon tumorigenesis and demonstrated that the interaction between dietary polyunsaturated fats and dietary cholesterol and/or tissue cholesterol may promote tumorigenesis when compared with dietary saturated fat and cholesterol in the animal model. In another study, Broitman et al. (82) found that the effects of dietary cholesterol on colon carcinogenesis appear: (a) to be independent of the putative effects of bile acids as promoters of colon carcinogenesis; and (b) to act during the initiation stage of MNU-induced tumorigenesis rather than the promotional stage.

The mutagenic, as well as colon tumor-promoting, activity of cholesterol and its metabolites have been tested by Smith et al. (83) and Reddy and Watanabe (74). Naturally air-aged commercial samples of cholesterol contain components that are mutagenic towards *Salmonella typhimirium* TA1537, TA1538, and TA98; however, cholesterol epoxide is nonmutagenic towards these strains (83). Our recent studies also indicate that cholesterol, cholesterol epoxide, and triol do not exhibit any colon tumor-promoting activity in both germfree and conventional rat models (74). These observations suggest that cholesterol or its metabolites as produced by colonic bacteria are not detectable either as carcinogens or as promoters for the colonic mucosa; but cholesterol, added to the diet, enhances colon tumorigenesis in the animal model.

METABOLIC INDICATORS OF COLON CANCER

It is clear that the examination of high-risk populations for the development of colon cancer offers unusual opportunities for the identification of carcinogens, co-carcinogens, and promoters. A substantial endeavor is required to obtain the necessary data to assure that individuals at high risk can be identified with precision. At the present time, there is evidence that these populations at high risk excrete increased levels of mutagens (presumptive carcinogens) as well as promoters that are diet-related. Thus, diet provides either directly or after metabolism a mutagen(s) and/or co-carcinogens and promoters capable of affecting the colon. The nature and magnitude of the data encourage optimism that these compounds could be used as markers (indicators) which distinguish low- and high-risk individuals for colon cancer, but without any clinical evidence of cancer. These observations also offer a significant strategy in the overall effort at preventing colon cancer in man.

Evidence has accumulated that populations at high risk for the development of colon cancer excrete in their feces high levels of mutagens, secondary bile acids such as deoxycholic acid and lithocholic acid, and bacterial β-glucuronidase and 7α-dehydroxylase activities. Many exogenous and endogenous substances, including important tumorigenic metabolites, are excreted via bile as glucuronide conjugates and then hydrolyzed by bacterial β-glucuronidase into active metabolites. 7α-Dehydroxylase metabolizes primary bile acids such as cholic acid and chenodeoxycholic acid to deoxycholic acid and lithocholic acid, respectively, which have tumor-promoting activity in the colon. It is important to recognize that the above compounds, whether they are carcinogens, co-carcinogens, or tumor promoters, and the enzymes which produce these compounds, play an essential role in human colon carcinogenesis and may be particularly used as tools to accomplish primary prevention.

CONCLUSIONS

During the last decade, some progress has been made in the understanding of the role played by nutrition and specific nutritional components in colon cancer, as well as dietary habits in various parts of the world. The epidemiologic studies on the relationship of lipids and fiber to colon cancer have been strengthened by animal model studies. Populations with high incidences of cancer of the colon are characterized by consumption of high levels of dietary fat. Furthermore, dietary fat may be a risk factor for colon cancer in the absence of factors that are protective, such as use of high fibrous foods and cereals and whole grains. Thus, alteration of dietary habits leading to a lower intake of fat and higher intake of certain fibers would be indicated to decrease the risk of this important cancer.

Most human cancers in reality result from a complex interaction of carcinogens, co-carcinogens, and tumor promoters. However, the understanding of post-initiating events appears to offer some promise. Most nutritional or dietary factors act at the promotional phase of carcinogenesis. Because promotion is a reversible process, in contrast to the rapid, irreversible process of initiation by carcinogens, manipulation of promotion would seem to be the best method of colon cancer prevention.

ACKNOWLEDGEMENTS

This research was supported in part from grants CA-16382, CA-32617, CA-29602, and CA-17613, and contracts CP-85659 and CP-05721 from the National Cancer Institute. The author acknowledges the expert assistance of Ms. Arlene Banow in preparation of the manuscript.

REFERENCES

1. Committee on Diet, Nutrition, and Cancer of the National Academy of Sciences. Diet, Nutrition, and Cancer. National Academy Press, Washington D.C., 1982.
2. Reddy, B.S., L. Cohen, G.D. McCoy, P. Hill, J.H. Weisburger, and E.L. Wynder (1980) Adv. in Cancer Res. 32:237.
3. Armstrong, D., and R. Doll (1975) Int. J. Cancer 15:617.
4. Burkitt, D.P. (1978) Am. J. Clin. Nutr. 31:S58.
5. Correa, P., and W. Haenszel (1978) Adv. Cancer Res. 26:1.

6. Graham, S., and C. Mettlin (1979) Am. J. Epidemiol. 109:1.
7. Wynder, E.L., T. Kajitani, S. Ishekawa, H. Dodo, and A. Takano (1969) Cancer 23:1219.
8. West, D.K. (1980) Banbury Report #4, Cancer Incidence in Defined Populations, J. Cairns, J.L. Lyon, and M. Skolnick, eds. Cold Spring Harbor Laboratory, New York, p. 31.
9. Reddy, B.S., A.R. Hedges, K. Laakso, and E.L. Wynder (1978) Cancer 42:2832.
10. Doll, R. (1980) Colorectal Cancer: Prevention, Epidemiology and Screening. S. Winawer, D. Schottenfeld, and P. Sherlock, eds. Raven Press, New York, p. 3.
11. Phillips, R.L., L. Garfinkel, J.W. Kuzma, W.L. Beeson, T. Lotz, and B. Brin (1980) J. Natl. Cancer Inst. 65:1097.
12. MacMahon, B., G. Copley, J.F. Fraumeni, Jr., P. Greenwald, and W. Haenszel, eds. J. Natl. Cancer Inst. 65:1055.
13. Haenszel, W., J.W. Berg, M. Segi, M. Kurihawa, and F.B. Locke (1973) J. Natl. Cancer Inst. 51:1765.
14. Staszewski, J., M.G. McCall, and N.S. Stenhouse (1971) Br. J. Cancer 25:599.
15. Hirayama, T. (1978) Prev. Med. 7:173.
16. Jussawalla, D.J., and D.K. Jain (1976) Cancer Incidence in Greater Bombay 1970-1972. Bombay Cancer Registry, India.
17. Wynder, E.L. (1979) Cancer 43:1955.
18. Wynder, E.L., and B.S. Reddy (1973) J. Natl. Cancer Inst. 50:1099.
19. Carroll, K.K., and H.T. Khor (1975) Prog. Biochem. Pharmacol. 10:308.
20. Wynder, E.L., T. Kajitani, S. Ishikawa, H. Dodo, and A. Takano (1969) Cancer 23:1210.
21. Jain, M., G.M. Cook, F.G. Davis, M.G. Grace, G.R. Howe, and A.B. Miller (1980) Int. J. Cancer 26:757.
22. Burkitt, D.P. (1971) J. Natl. Cancer Inst. 47:913.
23. International Agency for Research on Cancer Intestinal Microecology Group (1977) Lancet 2:207.
24. Modan, B., V. Barrel, F. Lubin, M. Modan, R.A. Greenberg, and S. Graham (1975) J. Natl. Cancer Inst. 55:15.
25. Dales, L.G., G.D. Friedman, H.K. Wry, S. Grossman, and S.R. Williams (1979) Am. J. Epidemiol. 109:132.
26. Bjelke, E. (1974) Scand. J. Gastroent. 9: (Suppl. 31), 1.
27. Graham, S., H. Dayal, M. Swanson, A. Mittelman, and G. Wilkinson (1978) J. Natl. Cancer Inst. 61:709.
28. Aries, V., J.S. Crowther, B.S. Drasar, M.J. Hill, and R.E.O. Williams (1969) Gut 10:334.
29. Reddy, B.S., J.H. Weisburger, and E.L. Wynder (1978) Carcinogenesis 2:453.
30. Reddy, B.S. (1981) Cancer Res. 41:3700.
31. Reddy, B.S. (1981) Cancer Res. 41:3766.
32. Talbot, J.M. (1980) Fed. Amer. Soc. for Exper. Biol.
33. Nigro, N.D., A.W. Bull, B.A. Klopfer, M.S. Pak, and R.L. Campbell (1979) J. Natl. Cancer Inst. 62:1097.

34. Spiller, G.A. (1979) Am. J. Clin. Nutr. 31 (suppl.) S231.
35. Kern, F., H.J. Birkner, and V.S. Ostrower (1979) Am. J. Clin. Nutr. 31:S175.
36. Cummings, J.H., M.J. Hill, T. Jivraj, H. Houston, W.J. Branch, and D.J.A. Jenkins (1979) Am. J. Clin. Nutr. 32:2086.
37. Cummings, J.H., H.S. Wiggins, D.J.A. Jenkins, H. Houston, T. Jivraj, B.S. Drasar, and M.J. Hill (1978) J. Clin. Invest. 61:953.
38. Kay, R.M., and A.S. Truswell (1977) Br. J. Nutr. 37:227.
39. Sugimura, T. (1982) Cancer 49:1970.
40. Weisburger, J.H., B.S. Reddy, N.E. Spingarn, and E.L. Wynder Colorectal Cancer: Prevention, Epidemiology and Screening (1980) S. Winawer, D. Schottenfeld, and P. Sherlock, eds. Raven Press, New York, 19.
41. Ehrich, M., J.E. Ashell, R.L. Tassell, T.D. Wilkins, A.R.P. Walker, and N.J. Richardson (1979) Mutat. Res. 64:231.
42. Bruce, W.R., A.J. Varghese, R. Furrer, and P.C. Land (1977) Origins of Human Cancer 4:1651. Cold Spring Harbor Laboratory, New York, 4:1651.
43. Kuhnlein, U., D. Bergstrom, and H. Kuhnlein (1981) Mutat. Res. 85:1.
44. Mower, H.F., D. Ichinotsubo, L.W. Wang, M. Mandel, C. Stemmerman, A. Nomura, and L. Heilburn (1982) Cancer Res. 42:1164.
45. Rao, T.K., J.A. Young, C.E. Weeis, T.J. Slaga, and J.L. Epler (1979) Environ. Mutagenesis 1:105.
46. Nagao, M. T. Yahagi, T. Kawachi, K. Kosuge, K. Tsuji, S. Wakabayashi, S. Mizusaki, and T. Matsumoto (1977) Proc. Japan. Acad. 53:95.
47. Matsumoto, T., D. Yoshida, and S. Mizusaki (1972) Mutat. Res. 56:85.
48. Reddy, B.S., C. Sharma, and E.L. Wynder (1980) Cancer Lett. 10:123.
49. Hill, M.J., B.S. Drasar, V.C. Aries, J.S. Crowther, G.B. Hawksworth, and R.E.O. Williams (1971) Lancet 1:95.
50. Goldin, B.R., and S.L. Gorbach (1976) J. Natl. Cancer Inst. 57:371.
51. Campbell, T.C. (1979) Adv. Nutr. Res. 2:29.
52. Maskens, A.P. Gastrointestinal Tract Cancer (1978) M. Lipkin and R. Good, eds. Plenum Press, New York, p. 361.
53. Deschner, E.E., F.C. Long, and A.P. Maskens (1979) Cancer Lett. 8:23.
54. Maskens, A.P. (1981) Cancer Res. 41:1240.
55. Nigro, N.D., D.V. Singh, R.L. Campbell, and M.S. Pak (1975) J. Natl. Cancer Inst. 54:429.
56. Bansal, B.R., J.E. Rhoads, Jr., and S.C. Bansal (1978) Cancer Res. 38:3293.
57. Rogers, A.E., and P.M. Newberne (1975) Cancer Res. 35:3427.
58. Reddy, B.S., T. Narisawa, D. Vukusich, J.H. Weisburger, and E.L. Wynder (1976) Proc. Soc. Exp. Biol. Med. 151:237.

59. Broitman, S.A., J.J. Vitale, E. Vavrousek-Jakuba, and L.S. Gottlieb (1977) Cancer 40:2455.
60. Reddy, B.S., S. Mangat, J.H. Weisburger, and E.L. Wynder (1977) Cancer Res. 37:3533.
61. Reddy, B.S., and T. Ohmori (1981) Cancer Res. 41:1363.
62. Bull, A.W., B.K. Soullier, P.S. Wilson, M.T. Hayden, and N.D. Nigro (1979) Cancer Res. 39:4956.
63. Diamond, L., T.G. O'Brien, and W.M. Baird (1980) Adv. Cancer Res. 32:1.
64. Wilson, R.B., D.P. Hutcheson, and L. Wideman (1977) Am. J. Clin. Nutr. 30:176.
65. Freeman, H.J., G.A. Spiller, and Y.S. Kim (1978) Cancer Res. 38:2912.
66. Watanabe, K., B.S. Reddy, J.H. Weisburger, and D. Kritchevsky (1979) J. Natl. Cancer Inst. 63:141.
67. Reddy, B.S., H. Mori, and M. Nicolais (1981) J. Natl. Cancer Inst. 66:553.
68. Linos, D.A., C.M. Beard, W.M. O'Fallon, M.B. Dockerty, R.W. Beart, Jr., and L.T. Kurland (1981) Lancet 2:379.
69. Chomchai, C., T. Bhadrachari, and N.D. Nigro (1974) Dis. Colon Rectum 17:310.
70. Narisawa, T., N.E. Magadia, J.H. Weisburger, and E.L. Wynder (1974) J. Natl. Cancer Inst. 53:1093.
71. Reddy, B.S., T. Narisawa, J.H. Weisburger, and E.L. Wynder (1976) J. Natl. Cancer Inst. 56:441.
72. Reddy, B.S., C.W. Martin, and E.L. Wynder (1977) Cancer Res. 37:1697.
73. Reddy, B.S., K. Watanabe, J.H. Weisburger, and E.L. Wynder (1977) Cancer Res. 37:3238.
74. Reddy, B.S., and K. Watanabe (1979) Cancer Res. 39:1521.
75. Cohen, B.I., R.F. Raicht, E.E. Deschner, M. Takahashi, A.N. Sarwal, and E. Fazzini (1980) J. Natl. Cancer Inst. 64:573.
76. Bagheri, S.A., M.G. Bolt, J.L. Boyer, and R.H. Palmer (1978) Gastroenterol. 7:188.
77. Lipkin, M. (1975) Cancer 36:2319.
78. Takano, S., M. Matsushima, E. Erturk, and G.T. Bryan (1981) Cancer Res. 41:624.
79. Williams, R.R., P.D. Sorlie, M. Feinleib, P.M. McNamara, W.B. Kannel, and T.R. Dawber (1981) JAMA 245:247.
80. Reddy, B.S., and E.L. Wynder (1977) Cancer 39:2533.
81. Cruse, J.P., M.R. Lewin, and C.G. Clark (1979) Lancet 1:152.
82. Broitman, S.A., L.S. Gottlieb, and J.J. Vitale (1982) Am. J. Clin. Nutr. 35:842.
83. Smith, L.L., V.B. Smart, and G.A.S. Ansari (1979) Mutat. Res. 68:23.

TUMOR ANTIGENS AS MARKERS FOR CARCINOGENS

Margaret L. Kripke

Basic Research Program
Litton Bionetics, Inc.
Frederick Cancer Research Facility
Frederick, Maryland 21701

INTRODUCTION

Many tumors induced in laboratory animals by chemical and physical agents express antigens not found on the normal tissue from which the tumors originated. Antigens of this type that are detected by their ability to induce increased resistance to tumor challenge in animals of the same inbred strain are called tumor-specific transplantation antigens (TSTA). After Foley (1) and Prehn and Main (2) first demonstrated the existence of such antigens on murine tumors induced by the polycyclic hydrocarbon 3-methylcholanthrene, many studies of TSTA were undertaken. From these studies, two important conclusions were drawn concerning the TSTA found on chemically and physically induced tumors: First, these antigens appear to be individually specific; that is, immunization with one tumor does not confer protection against challenge with a second tumor, even when the second tumor is induced by the same carcinogen in the same animal (3). Second, the degree and strength of antigenicity of tumors induced by a given carcinogen can vary considerably; however, different carcinogens appear to have different ranges of variability. For example, strong carcinogens, which induce tumors in a high proportion of animals in a short period of time, tend to produce highly antigenic tumors, whereas weak carcinogens tend to produce tumors that are weakly antigenic. There is, however, still a considerable range of antigenicity within these categories (4).

These variations in the strength and specificity of TSTA form the basis of the view that these tumor antigens represent random molecular alterations of some surface determinants on the tumor cells which, because of their heritable nature, ultimately must represent an alteration at the level of the cellular genome.

Because of the seemingly random variations in specificity of these antigens, there has been little reason to consider using TSTA as potential markers for carcinogens. However, recent studies on skin cancers induced in mice by ultraviolet (UV) radiation suggest that, at least in this system, a common carcinogen-associated antigen may be present on the tumor cells in addition to the classic, individually specific antigens. This common antigen was not recognized in the past because it usually cannot be detected by conventional *in vivo* cross-protection tests. Instead, it is detected by means of its interaction with immune regulatory cells, in particular, with suppressor T lymphocytes.

TRANSPLANTATION STUDIES WITH UV RADIATION-INDUCED TUMORS

The discovery of a common UV radiation-associated antigen stemmed from studies on the relationship between host immunity and tumor development during UV radiation carcinogenesis. Many primary fibrosarcomas and squamous cell carcinomas induced in mouse skin by chronic UV irradiation are immunologically rejected when transplanted to normal syngeneic recipients (5,6). The resulting immunity is specific for the particular tumor transplanted. This finding that the UV radiation-induced skin cancers were so antigenic that they were rejected by normal animals raised the intriguing question of how these tumors avoid immunological destruction in the primary host.

Studies addressing this question produced the unexpected finding that exposing mice to a short course of UV radiation rendered them unable to reject transplants of these highly antigenic tumors, even when the tumors were implanted into unirradiated sites. This finding demonstrated that exposing the animals to UV radiation produced a systemic alteration that abrogated their ability to reject these cancers. Furthermore, this alteration occurred before primary skin cancers could be detected, indicating that the alteration was not related to the growth of a primary tumor (7).

NATURE OF THE SYSTEMIC ALTERATION IN UV-IRRADIATED MICE

The specificity of this alteration in tumor rejection exhibited by UV-irradiated mice was determined by studies on the ability of these animals to carry out other immune responses. These studies demonstrated that many immune reactions proceed in normal fashion in UV-irradiated mice, even though the mice are incapable of rejecting UV radiation-induced tumors (7-10). Most revealing was the finding that these mice are able to reject transplanted allogeneic tumors and even syngeneic tumors induced by chemicals or viruses (11). This finding indicates that the systemic alteration is selective for only certain antigens and that it does not result from generalized immunosuppression.

The systemic alteration exhibited by UV-irradiated mice was shown to be immunological in nature because it could be transferred to secondary hosts with lymphoid cells. In these experiments, spleen and lymph node cells from normal or UV-irradiated mice were used to reconstitute lethally X-irradiated syngeneic mice. Animals that received injections of normal lymphoid cells were able to reject transplants of UV radiation-induced tumors, whereas those reconstituted with lymphoid cells from UV-irradiated donors could not reject such transplants. Repopulating the X-irradiated mice with cells from both UV-irradiated and normal donors also rendered them susceptible to tumor challenge, suggesting that a suppressor mechanism was involved (12).

Studies by Daynes and his colleagues demonstrated that the thymus-dependent (T) lymphocytes in the lymphoid cell suspension were responsible for the suppressive activity (13). After exposing mice to a sufficient dose of UV radiation, these suppressor T lymphocytes can be found in the spleen and lymph nodes, and they are capable of inhibiting the immunological rejection of syngeneic UV radiation-induced tumors (14,15). Recent studies have shown, in addition, that the UV radiation-induced suppressor T lymphocytes are important in determining the outcome of UV radiation carcinogenesis in the primary host (16).

ANTIGENIC SPECIFICITY OF UV RADIATION-INDUCED SUPPRESSOR CELLS

The activity of the UV radiation-induced suppressor T lymphocytes appears to be restricted to preventing an immune response against UV radiation-induced tumors. This conclusion is based on a few experiments that tested the ability of X-irradiated, reconstituted mice to reject various syngeneic and allogeneic tumors (12,15) and on many experiments that tested various syngeneic tumors for preferential growth in UV-irradiated mice (11,17). In both sets of experiments, the UV radiation-induced suppression of tumor rejection was limited to tumors induced by UV radiation. No evidence of suppressive activity was detected in UV-irradiated mice challenged with allogeneic tumors, syngeneic skin tumors induced by methylcholanthrene, psoralen plus long-wave UV radiation, or Moloney sarcoma virus, or with various spontaneous tumors.

Certain tumors induced by *in vitro* transformation with UV radiation also appear to express the antigen that is recognized by the UV radiation-induced suppressor cells. Preliminary experiments suggest that some murine epidermal cells transformed *in vitro* by UV radiation exhibit preferential growth in UV-irradiated syngeneic mice, relative to their growth in normal hosts (18). In addition, sarcomas produced by *in vitro* irradiation of C3H 10T1/2 fibroblasts with 254-nm UV radiation appear to be recognized by the UV radiation-induced suppressor cells (Fisher, Chan, and Kripke, unpublished data).

The induction of immune regulatory cells requires antigenic stimulation. For this reason, the appearance of suppressor T lymphocytes in the lymphoid organs of UV-irradiated mice suggests that a new antigen has been induced or uncovered by exposing the mouse skin to UV radiation. The finding that the suppressor T lymphocytes interfere with the rejection of all UV radiation-induced tumors suggests, in addition, that this new antigen is present on all such tumors. Thus, the available information suggests that these tumors express at least two types of antigens: the individually specific TSTA against which immune effector cells are directed, and a common, UV radiation-associated antigen against which the immune regulatory cells (suppressor and helper T lymphocytes) are directed (Figure 1). Since the common antigen is the target primarily of the regulatory cells of the immune system, its presence has been difficult to demonstrate by conventional immunization and challenge protocols. Spellman and Daynes (19) reported the induction of transplantation immunity against this common antigen, but their study could not be confirmed by others (reviewed in ref. 18). However, recent studies by Schreiber (20) have demonstrated that in order to detect the common antigen by transplantation immunity, the individually specific TSTA must first be removed by immunoselection.

CONCLUSIONS

These studies suggest that the UV-irradiated mouse could be used as a diagnostic "reagent" for tumors induced by UV radiation. Since only tumors of this etiology appear to be recognized by the UV radiation-induced suppressor cells, the tumors could be identified by their ability to grow preferentially in UV-irradiated mice, relative to their growth in normal animals. Thus, in theory, a test could be developed for identification of these tumors based on their expression of a common UV radiation-associated antigen.

Currently, we do not know whether the UV radiation-associated antigen arises during neoplastic transformation of the target cells or whether normal cells that have been exposed to UV radiation but that are not tumorigenic also express the antigen. We also do not know whether carcinogens other than UV radiation also induce common, carcinogen-associated antigens. At present, these points are a matter of conjecture, but it seems quite possible that other carcinogen-associated antigens exist and that they are independent of the transformation event. This speculation is based on the finding of Boon and his co-workers (21,22), recently confirmed by Frost et al. (23), that treatment of the cells from spontaneously arising murine tumors with certain carcinogens *in vitro* induced new antigens on

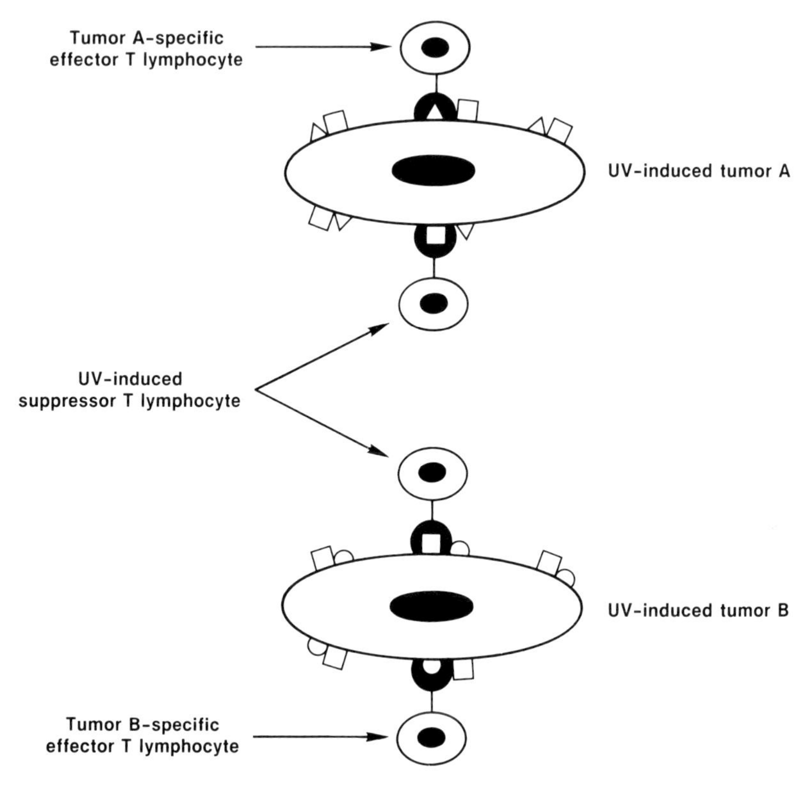

Fig. 1. Model of the antigenic makeup of UV radiation-induced tumors with individually specific determinants and common regulatory determinants.

these already transformed cells. Support for this view is also found in the sporadic reports of cross-reactions among chemically induced tumors (24-26). In most instances where this was observed, the tumors had been passaged in animals or in culture for extended periods of time, procedures that may have permitted the loss of the individually specific TSTA and consequently the expression of the common antigen. Previous studies, which relied on transplantation immunity, may not have revealed the presence of these common carcinogen-associated antigens because they function primarily as the targets of immune regulatory cells rather than immune effector cells. Even though the model presented for the antigenic makeup of UV-radiation induced tumors may not be accurate in every detail, in all likelihood the exploration of the general applicability of the model to other types of carcinogen-induced tumors will be profitable.

ACKNOWLEDGEMENTS

Research sponsored by the National Cancer Institute, DHHS,

under contract No. NO1-CO-23909 with Litton Bionetics, Inc. The contents of this publication do not necessarily reflect the views or policies of the Department of Health and Human Services, nor does mention of trade names, commercial products, or organizations imply endorsement by the U.S. Government.

REFERENCES

1. Foley, E.J. (1953) Antigenic properties of methylcholanthrene-induced tumors in mice of the strain of origin. Cancer Res. 13:835-837.
2. Prehn, R.T., and J.M. Main (1957) Immunity to methylcholanthrene-induced sarcomas. J. Natl. Cancer Inst. 18:769-778.
3. Old, L.G., E.A. Boyse, D.A. Clarke, and E.A. Carswell (1962) Antigenic properties of chemically induced tumors. Ann. N.Y. Acad. Sci. 101:80-106.
4. Stjernswärd, J. (1969) Immunosuppression by carcinogens. Antibiot. Chemother. 15:213-233.
5. Kripke, M.L. (1974) Antigenicity of murine skin tumors induced by ultraviolet light. J. Natl. Cancer Inst. 53:1333-1336.
6. Kripke, M.L. (1977) Latency, histology, and antigenicity of tumors induced by ultraviolet light in three inbred mouse strains. Cancer Res. 37:1395-1400.
7. Kripke, M.L., and M.S. Fisher (1976) Immunologic parameters of ultraviolet carcinogenesis. J. Natl. Cancer Inst. 57:211-215.
8. Kripke, M.L., J.S. Lofgreen, J. Beard, J.M. Jessup, and M.S. Fisher (1977) In vivo immune responses of mice during carcinogenesis by ultraviolet radiation. J. Natl. Cancer Inst. 59:1227-1230.
9. Norbury, K.C., M.L. Kripke, and M.B. Budmen (1977) In vitro reactivity of macrophages and lymphocytes from UV-irradiated mice. J. Natl. Cancer Inst. 59:1231-1235.
10. Spellman, C.W., J.G. Woodward, and R.A. Daynes (1977) Modification of immunologic potential by ultraviolet radiation. I. Immune status of short-term UV-irradiated mice. Transplantation 24:112-119.
11. Kripke, M.L., R.M. Thorn, P.H. Lill, C.I. Civin, and M.S. Fisher (1979) Further characterization of immunologic unresponsiveness induced in mice by UV radiation: Growth and induction of non-UV-induced tumors in UV-irradiated mice. Transplantation 28:212-217.
12. Fisher, M.S., and M.L. Kripke (1977) Systemic alteration induced in mice by ultraviolet light irradiation and its relationship to ultraviolet carcinogenesis. Proc. Natl. Acad. Sci. USA 74:1688-1692.
13. Spellman, C.W., and R.A. Daynes (1977) Modification of immunologic potential by ultraviolet radiation. II. Generation of suppressor cells in short-term UV-irradiated mice. Transplantation 24:120-126.

14. Spellman, C.W., and R.A. Daynes (1978) Properties of ultraviolet light-induced suppressor lymphocytes within a syngeneic tumor system. Cell. Immunol. 36:383-387.
15. Fisher, M.S., and M.L. Kripke (1978) Further studies on the tumor-specific suppressor cells induced by ultraviolet radiation. J. Immunol. 121:1139-1144.
16. Fisher, M.S., and M.L. Kripke (1982) Suppressor T lymphocytes control the development of primary skin cancers in ultraviolet-irradiated mice. Science 216:1133-1135.
17. Kripke, M.L., W.L. Morison, and J.A. Parrish (1982) Induction and transplantation of murine skin cancers induced by methoxsalen plus ultraviolet (320-400 nm) radiation. J. Natl. Cancer Inst. 68:685-690.
18. Kripke, M.L. (1981) Immunologic mechanisms in UV radiation carcinogenesis. Adv. Cancer Res. 34:69-106.
19. Spellman, C.W., and R.A. Daynes (1978) Ultraviolet light-induced murine suppressor lymphocytes dictate specificity of anti-ultraviolet tumor immune responses. Cell. Immunol. 38:25-34.
20. Schreiber, H. Proceedings of Workshop on Photoimmunology, Albuquerque, N.M., June, 1982, Plenum Press (in press).
21. Boon, T., and O. Kellerman (1977) Rejection by syngeneic mice of cell variants obtained by mutagenesis of a malignant teratocarcinoma line. Proc. Natl. Acad. Sci., USA 74:272-275.
22. Boon, T., and A. Van Pel. Teratocarcinoma cell variants rejected by syngeneic mice: Protection of mice immunized with these variants and against the original malignant cell line. Proc. Natl. Acad. Sci., USA 75:1519-1523.
23. Frost, D.P., R.S. Kerbel, E. Bauer, R. Tartamella-Biondo, and W. Cefalu (1982) Mutagen treatment as a means for selecting immunogenic variants from otherwise poorly immunogenic malignant murine tumors. Cancer Res. (in press).
24. Reiner, J., and C.M. Southam (1967) Evidence of common antigenic properties in chemically induced sarcomas of mice. Cancer Res. 27:1243-1247.
25. Leffell, M.S., and J.H. Coggin, Jr. (1977) Common transplantation antigens on methylcholanthrene-induced murine sarcomas detected by three assays of tumor rejection. Cancer Res. 37:4112-4119.
26. Hellstrom, K.E., I. Hellstrom, and J.P. Brown (1978) Unique and common tumor-specific transplantation antigens of chemically induced mouse sarcomas. Int. J. Cancer 21:317-322.

CHAIRMEN'S OVERVIEW ON *IN VITRO* TESTS

T. Ming Chu and David Brusick*

Roswell Park Memorial Institute
666 Elm Street
Buffalo, New York 14263

For all practical purposes, *in vitro* cell culture provides inexpensive and short-term testing for carcinogen. In addition to its technical simplicity and relative reliability, frequently it can give much valuable information. The *in vitro* assay generally has been regarded as an effective approach to the identification of chemical carcinogens and the elucidation of action mechanisms. Although *in vivo* animal bioassay often provides definitive endpoints, such as tissue pathology, *in vitro* testing serves as an invaluable preliminary tool.

At the present stage of its development of carcinogen testing, it is an acceptable hypothesis that a chemical which damages DNA and/or its transcriptional process is most likely a carcinogen. Short-term cell culture systems usually permit the detection of the endpoint, which includes change or deletion of cellular DNA, or one of its consequences, such as mutagenesis, chromosomal alteration, and the final malignant transformation. Recent technological development has resulted in more than 100 short-term cell culture assay systems. Most of the available assays are designed as detection or confirmation tests, or both when a series of tests are combined which are more effective than single assay alone. Generally these systems involve the use of mutant bacteria and mammalian cells. In the latter, culture systems mimicking the human metabolic environment commonly are used. Assays then are employed to observe the effect, if any, of a candidate chemical on the damage of DNA at nucleotide level or chromosome level or on other endpoint markers, although the ultimate malignant transformation also has been used

*Litton Bionetics, Inc., 5516 Nicholson Lane, Kensington, Maryland 20895

increasingly as a criterion of these *in vitro* testings. Results, either positive or negative, can be used as complements to those of long-term *in vivo* rodent tests, and often have served as a decision-making parameter for further stages of carcinogen testing.

It is with a great deal of pleasure that we introduce four of the most distinguished researchers as our speakers in this session. Dr. Andrew Sivak will discuss some new cellular approaches to the identification of tumor promoters by reviewing the merits and problems associated with the existing procedures. His presentation should reveal additional new information on this important area of carcinogen testing. Dr. J. Justin McCormick reviews some possible difficulties in the Ames test and proposes alternative promising procedures which would resolve and complement this most commonly used *in vitro* microbial assay. Dr. Gary Williams will present his decision-point approach using a battery of assays for the detection and confirmation of genotoxic and epigenetic carcinogens in cell culture, which should provide reliable and inexpensive *in vitro* assays for carcinogen. Dr. David Brusick, chairman of our session, will share with us his data on comparative evaluation of a microbial short-term test, the Ames test, and *in vivo* animal bioassays obtained from testing over 60 chemicals. His data should validate the value of *in vitro* test for carcinogen.

BIOCHEMISTRY AND BIOLOGY OF 2-ACETYL-AMINOFLUORENE IN PRIMARY CULTURES OF ADULT RAT HEPATOCYTES

H.L. Leffert,* K.S. Koch, S. Sell,[1] H. Skelly, and W.T. Shier[2]

Division of Pharmacology (M-013 H)
Department of Medicine
University of California, San Diego
La Jolla, California 92093

INTRODUCTION

Most hepatocarcinogenesis models assume that chemical carcinogens (or their reactive metabolites) bind to hepatocyte DNA and, from mutations caused by this event, "initiate" malignant transformation (1). Recent studies of oncogene transfection (2), structure (3), and expression in normal regenerating rat liver (4) tend to strengthen - but not prove (5) - these concepts. For example, liver stem cells might instead be tumor progenitors and macromolecules besides DNA might instead be initiating "targets" (6). Nonetheless, if the "hepatocyte DNA target" point of view prevails, long-term primary cultures of such adult cells will be useful to analyze the molecular and cellular basis of chemical carcinogenesis in normal epithelial systems. We have been working since 1970 to develop a system of this kind. We highlight, here, some of our results along these lines (see also refs. 7-11). Detailed accounts, mainly with 2-acetyl-aminofluorene (AAF), will appear elsewhere (12-15).

---

[1]Department of Pathology and Laboratory Medicine, University of Texas Medical School, Houston, Texas 77030;
[2]Department of Pharmacognosy, College of Pharmacy, University of Minnesota, Minneapolis, Minnesota 55455.

* To whom correspondence should be addressed.

GENERAL PROPERTIES OF THE SYSTEM

Figure 1 shows the morphological appearance of the cultures over an 18-day period in vitro. The cells, obtained from normal adult rat livers perfused with collagenase and hormone-supplemented buffers, are plated into uncoated plastic tissue culture dishes containing arginine-free medium (7). Full details have been published elsewhere (7,16,17). If weekly media changes are made beginning at 10-12 days post-plating, viability is maintained beyond 30 days (10,17).

Hepatocytes in these cultures are structurally differentiated and capable of proliferating (7-10,16-18). A large number of prominent liver-specific functions are expressed, many of which are growth-state dependent (18-20). Several are listed in Table 1. In particular, the cells activate pro-carcinogens like AAF (7-9) and

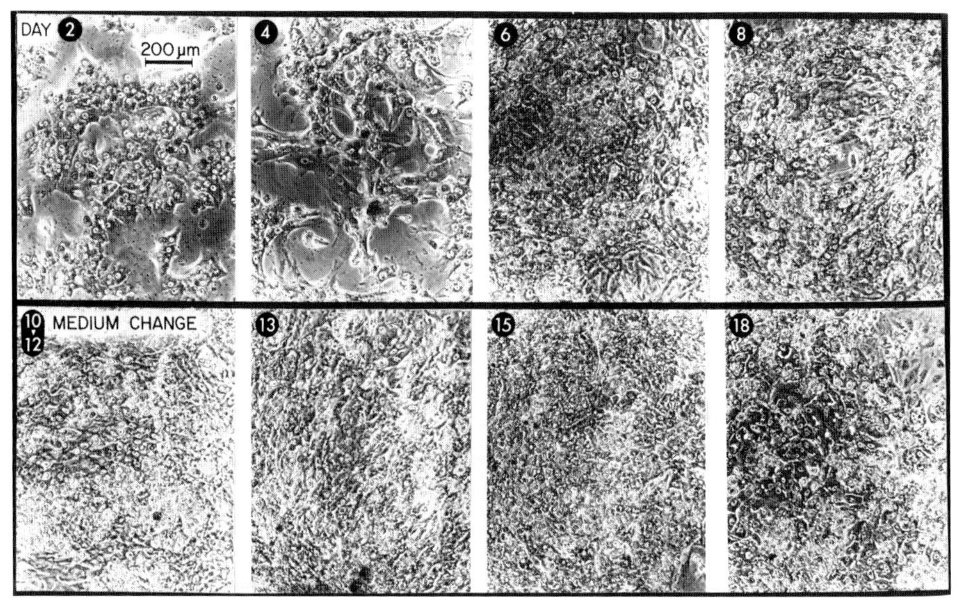

Fig. 1. Phase micrographs of adult rat hepatocytes in primary culture. The cells were plated ($1x10^6$ cells per 2 ml per 35 dish) into medium containing 15% (v/v) dialyzed fetal bovine serum, and 10 μg/ml each of insulin, hydrocortisone, and inosine as described elsewhere (7-9). On day 12, and thereafter, cultures were fluid changed into fresh serum-free medium supplemented with 50 ng/ml each of EGF (epidermal growth factor), insulin, and glucagon (26-29). On various days post-plating (encircled numbers, upper left corners of each panel), the cells were fixed for photomicroscopy (6).

Table 1. Differentiated properties of rat hepatocytes in primary "monolayer" cultures.

Property		Fetal	Adult
ENZYMES	Alcohol dehydrogenase [EC 1.1.1.1]	-	+[b]
	Gluconeogenic	+	+[b]
	Glutathione S-transferase B[EC 2.5.1.18]	U[a]	+[b]
	Glycogenolytic	+	+
	Microsomal $P_{448-450}$	+	+
	Pyruvate kinase [EC 2.7.1.40]		
	Fetal (type III)	+	+[b]
	Adult (type I)	-	+[b]
	Tyrosine aminotransferase [EC 2.6.1.5]	+	+
	Urea "cycle" (ornithine transcarbamylase [EC 2.1.3.3])	+	+
SECRETORY PROTEINS	Albumin	+	+
	$Alpha_1$-fetoprotein	+[b]	+[b]
	Fibrinogen	U	+
	Haptoglobin	+	+
	Hemopexin	+	+
	Transferrin	U	+
	Very low density lipoprotein	-	+[b]
ULTRASTRUCTURE	Bile canaliculi (vectorial transport)	+	+
	Desmosomes	+	+
	Glycogen granules	+	+
	Peroxisomes	+	+
	Tight junctions	+	+

[a] Unknown
[b] Growth state dependent (18 20).

3-methylcholanthrene (see below). Because this activation occurs in proliferation-competent hepatocytes (7), the culture system provides a model to study the early and, potentially, the late events in the interactions between carcinogens and epithelial cells.

CHEMICAL CARCINOGEN UPTAKE, METABOLISM, AND BINDING

The Acetyl-aminofluorene enters hepatocytes immediately by a noncarrier-mediated process. Uptake is probably determined by partitioning coefficients of AAF with different membrane lipids. By the criterion of the quantity of water-soluble, hydroxylated AAF metabolites released into the medium per 24 hr, measured by thin layer chromatography in two different solvent mixtures, two AAF-metabolizing systems are present with K_m [apparent] values of $1\text{-}3 \times 10^{-7}$M (Metabolism System 1) and $3\text{-}4 \times 10^{-5}$M (Metabolism System 2). Conversion of AAF to aminofluorene (AF), presumably by a deacetylase, occurs only through a high K_m system (12). The kinetic constants are summarized in Table 2.

The low K_m system may be situated in or associated with the nucleus. AAF metabolism by rat liver nuclei (21) has been confirmed in this laboratory with extensively washed nuclei. We find that nuclear metabolites, formed by a strictly NADPH-dependent reaction, bind covalently to nuclear macromolecules (12). This is shown in Table 3. Nuclear binding is virtually abolished by boiling for 2 min. Significant inhibition of metabolite binding to DNA and protein also is seen after pretreatment of nuclei with DNase and pronase. Nuclear RNA-containing structures available for binding to AAF metabolites seem to be more resistant to enzymatic disruption. The reasons for this are unclear. Interestingly, 0.5% (v/v) Triton X-100 fails to remove all of the nuclear metabolism-binding capacity, suggestive of a tight association between nuclear structures required for AAF metabolism and/or metabolite binding (see below).

Table 2. Apparent "Michealis" constants for two AAF-metabolizing systems in primary adult rat hepatocyte cultures.

Soluble Metabolite Formed from AAF	$K_{m(apparent)}$, M	
	Low	High
1-OH-AAF	2.0×10^{-7}	4.0×10^{-5}
3-OH-AAF	1.0×10^{-7}	3.0×10^{-5}
5-OH-AAF	3.0×10^{-7}	4.0×10^{-5}
7-OH-AAF	3.0×10^{-7}	4.0×10^{-5}
N-OH-AAF	2.0×10^{-7}	4.0×10^{-5}
Aminofluorene	---	3.0×10^{-5}

Three day old cultures (3 to 4×10^5 cells/dish) were pulsed 24 hr with $[^{14}C]$AAF (2×10^{-8}M to 2×10^{-4}M; sp. act. 100 dpm/pmol). Water soluble metabolites excreted into culture fluids were extracted after treatment with β-glucuronidase and α-amylase, and quantitated by thin layer chromatograms developed in chloroform: methanol (97:3) and benzene:acetone (4:1) as previously described (refs. 7,9). Apparent K_m values were estimated from two-component curves obtained by Lineweaver-Burk double reciprocol plots (12).

Table 3. Covalent binding of [^{14}C]AAF metabolites to nuclei from primary adult rat hepatocyte cultures.

Pre-treatment	Cofactor	Fluorene Residues Bound pmoles/mg protein/60 min (% maximal)
None	--	0.079 (24)
	NADP	0.080 (24)
	NADPH	0.328 (100)
Wash, x2 (Sucrose step-gradient)	--	0.070 (21)
	NADPH	0.320 (98)
Boil, 2 min	--	0.042 (13)
	NADPH	0.100 (30)
DNase	--	0.085 (26)
	NADPH	0.228 (70)
RNase	--	0.057 (17)
	NADPH	0.300 (92)
Pronase	--	0.042 (13)
	NADPH	0.128 (39)
Triton X-100, 0.5% (v/v)	--	0.090 (27)
	NADPH	0.164 (50)

Nuclei were prepared from washed monolayers by standard procedures and subjected to various physicochemical or enzymatic (50 μg enzyme/ml) pretreatments. The nuclei were recovered, washed, and incubated (∿1.4 mg DNA/ml) with or without 1 mg cofactor/ml and [^{14}C]AAF (0.65×10^{-7} M; sp. act. 100 dpm/pmol) at 37°C for 1 hr. The reaction was stopped by bringing the mixture into 10% (v/v) trichloroacetic acid at 100°C for 5 min. Insoluble material was precipitated onto Whatman GF/C filters, which were then washed, dried, and counted to ±1% errors. NADPH-independent binding averaged ∿20% of cofactor-dependent binding.

Figure 2 shows indirect evidence of hepatocyte metabolism of [^{14}C]3-methylcholanthrene (3-MC). The cultures were incubated with the chemical, and after preparing soluble detergent-extract mixtures, radioactive proteins were immunoprecipitated with monospecific polyclonal rabbit antiserum to rat glutathione-S-transferase B [E.C. 2.5.1.18], a protein whose subunits bind covalently to 3-MC metabolites (22). The presence of bound adducts on precipitated proteins of M_r [apparent] ∿26,000 and 24,000 daltons, revealed on SDS-PAGE gels, strongly suggests that microsomal P_{448} is present in these cultures.

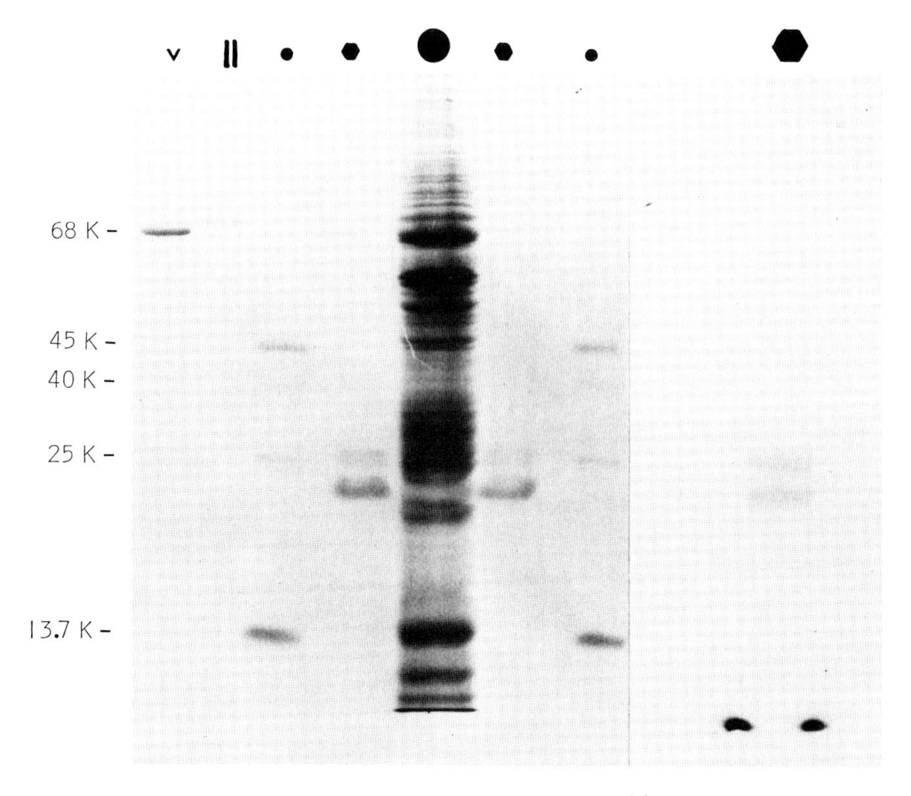

Fig. 2. Covalent binding of metabolites of [^{14}C]3-methylcholanthrene to cellular proteins in primary adult rat hepatocyte cultures. Nine day old cultures (~0.6x10^5 cells/dish) were incubated with [^{14}C]3-methylcholanthrene (5μCi/ml; sp. act. 40-60 mCi/mmol) for 24 hr. Two dishes were washed and the "monolayers" dissolved into a buffer solution containing 0.2% (v/v) Triton X-100 (18). The pooled lysates were incubated with rabbit antiserum containing precipitating antibodies to "ligandin" (glutathione S-transferase B) overnight at 4°C. The precipitate that formed was washed extensively, dissolved, and boiled 2 min in β-mercaptoethanol containing SDS-buffer. About 200 specifically bound cpm were electrophoresed on a 10% (w/v) polyacrylamide SDS-slab gel - which was then subjected to fluorography for 5 days under standard conditions (large ⬢ track at extreme right). Coomassie Blue stained tracks on the gel are, from left to right, rat albumin (v; 68K), rat α_1-fetoprotein (‖ ; 70K), authentic molecular weight marker proteins (small ●), authentic ligandin (small ⬢), and the non-precipitated cellular lysate (large ●). Positions of molecular weights are given at left.

Table 4. Two classes of "AAF" binding sites for AAF-metabolites in primary adult rat hepatocyte cultures.

			SITE I		SITE II	
Diet	Strain	Number of Experiments	K_D (apparent)*	Binding Capacity†	K_D (apparent)*	Binding Capacity†
Normal	Fischer/344	2	7×10^{-6} M	6.0	2×10^{-3} M	400
	Sprague-Dawley	3	5×10^{-6} M	5.0	1×10^{-3} M	300
Lipotrope Deficient	Sprague-Dawley	2	5×10^{-6} M	5.0	1×10^{-3} M	100

* Corrected intracellular "free" [^{14}C]AAF concentration after 24 hr labelling.

† Picomoles [^{14}C]fluorene residues per 10^6 cells after 24 hr labelling.

Three day old cultures (from various animal donors) were incubated with varying concentrations of [^{14}C]AAF as described under Table 2. After 24 hr, the monolayers were washed with cold buffer and extracted overnight at 4°C with 2 ml 5% (v/v) trichloroacetic acid. These extracts were used to measure and to quantitate cellular levels of AAF ("free AAF"). The remaining monolayers were solubilized and acid-insoluble, heat-stable radioactive material collected onto filters as described under Table 3. These counts were taken to represent "bound AAF". Using Scatchard analysis-like assumptions, apparent dissociation constants (K_D), and binding site "capacities" were calculated from curves relating [bound/free] vs. [bound] AAF (13). Intracellular water-spaces were found to be ~2 pl H_2O/cell and it was assumed that free AAF distributed in this volume. We further assumed that: 1) All cells are targets; 2) ~20 pg DNA/cell; 3) 1 genome ~6×10^{12} daltons ≈4×10^9 dG residues; and 4) fluorene residues bind exclusively to dG binding sites in a ratio of 1:1.

Table 4 summarizes results of experiments with the adult cells suggesting the existence of two major classes of binding sites for metabolites of AAF, having K_D[apparent] values of 6 μM (Binding Site 1) and 2 mM (Binding Site 2). Direct evidence indicates that Site 1 is nuclear DNA. As ambient AAF levels are raised in culture, the binding of fluorene residues to DNA saturates at an amount predicted from Scatchard-like curves of 6 pmoles per 10^6 cell DNA equivalents (13). DNA-repair processes might explain this saturation phenomenon in living cells. More subtle structural properties of chromatin could also be involved. In addition, nuclear binding of AAF metabolites to DNA may be non-random with preferences for newly-synthesized, high molecular weight DNA. Direct evidence for this comes from studies with purified DNA obtained from growing cells labelled with [^{3}H]dT or [^{3}H]AAF, and treated with DNA site- specific restriction endonucleases (13). This is shown in Figure 3. Notably, studies of covalent binding of AAF metabolites with nuclear macromolecules show that very low ambient AAF levels (10^{-8} to 10^{-7} M) give rise to nuclear-DNA adducts (Tab. 3). At these doses, intracellular AAF levels are less than the K_m for Metabolism System 2.

Binding Site 2 seems to be comprised mainly of proteins. At low ambient AAF levels (10^{-8} to 10^{-6} M), little protein binding is detected. Only at higher AAF concentrations in the range of the K_m for Metabolism System 2 (>10^{-5} M) is significant protein binding revealed (13). This is shown in Figure 4. Fluorograms of cellular proteins analyzed by SDS-PAGE under denaturing conditions show radioactive AAF metabolites bound predominantly to two cellular (56,000 and 31,000 dalton) proteins and to secreted albumin (68,000 daltons). Neither major nor minor cellular "targets" detected on this gel are principal proteins with respect to Coomassie staining (13). These studies involve the use of a high specific activity

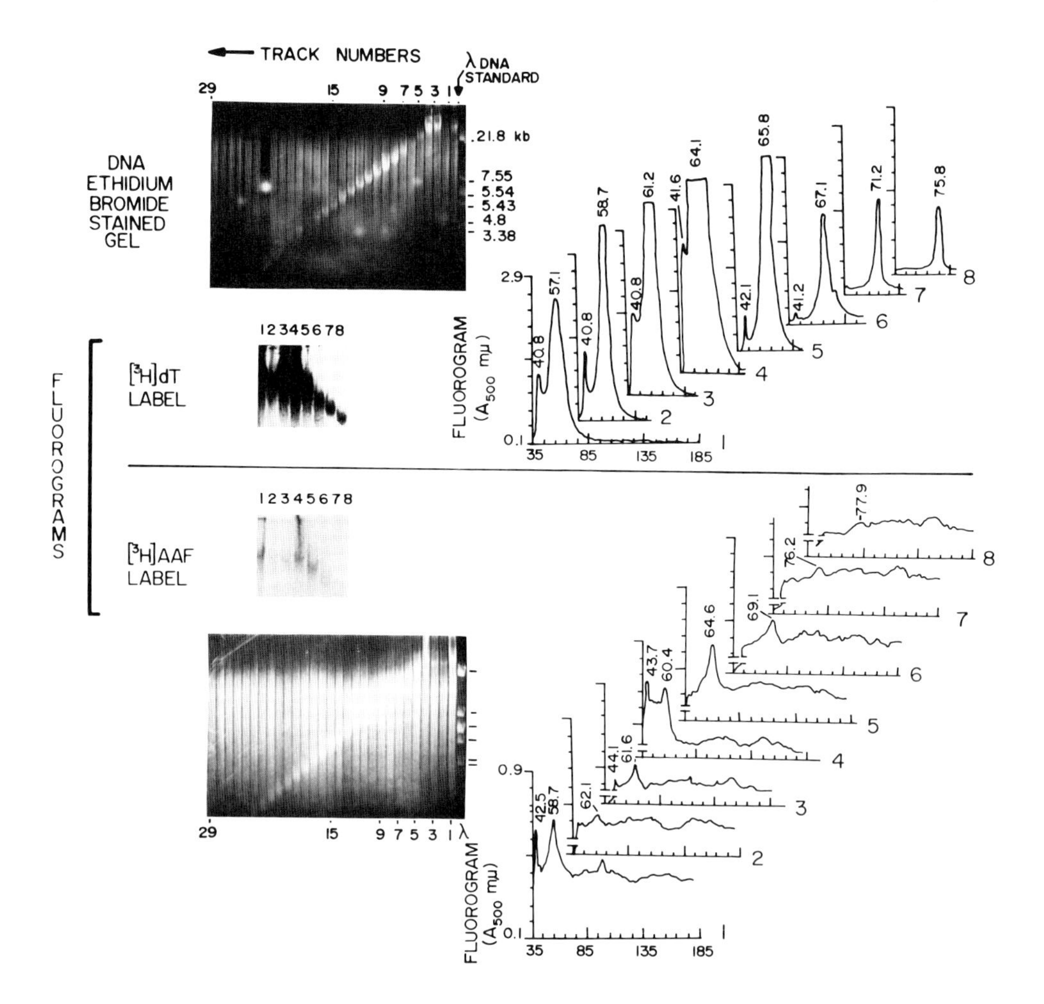

[^{3}H]AAF probe (see also Fig. 3) without which considerably longer gel exposure times would be necessary. Cells from lipotrope-deficient rats show Site 2 binding capacity defects with "normal" Site 1 constants (Table 4; see also refs. 8,9).

WORKING HYPOTHESIS

Cultured hepatocytes might preferentially bind metabolites of AAF to "targets" of different affinities depending on external AAF concentrations presented to the cells. At low AAF levels, DNA is a preferred target (because of its lower "K_D" and because of a nuclear AAF-metabolizing system with a low K_m), whereas, at high AAF levels, cytoplasmic protein targets become available, partly because of the concomitant operation of a high K_m cytoplasmic AAF-metabolizing system and its putative proximity to or equivalence with low affinity target proteins. A prediction of this hypothesis is that the ratio of AAF metabolites bound to nuclear sites compared to those

Fig. 3. Direct evidence of covalent binding of AAF metabolites to nuclear DNA in primary adult rat hepatocyte cultures. Three day old cultures were divided into two groups: one was incubated 24 hr with [^{3}H]AAF ($2x10^{-5}$ M; 4000 dpm/pmol), and the other pulsed 24 hr with [^{3}H]thymidine ($3x10^{-6}$ M unlabelled TdR; 1.25 μCi [^{3}H]dT/ml, sp. act. 20 Ci/mmol). Nuclei from washed monolayers (a total of $1\text{-}2x10^{-7}$ cells/group) were then prepared, incubated with 100 μg carrier calf thymus DNA, and highly purified DNA obtained ($A_{260/280}$ ∿1/0.5; $S_{20,w}$ = 21-23; and ρ_B ∿ 1.702 in CsCl). The DNA mixture from each group was first incubated at 37°C for 24 hr with Hind III and electrophoresed on 0.7% (w/v) agarose flat-bed gels. DNA from these gels (29 tracks, with #1 at the top and #29 at the bottom) was eluted with NaI, recovered, and incubated with EcoRI (along with λ viral DNA as a standard) as above, and electrophoresed again on agarose gels. The second gels are shown above, after staining with ethidium bromide. The direction of track numbers (i.e., high to low MW) goes from right to left. The gels were subjected to fluorography (216 days); the results are shown for tracks #1-8 (left to right). It should be noted that, for the [^{3}H]dT gel (top), the band intensifies on the fluorogram were directly proportional to the total radioactivity loaded per track. Under these conditions, labelled bands on the [^{3}H]AAF fluorogram (bottom) were not seen beyond track #8 (see ref. 13).

bound to cytoplasmic sites should decrease as ambient AAF levels are increased. Direct evidence supports this prediction, as shown elsewhere (10).

The results also suggest that "normal" binding of AAF metabolites to DNA occurs in hepatocytes from lipotrope-deficient rats, whereas a more prominent "defect" exists with respect to protein binding. The biochemical nature of this defect and its role, if any, in promotion of hepatocarcinogenesis - an effect attributed to lipotrope deficiency - remain unknown.

EFFECTS OF AAF ON RE-INITIATION OF HEPATOCYTE DNA SYNTHESIS IN DEFINED MEDIUM

Perhaps of even more interest to the carcinogenesis problem is the finding that at exogenous AAF levels (ca. 8×10^{-7} M), which half-saturate DNA binding sites (13), neither the initiation nor the

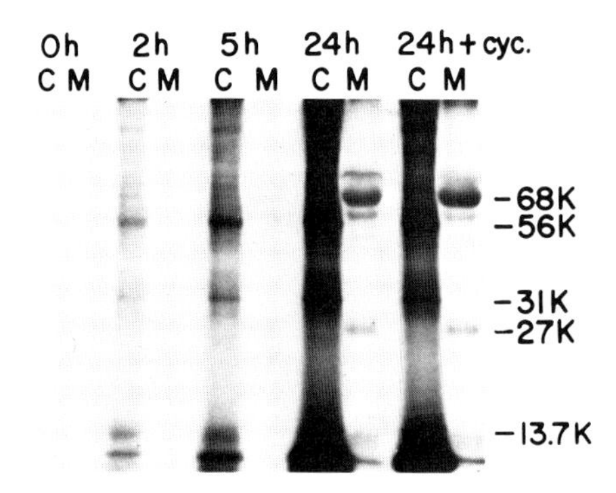

Fig. 4. Covalent binding of metabolites of [^{3}H]AAF to cellular and secreted proteins in primary adult rat hepatocyte cultures. Three day old cultures ($3\text{-}4 \times 10^5$ cells/dish) were pulsed for varying times (0,2,5, and 24 hr) with [^{3}H]AAF (2×10^{-5} M; sp. act. ~4000 dpm/pmol). Some cultures, harvested at 24 hr, also received 5 μg cycloheximide/ml at zero time. At the indicated times, culture fluids (with particulate matter removed from them by centrifugation) and cellular extracts were obtained from washed monolayers and dissolved into SDS-gel running buffer. All samples (C = cell extracts; M = culture media) were boiled 2 min in buffer containing SDS and β-mercaptoethanol, and subjected to SDS-PAGE as described under Fig. 2. The gel was dried and subjected to fluorography for 93 days. MW markers also were run (not shown) to give relative MW values (at right) of radioactive proteins shown. If the samples were incubated with 50 μg pronase/ml prior to SDS-PAGE, none of the bands revealed were present after fluorography (13).

continuation of hepatocyte DNA synthesis are impaired (14). In fact, hepatocyte DNA synthesis is blocked only if 100-fold higher AAF doses are given (K_i[apparent] = $6x10^{-5}$ M). This is shown in Figure 5. These results suggest that inhibition of DNA synthesis is indirect, involving Metabolism System 2 and Binding Site 2. Such

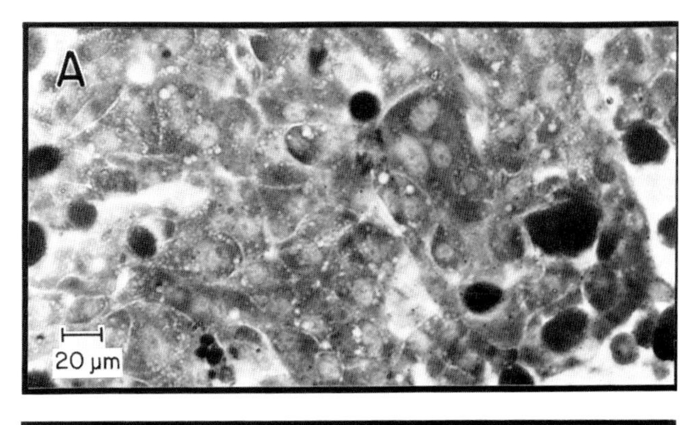

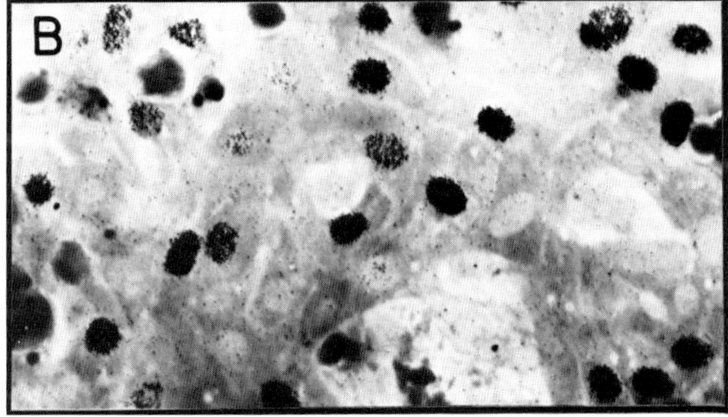

Fig. 5. Direct evidence of inhibition of DNA synthesis by AAF in primary rat hepatocyte cultures. Twelve day old cultures (∿$0.8x10^5$ cells/dish) were used for growth reinitiation assays exactly as described elsewhere (26-29). Control cultures (panel B) were shifted into fresh medium containing 50 ng/ml each of EGF, insulin, and glucagon plus the vehicle used to dissolve AAF (i.e. 1% [v/v] ethanol). Experimental cultures were incubated under identical conditions together with $2x10^{-4}$ M AAF. [^{3}H]dT (1.25 μCi/ml) was added at 12 hr. At 24 hr (day 13), the cultures were fixed, subjected to radioautography, and stained with crystal violet (26). Significant numbers of DNA synthesizing hepatocytes are seen in the hormone-stimulated, untreated cultures (labelling index ≅35-40%; panel B); no labelled hepatocytes are seen in AAF treated cultures (panel A). Control studies rule out artifacts of potential inhibition of [^{3}H]dT transport or conversion to dTTP by AAF (see ref. 14).

findings are strong evidence that procarcinogen feeding <u>in vivo</u> inhibits hepatocyte proliferation by a direct interaction of AAF with the liver (15). Further studies indicate that different AAF metabolites, including the ring-hydroxylated ones (see Tab. 2), also block DNA synthesis but with varying potencies that are less than AAF or N-OH-AAF (14). Such findings suggest that the so-called "non-carcinogenic" detoxified metabolites re-enter the cultured cells, where they may become N-hydroxylated and further converted into reactive electrophiles.

CONCLUSIONS AND FUTURE PROSPECTS

The observations have two important corollaries. First, if hepatocyte AAF-DNA-adducts initiate chemical hepatocarcinogenesis then inducing-doses of such chemicals may be far below those causing detectable DNA repair synthesis in nonreplicating cultures (23) or mutagenesis in bacterial test systems (24). Second, because hepatocyte proliferation can occur in cells with DNA which contain bound-adducts, altered cells may arise (if such events occur) under conditions that do not necessitate growth-inhibition of the bulk-parenchymal population. We have not yet determined whether the sites of AAF-DNA binding (Fig. 3) reside in specific genes or control regions, or whether they coincide with cellular proto-oncogenes. We have attempted to transform normal adult hepatocytes with hepatoma DNA, but we have been unsuccessful thus far.

A minority cell type in the adult rat liver - the "oval" cell - may also be a hepatoma precursor (6). Oval cell proliferation is facilitated by conditions in which parenchymal cell proliferation or function is depressed (for example, by AAF feeding) and by nutritional conditions which promote the carcinogenic process (25) and might reduce the methyl content of liver cell DNA (lipotrope deficiency). Oval cells have not been observed to participate in normal rat liver regeneration. There is no evidence yet that such cells are recovered from normal tissues under routine hepatocyte plating conditions. Thus, special methods are needed to obtain large numbers of oval cells for <u>in vitro</u> work (6). The biochemistry of AAF and of its interactions with mature hepatocytes, and the expression of oncogenes in these fascinating rat liver cell populations (and their role, if any, in hepatocarcinogenesis) remain to be determined.

ACKNOWLEDGMENTS

This work was supported by grants from the USPHS (CA29540, AM 28215, AM 28392). We thank I. Arias for providing us with ligandin antibody and S. Dutky for computerized typing assistance. This is paper No. 11 in a series. Papers 8-10 are refs. 27-29, respectively.

REFERENCES

1. Heidelberger, C. (1975) Chemical carcinogenesis. Annu. Rev. Biochem. 44:79-121.
2. Shilo, B.Z., and R.W. Weinberg (1981) Unique transforming gene in carcinogen-transformed mouse cells. Nature 289:607-609.
3. Tabin, C.J., S.M. Bradley, C.I. Bargmann, R.A. Weinberg, A.G. Papageorge, E.M. Scolnick, R. Dhar, D.R. Lowy, and E. Chang (1982) Mechanism of activation of a human oncogene. Nature 300:143-149.
4. Fausto, N., M. Goyette, C. Petropoulos, and P. Shank (1982) Oncogene expression during liver regeneration. J. Cell Biol. 95:476a.
5. Logan, J., and J. Cairns (1982) The secrets of cancer. Nature 300:104-105.
6. Sell, S., and H.L. Leffert (1982) An evaluation of cellular lineages in the pathogenesis of experimental hepatocellular carcinoma. Hepatology 2:77-86.
7. Leffert, H.L., T. Moran, R. Boorstein, and K.S. Koch (1977) Procarcinogen activation and hormonal control of cell proliferation in differentiated primary adult rat liver cell cultures. Nature 267:58-61.
8. Leffert, H.L., and K.S. Koch (1978) Proliferation of hepatocytes. In Hepatotrophic Factors, CIBA Foundation Symposium, No. 55, pp. 61-94.
9. Leffert, H.L., K.S. Koch, B. Rubalcava, S. Sell, T. Moran, and R. Boorstein (1978) Hepatocyte growth control: In vitro approach to problems of liver regeneration and function. In Gene Expression and Regulation of Cultured Cells, Third Decennial Review Conference, TCA, Natl. Canc. Inst. Monogr., 48:87-101.
10. Koch, K.S., and H.L. Leffert (1980) Growth control of differentiated adult rat hepatocytes in primary culture. Ann. N.Y. Acad. Sci. 349:111-127.
11. Leffert, H.L., W.T. Shier, H. Skelly, and T. Moran (1980) Biochemical processing and anti-proliferative effects of an hepatoprocarcinogen in primary adult rat hepatocyte cultures. Proc. Amer. Assoc. Cancer Res. 21:114.
12. Leffert, H.L., W.T. Shier, H. Skelly, and T. Moran. Adult rat hepatocytes in primary culture. XII. 2-Acetyl-aminofluorene (AAF): Two classes of low and high affinity AAF-metabolite forming systems (in preparation).
13. Leffert, H.L., W.T. Shier, H. Skelly, T. Moran, and R. Gutel. Adult rat hepatocytes in primary culture. XIII. 2-Acetyl-aminofluorene (AAF): Two classes of low and high affinity AAF-binding sites (in preparation).
14. Leffert, H.L., H. Skelly, and T. Moran. Adult rat hepatocytes in primary culture. XIV. 2-Acetyl-aminofluorene (AAF): Direct evidence for the inhibition of hepatocyte DNA synthesis-initiation and replication under chemically defined conditions (in preparation).

15. Leffert, H.L., S. Sell, H. Skelly, and T. Moran. Adult rat hepatocytes in primary culture. XV. 2-Acetyl-aminofluorene (AAF): Multicycle characterization of short-term proliferative and functional properties of cells obtained from rats fed a carcinogenic diet (in preparation).
16. Leffert, H.L., K.S. Koch, T. Moran, and M. Williams (1979) [47] Liver cells. In Methods in Enzymology (S.P. Colowick and N.O. Kaplan, eds.), Vol. 58 (W. Jakoby and I. Pastan, eds.) pp. 536-544.
17. Leffert, H.L., K.S. Koch, and H. Skelly (1983) Methods for primary culture of hepatocytes. In Methods in Molecular and Cell Biology (G. Sato, D. Sirbasku, and D. Barnes, eds.), Alan R. Liss, Inc., New York (in press).
18. Leffert, H., T. Moran, S. Sell, H. Skelly, K. Ibsen, M. Mueller, and I. Arias (1978) Growth state dependent phenotypes of adult hepatocytes in primary monolayer culture. Proc. Natl. Acad. Sci., USA 75:1834-1838.
19. Leffert, H.L. and K.S. Koch (1979) Regulation of growth of hepatocytes by sodium ions. Chapter 6. In Progress in Liver Diseases, Vol. VI, pp. 123-134.
20. Lad, P.J., W.T. Shier, H. Skelly, B. de Hemptinne, and H.L. Leffert (1982) Adult rat hepatocytes in primary culture. VI. Developmental changes in alcohol dehydrogenase activity and ethanol conversion during the growth cycle. Alcoholism: Clin. Exp. Res. 6:64-71.
21. Kawajiri, K., H. Yonekawa, E. Hara, and Y. Tagashira (1979) Activation of 2'-acetyl-aminofluorene in the nuclei of rat liver. Cancer Res. 39:1089-1093.
22. Litwack, G., B. Ketterer, and I.M. Arias (1971) Ligandin: A hepatic protein which binds steroids, bilirubin, carcinogens, and a number of exogenous organic anions. Nature 234:466-467.
23. Williams, G.M. (1977) Detection of chemical carcinogens by unscheduled DNA synthesis in rat liver primary cell cultures. Cancer Res. 37:1845-1851.
24. Ames, B.M., E.G. Gurney, and H. Bartsch (1972) Carcinogens as frameshift mutagens: Metabolites and derivatives of 2-acetyl-aminofluorene and other aromatic amine carcinogens. Proc. Natl. Acad. Sci., USA 69:3128-3132.
25. Rogers, A.E. (1975) Variable effects of a lipotrope-deficient diet, high fat on chemical carcinogenesis in rats. Cancer Res. 35:2469-2475.
26. Koch, K.S., and H.L. Leffert (1979) Increased sodium ion influx is necessary to initiate rat hepatocyte proliferation. Cell 18:153-163.
27. Leffert, H.L., and K.S. Koch (1982) Hepatocyte growth regulation by hormones in chemically-defined medium: A two-signal hypothesis. Cold Spring Harbor Confer. Cell Prolif. 9:597-613.
28. Leffert, H.L., K.S. Koch, M. Fehlmann, W. Heiser, P.J. Lad, and H. Skelly (1982) Amiloride blocks cell-free protein synthesis at levels attained inside cultured hepatocytes. Biochem. Biophys. Res. Commun. 108:738-745.

29. Koch, K.S., P. Shapiro, H. Skelly, and H.L. Leffert (1982) Rat hepatocyte proliferation is stimulated by insulin-like peptides in defined medium. Biochem. Biophys. Res. Commun. 109:1054-1060.

RELATING *IN VITRO* ASSAYS OF CARCINOGEN-INDUCED GENOTOXICITY TO TRANSFORMATION OF DIPLOID HUMAN FIBROBLASTS

J. Justin McCormick and Veronica M. Maher

Carcinogenis Laboratory - Fee Hall
Department of Microbiology and Department of Biochemistry
Michigan State University
East Lansing, Michigan 48824-1316

INTRODUCTION

As part of this EPA-sponsored discussion of the application of biological markers to carcinogen testing, we have been asked to consider some general aspects of the use of *in vitro* tests to quantitate the genotoxic effects of chemicals, as well as to describe some of our recent results on the transformation of diploid human fibroblasts by chemicals and radiation. The term genotoxicity is recent in origin and is used to designate the property of a chemical or radiation which results in damage to the genetic apparatus of a cell or virus. The fact that short term, *in vitro* assays designed to detect genotoxicity are commonly employed as a means of identifying potential cancer-causing agents indicates that it is now commonly assumed that DNA is the principal cellular target for carcinogenesis. That is the underlying basis, for example, of assays used to detect potential carcinogencity by screening chemicals for their mutagenic action in bacteria and for their ability to cause DNA damage and/or repair. This assumption does not necessarily underlie *in vitro* assays which use morphological transformation of mammalian cells as their end point for detecting carcinogens. However, as will be described, the induction of loss of anchorage dependence in diploid human fibroblasts, as well as in at least two other mammalian cell transformation systems, appears to be the result of a mutagenic event. Therefore, such assays can also be used to quantitate the genotoxicity of various environmental agents as well as their potential carcinogenicity.

BACTERIAL MUTAGENESIS ASSAY

The most widely used assay among the tests for genotoxicity is the Ames' Salmonella *his*$^-$ to *his*$^+$ reverse mutation assay (1). The simplicity of the assay and its ability to be rapidly standardized have been two of its most attractive features. Many workers without formal training in mutagenesis have found that with a little practice, they could successfully utilize this assay. Another strong advantage of this assay is its high sensitivity. Strains TA98 and TA100, containing the plasmid pKM101, are extremely sensitive to the mutagenic action of many compounds. For example, in strain TA98, 1,8-dinitropyrene has been reported to give 254,000 revertants per nmole (2) and is one of the most mutagenic compounds ever reported in Salmonella.

One of the important questions raised is whether or not results obtained with Salmonella are predictive of the response of mammalian cells to the same compounds. That is, will agents that give high mutagenicity in Salmonella similarly yield high mutagenicity in assays carried out in mammalian cells? A review of the literature reveals that results with the Ames' strains are generally reliable _qualitative_ predictors of the mutagenicity for organic compounds, but the quantitation observed does not always correlate with the _quantitative_ results obtained in mammalian cell mutagenicity assays or in whole animal tumorigenicity assays. For example, as indicated above, nitro aromatic compounds such as 1,6- or 1,8-dinitropyrene are among the most mutagenic compounds ever tested in Salmonella. Yet in mammalian cells (3,4) and in a tumor assay (5), nitro aromatic compounds demonstrate only modest activity. Such differences are to be expected if the bacteria possess large amounts of highly efficient nitro reductase activity needed to convert such compounds into reactive derivatives, whereas mammalian cells are less efficient at such conversions.

Providing Metabolic Activation of Parent Carcinogens

In electing to use S9 fractions prepared from rat livers as the source of the metabolizing enzymes needed to activate parent compounds, Ames hoped to mimic as closely as possible the way test chemicals would be activated by the cells of intact rodents. However, recent evidence shows that the metabolism of several polycyclic aromatic hydrocarbons by S9 fractions from rat liver differs qualitatively as well as quantitatively from their metabolism by the intact cells of mouse skin, a widely-studied target organ for measuring carcinogenesis induced by this class of carcinogens. In intact mammalian cells, polycyclic aromatic hydrocarbons containing a "bay region" are metabolized into diol epoxides by a multi-stepped process involving a series of enzymes (6). Therefore, it is to be expected that differences in the amount of activity of these various enzymes in an S9 preparation, or the availability of co-factors or of competing enzymatic pathways would influence the amount of particular products formed. The predicted result would be that single oxidative metabolites would predominate rather than the diol epoxide of such hydrocarbons. Therefore, it was surprising that Santella et al. (7) found that the principal DNA adduct formed in _Salmonella typhimurium_ cells exposed to benzo(a)pyrene in the presence of S9 fractions of rat liver was identical to that formed in intact human cells as well as in other mammalian cells, i.e., that formed by the anti isomer of the 7,8-diol-9,10-epoxide.

However, Bigger et al. (8) showed that, in contrast to the results obtained by Santella et al. (7) with benzo(a)pyrene, when 7,12-dimethylbenz(a)anthracene (DMBA) is metabolized by a S9 fraction prepared from rat liver, the types of adducts found in the calf thymus DNA used to trap the reactive DMBA derivatives produced

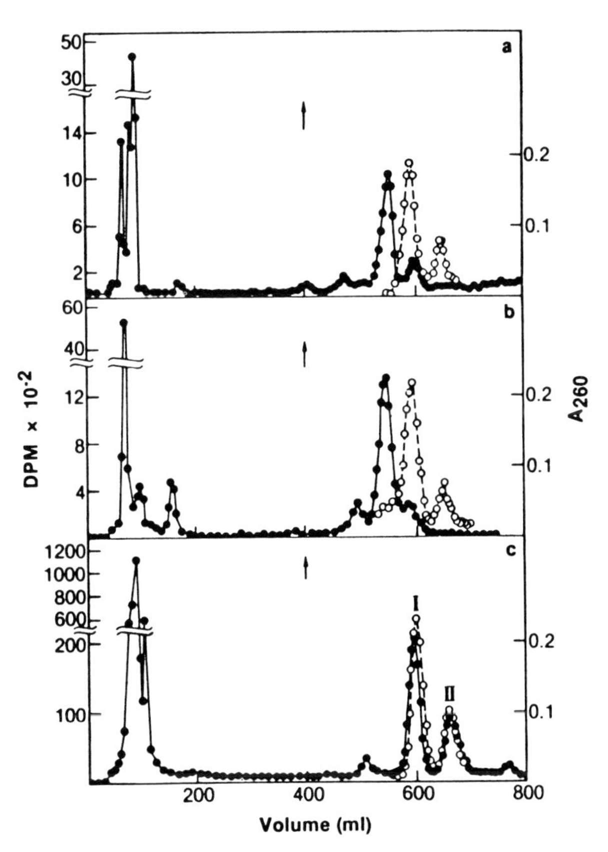

Fig. 1. Comparison of Sephadex LH20 chromatography of DMBA 5,6-oxide-nucleoside products (0---0) formed by enzymic digestion of calf thymus DNA which had been reacted with DMBA 5,6-oxide in vitro with enzymically digested DMBA-DNA products from: (a) mouse skin from animals treated 24 hours with [^{3}H]DMBA; (b) mouse embryo cell cultures treated 24 hours with [^{3}H]DMBA; and (c) calf thymus DNA treated with [^{3}H]DMBA for 2 hours in the presence of Aroclor-stimulated rat microsomes. 0---0, DPM; 0---0, absorbance at 260 nm; ↑, position for elution of added 4-(p-nitrobenzyl)-pyridine marker. Taken from ref. 8 with permission.

depends upon the ratio of DMBA to S9 in the incubation mixture (see Figure 1). With higher concentrations of DMBA, new species of DNA adducts, not present in DMBA-treated mouse skin, appear. Prominent among the new species are adducts of the K-region epoxide of DMBA. Since the carcinogencity of K-region epoxides of polycyclic aromatic hydrocarbons is not correlated with that of the parent compounds, it is now commonly assumed that such metabolites are not the metabolic precursors that cause carcinogenicity. In contrast, when Bigger et al. (9) varied the concentration of DMBA that mouse cells were exposed to by 40-fold, they found no qualitative change in the kinds of carcinogen adducts produced in the DNA of these cells (Figure 2). Since Salmonella tests or other tests using the S9 fraction for activation are usually carried out at the highest possible concentration of the test chemicals, these results of Bigger et al. (9) suggest that the adducts formed in the DNA of the test cells may be different from those found in the target tissues of the animals used for carcinogenicity studies. This result is not surprising in view of the need for a 3-step activation into diol epoxide derivative, but it underscores the fact that even when one is working with test compounds taken from the same general class of chemicals, one cannot assume that the reactive derivatives formed by the S9 fraction of rat liver or other sources of metabolism will always correspond to those formed within intact metabolizing cells.

This problem should be less serious when one is working with chemicals which need only a single enzymatic step for activation. However, the Ames' test including the rat liver S9 fractions is most frequently employed to assay the mutagenicity of chemicals for which the metabolic activation pathways are unknown and no reference material to indicate the "correct" DNA adduct profile is available. One obvious solution is to utilize mammalian or human cells rather than homogenates as the activation system for Salmonella. Studies developing such protocols, especially with rat hepatocytes as the metabolizing cells, are underway, but the data base is still small.

Limitations of the Assay

There are compounds such as hydrogen peroxide, nalidixic acid, bleomycin, actinomycin D, and methotrexate that are negative in the usual Ames' tester strains but are mutagenic in *Bacillus subtilis* multigene assay (10) as well as other tests. The failure of the Ames' strains to pick up such mutagens may reflect the fact that these strains only exhibit reverse mutations or may be related to less specific causes such as the difference in species. Another limitation in the Ames' Salmonella assay is the lack of a toxicity assay suitable for measuring the killing of the test strains by the mutagens. Such an assay is important because if a toxic compound is only weakly mutagenic, one may see fewer *his*$^+$ revertant colonies per plate at higher concentrations of the test chemical because of toxicity. Appropriate cytotoxicity data would indicate whether this decrease was caused by the killing of some of the mutant colonies by the test chemical. If so, an appropriate correction could be made

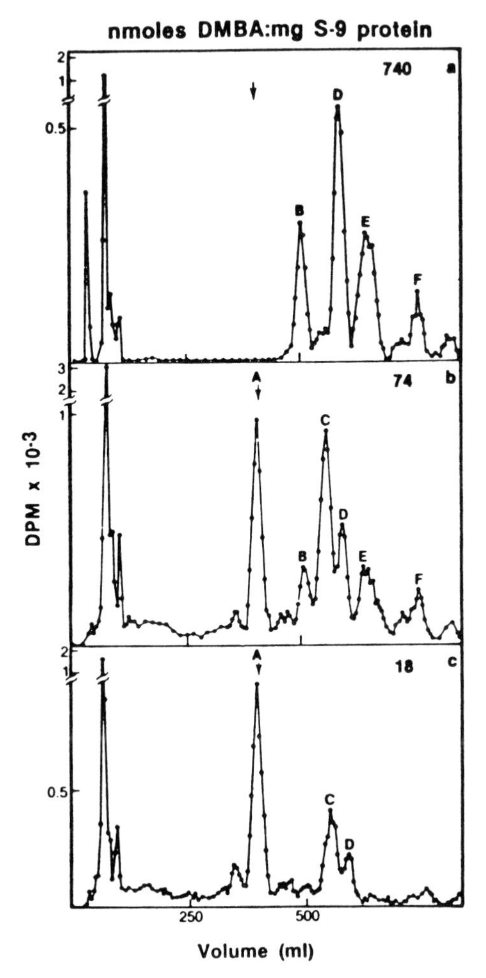

Fig. 2. Sephadex LH-20 column chromatography of DMBA-deoxyribonucleoside adducts formed by enzymatic digestion of calf-thymus DNA which had been treated in the presence of rat liver S9 fraction with various doses (a-c) of [^{3}H]DMBA. The single-headed arrow denotes the position of elution of an added UV-absorbing marker, 4-(p-nitrobenzyl)pyridine. Taken from ref. 9 with permission.

in the mutation frequencies. A method to carry out such a cytotoxicity assay by means of newly created Salmonella strains isogenic with the Ames' tester strains has recently been published by Waleh et al. (11). An example of how useful such a procedure is was reported by Kaden et al (12). They treated Salmonella cells with fluoranthene, cyclopenta(c,d)pyrene, and benzo(a)pyrene and found no numerical increase in the number of mutants observed in a forward mutation assay (loss of a functional hypoxanthine(guanine)phosphoribosyltransferase). However, when they corrected for the toxic effects of the compounds, they found the results shown in Figure 3. Clearly, wider use of toxicity assays with Salmonella assays may be valuable. Nevertheless, considering the problems of providing for metabolic activation discussed above, as well as the pharmacodynamic factors which can be expected to occur in the body of intact animals, the correlations which have been obtained between the mutagenicity of chemicals as assayed with the standard Ames' test and their carcinogenicity assayed in intact animals is very impressive.

IN VIVO - *IN VITRO* TEST FOR DNA DAMAGE AND REPAIR

An interesting example of a compound which is a strong hepatocarcinogen in rats, but which has proved negative in most tests which assay genotoxicity, is commercial grade dinitrotoluene (DNT), a mixture of the 2,4- and 2,6-DNT isomers. In the Ames' Salmonella assay, 2,4-DNT isomer was reported to be negative (13,14), but later, when retested in the light of the strong tumor response, was shown to give a weak response (15). When this compound or its isomers were assayed for induction of forward mutations to 6-thioguanine resistance in a CHO cell line, in the presence or absence of metabolic activation they proved negative (16). Neither were these compounds, the 2,4- and 3,5-DNT isomers, able to induce mitotic recombination of *Saccharomyces cerevisiae* (14). Similarly, assays of 2,4-DNT, as well as other isomers, for mutagenicity in mice by the dominant lethal test, as well as by the sperm morphology and/or recessive spot test were negative (17). Thus, this is a case of a strong liver carcinogen giving a negative, or at most a weak positive, response in short-term assays of genotoxicity.

However, when Mirsalis and Butterworth (18) tested DNA isomers using an *in vivo* - *in vitro* assay they have developed (19), they found an excellent correlation between carcinogenicity and induction of unscheduled DNA synthesis (UDS) in primary rat hepatocytes. Male rats were exposed to 2,4-DNT or 2,6-DNT and then UDS was measured in primary hepatocyte cultures prepared from them. There was a strong positive UDS response with 2,6-DNT and only a weak response with 2,4-DNT. Female rats gave only a modest UDS response with either isomer (18). These results parallel those from tumorigenicity studies (20). When similar studies were undertaken with germ-free rats, exposure to DNT no longer caused UDS in hepatocytes (21). If two weeks before exposure to DNT such germ-free animals were allowed to acquire eight anaerobic bacterial strains, similar to those found in normal gut microflora, the UDS response of the hepatocytes was positive. These results suggest that the bacteria flora of the gut perform some necessary metabolic step to convert DNT into a genotoxic form.

Metabolism studies by other workers (22) have shown that axenic rats treated with 2,4-DNT have markedly reduced levels of two of the

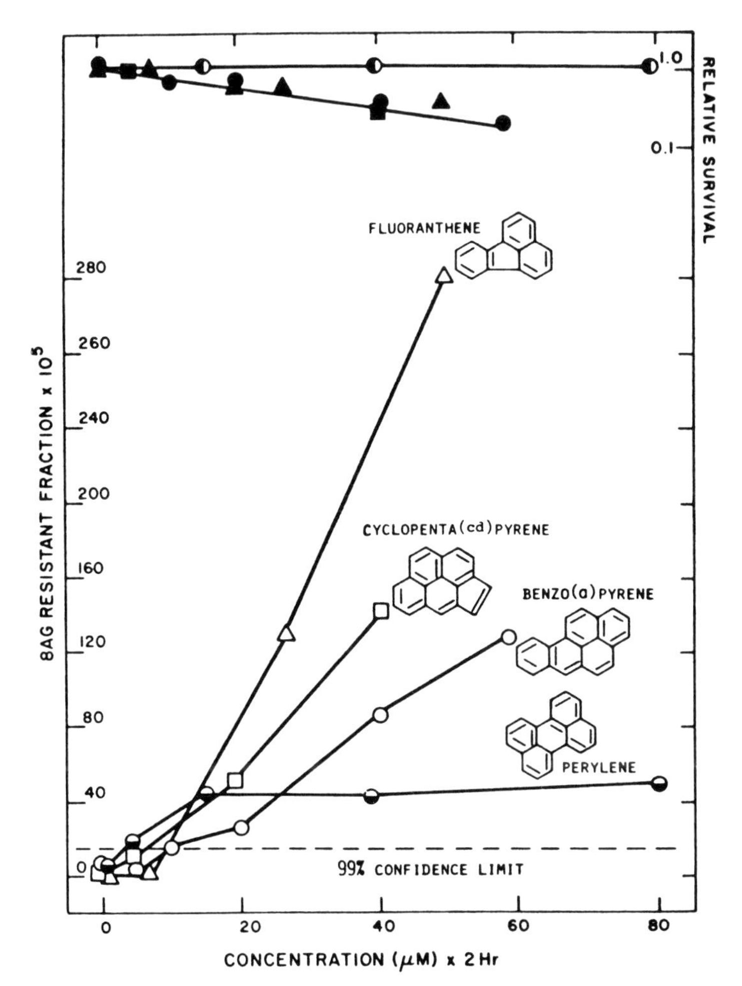

Fig. 3. Concentration-dependent mutagenicity (open symbols) and toxicity (closed symbols) of cyclopenta(c,d)pyrene (□,■), fluoranthene (△,▲), benzo(a)pyrene (○,●), and perylene (◐,◒), to S. typhimurium. All points were assayed in the presence of aroclor-induced postmitochondrial supernatant. Each point represents the average of 2 independent determinations. 8AG, 8-azaguanine. Taken from ref. 12 with permission.

four major DNT metabolites in their urine. Studies with rat intestinal flora in vitro showed that various reduced metabolites of 2,4-DNT were produced (23). Such metabolites are not produced in hepatocytes in culture at physiological oxygen concentrations (24). Taken together, these data indicate that reductive metabolism is required to convert parent DNT into reactive species that are genotoxic or to produce metabolic intermediates that require further metabolism in order to produce the ultimate genotoxic agents. Since the S9 fraction or feeder layer cells usually used as metabolic activation systems in short-term assays primarily carry out oxidative rather than reductive metabolism of such compounds, it is not surprising that these latter assays gave negative results with this particular class of agents. Nevertheless, this important finding emphasizes the need for caution in interpreting results of *in vitro* tests.

TRANSFORMATION OF DIPLOID HUMAN FIBROBLASTS AS AN ASSAY FOR GENOTOXICITY

Mammalian cell mutagenesis studies. Quantitative forward mutagenesis assays based on mammalian cells, including human cells, also have been developed over the past decade (see ref. 25 for a review) and are now commonly employed as tests for genotoxicity. Such systems involving mammalian cells have the potential to detect classes of compounds which cannot exert their genotoxic effects in bacteria. Examples are hormones and mitotic spindle poisons. However, these assays are far more labor intensive than bacterial tests such as that of Ames, take longer to complete, and demand a greater degree of sophistication on the part of the person carrying out the mutagenesis tests. Furthermore, because most of the target cells employed lack the ability to metabolize parent compounds into reactive derivatives, such assays require the same kinds of provisions for exogeneous activation as Salmonella and so suffer from similar problems. For this reason the data base for these mammalian or human cell systems is much smaller than that which has accrued for Salmonella. However, it continues to increase and quantitative mutagenesis studies conducted in these "higher" cells contribute to the tier approach recommended for use in testing unknown compounds for potential genotoxicity.

Transformation Studies with Diploid Human Cells

For the past few years, we have been developing a system for the quantitation of human fibroblast transformation induced by chemical carcinogens or radiation (26-29). This area has long been a problem. For a number of years, epidemiological data indicated that a large percentage of human cancer may be caused by exposure to environmental agents (30). At the same time, human cells in culture proved refractory to transformation by chemical agents or radiation but could be transformed by SV40 virus and by certain RNA tumor viruses. In 1977, Kakunaga was able to demonstrate that normal human fibroblasts could be transformed by chemical agents (31). Since that time we have been working on techniques to quantitate human cell transformation in order to make use of an *in vitro* transformation system to investigate the number and kinds of steps involved in carcinogenesis.

Transformation as the result of a mutagenic event. In the past, most workers who were developing transformation assays have assumed either explicitly or implicitly that the transformation event(s) are not the result of mutations. Consequently, most transformation assays were not designed to simultaneously yield information on transformation frequencies and mutation frequencies using the same treated cell populations. Furthermore, the protocols used often prevented one from recognizing whether mutation(s) were involved in transformation. Since we have had extensive experience developing mutation assays for diploid human fibroblasts before we began development of a transformation assay, and since we believed that the transformation process was likely to involve one or more mutagenic events, we deliberately designed a human cell transformation assay modeled after our human cell mutagenesis assay (26). We were also aware that the induction of anchorage-independent growth in embryonic mouse cells had been shown by fluctuation analysis to occur in a pattern consistent with it arising as a result of a mutational event (32). This fact also encouraged us to consider that the induction of anchorage independent growth in human cells by carcinogenic agents might result from a mutational event.

For example, we expected the process of transformation of human fibroblasts to have in common with the process of mutagenesis: a) an expression period, i.e, a period of time between carcinogen damage of cells and their ability to express the new (i.e., transformed) phenotype; b) a linear increase in the frequency of transformants with an increase in carcinogen dose; c) a concentration dependence for the carcinogenic agent which resembled that required for induction of mutations so that strong mutagens would usually be strong transforming agents; d) a low, but measurable frequency of transformed cells in non-carcinogen treated cell populations just as one finds a low, but measurable, frequency of mutant cells in such populations; e) a higher frequency of transformation per dose in DNA repair-deficient cells than in normal cells, as we had shown for the induction of mutants in XP cells; and f) a cell cycle dependence similar to that which occurs for mutation induction, so that populations of cells treated just before S phase would show a higher frequency of transformation than cell populations treated with the same dose of the agent far from S phase. Our results described below and elsewhere (27,28) confirmed each of these predictions.

Our assay, unlike that of Kakunaga (31), is based on the induction of ability of the cells to grow in an anchorage-independent manner following carcinogen treatment. Human fibroblasts are treated with reactive derivatives of chemical carcinogens or radiation (UV or ionizing) using doses which lower the survival, as judged by cloning, to between 40% and 10% of the control population. The surviving cells (10^6 or more) are kept in exponential growth for 8 to 10 population doublings to allow full expression of the anchorage-independent phenotype (Figure 4) and are then plated as single

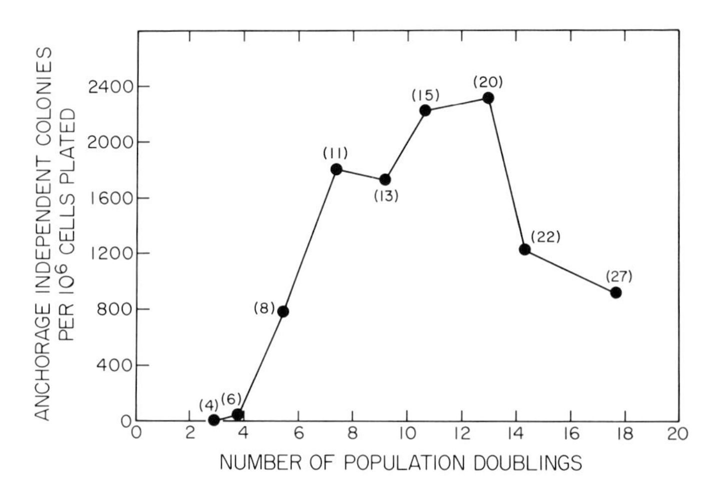

Fig. 4. Kinetics of expression of anchorage independence. Cells treated with propane sultone (21.5 ug/ml) for 14 hrs after the cells entered S phase. The surviving population (∿28%) as well as the control population were trypsinized, diluted 1:2, replated, and kept in exponential growth by continuous cubculturing as required. On each day of the days indicated in parenthesis, 2 to 5 x 10^6 cells were trypsinized, pooled, and counted to determine the number of population doublings which occurred. Then, for each determination, 10^6 cells were assayed for anchorage independent growth (5 x 10^4 cells/dish), and the rest were replated for continued propagation and subsequent assay. Taken from ref. 27 with permission.

cells into medium containing 0.33% agar (28). Colonies of anchorage-independent cells develop to a size which can be counted (∿ 0.2 mm) after 3 to 4 weeks. Large-sized colonies, isolated from the agar by use of a Pasteur pipette, must be separated from the more numerous non-transformed cells present in the surrounding agar, since the latter remain viable even though they do not replicate in agar. The anchorage independent cells are then pooled and allowed to replicate into large populations.

Representatives of these anchorage-independent cells have been repeatedly tested for tumorigenicity by injecting 10^7 cells s.c. into sublethally X-irradiated athymic mice. No tumors develop in animals injected with non-transformed, control populations of anchorage dependent cells (>$2x10^7$/injection). In contrast, the anchorage-independent cells grow subcutaneously in these mice and form nodules 0.7 to 2 cm in diameter in 8 to 16 days. Representative examples of such nodules were surgically removed for examination by pathologists and in order to return cells to culture.

Reports from the pathologists who examined sections prepared from these nodules indicated that they were fibrosarcomas or undifferentiated sarcomas. Those designated fibrosarcomas were indistinguishable from the slides prepared from sections of nodules formed by injecting HT1080 cells, a malignant cell line derived from a human fibrosarcoma. However, in the majority of cases, if the nodules were not removed when they attained a diameter of 0.7 to 1.0 cm, they were seen to regress in size. Pathology examination of sections through the latter nodules showed heavy infiltration by lymphocytes. The cells derived from these regressing tumors which have been examined for karyotypes appear to be diploid. When reinjected into X-irradiated mice at 10^7 cells/injection they tend to produce regressing tumors.

However, in carrying out the above protocols with anchorage-independent cells obtained from agar colonies, we obtained two nodules which attained a size which exceeded the average size cited above. After one month these two tumors were excised. Half of each tumor was sent to pathology and the other half returned to culture. These nodules also proved to be fibrosarcomas. Karyotype examination of these latter two cell strains showed that they were composed of heteroploid cells. Each strain had its own model number of chromosomes. When these cells were reinjected into the animals, they again formed non-regressing tumors which threatened to kill the animals. Because of their characteristics these cells seem to have spontaneously undergone additional changes. Our current hypothesis is that these latter two cell strains are fully transformed whereas the anchorage-independent cells are only partially transformed.

Further evidence that loss of anchorage dependence with its subsequent tumorigenicity is induced as a genetic event in human diploid cells derives from a recent study in our laboratory (28) comparing the frequency of induction in normal human cells and in cells derived from two excision repair-deficient xeroderma pigmentosum patients, XP7BE (complementation group D) and XP12BE from group A. [Note that these XP cells are not malignant; the biopsies from which they are derived are always obtained from non-sunlight exposed areas of the skin. This is confirmed by the fact that injection of $>2 \times 10^7$ cells into athymic mice did not produce tumors (28).] The results are shown in Figure 5.

The survival data (top panel) show that the XP cells are significantly more sensitive than NF cells to the lethal effect of UV; e.g., a dose of 0.5 or 1.1 J/m^2 decreased their survival to 20%, whereas 7.5 J/m^2 was required to cause the same decrease in NF cells. The mutagenicity data (middle panel) show that the XP cells are significantly more sensitive than normal cells to the mutagenic action of UV. The mutant frequencies were corrected for the cloning efficiency of the cells because we showed that, under the selection conditions, their cloning efficiency is similar to that of cells

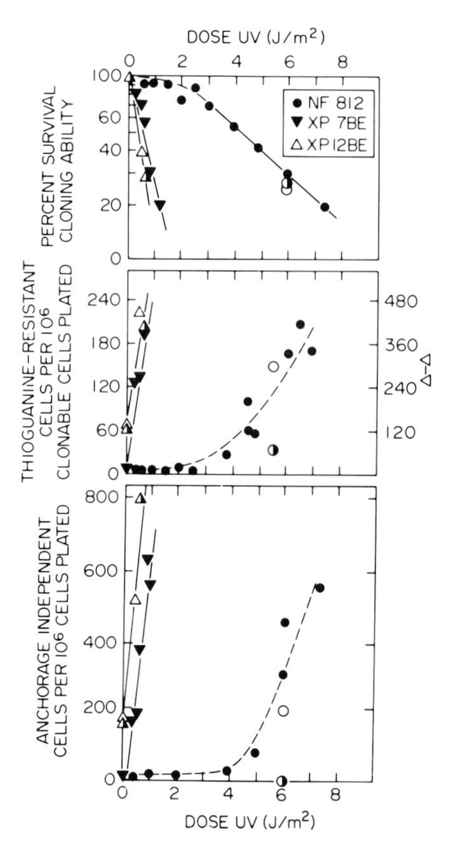

Fig. 5. Cytotoxicity, mutagenicity, and transforming ability of UV radiation in normal (circles) and XP cells (XP12BE, triangles, XP7BE, inverted triangles). The frequency of thioguanine resistant cells was assayed after 6 doublings; that of anchorage-independent cells after 9 to 11 doublings. The former were corrected for cloning efficiency on plastic. Solid symbols, populations irradiated in exponential growth; open symbols, cells synchronized by release after confluence and irradiated shortly before onset of S phase; half-solid symbols, cells irradiated 18-20 hrs prior to S phase. See text for details. Some of these data were taken from ref. 36.

plated at cloning density in the absence of selection (33). These results with XP7BE confirm those found earlier by Glover et al. (34) using a different genetic marker and support our earlier findings with XP2BE, complementation group C (35).

The transformation frequency data (bottom panel) indicate that the XP cells are also significantly more sensitive than normal cells to UV-induced loss of anchorage dependence. These particular data were obtained by assaying the progeny of the irradiated cells after an expression period of about nine population doublings. However, in the majority of cases, the frequency of anchorage-independent cells was assayed twice - i.e., after 9 and 11 doublings - and yielded approximately the same frequencies both times. (Note that these data are the observed frequencies of agar colonies counted directly without a correction factor.) The data indicate that to achieve a particular degree of cell killing, mutagenesis, and transformation, NF cells have to be exposed to 8- to 10-fold higher doses of UV radiation than XP cells. As discussed above, this is the result expected if induction of anchorage independence as well as thioguanine resistance results ultimately from DNA damage remaining unexcised in the cell at some critical time after irradiation and if, because of the difference in their respective rates of excision repair, the average number of lesions remaining at this critical time is approximately equal in the three populations.

Effect of Time for Excision Repair Before S Phase on the Frequency of Anchorage Independence in Repair-Proficient Normal Cells

We showed elsewhere (36) that if synchronous populations of NF cells are irradiated shortly before the onset of S phase, the frequency of thioguanine resistant cells induced is $\sim$8-fold higher than in the same cells treated 18 hrs prior to S phase. No such difference is observed when the target cells are virtually incapable of excision repair. To determine if a similar cell cycle effect occurred for induction of transformation, we irradiated NF and XP12BE cells in early and late G phase and assayed their progeny for frequency of anchorage-independent cells. The results are included in Fig. 5 as open and half-solid symbols. The transformation frequencies of the repair-proficient normal cells showed a strong cell cycle dependence, i.e., cells irradiated with 6 J/m^2 $\sim$ 3 hrs prior to onset of S phase yielded 200 anchorage-independent cells per 10^6 cells plated; the control cells in this experiment also gave no colonies out of 2 x 10^6. In contrast, the frequency of anchorage-independent cells in XP12BE population irradiated in early G_1 phase did not decrease; in fact it was somewhat higher. In the corresponding mutation experiment, the frequencies were equal. In the mutagenesis experiments with normal cells from which the data in the middle panel were taken, the frequency of mutant cells did not decrease completely to the background level. However, in this mutagenesis experiment, the cells irradiated in G_1 phase had somewhat less time for excision repair before onset of S phase than was available in the transformation experiments. The fact that allowing substantial time for excision before DNA synthesis eliminated the potentially mutagenic and transforming effect of UV radiation in normal cells, but not in XP12BE cells, suggests that DNA synthesis on a template still containing unexcised lesions is the cellular event responsible for permanently "fixing" the mutations and transformation.

In addition to the information that can be gained by analyzing the 8- to 10-fold difference between XP and NF cells in Fig. 5, it is useful to compare the frequencies of anchorage-independent cells with those of thioguanine-resistant cells. The similarity of the dose-responses for the two phenotypes is obvious and supports the idea that acquisition of anchorage independence (transformation) in human cells occurs as the result of a single mutation event. The frequencies of mutations and transformation differ by a factor of 2.5. If we assume that anchorage independence results from a mutation, we might speculate about the size of the DNA target involved. The observed number of thioguanine-resistant colonies (selected on plastic at a density equivalent to 10^4 cells per 60 mm dish) has been corrected for the cells' intrinsic cloning efficiency (i.e., 30-50%). The number of anchorage-independent cells (selected at a density of 10^5 cells per 3 ml of soft agar), on the other hand, was determined directly from the observed number of agar colonies without correction. There is no simple way to determine the "intrinsic cloning efficiency" under these conditions in which 10^5 cells are plated per dish. If the cloning efficiency in agar were less than 100%, the frequencies would have to be increased accordingly. Nevertheless, the relationship between the frequencies observed for thioguanine resistance as well as anchorage independence in the three strains that differ in DNA repair capacity supports the hypothesis that mutations are involved in transformation.

If our interpretation of the induction of anchorage independence in human fibroblasts is correct, one has in this assay combined a mutagenesis/transformation assay. The fact that the cells become only partially transformed by a single carcinogen treatment is consistent with the multi-stepped nature of the carcinogenesis process. It is somewhat similar to the results obtained in the only commonly used transformation assay that utilizes diploid cells, the Syrian hamster embryo system (37-39). In that assay one observes morphological transformation within a few days after treatment with a carcinogen. However, cells capable of forming malignant tumors are not present until 35 to 70 generations later (38). Treatment of these hamster cells with a carcinogen seems to act to immortalize a small proportion of cells and it is among the progeny of these that the fully transformed cells arise (40). The mouse cell lines used in transformation assays, e.g., Balb C or C3H10T½, seem to represent cells that have already undergone partial transformation and, by treatment with a carcinogen, become fully transformed cells.

CONCLUSION

In our discussion of selected aspects of the widely used Ames' Salmonella mutagenicity assay as these are related to the predictive value of this test, we did not intend to give an overview of the assay. A fine review of the assay as it has been applied to carcinogenicity screening has recently been prepared by Bartsch et al.

(41). Our purpose was rather to point out some of the complicating factors which have been encountered in order to emphasize the caution which must be taken in using this genotoxicity assay to predict effects in higher organisms. We gave a short description of the ingenious use of whole animals to provide needed metabolism of parent carcinogens (which in actuality were metabolically activated by anaerobic bacteria) to illustrate the degree of sophistication which even simple tests of DNA repair can require. Finally, we described a new assay for induction of genotoxicity, namely, the _in vitro_ transformation of diploid human fibroblasts to anchorage independence. The latter studies were undertaken primarily to be able to dissect the steps necessary for carcinogenesis _in vivo_. However, with certain modifications the assay also can be used to determine whether agents can cause genotoxicity and to quantitate the DNA damage induced.

ACKNOWLEDGEMENT

The research summarized in this report was supported in part by Contract ES-78-4659 from the Department of Energy and by Grants CA21247, CA21253, CA21289, and ES 07076 from the Department of Health and Human Services, N.I.H. Additional financial assistance was provided by the Michigan Osteopathic College Foundation.

REFERENCES

1. Ames, B.N., J. McCann, and E. Yamasaki (1975) Methods for detecting carcinogens and mutagens with the Salmonella/mammalian microsome mutagenicity test. _Mutat. Res._ 31:347-364.
2. Mermelstein, R., D.K. Kiriazides, M. Butler, E.C. McCoy, and H.S. Rosenkranz (1981) The extraordinary mutagenicity of nitropyrenes in bacteria. _Mutat. Res._ 89:187-196.
3. Cole, J., C.F. Arlett, J. Lowe, and B.A. Bridges (1982) The mutagenic potency of 1,8-dinitropyrene in cultured mouse lymphoma cells. _Mutat. Res._ 93:213-220.
4. Nakayasu, M., H. Sakamoto, K. Wakabayashi, M. Terada, T. Sugimura, and H.S. Rosenkranz (1982) Potent mutagenic activity of nitropyrenes on Chinese hamster lung cells with diphtheria toxin resistance as a selective marker. _Carcinogenesis_ 3:917-922.
5. Ohgaki, H., N. Matsukura, K. Morino, T. Kawachi, T. Sugimura, K. Morita, H. Tokiwa, and T. Hirota (1982) Carcinogenicity in rats of the mutagenic compounds 1-nitropyrene and 3-nitrofluoranthene. _Cancer Letters_ 15:1-7.
6. Jerina, D.M., J.M. Sayer, D.R. Thakker, H. Yagi, W. Levin, A.W. Wood, and A.H. Conney (1980) Carcinogenicity of polycyclic aromatic hydrocarbons: The bay-region theory. In _Carcinogenesis: Fundamental Mechanisms and Environmental Effects_, B. Pullman, P.O.P. Ts'o, and H. Gelboin, eds. D. Reidel Publishing Co., Dordrecht, Holland, pp. 1-12.

7. Santella, R.M., D. Grunberger, and I.B. Weinstein (1979) DNA-benzo(a)pyrene addults formed in a Salmonella typhimurium mutagenesis assay system. Mutat. Res. 61:181-189.
8. Bigger, C.A.H., J.E. Tomaszewski, and A. Dipple (1978) Differences between products of binding of 7,12-dimethylbenz(a)-anthracene to DNA in mouse skin and in a rat liver microsomal system. Biochem. and Biophys. Res. Commun. 80:229-235.
9. Bigger, C.A.H., R.C. Moschel, and A. Dipple (1981) Differential effect of substrate concentrate on metabolic activation: Comparison of intact tissues, intact cells and homogenates. In Chemical Analysis and Biological Fate: Polynuclear Aromatic Hydrocarbons. Fifth International Symposium. M. Cooke and A.J. Dennis, eds. Battelle Press, Columbus, Ohio, pp. 209-219.
10. Sacks, L.E., and J.T. MacGregor (1982) The B. subtilis multigene sporulation test for mutagens: Detection of mutagens inactive in the Salmonella his reversion test. Mutat. Res. 95:191-202.
11. Waleh, N.S., S.J. Rapport, and K. Mortelmans (1982) Development of a toxicity test to be coupled to the Ames Salmonella assay and the method of construction of the required strains. Mutat. Res. 97:247-256.
12. Kaden, D.A., R.A. Hites, and W.G. Thilly (1979) Mutagenicity of soot and associated polycyclic aromatic hydrocarbons to Salmonella typhimurium. Cancer Res. 39:4152-4159.
13. Chium, C.W., L.H. Lee, C.Y. Wang, and G.T. Bryan (1978) Mutagenicity of some commercially available nitro compounds for Salmonella typhimurium. Mutat. Res. 58:11-22.
14. Simmon, V.F., A.F. Eckford, R. Griffin, R. Spanggord, and G.W. Newell (1977) Munitions waste water treatments: Does chlorination or ozonation of individual components produce microbial mutagens? Toxicol. Appl. Pharmacol. 41:197.
15. Couch, D.B., P. Flowe, and D. Ragan (1979) The mutagenicity of dinitrotoluenes in Salmonella typhimurium. Environ. Mutagen. 1:168.
16. Abernethy, D.J., and D.B. Couch (1982) Cytotoxicity and mutagenicity of dinitrotoluenes in Chinese hamster ovary cells. Mutat. Res. 103:53-59.
17. Soares, E.R., and L.F. Lock (1980) Lack of an indication of mutagenic effects of dinitrotoluenes and diaminotoluenes in mice. Environ. Mutagen. 2:111-124.
18. Mirsalis, J.C., and B.E. Butterworth (1982) Induction of unscheduled DNA synthesis in rat hepatocytes following in vivo treatment with dinitrotoluene. Carcinogenesis 3:241-245.
19. Mirsalis, J.C., and B.E. Butterworth (1980) Detection of unscheduled DNA synthesis in hepatocytes isolated from rats treated with genotoxic agents: An in vivo - in vitro assay for potential carcinogens and mutagens. Carcinogenesis 1:621-625.
20. Chemical Industry Institute of Toxicology, a twenty-four month toxicology study in Fischer-344 rats given dinitrotoluene, 12 month report (1978) CIIT Docket #327N8.

21. Mirsalis, J.C., T.E. Hamm, Jr., J.M. Sherrill, and B.E. Butterworth (1982) Role of gut flora in the genotoxicity of dinitrotoluene. Nature 295:322-323.
22. Rickert, E.E., R.M. Long, S. Krakowka, and J.G. Dent (1981) Metabolism and excretion of 2,4- ^{14}C dinitrotoluene in conventional and anenic Fischer-344 rats. Toxic Appl. Pharmac. 59:574-579.
23. Guest, D., S.R. Schnell, and J.G. Dent (1981) The Toxicologist 1:110.
24. Hollstein, M., J. McCann, F.A. Angelosanto, and W.W. Nichols (1981) Metabolism of 2,4-dinitro ^{14}C toluene by freshly isolated Fischer-344 rat primary hepatocytes. Drug Metab. Disposit. 9:10-14.
25. Maher, V.M., and J.J. McCormick (1982) Measurement of mutations in somatic cells in culture. In Mutagenicity: New Horizons in Genetic Toxicology, J.A. Heddle, ed. Academic Press, New York, pp. 215-240.
26. McCormick, J.J., K.C. Silinskas, and V.M. Maher (1980) Transformation of diploid human fibroblasts by chemical carcinogens. In Carcinogenesis, Fundamental Mechanisms and Environmental Effects, B. Pullman, P.O.P. Ts'o, and H. Gelboin, eds. D. Reidel Publ. Co., Dordrecht, Holland, pp. 491-498.
27. Silinskas, K.C., S.A. Kateley, J.E. Tower, V.M. Maher, and J.J. McCormick (1981) Induction of anchorage independent growth in human fibroblasts by propane sultone. Cancer Res. 41:1620-1627.
28. Maher, V.M., L.A. Rowan, K.C. Silinskas, S.A. Kateley, and J.J. McCormick (1982) Frequency of UV-induced neoplastic transformation of diploid human fibroblasts is higher in xeroderma pigmentosum cells than in normal cells. Proc. Natl. Acad. Sci., USA 79:2613-2617.
29. McCormick, J.J., L.A. Rowan, S.A. Kateley, M.M. Moon, and V.M. Maher. Cells from xeroderma pigmentosum variants are abnormally sensitive to transformation (submitted for publication, 1983).
30. Higginson, J., and C.S. Muri (1973) Epidemiology of Cancer. In Cancer Medicine, J.F. Holland and E. Frei III, eds. Lea and Febiger, Philadelphia, Pennsylvania, pp. 241-306.
31. Kakunaga, T. (1977) The transformation of human diploid cells by chemical carcinogens. In Origins of Human Cancer, H.H. Hiatt, J.D. Watson, and J.A. Winsten, eds. Cold Spring Harbor Laboratory Press, Cold Spring Harbor, New York, pp. 1537-1548.
32. Bellett, A.J.D., and H.B. Younghusband (1979) Spontaneous, mutagen-induced and adenovirus-induced anchorage independent tumorigenic variants of mouse cells. J. Cell Physiol. 101:33-48.
33. Maher, V.M., J.J. McCormick, P.L. Grover, and P. Sims (1977) Effect of DNA repair on the cytotoxicity and mutagenicity of polycyclic hydrocarbon derivatives in normal and xeroderma pigmentosum human fibroblasts. Mutat. Res. 43:117-138.

34. Glover, T.W., C.-C. Chang, J.E. Trosko, and L.S.-L. Li (1979) Ultraviolet light induction of diphtheria toxin-resistant mutants in normal and xeroderma pigmentosum human fibroblasts. Proc. Natl. Acad. Sci., USA 76:3982-3986.
35. Maher, V.M., D.J. Dorney, A.L. Mendrala, B. Konze-Thomas, and J.J. McCormick (1979) DNA excision-repair processes in human cells can eliminate the cytotoxic and mutagenic consequences of ultraviolet radiation. Mutat. Res. 62:311-323.
36. Konze-Thomas, B., R.M. Hazard, V.M. Maher, and J.J. McCormick (1982) Extent of excision repair before DNA synthesis determines the mutagenic but not the lethal effect of UV radiation. Mutat. Res. 94:421-434.
37. Pienta, R.J., J.A. Poiley, and W.B. Lebherz (1977) Morphological transformation by early passage golden Syrian hamster embryo cells derived from cryopreserved primary cultures as a reliable in vitro bioassay for identifying diverse carcinogens. Int. J. Cancer 19:642-655.
38. Barrett, J.C., B.D. Crawford, and P.O.P. Ts'o (1980) The role of somatic mutation in a multi-stage model of carcinogenesis. In Mammalian Cell Transformation by Chemical Carcinogens, N. Mishra, V. Dunkel, and M. Mehlman, eds. Senate Press Inc., Princeton, New Jersey, pp. 467-501.
39. Barrett, J.C., and P.O.P. Ts'o (1978) Evidence of the progressive nature of neoplastic transformation in vitro. Proc. Natl. Acad. Sci., USA 75:3761-3765.
40. Newbold, R.F., R.W. Overell, and J.R. Connell (1982) Induction of immortality is an early event in malignant transformation of mammalian cells by carcinogens. Nature 299:633-635.
41. Bartsch, H., T. Kuroki, M. Roberfroid, and C. Malaveille (1982) Metabolic activation systems in vitro for carcinogen/mutagen screening tests. In Chemical Mutagens, F.J. de Serres and A. Hollaender, eds. Plenum Press, New York, pp. 95-161.

EVALUATION OF CHRONIC RODENT BIOASSAYS AND AMES ASSAY TESTS AS ACCURATE MODELS FOR PREDICTING HUMAN CARCINOGENS

David Brusick

Litton Bionetics, Inc.
5516 Nicholson Lane
Kensington, Maryland 20895

INTRODUCTION

The question "What percentages of mammalian carcinogens and noncarcinogens can be predicted by bacterial mutation tests?" probably cannot be answered for the following two reasons:

1.) There seems to be no universal agreement upon what criteria are absolutely necessary to unequivocally establish a carcinogen for mammals. Consequently, the bull's eye of the target for which the bacterial tests must aim continually shifts, causing the results of comparisons conducted at one point in time to be adjusted at a different point in time.

2.) Carcinogen results are not obtained for "mammals", they are obtained in unique mammalian species which often differ among themselves as to whether a particular chemical is a carcinogen or not. Bacterial tests might hopefully match the results in a single mammalian model but cannot be expected to match the combined results from several mammalian species which fail to agree among themselves as to the correct answer.

The dilemma identified in the previous two points is best shown via an analysis of the correlation between mutation data from the "standard" Ames test and the "standard" rodent lifetime cancer bioassay. The data used in this analysis is derived from 200 studies supported by the National Cancer Institute (NCI) (3). The comparison includes a sample of approximately 25% of the total data base and is probably representative of the complete set of responses.

The Bacterial Mutagenesis Study

This portion of the data base was developed from the results of a detailed collaborative study of the Ames test coordinated by Dr. V.C. Dunkel. A portion of this study, conducted in four laboratories, involved a set of chemicals not among the 200 reported by NCI and preliminary results have been reported by Dunkel in 1979 (4). The initial group of chemicals were tested to provide a summary of data from studies conducted to standardize test methods, the evaluation of factors involved in inter-laboratory variability, and the criteria involved in data evaluation. The set of chemicals reported here were chosen on compound availability and stability since many of the cancer bioassays had been conducted several years earlier.

The investigation was designed to be a study of how the results from the Ames test compared with those in the NCI animal data base. All four laboratories tested the same compounds under code, and the data were evaluated and reported by code numbers. Only after all four laboratories had completed a set of chemicals and submitted their evaluations to the study coordinator were the chemicals identified. The test materials used in the Ames test were from the same lots used in the rodent carcinogenicity studies, and the rat and mouse S9 preparations were made from the two rodent strains used in the NCI bioassay (Fischer 344 rats and B6C3F1 mice).

The protocol employed was extensive with redundancies designed into the procedures. Table 1 summarizes the basic parameters of the test methods used in each of the four laboratories for each compound. A single chemical, for example, would have a total of 84 plates per dose level per strain tested (3 plates x 1 nonactivation group x 6 activation groups x 4 laboratories), therefore, the minimal composite data from the Ames assay for each chemical consisted of a minimum of 3,360 plates (84 plates per dose group x a minimum of 6 dose levels and 2 control groups x 5 strains). Consequently, the resolving power of this protocol was extremely high and a composite positive or negative response carried substantial reliability.

The criteria used to decide whether the data for a chemical were positive consisted of the demonstration of a dose response effect with a concomitant two-fold increase over the solvent control value in at least one Salmonella tester strain for at least one treatment condition. Uniformity among laboratories with respect to strain and activation system specificity was very high (approximately 95% concordance in all four laboratories). For most compounds, the dose range showing activity was also similar among the four laboratories.

Thus, the Ames test data base represented a relatively good indication of the type of data which can be expected from multiple

Table 1. Example of the design of the NCI Ames test. A total of five strains; TA-1535, TA-1537, TA-1538, TA-98, and TA-100 were used for each chemical tested and each chemical was tested in all four laboratories. The number of dose levels ranged from 6 to 8 with 6 or 7 being typical. As shown below, the number of plates involved in the composite data for each chemical tested in the study ranged from 3,360 to 3,780. Concurrent controls were conducted in each test.

Strain	Treatment	Non-Activation	Mouse S9 Ind-uced	Mouse S9 Non-Ind-uced	Rat S9 Ind-uced	Rat S9 Non-Ind-uced	Hamster S9 Ind-uced	Hamster S9 Non-Ind-uced	Total
TA-1535	Solvent	3	3	3	3	3	3	3	21
	Positive Control	3	3	3	3	3	3	3	21
	Test Agent								
	1	3	3	3	3	3	3	3	21
	2	3	3	3	3	3	3	3	21
	3	3	3	3	3	3	3	3	21
	4	3	3	3	3	3	3	3	21
	5	3	3	3	3	3	3	3	21
	6	3	3	3	3	3	3	3	21
	7	3	3	3	3	3	3	3	21
								TOTAL	189*

*To calculate the total number of plates scored per chemical in all four laboratories, the plates per strain (189) is multiplied by 5 (for five strains) to obtain 945 then times 4 (for four laboratories) to obtain 3780.

laboratories when the test procedures are well standardized. Detailed results from this investigation are being prepared for publication (V.C. Dunkel, pers. commun.).

The Rodent Carcinogenesis Study

The typical bioassay procedure used to generate the results presented in Table 2 are described by Chu et al. (3). The chemicals listed in this table are from among those tested employing two rodent species and conducted by NCI between 1972 and 1981.

The protocols generally involved the mouse B6C3F1 hybrid and the Fischer 344 rat. Occasionally, other species were used but their data do not constitute a substantial portion of the overall body of data. The test chemical was administered continuously to groups of 50 male and female mice (18-24 months) and 50 male and female rats (20-24 months). Generally, a control group (untreated or vehicle) and two dose groups were included in a study. The high dose administered to the animals was the maximum tolerated dose (MTD) defined as "the highest dose that causes no more than a 10% weight decrement as compared to the appropriate control groups, and does not produce mortality, clinical signs of toxicity, or

Table 2. Composition of the chemical sample employed.

Chemical	Rat	Mouse	Ames Test	Cancer Evidence Category
2-Aminoanthraquinone	+	+	+	Sufficient (2 species)
o-Anisidine HCl	+	+	+	Sufficient (2 species)
4-Chloro-o-phenylenediamine	+	+	+	Sufficient (2 species)
p-Cresidine	+	+	+	Sufficient (2 species)
Cupferron	+	+	+	Sufficient (2 species)
2,4-Diaminoanisole SO_4	+	+	+	Sufficient (2 species)
1,2-Dibromoethane	+	+	+	Sufficient (2 species)
Hydrazobenzene	+	+	+	Sufficient (2 species)
4,4'-Methylene bis (N,N-dimethyl) benzenamine	+	+	+	Sufficient (2 species)
Michler's Ketone	+	+	+	Sufficient (2 species)
1,5-Naphthalenediamine	+	+	+	Sufficient (2 species)
5-Nitroacenaphthene	+	+	+	Sufficient (2 species)
Nitrofen	+	+	+	Sufficient (2 species)
2-Aminoanthraquinone	+	+	+	Sufficient (2 species)
Nitrilotriacetic acid, trisodium salt (NTA)	+	+	-	Sufficient (2 species)
Reserpine	+	+	-	Sufficient (2 species)
Tris (2,3-dibromopropyl) PO_4	+	+	+	Sufficient (2 species)
4-Amino-2-nitrophenol	+	-	+	Sufficient (1 species)
3-(Chloromethyl) pyridine HCl	+/-	+	+	Sufficient (1 species)
5-Nitro-o-toluidine	-	+	+	Sufficient (1 species)
p-Quinone dioxime	+	-	+	Sufficient (1 species)
Cinnamyl anthranilate	-	+	-	Sufficient (1 species)
m-Cresidine	+	-	-	Sufficient (1 species)
6-Nitrobenzimidazole	-	+	±	Sufficient (1 species)
N-Nitrosodiphenylamine	-	+	-	Sufficient (1 species)
Dapsone	+	-	-	Sufficient (1 species)
3,3'-Dimethoxybenzidine-4,4'-diisocyanate	+	-	+	Sufficient (1 species)
3-Amino-4-ethoxyacetanimide	-	+	+	Limited Evidence
4-Chloro-m-phenylenediamine	+	+	+	Limited Evidence
Diaminozide	+	+/-	-	Limited Evidence
2-Nitro-p-phenylenediamine	-	+	+	Limited Evidence
2,4-Dinitrotoluene	+	-	+	Limited Evidence
Pivalolactone	+	-	+	Limited Evidence
p-Anisidine HCl	+/-	-	+	Limited Evidence
1H-benzotriazole	+/-	+/-	+	Limited Evidence
Acetohexamide	+/-	-	-	Limited Evidence
p-Chloroaniline	+/-	+/-	+	Limited Evidence
Proflavine	+/-	+/-	+	Limited Evidence
Styrene	+/-	-	-	Limited Evidence
Aldicarb	-	-	-	Without Evidence
Anilazine	-	-	-	Without Evidence
Coumaphos	-	-	-	Without Evidence
Diazinon	-	-	-	Without Evidence
N,N'-Dicylohexylthiourea	-	-	-	Without Evidence
2,4-Dimethoxyaniline HCl	-	-	+	Without Evidence
4'-(Chloroacetyl) acetanilide	-	-	+	Without Evidence
APC Mixture	-	-	-	Without Evidence
3-Nitropropionic acid	-	-	+	Without Evidence
Ethylenediaminete-traacetate trihydrate (EDTA)	-	-	-	Without Evidence
Fluometuron	-	-	-	Without Evidence
Lithocholic acid	-	-	-	Without Evidence
4-Nitro-o-phenylenediamine	-	-	+	Without Evidence
Triphenyltin hydroxide	-	-	-	Without Evidence
Sulfisoxazole	-	-	-	Without Evidence
1-Phenyl-3-methyl-5-pyrazolone	-	-	-	Without Evidence
p-Phenylenediamine dihydrochloride	-	-	+	Without Evidence
2,3,5,6-Tetrachloro-4-nitroanisole	-	-	-	Without Evidence
2,5-Toluenediamine SO_4	-	-	+	Without Evidence
1-Nitronaphthalene	-	-	+	Without Evidence
Titanium dioxide	-	-	-	Without Evidence

Citation: Animal data derived from Chu, K., Cueto, C., and Ward, J.M.: Factors in the evaluation of 200 National Cancer Institute carcinogen bioassays. J. of Toxicol. and Environ. Health, 8:251-280, 1981.

pathological lesions that could be predicted to shorten the animal's natural life span". The low dose was one half the MTD. Analysis of pathology was performed on 25-30 tissues from all animals. Observations were made and clinical data collected over the period of the study.

An evaluation of the test results involved the integration of toxicological, pathological, and statistical data.

- Analysis of pathology consisted of identification and enumeration of the tumors, identification of the target organ(s), and a determination of whether tumors identified were rare in control animals. In addition, the progression of the lesions (metastatic or benign) may influence the final assessment of the biological activity of the chemical.

- Statistical analysis included comparison of tumor incidence in the treated groups with that in the control group, and also the time to tumor detection values for both groups. Study results might be subclassified and analyzed by target organ or sex.

- Evaluation of toxicological findings included documentation of dose administration, necropsy finding, clinical chemistry results, and routine in-life observations for toxic manifestations.

The data for the 200 chemicals were evaluated using the following premises:

1) A carcinogen was any agent that induced malignant and benign tumors in either species over control values.

2) A chemical producing only benign tumors was considered a suspect carcinogen.

3) The weight of evidence for positive responses varied from "sufficient results from 2 species", "sufficient evidence from one species", "limited evidence in two species", or "marginal evidence in one species". Negative responses included all categories without sufficient evidence of tumor induction.

Other inferences from the evaluation were that negative results do not necessarily mean that the test chemical is not a carcinogen and that any substance that is designated a carcinogen or suspect carcinogen represents a potential risk to humans.

CORRELATION OF RESULTS FOR RODENTS AND BACTERIAL TESTS

The mutagenicity and carcinogenicity results for 60 chemicals are summarized in Table 2. Levels of concordance for positive and negative responses were determined for rodent to rodent and rodent to bacteria. The evaluation by species showed a 72% agreement between the mouse and the rat, a 69% between the rat and the Ames test, a 71% between the mouse and the Ames test, and a 74% agreement between the Ames test and the two rodent species combined. The 30% nonagreement indicates that neither rodent species is better at predicting carcinogenicity in this sample of chemicals in the other species than is the Ames test.

The predictive power of a test is categorized by several parameters, including:

1.) Test System Sensitivity - Defined as the number of true positives* correctly identified by the model test.

2.) Test System Specificity - Defined as the number of true negative* compounds correctly identified by the model test.

3.) Test System Predictive Value - Defined as the proportion of positive responses generated by the model system which were true positives*.

4.) False Positive - The number of noncorrelated positive responses in the model system.

5.) False Negative - The number of noncorrelated negative responses

Generating the values for these five parameters involves acknowledging the most critical assumptions involved in this entire analysis, which is that a standard set of true responses exists against which the performance of a model system can be accurately compared. For this exercise the standard is presumed to be the rodent. A critical question, however, is which rodent is the true standard, the mouse or the rat? The answer to this question is, of course, dependent upon which species best reflects the hypothetical response pattern of humans for the same set of 60 chemicals. Since the two species agree only 70% of the time, the human response pattern, if known, could not possibly correlate equally with both. Therefore, only one of the two species could <u>most</u> accurately represent the true standard.

In order to look at this aspect of the analysis in more detail, the sensitivity, specificity, predictive value and number of apparent false responses were calculated assuming each rodent species as the most predictive standard (Table 3). The results indicate that regardless of which animal species is selected as the standard, the

---

* The terms true negative and true positives refer to the responses in the designated bioassay standard.

Table 3. Correlation levels between species assuming either the rat or the mouse as the best standard for detecting human carcinogens.

Presumed True Standard[a]	Proposed Predictive Model	Model Assay Sensitivity	Model Assay Specificity	Model Assay Predictive Value	False Positives	False Negatives
Rat Species	Mouse Species	70%	75%	75%	7	9
Rat Species	Ames Test	78%	58%	69%	11	7
Mouse Species	Rat Species	75%	70%	70%	9	7
Mouse Species	Ames Test	85%	58%	64%	13	4

[a]Species assumed to identify carcinogens and noncarcinogens most similar to humans.

Ames test is more accurate than the remaining rodent in identifying true carcinogens. The Ames test is less accurate, however, in correctly identifying noncarcinogens (it produces a slightly higher degree of false positives).

An alternate interpretation approach is to make the assumption that combined results from both rodent models are needed to match the human response pattern to the 60 chemicals. This assumption appears to be illogical because: (a) both a "+" and "-" response cannot be the correct prediction of the human response for single chemicals, and (b) if only the combined "+" responses are considered as indication of true human risk, then the human species must be viewed as having a broader range of susceptibility than either the mouse or the rat to carcinogens because all 37 positives from the two rodent species (Table 2) would be required to best reflect the response pattern of the single human species.

But, in spite of logic to the contrary, one can construct the values for the same parameters as shown in Table 3 for the two rodent species combined as the standard (Table 4). Again, the Ames test performs reasonably well and is slightly better than the mouse in detecting the true carcinogens from the combined rodent data set. The rat species is significantly better than either the Ames test or the mouse species.

In summary then, logic would seem to dictate that since the mouse and rat responses differed from each other for approximately 30% of the chemicals evaluated, one species or the other, but not both, will best reflect the response pattern of humans (if such human data were available). In the event that either the mouse or the rat species is, in fact, the appropriate standard, the Ames test appears to provide a comparable (at least for this group of 60 chemicals) detection of carcinogens. It is possible, as well, that neither rodent species is an adequate standard for humans. This possibility is addressed in the next section.

Table 4. Correlation levels between species assuming that both positives from both the mouse and rat are necessary to best reflect the human response pattern to carcinogens.

Presumed True Standard	Proposed Predictive Model	Model Assay Sensitivity	Model Assay Predictive Value	False Negatives
Rat/Mouse	Rat	85%	100%*	5
Rat/Mouse	Ames Test	78%	81%	8
Rat/Mouse	Mouse	73%	100%*	10

Notes: Specificity cannot be calculated for this comparison because rodent negatives cannot be true responses as some chemicals have both a "+" and "-" rodent response, and this approach uses the "+" response considering the negative response not a true indication of carcinogenicity. False positives cannot be calculated using this approach since all rodent positives are considered true carcinogens.

*The predictive values in these two instances are a compulsory 100% because all rodent positives are by definition of the standard true carcinogens.

Other philosophical and interpretational difficulties involving the rodent bioassay arise if an assumption is made that the combined response pattern of the two rodent species is the appropriate standard against which results of the Ames test must be correlated. For example, this assumption could indicate that the human species is more susceptible than either the mouse or the rat (since neither species alone can detect all true carcinogens among the 60 chemicals), and would also indicate that neither species is capable of accurately defining true noncarcinogens for humans since a positive response in one species is taken as evidence for presumptive human risk even when the other species showed a negative response.

Some chemicals, not carcinogenic in the mouse and rat, are carcinogenic when tested in a third species (e.g., hamster, dog). Customarily, such chemicals would be viewed as possible human carcinogens. However, addition of a third species to the bioassay procedure might produce some confounding consequences. For example:

1.) Chemicals which induce tumors in the third species but not the mouse and rat would further erode confidence that rodent tests can identify true negatives. In fact, the concept of a "true negative" would not be possible until all nonhuman mammalian species had been exhausted as test organisms. If one cannot demonstrate true negatives, then the term "false negative" also has no meaning. Thus, for purposes of review, only true positives and false positives exist.

2.) Because the Ames test has a higher proportion of "false positives", its correlation against a three species data base should improve relative to each of the rodent species examined individually. Thus, additional rodent species added to the bioassay data base may not reduce the overall uncertainties of model system predictability.

CORRELATION OF RESULTS FOR HUMAN, RODENT AND BACTERIAL TESTS

If the findings of the previous section are typical of the larger data bases they represent, assessing chemicals for their possible carcinogenic activity to humans appears to be more complex than current practices in cancer bioassays indicate.

Use of the human data base might be of some value in comparing the predictive powers of various rodent and bacteria species. The final answer cannot be obtained in this fashion, however, because the human data base consists only of carcinogens (true positives) and no noncarcinogens. Thus, calculated values for test system specificity, predictive coefficient, and false positive responses cannot be generated. The only available data for comparison, then, are the sensitivity of the test system and estimates of false negative responses. A comparison of the human, rodent, and bacteria results for 14 human carcinogens is shown in Table 5.

The chemicals listed in Table 5 are restricted to those agents with sufficient human evidence to be considered by the International Agency for Research on Cancer (IARC) as established human

Table 5. Chemicals evaluated as carcinogenic for humans and their responses in rodent and bacterial predictive assays.

Chemical	Human	Rat Bioassay	Mouse Bioassay	Ames Test
4-Aminobiphenyl	+	+	+	+
Arsenic	+	-	-	-
Asbestos	+	+	+	-
Benzene	+	-	-	-
Benzidine	+	+	+	+
Bis(chloromethyl)ether	+	+	+	+
Chromium and some chromium compounds	+	+	-	+
Cyclophosphamide	+	+	+	+
Diethylstilbestrol	+	+	+	-
Melphalan	+	+	+	+
Mustard Gas	+	No Data	+	+
2-Naphthylamine	+	-	+	+
Soot, tars	+	-	+	+
Vinyl chloride	+	+	+	+

Sources for:

- The Rodent Bioassay Responses

  Handbook of Identified Carcinogens and Noncarcinogens, Vol. I and II, J.V. Soderman, ed., CRC Press, Boca Raton, Florida, 1982.

- The Chemical Selection and Mutation Results

  Bartsch H., Tomatis, L. and Malaveille, C.: Mutagenicity and Carcinogenicity of Environmental Chemicals, Regulatory Toxicology and Pharmacology, 2:94-105, 1982.

Not included are; manufacture of auramine, boot and shoe manufacture, furniture and cabinet making, hematite mining, manufacture of isopropyl alcohol, nickel refining, conjugated estrogens. These processes do not involve discrete chemical entities that can be tested.

Table 6. Accuracy of rodent and bacterial predictive tests to identify human carcinogens.

Standard	Model	Model Sensitivity	False Negatives
Human	Rat	69% (9/13)*	4
Human	Mouse	79% (11/14)	3
Human	Ames Test	71% (10/14)	4
Human	Combined Mouse/Rat	86% (12/14)	2

*Data not available for one compound

carcinogens (2). The table does not include manufacturing or mining processes also associated with induction of tumors in humans. The calculated values for sensitivity and false negative responses are presented in Table 6.

The outcome of the comparison of such a limited data base does address several points raised in the analysis of the NCI rodent data base.

1.) Responses from the mouse, rat, and Ames test are not strikingly dissimilar nor are any of the three model systems accurate predictors of human responses. Used alone, none of the three model systems is adequate to protect the human from carcinogens.

2.) By combining the "+" responses from both rodent species, the predictive power of these models reaches a respectable 86%; however, by replacing either the mouse or the rat bioassay with the Ames test the same 86% sensitivity is achieved.

3.) The assumption that humans may have a broader susceptibility to carcinogens than either rodent species appears to be supported.

4.) The Ames test alone does not provide adequate predictive power for a set of known human carcinogens. However, if the Ames test was combined with two additional *in vitro* methods, cell transformation and cytogenetic analysis, all four false negative responses in Table 5 would be eliminated (1,5,6), and the sensitivity value for this battery would be 100%.

CONCLUSIONS

An evaluation of two carcinogen data bases and corresponding Ames test data bases supports a view that the predictive power to identify true chemical carcinogens is similar for the rat model, the mouse model and the Ames test model. The similarity in accurate carcinogen identification is true whether the chemicals are considered to be rodent carcinogens or human carcinogens.

Further analysis of these data bases indicated that a combined mouse/rat model was necessary to provide adequate correlation with the current human carcinogen data base. However, by combining "+" responses in order to maximize the test system sensitivity, "-" responses from either species becomes insignificant in assessing chemical safety.

Although highly comparable to either of the two rodent species when evaluated separately, the Ames test alone does not provide sufficient accuracy to be a reliable predictor of either rodent or human carcinogens. Combined with either rodent bioassay or with one or two other *in vitro* tests, the technique is equivalent to or more reliable than the combined mouse/rat model in predicting human carcinogens.

REFERENCES

1. McLachlan, J.A., A. Wong, G.H. Degen, and J.C. Barrett (1982) Morphological and neoplastic transformation of Syrian hamster embryo fibroblasts by diethylstilbestrol and its analogs. *Cancer Res.* 42:3040-3045.
2. Bartsch, H., L. Tomatis and C. Malaveille (1982) Mutagenicity and carcinogenicity of environmental chemicals. *Regulatory Toxicol. and Pharmacol.* 2:94-105.
3. Chu, K., C. Cueto, and J.M. Ward (1981) Factors in the evaluation of 200 National Cancer Institute Carcinogen Bioassays. *J. Toxicol. and Environ. Health* 8:251-280.
4. Dunkel, V.C. (1979) Collaborative studies on the *Salmonella*/microsome mutagenicity assay. *J. Assoc. Off. Anal. Chem.* 62:874-882.
5. Nakamuro, K. and Y. Sayato (1981) Comparative studies of chromosomal aberration induced by trivalent and pentavalent arsenic. *Mut. Res.* 88:73-80.
6. Siou, G., L. Conan, and M. el Haitem (1981) Evaluation of the clastogenic action of benzene by oral administration with 2 cytogenetic techniques in mouse and Chinese hamster. *Mut. Res.* 90:273-278.

INDICES FOR IDENTIFICATION OF GENOTOXIC AND EPIGENETIC CARCINOGENS IN CELL CULTURE

Gary M. Williams

Naylor Dana Institute for Disease Prevention
American Health Foundation
Valhalla, New York 10595

GENOTOXIC AND EPIGENETIC CARCINOGENS

The induction of cancer by chemicals is a process consisting of a series of steps, which comprise two distinct sequences: the conversion of the normal cell to a neoplastic cell, and the progression of the neoplastic cell to formation of a tumor. Chemicals are involved in this process at several points in both of the sequences, primarily as "initiating" agents which produce neoplastic conversion, most likely as a result of reaction with DNA, and as promoting or enhancing agents which facilitate the development of altered cells into tumors. Effects by chemicals in both sequences can lead to an increase of cancer in animals and, consequently, chemicals can be carcinogenic by a variety of mechanisms.

As an initial effort in making distinctions among carcinogens with different mechanisms of action, a proposal was advanced to categorize carcinogens into two principal types: genotoxic carcinogens that are capable of reacting with and damaging DNA, and epigenetic agents which do not damage DNA but produce other biological effects which result in the production of tumors (34,35). Specific classes of carcinogens have been assigned to these two categories (Table 1) based upon the established capacity of representatives to produce either DNA damage or other biological effects (29).

A carcinogen can be determined to be genotoxic either by the demonstration through biochemical techniques that it damages DNA or by the finding that it is active in tests which reliably measure DNA damage (36,43). Carcinogens with genotoxic properties consist mainly of organic compounds which, either in their parent form or after biotransformation by enzyme systems, give rise to electrophilic

Table 1. Classification of carcinogenic chemicals.[a]

Category and Class	Example
A. Genotoxic Carcinogens	
1. Activation-independent	alkylating agents, e.g., methyl methanesulfonate
2. Activation-dependent	polycyclic aromatic hydrocarbon, Nitrosamine
3a. Inorganic[b]	metal
B. Epigenetic Carcinogens	
3b. Inorganic[b]	metal
4. Solid state	plastics
5. Cytotoxic	nitrilotriacetic acid
6. Hormone-modifying	amitrole, estrogen
7. Immunosuppressor	purine analogs
8. Co-carcinogen	phorbol ester, ethanol
9. Promoter	saccharin, organochlorine pesticides

[a]Data based on Weisburger and Williams (29).
[b]Some are tentatively categorized as genotoxic because of evidence for damage of DNA; others may operate through epigenetic mechanisms such as alterations in fidelity of DNA polymerases.

reactants that form adducts on the bases in DNA. Some metal carcinogens have also displayed activity in short-term tests indicative of genotoxicity. *In vivo*, genotoxic carcinogens generally produce tumors in more than one organ and in high yield after a relatively short latent period.

In addition to genotoxins, which react with DNA, other agents may indirectly affect DNA (43). Certain metal carcinogens can alter the accuracy with which DNA polymerases replicate DNA thereby giving rise to abnormal DNA (25). Also, toxic agents generate reactive oxygen species which may damage DNA (14,18). Similarly, Shank and co-workers (4) have shown that certain agents can produce aberrant methylation of DNA under toxic conditions. In addition, production of other types of genetic effects, such as aneuploidy, which do not involve direct DNA damage may be important in carcinogenesis (21). These types of effects require consideration, but still may be distinct from genotoxicity as regards the conditions under which they occur.

Epigenetic carcinogens are defined as being nongenotoxic and producing a biological effect that could account for their carcinogenicity. Thus, the first step in identifying an epigenetic carcinogen is to establish that it does not damage DNA. This may be done using biochemical studies or short-term tests for genotoxicity. Then, evidence must be developed on other biological effects.

The *in vivo* carcinogenic effects of a chemical may provide indications of possible epigenetic effects. For example, an agent that produces tumors only in hormonally responsive tissues may do so as a consequence of perturbation of the endocrine system. An example of this is amitrole (Table 1) which interferes with thyroid function leading to overproduction of pituitary thyroid-stimulating hormone, which exerts a persistent trophic effect on the thyroid leading to the development of thyroid tumors. Another effect suggestive of an epigenetic action is the production of tumors only in a single organ.

As an example of such organ specificity, nitrilotriacetic acid has been negative in a variety of short-term tests and produces tumors only in the urinary tract. Extensive research by Anderson and co-workers (2) has shown that this selective effect occurs only at doses that are toxic to the kidney and, therefore, this agent is proposed to be tumorigenic through its cytotoxic effects (Tab. 1). Likewise, saccharin, which produces only bladder cancer in low yield, is nongenotoxic (3). In this case, saccharin produces a promoting action in the bladder (6,10), which could account for its carcinogenicity. Therefore, saccharin is proposed to be an epigenetic carcinogen of the promoter class (Tab. 1). Similarly, several organochlorine pesticides have been found to produce tumors exclusively or predominantly in the livers of mice and rats (39). These have been generally nongenotoxic in several different types of short-term tests (36) and importantly do not elicit DNA repair in hepatocytes from mice, rats, or hamsters (13). One organochlorine pesticide, dichloro-diphenyl-trichloro-ethane (DDT), has been reported to be a liver tumor promoter (20) and several have shown *in vitro* effects with this chemical indicative of a promoting action (26,27,45). These organochlorine compounds, therefore, appear to belong to the promoting class also (Tab. 1). Thus, the evidence of epigenetic mechanisms is usually derived from whole animal studies, but, *in vitro* systems to be discussed below are now becoming available for identifying some epigenetic effects.

CELL CULTURE INDICES FOR GENOTOXIC CARCINOGENS

Most *in vitro* test systems measure DNA damage or one of its consequences such as mutagenesis, chromosomal effects, or transformation. The different genetic end points that can be measured and the levels of biological complexity at which they can be assessed

Table 2. Cell culture systems for detecting genotoxins

Biological System	Genetic Effect		
	DNA	Gene	Chromosome
Bacteria	+	+	-
Fungi	+	+	+
Plants	i	+	+
Mammalian Cells	+	+	+

+ = generally used; i = infrequently used.

are shown in Table 2. There are ten available combinations of different biological systems and end points. Largely, as a reflection of this, over one hundred different in vitro systems have been described for measuring and detecting mutagens and carcinogens.

Genotoxicity, as used in the present context, refers specifically to DNA damage and, therefore, the most definitive tests for genotoxicity are those which measure DNA damage or the resulting DNA repair. The most useful screening systems for DNA damage are those that have as their indicator a general consequence of DNA damage such as DNA fragmentation or DNA repair (35). A validated test for DNA damage is the hepatocyte primary culture (HPC) or HPC/DNA repair test (33,38) which offers the advantage of intact cell metabolism in a cell type with broad metabolic capability. The HPC/DNA repair test is highly sensitive, and positive results in it correlate extremely well with carcinogenicity (40,41,44).

Tests for mutagenicity in bacteria, particularly the Ames Salmonella/microsome test (8,9), are the most widely applied screening tests for carcinogens. These tests are highly reliable indicators of genotoxicity and correlate well with carcinogenicity. Discrepancies arise mainly because of the artifactual metabolism performed by the subcellular fractions used for metabolic activation. Mammalian cell mutagenicity tests offer the advantage of assessing genetic effects at a higher level of genetic organization. They also are reliable indicators of genotoxicity and yield results that correlate well with carcinogenicity.

Chromosome tests monitor for genetic effects at the highest level of genetic organization. Unfortunately, most systems have not been thoroughly evaluated and discrepancies with carcinogenicity exist. Although chromosome effects and sister chromatid exchange can arise from DNA damage, it is possible that other effects result in their expression.

Transformation systems using fibroblasts and epithelial cells have been described (16). In theory, transformation *in vitro* has the potential to mimic neoplastic conversion *in vivo* and, thus, could be produced by the wide variety of chemicals that are carcinogenic. Indeed, transformation has been reported for agents that are not active in other *in vitro* systems. This sensitivity represents an advantage of transformation, but it makes the end point of limited reliability for identification of genotoxins.

No single *in vitro* test has proved capable of detecting all genotoxic carcinogens. Therefore, a series of tests is generally recommended. A battery that incorporates the end points of genotoxicty in Tab. 2 is shown in Table 3. This battery was formulated as part of a systematic approach to carcinogen detection referred to as the Decision Point Approach (30,46). Available results with chemicals examined in the tests recommended in this battery show an excellent correlation with carcinogenicity (46). In addition, Brusick has proposed a way to quantify the results in multiple test batteries such as this (5,48). Thus, it is concluded that adequate culture systems are available for identifying genotoxic carcinogens.

CELL CULTURE INDICES FOR EPIGENETIC CARCINOGENS

Since epigenetic carcinogens are suspected to operate by a variety of different biological effects, it is unlikely that any single cell culture system would be responsive to all types of epigenetic agents.

For those agents suspected to operate through production of chronic toxicity, a variety of systems are available for measuring cytotoxic effects (1,7,15,32). Thus far, these tests have usually been applied to the assessment of toxicity rather than the study of cytotoxic effects of carcinogens with the exception of studies on asbestos (24).

Co-carcinogenic effects could in principle be measured *in vitro* by determination of the enhancement by a test agent of transformation induced by a genotoxic carcinogen. Another approach would be to measure co-mutagenic effects, as has been done in bacterial mutagenicity assays. In a mammalian cell culture system, asbestos has been shown to enhance the mutagenicity of benzo(a)pyrene (23).

Tumor promoters have been reported to produce a variety of effects in cell culture (Table 4), including enhanced transformation by exposure to promoters after treatment with a carcinogen (12,17, 22). However, most indices have been tested only for phorbol-type compounds, leaving the general applicability of these effects in question as end points for the identification of promoters. Recently, a new *in vitro* approach to the study of tumor promotion has become available through the work of Yotti et al. (49) and Murray and Fitzgerald (19), who reported inhibition of intercellular com-

Table 3. Decision point approach: Battery for genotoxic carcinogens.[a]

1. Hepatocyte DNA repair
2. Salmonella mutagenesis
3. Mammalian cell mutagenesis
4. Sister chromatid exchange
5. Cell transformation[b]

[a] Data from Weisburger and Williams (30).
[b] Optional supplement to items 1-4.

Table 4. Cell cultures indices for promoters.

Enhancement of neoplastic phenotype
Increased membrane permeability
Increased production of plasminogen activator
Inhibition of differentiation
Increased ornithine decarboxylase activity
Increased transformation
Production of sister chromatid exchange or aneuploidy
Inhibition of intercellular communication

munication between cultured fibroblasts by tumor promoters. This effect has now been demonstrated for a variety of promoters (27,28) and extended to the use of metabolically competent liver cells by Williams and co-workers (26,36,45).

Inhibition of intercellular communication by tumor promoters is potentially important because *in vivo* it could serve to release dormant tumor cells from growth control by surrounding normal cells. This action would be part of the second sequence of steps in tumor development: the progression of neoplastic cells to formation of tumors. The production of liver tumors in mice and rats by nongenotoxic organochlorine pesticides may result from the promoting effect of these compounds on pre-existing cells with an abnormal genotype (37). Importantly, if it is established that a membrane effect is one basis for tumor promotion, this would represent a true epigenetic effect.

Thus, culture systems are available for detecting epigenetic effects. However, unlike tests for genotoxins, none of these has been established to be predictive of carcinogenicity.

SIGNIFICANCE OF GENOTOXIC AND EPIGENETIC CARCINOGENS

Cell culture systems provide several indices for distinguishing between carcinogens that produce their effects by different modes of action and for detecting these agents.

The identification of genotoxicity is a significant finding in the evaluation of the potential carcinogenicity of a chemical. The correlation between genotoxicity and carcinogenicity is very high (46) and, consequently, genotoxicity in a battery of short-term tests, including measurement of DNA damage, mutagenesis, and chromosomal effects may be taken as presumptive evidence of carcinogenicity (30,46). This suggestion is supported by the findings of a recent working group of the International Agency for Research on Cancer (IARC) (11) that, for forty-one chemicals classified as genotoxic, the results of animal studies on 88% were judged as showing sufficient or limited evidence of carcinogenicity (43). Other combinations of test results may be equally significant, as for example, positive results in both the hepatocyte primary culture/DNA repair test and the Ames Salmonella/microsome test have always been predictive of carcinogenic activity (38,44). Thus, a battery of short-term tests for genotoxicity is a reliable means for detecting one type of chemical carcinogen, and obviates the need for further testing since genotoxins are virtually certain to be carcinogenic under some conditions and, regardless, represent an unequivocal toxic hazard.

The absence of activity in short-term tests does not preclude carcinogenicity through nongenotoxic mechanisms in chronic studies. Previously, the lack of activity of carcinogens such as asbestos, hormones, and organochlorine pesticides in tests for genotoxicity was interpreted by some as indicating a lack of sensitivity of the tests. According to the concept of genotoxic and epigenetic carcinogens, however, such negative results can be a true reflection of the biological activity of certain chemicals and provide information on their mechanism of carcinogenicity. Thus, it is important that such negative results for nongenotoxic agents not be used to calculate misleading low percentages of correlation between short-term test results and carcinogenicity. For a compound that is inactive in short-term tests for genotoxicity, further evaluation is primarily by chronic *in vivo* studies, but should also include short-term tests for epigenetic effects.

Another major implication of the concept that carcinogens operate through different mechanisms is that the hazard evaluation for each agent must take this into account (31,47). Genotoxic carinogens vary greatly in their potency, but as a group, their carcinogenic effects can pose extreme hazards, including activity with a single exposure, activity at low doses, and transplacental and enhanced neonatal effects (43). In contrast, for at least one class of epigenetic agents, tumor promoters, the characteristics of their carcinogenic and promoting effects are quite different. Therefore, these two types of agents represent different kinds of hazards to human health (42,47).

In conclusion, the distinction between different types of carcinogens has a solid scientific base and leads to new approaches to the evaluation of data from short-term tests and carcinogenicity bioassays.

ACKNOWLEDGEMENTS

Many of the concepts presented here were developed through collaboration with Dr. John H. Weisburger and co-workers at the Naylor Dana Institute. During the performance of most of the work reported in this paper, the author was supported by grant CA 17613 from the National Cancer Institute.

REFERENCES

1. Acosta, D., D.C. Anuforo, and R.V. Smith (1980) Cytotoxicity of acetaminophen and papaverine in primary cultures of rat hepatocytes. Toxicol. and Appl. Pharm. 53:306-314.
2. Anderson, R.L., C.L. Alden, and J.A. Merski (1982) The effects of nitrilotriacetate on cation disposition and urinary tract toxicity. Fd. and Cosmet. Toxicol. 20:105-122.
3. Ashby, J., J.A. Styles, D. Anderson, and D. Paton (1978) Saccharin: A possible example of an epigenetic carcinogen/mutagen. Fd. and Cosmet. Toxicol. 16:95-103.
4. Barrows, L.R., and R.C. Shank (1981) Aberrant methylation of liver DNA in rats during hepatotoxicity. Toxicol. and Appl. Pharmacol. 60:334-345.
5. Brusick, D. (1981) Unified scoring system and activity definitions for results from in vitro and submammalian mutagenesis test batteries. In Health Risk Analysis: Proceedings of the Third Life Sciences Symposium, C.R. Richmond, P.J. Walsh, and E.D. Copenhaver, eds. The Franklin Press, Philadelphia, PA, pp. 273-286.
6. Cohen, S.M., M. Arai, J.B. Jacobs, and G.H. Friedell (1979) Promoting effect of saccharin and DL-tryptophan in urinary bladder carcinogenesis. Cancer Res. 39:1207-1217.
7. Ekwall, B. (1980) Screening of toxic compounds in tissue culture. Toxicol. 17:127-142.
8. Haroun, L., and B.N. Ames (1981) The Salmonella mutagenicity test: An overview. In Short-term Tests for Chemical Carcinogens, H.F. Stitch and R.H.C. San, eds. Springer-Verlag, New York, NY, pp. 108-119.
9. Hollstein, M., J. McCann, F.A. Angelosanto, and W.W. Nichols (1979) Short-term tests for carcinogens and mutagens. Mutat. Res. 65:133-226.

10. Hooson, J., R.M. Hicks, P. Grasso, and J. Chowaniec (1980) Ortho-toloune sulphonamide and saccharin in the promotion of bladder cancer in the rat. Br. J. Cancer 42:129-147.
11. International Agency for Research on Cancer (1983) IARC Monographs on the Evaluation of the Carcinogenic Risk of Chemicals to Humans, Supplement 4. IARC, Lyon, France.
12. Lanse, C., A. Gentil, and I. Chouroulinkov (1974) Two-stage malignant transformation of rat fibroblasts in tissue culture. Nature 247:490-491.
13. Maslansky, C.J., and G.M. Williams (1981) Evidence for an epigenetic mode of action on organochlorine pesticide hepatocarcinogenicity: A lack of genotoxicity in rat, mouse, and hamster hepatocytes. J. Toxicol. Environ. Health 8:463-477.
14. Mason, R.P., and C.F. Chingnell (1982) Free radicals in pharmacology and toxicology. Pharmacol. Rev. 33:189-212.
15. McQueen, C.A., and G.M. Williams (1982) Cytotoxicity of xenobiotics in adult rat hepatocytes in primary culture. Fund. and Appl. Toxicol. 2:139-144.
16. Mishra, N., V. Dunkel, and M. Mehlman, eds. (1981) Advances in modern environmental toxicology, Vol. 1. In Mammalian Cell Transformation by Chemical Carcinogens. Senate Press, Inc., Princeton, New Jersey.
17. Mondal, S., and C. Heidelberger (1976) Transformation of C3H/10T1/2C18 mouse embryo-fibroblasts by ultraviolet irradiation and a phorbol ester. Nature 260:710-711.
18. Moody, C.S., and H.M. Hassan (1982) Mutagenicity of oxygen free radicals. Proc. Natl. Acad. Sci., USA 79:2855-2859.
19. Murray, A.W., and D.J. Fitzgerald (1979) Tumor promoters inhibit metabolic cooperation in cocultures of epidermal and 3T3 cells. Biochem. Biophys. Res. Commun. 91:395-401.
20. Peraino, C., R.J.M. Fry, and D.D. Grube (1978) Drug-induced enhancement of hepatic tumorigenesis. In Carcinogenesis, Mechanisms of Tumor Promotion and Cocarcinogenesis, T.J. Slaga, A. Sivak, and R.K. Boutwell, eds. Raven Press, New York, NY, pp. 421-432.
21. Parry, J.M., E.M. Parry, and J.C. Barrett (1981) Tumour promoters induce mitotic aneuploidy in yeast. Nature 294:263-265.
22. Poiley, J.A., R. Raineri, and R.J. Pienta (1979) Two-stage malignant transformation in hamster embryo cells. Brit. J. Cancer 39:8-14.
23. Reiss, B., C. Tong, S. Telang, and G.M. Williams. (1983) Enhancement of benzo(a)pyrene mutagenicity by chrysotile asbestos in rat liver epithelial cells. Environ. Res. 31:100-104.
24. Reiss, B., S. Solomon, and G.M. Williams (1980) Comparative toxicities of different forms of asbestos in a cell culture assay. Environ. Res. 22:109-129.
25. Sirover, M.A., and L.A. Loeb (1976) Metal-induced infidelity during DNA synthesis. Proc. Natl. Acad. Sci., USA 73:2331-2335.

26. Telang, S., C. Tong, and G.M. Williams (1982) Epigenetic membrane effects of a possible tumor promoting type on cultured liver cells by the nongenotoxic organochlorine pesticides chlordane and heptachlor. Carcinogenesis 3:1175-1178.
27. Trosko, J.E., L.P. Yotti, B. Dawson, and C.C. Chang (1981) In vitro assay for tumor promoters. In Short-Term Tests for Chemical Carcinogens, H.F. Stich and R.H.C. San, eds. Springer-Verlag, Inc., New York, pp. 420-427.
28. Umeda, M., K. Noda, and T. Ono (1980) Inhibition of metabolic cooperation in Chinese hamster cells by various chemicals including tumor promoters. Gann 71:614-620.
29. Weisburger, J.H., and G.M. Williams (1980) Chemical carcinogens. In Toxicology: The Basic Science of Poisons, 2nd Edition, J. Doull, C.D. Klaasen, and M.O. Amdur, eds. Macmillan Publ. Co., Inc., New York, pp. 84-138.
30. Weisburger, J.H., and G.M. Williams (1981) Carcinogen testing: Current problems and new approaches. Science 214:401-407.
31. Weisburger, J.H., and G.M. Williams (1981) Basic requirement for health risk analysis: The decision point approach for systematic carcinogen testing. In Proceedings of the Third Life Sciences Symposium on Health Risk Analysis, C.R. Richmond, P.J. Walsh, and E.D. Copenhaver, eds. Franklin Press, Philadelphia, PA, pp. 249-271.
32. Wiebkin, P., J.R. Fry, and J.W. Bridges (1978) Metabolism-mediated cytotoxicity of chemical carcinogens and non-carcinogens. Biochem. Pharm. 27:1849-1851.
33. Williams, G.M. (1977) The detection of chemical carcinogens by unscheduled DNA synthesis in rat liver primary cell cultures. Cancer Res. 37:1845-1851.
34. Williams, G.M. (1979) A comparison of in vivo and in vitro metabolic activation systems. In Critical Reviews in Toxicology-Strategies for Short-term Testing for Mutagens/Carcinogens, B. Butterworth, ed. C.R.C. Press, West Palm Beach, FL, pp. 96-97.
35. Williams, G.M. (1979) The status of in vitro test systems utilizing DNA damage and repair for the screening of chemical carcinogens. J. Assoc. Off. Anal. Chem. 62:857-863.
36. Williams, G.M. (1980) Classification of genotoxic and epigenetic hepatocarcinogens using liver culture assays. Ann. N.Y. Acad. Sci. 349:273-282.
37. Williams, G.M. (1980) The pathogenesis of rat liver cancer caused by chemical carcinogens. Biochim. Biophys. Acta 605: 167-189.
38. Williams, G.M. (1980) The detection of chemical mutagens/carcinogens by DNA repair and mutagenesis in liver cultures. In Chemical Mutagens, Vol. VI, F.J. de Serres and A. Hollaender, eds. Plenum Press, New York, pp. 61-79.
39. Williams, G.M. (1981) Liver carcinogenesis: The role for some chemicals of an epigenetic mechanism of liver tumor promotion involving modification of the cell membrane. Fd. Cosmet. Toxicol. 19:577-583.

40. Williams, G.M. (1981) The detection of genotoxic chemicals in the hepatocyte primary culture/DNA repair test. In Mutation Promotion and Transformation, N. Inui, T. Kuroki, M.-A. Yamada, and C. Heidelberger, eds. Gann Monograph on Cancer Research 27:47-57, University of Tokyo Press, Tokyo, Japan.
41. Williams, G.M. (1981) Liver culture indicators for the detection of chemical carcinogens. In Short Term Tests for Chemical Carcinogens, R.H.C. San and H.F. Stich, eds. Springer-Verlag, New York, pp. 581-609.
42. Williams, G.M. (1983) Epigenetic effects of liver tumor promoters and implications for health effects. Environ. Hlth. Perspect.
43. Williams, G.M. (1983) Genotoxic and epigenetic carcinogens: Their identification and significance. Ann. N.Y. Acad. Sci. 407:328-333.
44. Williams, G.M., M.F. Laspia, and V.C. Dunkel (1982) Reliability of the hepatocyte primary culture/DNA repair test in testing of coded carcinogens and noncarcinogens. Mutat. Res. 97:359-370.
45. Williams, G.M., S. Telang, and C. Tong (1981) Inhibition of intercellular communication between liver cells by the liver tumor promoter 1,1,1-trichloro-2,2-bis (P-chlorophenyl) ethane (DDT). Cancer Lett. 11:339-344.
46. Williams, G.M., and J.H. Weisburger (1981) Systematic carcinogen testing through the decision point approach. Ann. Rev. Pharm. and Toxicol. 21:393-416.
47. Williams, G.M., and J.H. Weisburger. Risk assessment of dietary carcinogens and tumor promoters. In Diet and Cancer: From Basic Research to Policy Implications, T.C. Campbell and D. Schottenfeld, eds. A.R. Liss, Inc., New York, New York, (in press).
48. Williams, G.M., J.H. Weisburger, and D. Brusick (1981) The role of genetic toxicology in a scheme of systematic carcinogen testing. In American Chemical Society Symposium Series: The Pesticide Chemist and Modern Toxicology, S.K. Bandal, G.J. Marco, L. Goldberg, and M.L. Leng, eds., pp. 57-87.
49. Yotti, L.P., C.C. Chang, and J.E. Trosko (1979) Elimination of metabolic cooperation in Chinese hamster cells by a tumor promoter. Science 206:1089-1091.

CHAIRMEN'S OVERVIEW ON ENZYME MARKERS

Doris Balinsky and Russell Hilf*

Department of Biochemistry and Biophysics
Iowa State University
Ames, Iowa

When looking for suitable biological markers of cell alteration as a response to the action of carcinogens, enzymes are obvious candidates for several reasons. Firstly, the enzyme complement of an adult, differentiated tissue is unique to that tissue, being optimal for its functioning. Thus, although many enzymes are ubiquitous, since they are essential for general cell metabolism, the activities in different tissues generally differ. Secondly, a specific adult differentiated tissue has some enzymes which are unique to that tissue, or which occur in much higher concentration in that tissue than in other tissues. Thirdly, many enzymes exist in multiple forms, or isozymes; again, the isozyme pattern may distinguish a particular adult differentiated tissue from other differentiated adult tissues. Fourthly, because most enzymes are highly specific towards their substrates, it is possible to assay an enzyme and its isozymes in a crude homogenate containing many other proteins. Tumors frequently have a different enzyme and isozyme profile to the tissue of origin, since their metabolic requirements are so different. A potential carcinogen might thus be expected to perturb the metabolic profile of an adult differentiated tissue, as reflected in the enzyme and isozyme complement, and this can easily be detected by suitable assay procedures.

The choice of suitable enzymes to use as biological markers may be a difficult one. One can start with a logical approach, namely to ask how tumor metabolism might be expected to differ from that of the adult differentiated tissue, and which enzymes would thus be involved. Or one can simply try a hit-and-miss approach, examining enzymes which have no other claim to possible usefulness except that

*Department of Biochemistry, University of Rochester School of Medicine and Dentistry, Rochester, New York 14642

they have a convenient assay procedure, or are already under investigation in a particular laboratory. Both approaches have been successfully used, since carcinogenesis may lead to a general altered programming of protein synthesis.

As will be discussed more fully later, the changed enzyme complement observed in tumors frequently resembles that of fetal tissues, placental tissue, regenerating adult tissues, or adult differentiated tissues other than the tissue of origin. No new enzymes or isozyme forms have been observed in tumors, despite extensive searches for them.

Studies on carcinogenesis are in general not carried out for purely academic reasons, but are designed to help us understand and ultimately conquer cancer in humans, preferably by prevention, but otherwise by cure. It is important to compare the enzyme status of human and animal tumors derived from various tissues.

In this session we will hear firstly from Dr. Kouri, who will tell us how some animals may be protected from the effects of carcinogens by a genetically-determined deficiency of the enzyme arylhydrocarbon hydroxylase, which metabolizes compounds like benzpyrene to potent carcinogens. These studies have been extended by Dr. Kouri and his colleagues to human cancer patients who appear to have higher levels of arylhydrocarbon hydroxylase than a group of non-cancer patients.

Dr. Bill Richards will then discuss the enzyme gammaglutamyltranspeptidase, an enzyme which is widely thought to be growth-related, and is used as a tumor marker. He will present data indicating that this enzyme may in fact be a marker of cell differentiation and aging rather than proliferation.

We will hear next from Mari Haddox about the regulation of polyamine biosynthesis. The biosynthesis and accumulation of polyamines appear to be growth related phenomena, and Dr. Haddox will tell us how ornithine decarboxylase, the rate-limiting step of polyamine biosynthesis, is regulated, and how its regulation is altered in malignancy.

I will compare the levels and isozyme patterns of several enzymes of nucleic acid and carbohydrate metabolism in human and animal tumors from various organs, and point out how the patterns may differ between humans and animals, and among different human tumors. I will also indicate how changes similar to those found in tumors also occur in other growing tissues, and in uninvolved tissues of tumor-bearing animals and humans.

Finally, Dr. Russell Hilf will discuss enzyme and isozyme patterns in human breast and experimental animal mammary tumors and point out how they differ from normal.

ARYL HYDROCARBON HYDROXYLASE ACTIVITY

"OF MICE AND HUMANS"

R.E. Kouri[a,d], R.A. Lubet[a], C.E. McKinney[a], G.M. Connolly[a], D.W. Nebert[b], and T.L. McLemore[c]

[a]Division of Toxicology and Oncology
Microbiological Associates
Bethesda, Maryland 20816

INTRODUCTION

The fate of a particular chemical *in vivo* depends on a delicate balance between those enzymes capable of potentiating the effects of the chemical and those enzymes capable of detoxifying it to non-reactive intermediates. Within a given tissue or cell, several potential pathways exist for the metabolism of most xenobiotics (see reviews 1,2). One recent approach to the analysis of these complex events has been to reduce the number of these series of reactions to a few individual steps. This approach takes advantage of the fact that there is very often a rate-limiting step involved in a specific pathway. That is, one step is usually much slower than all the rest and, hence, determines the average rate of the whole process. The questions then become:

1. Can we identify which of these steps is rate-limiting?
2. Are there naturally-occurring variations in the activity of this particular step?
3. Is there a genetic basis for the regulation of these naturally-occurring differences?
4. Can these differences result from the action of a single gene or multiple genes?
5. Is there linkage between the presence or absence of this locus and a biological effect?

---

[b]Developmental Pharmacology Branch, NICHHD, NIH, Bethesda, Maryland 20205; [c]Southwest Foundation for Research and Education, San Antonio, Texas 78282; [d]Funded in part by Contract N01-HD92840 from NICHHD and contracts from the Council for Tobacco Research, USA.

These approaches have been used not only to analyze the rate-limiting step in monooxygenase metabolism [i.e., aryl hydrocarbon hydroxylase (AHH)], but also to correlate the levels of these enzymes to such biological endpoints as carcinogenesis, mutagenesis, and cytotoxicity (2,3).

In this chapter we review the data which suggest that: a) there are naturally-occurring variations in the steady-state levels of these microsomal monoxygenases; b) these observed naturally-occurring variations are under host-gene regulation by a small number of genes; and c) correlations exist between activities of these enzymes and susceptibility to cancer induced by these same chemicals in both animal model systems and humans.

ANIMAL STUDIES

The Ah Locus

The level of AHH activity is increased dramatically by pre-treatment with polycyclic aromatic hydrocarbons (PAH). This increase, or induction, is observed in the liver, lung, bowel, kidney, lymph nodes, skin, bone marrow, pigmented epithelium of the retina, brain, mammary gland, uterus, testes, and the ovary, of a variety of AHH-responsive mouse strains. It is absent, or markedly decreased, in these same tissues from AHH nonresponsive mouse strains (4,5). This "responsiveness" to aromatic hydrocarbons was originally designated the Ahh locus (6), and ultimately was designated the Ah locus (7). The allele Ah^b denotes the allele carried by the C57Bl/6(B6) inducible inbred strain, and Ah^d denotes the allele carried by the DBA/2(D2) non-inducible inbred strain. Numerous studies indicate that an important product of the Ah locus in mice is a cytosolic receptor (8,9) capable of binding to certain polycyclic aromatic inducers. This complex activates structural genes leading to increased levels of enzymes which metabolize the inducers and other polycyclic aromatic non-inducing compounds. In addition to innocuous products resulting from the metabolism, reactive metabolites may also be generated. Induction of one or more forms of cytochrome P-450 is associated with the induction of numerous monooxygenase activities (see reference 2). How so many substrates with very different chemical structures can be oxygenated by a single enzyme active site is not understood. The most likely possibility is that there are several forms of cytochrome P-450.

Regulation of Susceptibility to Chemically-Induced Cancers by the Ah Locus

Examples of how the Ah locus regulates susceptibility to chemically-induced cancers will now be discussed. In crosses between B6 (responsive) and D2 (nonresponsive) mice, AHH induction segregates as a single autosomal dominant gene (2). Table 1 presents the

Table 1. Genetic linkage between Ah^b allele and susceptibility to 3-methycholanthrene (MCA) - and dibenz(a,h)anthracene [DB(a,h)A]induced subcutaneous (SC) carcinogenesis[a,b].

STRAIN OFFSPRING	EXPRESSION AT Ah LOCUS[c]	SC CARCINOGENESIS					
		MCA			DB(a,h)A		
		TU/TR	LATENCY (DAYS)	CI	TU/TR	LATENCY (DAYS)	CI
C57B1/6(B6)	++	23/29	130	61	30/60	178	28
DBA/2(D2)	0	10.48	209	10	1/60	230	<1
B6D2F1	++	54/90	138	43	41/117	184	19
B6D2F1 x D2	++	15/20	136	55	63/101	248	25
	0	5/34	190	8	0/75	NA	0

[a] Mice at 4-6 weeks of age were treated SC with 150 μg of MCA in trioctanoin and observed daily for evidence of SC tumors over a 9 month period (see references 5 and 10 for procedures). DB(a,h)A-treated animals were given either 150 μg or 300 μg of chemical and observed over a 12-month period. Data are pooled for this presentation.

[b] Abbreviations are: TU/TR = Number of mice with SC fibrosarcomas per number of chemically-treated mice initially put on test; Latency is given in days, and CI = Carcinogenic Index; percent tumors ÷ average latency in daus x 100.

[c] Hepatic AHH levels 24 hours agter intraperitoneal (IP) treatment with 100 mg 3-methylcholanthrene [or dibenz(a,h)anthracene] per kg body weight are presented; ++ = ∿2500 pmoles 3-OHBP formed per min per mg microsomal protein, 0 = ∿3- pmoles 3-PHBP formed per min per mg microsomal protein. (See references 8 and 10 for details of assay).

typical hepatic AHH activities observed 24 hours following intraperitoneal (IP) treatment with either 3-methylcholanthrene (MCA) or dibenz (a,h)anthracene [DB(a,h)A] at 100 mg per kg body weight. For B6, B6D2F1, and D2 mice, the levels are ∿2,500, ∿2,100, and ∿300 units AHH activity/mg microsomal protein, respectively. [A unit of AHH = the formation of the fluorescent equivalent of 1.0 pmole 3-OH-benzo(a)pyrene per min at 37°C (see references 8 and 10 for details of assay)]. AHH levels in progeny from B6D2F1 x D2 backcrosses are also presented in Tab. 1. In progeny from this cross, it would be expected that ∿50% of the mice would express levels similar to the B6D2F1 parent and ∿50% would express levels similar to the D2 parent. The numbers of animals expressing the AHH responsive phenotype of the B6D2F1 parent (i.e., Ah^b allele) was 121 (20 animals from the MCA-treated study and 101 from the DB(a,h)-treated study). The number of animals expressing the AHH non-responsive phenotype (i.e., Ah^d allele) was 109 [34 animals from the MCA study and 75 from the DB(a,h)A study]. Thus, in these studies, the frequency of the Ah^b allele was 53% in the backcross progeny.

In two separate studies, backcross mice were treated subcutaneously (SC) with either MCA or DB(a,h)A in order to define the relationship between susceptibility to carcinogen-induced SC fibrosarcomas and expression of AHH levels as controlled by the *Ah* locus. MCA was given at a dose of 150 μg/0.02 ml trioctanoin as described previously (10,11). DB(a,h)A was given at either 150 μg or 300 μg/0.02 ml trioctanoin. For the purpose of this presentation, the data from both doses are pooled. SC carcinogenesis data are given in terms of the incidence of fibrosarcomas at the site of chemical inoculation, the average latency in days until the tumor reaches 1.0 cm in diameter, and the carcinogenic index (CI). The CI is used so that incidence and latency information can be pooled to generate one index which is indicative of level of carcinogenic activity observed (see reference 2 for discussion). Both the B6 and F1 animals were four to six times more susceptible to MCA-induced fibrosarcomas than the D2 mice. Of greatest importance is the fact that, in progeny from the B6D2F1 x D2 cross, the AHH-responsive progeny were also approximately seven times more susceptible to MCA carcinogenesis than the AHH non-responsive progeny.

The data using DB(a,h)A as the carcinogen are even more clear-cut. The CI from DB(a,h)A-treated B6 and F1 mice was 28 and 19, respectively (Tab. 1). This chemical induced only one tumor in the D2 strain. In backcross animals, a total of 63 DB(a,h)A-induced tumors were observed out of the total of 176 treated animals, and 100% of the tumors were found in AHH-responsive animals.

The results show very clearly that there is genetic linkage between expression of the AHH responsive phenotype (the Ah^b allele) and sensitivity to either MCA or DB(a,h)A-induced fibrosarcomas. This genetic linkage can be extended to include benzo(a)pyrene (11,12), and 7,12-dimethylbenz(a)anthracene (13), as well as other tumor endpoints such as skin carcinomas (12) and lung carcinomas (14). Thus, in animal model systems: a) there are naturally-occurring variations in the steady-state levels of certain microsomal monooxygenases (i.e., AHH); b) these variations are under host-gene regulation by a small number of genes (only one gene in the example presented); and c) there is a genetic correlation between both the ability to respond to and metabolize certain chemical carcinogens, and susceptibility to cancer induced by these chemicals. The question now becomes: can these same correlations be found in the human species?

HUMAN STUDIES

The human population varies in the levels of microsomal monooxygenases, as shown by *in vivo* assays of drug metabolism (15), and *in vitro* assays using such human tissues as liver (16), lung (17,18), colon (19), esophagus (20), placenta (21), monocytes (22), and lymphocytes (23-27). Twin and family studies have shown that genetic factors are primarily responsible, not only for maintaining the *in vivo* variations in rates of elimination of many commonly used drugs (15), but also for maintaining the observed variations in AHH levels in such human tissues as blood monocytes (22), and mitogen-activated lymphocytes (23-25). The data, however, are still somewhat equivocal because of the small numbers of individuals who have been sampled for such tissues as esophagus, colon, and liver, etc. in a given study, or because of methodologic problems surrounding the more accessible tissues such as the peripheral blood lymphocytes. The latter tissue has been used most frequently for determining AHH levels in the human population, but recent studies have shown that the levels of AHH activity are dependent on the degree of mitogen-induced activation, and this activation step is influenced by a variety of _in vivo_ and _in vitro_ factors which are difficult to define and control (28). Nutritional state (29), drug intake (30), age (31), and disease state (32) influenced the capacity of lymphocytes to respond to mitogens. Variations in AHH levels in lymphocytes also have been observed to occur seasonally in some specific geographic locations (33). A variety of _in vitro_ conditions can also influence the kinetics and absolute levels of both control and induced AHH levels in mitogen-stimulated lymphocytes. These include: the initial concentration of lymphocytes (26), the type and lot of serum supplement (26,28), and the type and lot of mitogen (34).

These problems make it mandatory that some basic methodologic questions be addressed before we can answer the questions of naturally-occurring variations in AHH levels in humans, the genetic control of these differences, or the role these AHH differences may play in cancer susceptibility. Our laboratory has recently completed a series of studies which have attempted to address the problems of quantitating AHH activity in mitogen-activated human lymphocytes. We have attempted to correct the potential points of variability by the following means:

- Use of human AB serum as the supplement in the culture medium.
- Use of only phytohemagglutinin (PHA)-M (the more crude form of lectin).
- Use of exactly 1 x 10^6 lymphocytes/ml culture medium to initiate the blastogenic assay.
- Use of benz(a)anthracene (BA) as the inducer, and provision of BA in the medium throughout the time of mitogen activation.
- Use of more than one time point for determination of peak AHH activity.
- Use of NADH-dependent cytochrome b_5 reductase activity as a basis for comparing AHH activity among individuals.
- Use of cryopreserved lymphocytes.

Descriptions of the studies which provided the rationale for these methods are in references 26, 27, and 35.

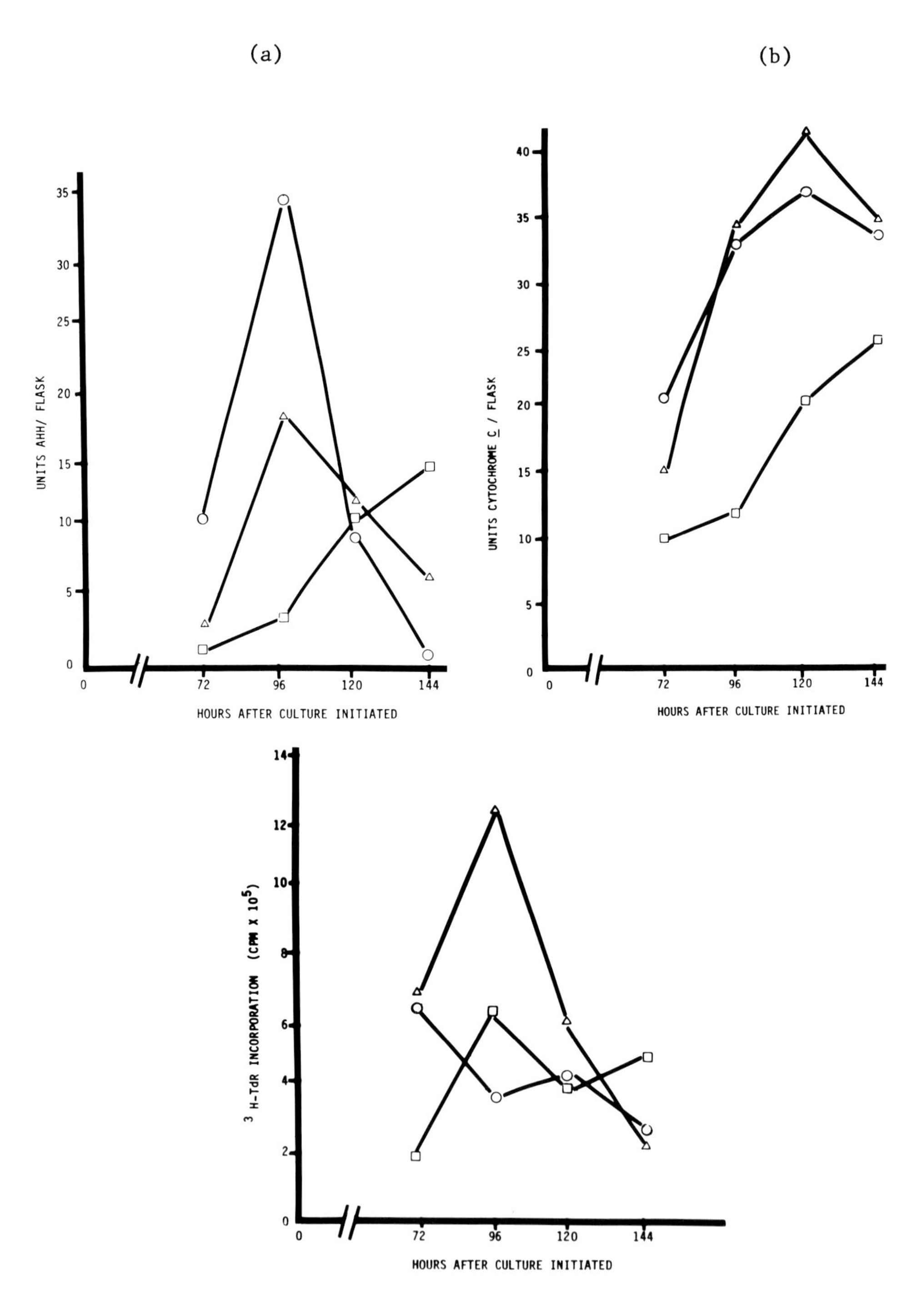
(a)
(b)
UNITS AHH/ FLASK
35
30
25
20
15
10
5
0
0
72
96
120
144
HOURS AFTER CULTURE INITIATED
UNITS CYTOCHROME C / FLASK
40
35
30
25
20
15
10
5
0
0
72
96
120
144
HOURS AFTER CULTURE INITIATED
3 H-TdR INCORPORATION (CPM X 10^5)
14
12
10
8
6
4
2
0
0
72
96
120
144
HOURS AFTER CULTURE INITIATED

(c)

AHH Levels Among Different Individuals

Using these "standard" conditions, the levels of BA-induced AHH activity, NADH-dependent cytochrome b_5 reductase activity, and the incorporation of ^{3}H-thymidine have been analyzed in a variety of individuals. The AHH assay was used as described previously, and measures the formation of the fluorescent equivalent of 3-hydroxybenzo(a)pyrene from benzo(a)pyrene in a 30 min *in vitro* incubation (26). The NADH-cytochrome b_5 reductase is measured using cytochrome *c* as the substrate and usually is called NADH-dependent cytochrome *c* reductase (Cyt *c*) (35). Data from this laboratory have shown that Cyt *c* is: a) microsomal; b) immunochemically similar to the NADH-cytochrome b_5 reductase and cytochrome b_5 found in liver microsomes, spleen microsomes, erythrocytes, and other mitochondrial membranes; c) distinct from NADH-cytochrome P-450 reductase; d) not influenced by mitogen activation; and e) a marker for total cells, viability of cells, and perhaps for microsomal content of lymphocytes. Levels of incorporation of ^{3}H-TdR in acid precipitable material were done as described previously (26). Lymphocytes were incubated with ^{3}H-TdR during the last 4 hr prior to cell collection and assay.

The time-dependent increases in BA-induced AHH, Cyt *c*, and ^{3}H-TdR incorporation for three individuals are shown in Figures 1a, 1b, 1c, respectively. AHH activity for the three individuals varied over the 144 hr period of incubation, with samples 47 and 48 expressing peak activity at 96 hr, and sample 45 showing peak activity at 120-144 hr in culture (see Fig. 1a). It should be pointed out that these AHH levels were easily detected, since AHH levels of non-BA treated lymphocytes, or non-mitogen activated lymphocytes, were always less than 2 units AHH/flask.

The development of NADH-dependent Cyt *c* activity in these frozen cells is shown in Figure 1b. The activity of this enzyme

←

Fig. 1. Kinetics of microsomal enzyme induction and mitogen activation in cryopreserved human lymphocytes. Lymphocytes were isolated from three plateletpheresis residue samples and cryopreserved as described elsewhere (27). Lymphocytes were thawed, exposed to PHA, and treated with BA according to previously described methods (27). Figure 1a depicts the kinetics of induction of BA-induced AHH levels in the PHA-activated lymphocytes from three different individuals. Figure 1b shows the kinetics of formation of Cyt *c* activity in these same lymphocytes. Figure 1c presents the level of ^{3}H-TdR incorporation (cpm x 10^5/flask) into these PHA-activated lymphocytes. Individuals are as follows: Sample 45 (□——□), Sample 47 (o——o), and Sample 48 (∇——∇).

increased for samples 47 and 48 for up to 120 hr in culture while that of sample 45 had not reached a peak until 144 hr in culture. The slower growth rate of sample 45 paralleled the more delayed time of appearance of peak AHH levels shown in Figure 1a. All blood samples showed an approximate doubling of the cell population over the period of 72-120 hr in culture.

The ability of these cells to incorporate ^{3}H-TdR is presented in Figure 1c. The level of incorporation of ^{3}H-TdR was similar to freshly cultured blood samples (see reference 27). Differences in the rate of incorporation among the 3 samples were observed, but these differences did not parallel rates of appearance of either AHH activity or the Cyt c activity.

These enzyme activities are retained for at least one year after cryopreservation. Sample 45 has been cultured and assayed 17 times over a period of one year with a mean AHH activity of 0.67 units/Cyt c and a coefficient of variation of 0.17.

Studies from our laboratories have shown that for ten normal human volunteers, the peak levels of AHH/Cyt c activity were similar when assayed in either freshly cultured, or in cryopreserved lymphocytes (27). The degree of reproducibility, however, was much better in assays utilizing frozen lymphocytes (27). The relative ranking of these 10 individuals, cultured fresh or after cryopreservation, is identical (27). As described previously for freshly cultured lymphocytes (26), the levels of AHH activity observed in these cryopreserved lymphocytes were unrelated to differences in degree of mitogen-induced blastogenesis.

The variation of peak AHH/Cyt c levels for four different individuals over an eleven month observation period is presented in Table 2. The four individuals were designated 101(R), 102(L), 103(C), and 104(A). Blood was drawn from all four individuals on day 14 ± 2 of the menstrual cycle of individual 103(C). The lymphocytes were cryopreserved from 3 to 14 months before all samples were thawed and assayed for both BA-induced AHH/Cyt c levels and cpm ^{3}H-TdR/Cyt c. There were significant variations in AHH/Cyt c levels for all four people. Average coefficient of variation for all four individuals was $\sim$0.35, with significantly lower levels observed in months 2 or 3 (Oct. and Nov.). The study was begun in September, thus the lower induced AHH levels could reflect some loss of viability after long-term cryopreservation. On the other hand, the data could reflect some seasonal variation (33), but both of these suggestions remain speculative at this time. It seems likely that the AHH levels in mitogen-activated lymphocytes vary considerably for a given individual; however, the variation between individuals is usually larger than the variation within a given individual. The intraclass correlation coefficient (r) for the relative ranking of these four individuals is: 0.70 ($p < .001$) -- that is, individuals 103(c) > 104(A) > 102(L) ≅ 101(R).

Table 2. Comparison of AHH/Cyt c activities in four individuals whose blood was collected monthly for eleven consecutive months[a,b,c]

INDIVIDUAL	MONTH OF BLOOD COLLECTION										
	1[d]	2	3	4	5	6	7	8	9	10	11
101(R) ♂	-	.19	.19	.34	.20	.32	51	.70	.46	.30	.39
102(L) ♂	.52	.37	.23	.25	.47	.22	.34	.48	.48	.49	.36
103(C) ♀	-	.43	.22	1.40	1.20	1.30	.95	1.60	1.30	1.50	1.40
104(A) ♀		1.20	.51	.30	.57	.65	1.10	1.10	.46	.52	.70

[a] Lymphocytes were isolated and frozen as described in reference 27. Cells were stored frozen from three months to fourteen months before thawing and assaying for benzo(a)anthracene-induced AHH levels and NADH-dependent Cyt c activity. Cells were assaysed at both 96 and 120 hr, and peak values are presented.

[b] Blood was drawn for all four donors on day 14 ± 2 of the menstrual cycle of individual 103 (C).

[c] Correlation coefficient (r) for relative ranking of these individuals was 0.70 ($p < 0.001$; DF = 3,36).

[d] September was the month when blood drawing was initiated and samples from individuals 101 (R) and 103 (C) were lost as the result of contamination.

AHH Levels in Monozygotic (MZ) and Dizygotic (DZ) Twins

The AHH/Cyt c levels in 9 pairs of MZ twins and 10 pairs of DZ twins are given in Table 3. Blood was collected once per month over a three-month period and lymphocytes were thawed and cultured as described previously (26,27). Data were analyzed for intraclass correlation coefficients (r), heritability, and between/within pair variance as described by Borresen et al. (25).

The AHH/Cyt c levels in cultured lymphocytes from MZ twins were highly correlated within the pairs of twins themselves ($r = 0.81$; $p<0.001$). The AHH/Cyt c levels in DZ twins were almost as highly correlated ($r = 0.58$; $p = 0.03$). That is, the within-twin variation in AHH/Cyt c levels for both MZ and DZ twins was much less than the between-twin variation. A significant intraclass correlation coefficient would be expected if AHH levels were either genetically or environmentally determined. The correlation coefficients were observed to be significantly greater than zero, and r_{mz} (0.81) was greater than r_{dz} (0.58). Testing for heritability by comparing "within pair" variance of DZ twins to "within pair" variance of MZ twins indicated that no significant differences existed. A much larger sample size would be required to see significant heritability given the data presented in Table 3 - at least 25 pairs of MZ and DZ twins. The data in Table 3 suggest that the trend is in the right direction for suggesting heritability, but attempts at delineating

Table 3. AHH/Cyt c activities in cryopreserved lymphocytes from twin pairs[a,b,c]

TWIN PAIR	SEX	MONOZYGOTIC TWINS INDIVIDUAL 1	INDIVIDUAL 2	SEX	DIZYGOTIC TWINS INDIVIDUAL 1	INDIVIDUAL 2
1	♂	0.24 ± 0.15	0.29 ± 0.12	♂	0.74 ± 0.22	0.66 ± 0.27
2	♀	0.30 ± 0.06	0.56 ± 0.12	♂	0.42 ± 0.18	0.48 ± 0.12
3	♂	0.40 ± 0.13	0.51 ± 0.20	♀	0.26 ± 0.07	0.54 ± 0.13
4	♀	0.41 ± 0.19	0.50 ± 0.18	♀	0.42 ± 0.18	0.39 ± 0.12
5	♀	0.84 ± 0.25	0.82 ± 0.27	♂	0.42 ± 0.15	0.49 ± 0.15
6	♀	0.42 ± 0.09	0.36 ± 0.08	♂	0.37 ± 0.10	0.55 ± 0.13
7	♀	0.56 ± 0.19	0.63 ± 0.14	♀	0.50 ± 0.16	0.50 ± 0.17
8	♀	0.30 ± 0.11	0.33 ± 0.19	♀	0.46 ± 0.19	0.42 ± 0.13
9	♀	0.44 ± 0.17	0.35 ± 0.04	♀	0.40 ± 0.14	0.35 ± 0.05
10				♀	0.65 ± 0.12	0.70 ± 0.43

[a] AHH activity is determined at peak time (either 96 or 120 hrs after culture) and the AHH/Cyt c ratio is calculated from this peak activity.

[b] The number of determinations is $N \geq 3$ in all cases. The average coefficient of variation (X/S) for all individuals was 0.33, with a range of 0.11 - 0.62.

[c] Intraclass correlation coefficients were: $r_{mz} = 0.81$ ($p=<0.001$) and $r_{dz} = 0.58$ ($p=0.03$). Heritability (h^2) for MZ = 0.81, for DZ = 1.16 and between/within pair variance ratio (F ratio) for AHH levels were: $F_{mz} = 8.69$ ($p<.001$) and $F_{dz} = 3.57$ ($p=0.03$).

the effect of environment on heritability will have to be done with much larger sample sizes. Family studies to try to address this issue are ongoing in our laboratory.

Variation in AHH/Cyt c Levels as a Function of the Menstrual Cycle

A comparison of peak AHH/Cyt c over the complete menstrual cycle of two females and the AHH/Cyt c levels of two age-matched males is presented in Figure 2. The data from both females and from both males are pooled for this presentation. Also given in this figure is a schematic representation of estrogen/progesterone levels during a normally-occurring menstrual cycle (36). The AHH/Cyt c levels in the two males were fairly reproducible and averaged ∿0.9 units AHH/Cyt c. The two females expressed a decline in AHH activity during the first 14-16 days of their cycle; post-ovulation, a significant increase in AHH/Cyt c activity was observed -- mean activity increased from ∿0.5 units AHH/Cyt c to ∿1.0 unit AHH/Cyt c. It is not known if these variations in AHH levels are the result of real changes in hydrocarbon metabolizing capacities of females or are the result of physiological changes which alter the type of

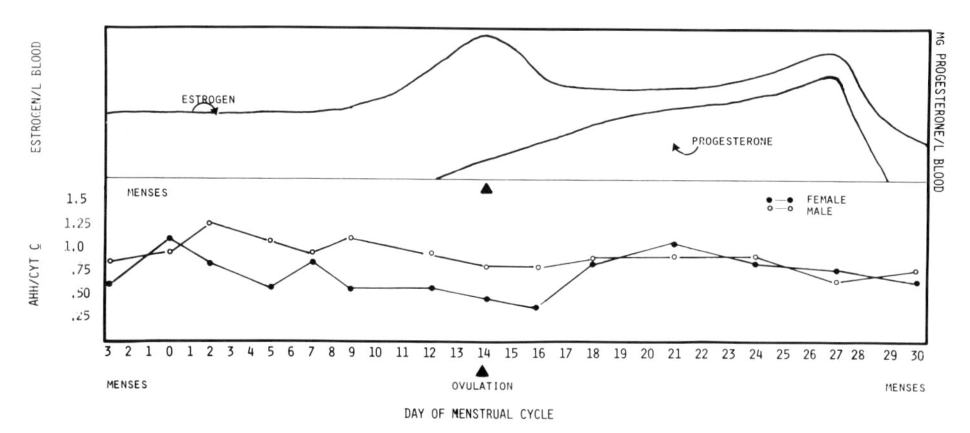

Fig. 2. Comparison of AHH levels as a function of the menstrual cycle of human females. Whole blood was drawn every other day throughout the course of one menstrual cycle for two females. Blood was drawn from two age-matched males for controls. Lymphocytes were isolated and cryopreserved as described previously (27). All samples were thawed, cultured, and assayed for AHH, Cyt c and ^{3}H-TdR incorporation as described previously (27). Data from females and from both males were pooled for this comparison. The top panel shows a typical hormone cycle during a normal menstrual cycle. [Adapted from Selkhurt (36)].

lymphocytes or responsiveness of lymphocytes in the *in vitro* blastogenesis assay. Attempts at mimicking this effect by adding estrogen/progesterone to the *in vitro* cultured lymphocytes failed to alter either control or BA-induced AHH levels (data not shown). Some of the variation observed in the studies depicted in Tables 2 and 3 could result from these cyclical variations in menstruating females. In fact, the twin data in Tab. 3 may be a case in point. Twins #2 of the MZ group were 54 year old females, one of whom was post-menopausal and one of whom was not. This one pair of twins yielded $\sim$70% of the total variability which was observed in all nine pairs of MZ twins. This could suggest that environment may play a major role in determining AHH levels in cultured lymphocytes.

AHH Levels in Lung Cancer Patients

In humans, reports from numerous laboratories have suggested that there is a relationship between higher levels of AHH activity in certain tissues and the occurrence of lung cancer (17,37-48), laryngeal cancer (49,50), and renal and ureter cancer (51). Susceptibility to leukemia may be linked to a lower capacity to metabolize chemical carcinogens, that is, lower AHH levels (52). On the other hand, there are also reports from a number of other laboratories suggesting a lack of relationship between high AHH levels in human-derived tissues and susceptibility to lung or laryngeal cancer (18,53-57). The reasons for these contradictory results probably reside in the methodology for assessing AHH activity, as described earlier in this chapter.

Using cryopreserved lymphocytes and the AHH assay described in detail in ref. 27, our laboratory has recently completed a collaborative project with the Veterans Hospital, Houston, Texas. Care was taken to insure that the individuals from whom blood was collected were on the same hospital diets for at least two days prior to phlebotomy, and the individuals were carefully matched for age, sex, and disease state. Details as to description of the patient population and methods for transferring blood samples between the laboratories are provided in reference 58.

The data from 51 individuals whose blood was collected and cryopreserved prior to any diagnosis are presented in panels B and C of Figure 3. These data are compared to the assay of 161 normal volunteers (see Panel A) whose assay for AHH/Cyt c levels was carried out at approximately the same time as the hospitalized patients.

Some clinical and experimental features of the fifty-one patients in panels B and C of Fig. 3 are as follows. Twenty-one patients were diagnosed as having primary lung cancers (panel C) while the remaining individuals expressed a variety of diseases, including chronic obstructive pulmonary disease, pneumonia, and pulmonary tuberculosis (panel B). The location of the tumor within the lung was also noted for eighteen of these cancers. Seven cancers were found in the right upper lung, five in the left upper lung, two were found in the right lower lung, three in the right middle lung, and one in the left lower lung. The age range of the patients with lung cancer was 47-78 years, with a mean of 59.7 years, while the age range for the non-cancer patients was 35-87 years, with a mean of 58.6 years. Nineteen of the 21 lung cancer patients were known cigarette smokers at the time of admission to the hospital. One individual with lung cancer had an unknown smoking history, and one was a pipe smoker. The patients averaged 67.5 "pack years" of smoking. (A pack-year is determined by multiplying the number of packs per day times the number of years the individual smoked.) Twenty-four of the thirty non-cancer patients also smoked (average of 55.8 pack years). Three individuals were non-smokers, one smoked cigars, and two had unknown smoking histories. Smoking history for lung cancer and non-lung cancer patients was not different ($p = 0.30$).

Peak AHH/Cyt c activities ranged from 0.02 to 1.87 units AHH/Cyt c for the fifty-one patients. The level of AHH/Cyt c in these fifty-one patients corresponded strikingly with diagnoses of lung cancer. Of the fourteen highest AHH/Cyt c activities observed, all were found in patients with lung cancer. The mean AHH/Cyt c

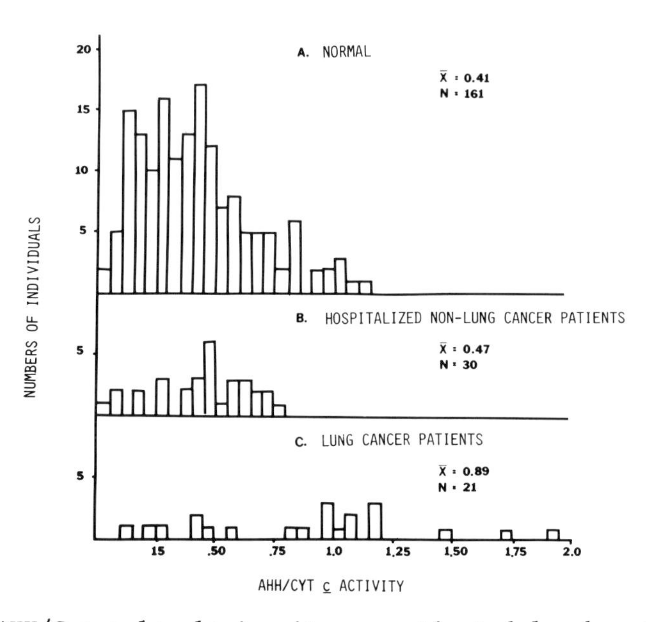

Fig. 3. AHH/Cyt c levels in mitogen-activated lymphocytes from normal individuals (Panel A), hospitalized non-lung cancer patients (Panel B), and hospitalized lung cancer patients (Panel C). Blood was collected from 161 individuals (Panel A) over a one yr period and lymphocytes were isolated and cryopreserved (27). Lymphocyte samples were given random numbers and thawed, cultured, and assayed for AHH activity four separate times over a two-week interval in order to complete the assays for all 161 individuals. AHH and Cyt c assays were as described (26,27). For Panels B and C, blood was collected from patients from the Veterans Administration Hospital, Houston, Texas and shipped to Microbiological Associates within 12 hr of collection. Individuals were in the hospital a minimum of two days, and no radiotherapy, chemotherapy, or other medication was given. A description of these individuals including all salient histopathological diagnoses are presented in reference 58. Of the twenty-one patients diagnosed with lung cancer, eleven had squamous cell carcinomas, seven had adenocarcinomas, and three had oat cell carcinomas.

activity for all the lung cancer patients was 0.89, while that of the non-lung cancer patients was 0.47 ($p < 0.001$). Within the group of cancer patients, there was no relationship between AHH/Cyt c levels and histologic type ($p = 0.13$) or the location of pulmonary tumors ($p = 0.80$). There was no relationship between AHH/Cyt c activity and age of the patient ($p = 0.45$), numbers of cigarettes smoked ($p = 0.80$), or family history of cancer ($p = 0.38$). For the non-lung cancer patients, the level of AHH/Cyt c activity observed was not correlated with age of the individual ($p = 0.45$), smoking history ($p = 0.35$), or cancer family history ($p = 0.71$).

The distribution of lymphocyte blastogenesis (as measured by cpm ^{3}H-Tdr/Cyt c) was uniform in these fifty-one patients (data not shown - see ref. 58 for discussion). No correlation was noted between AHH activities and blastogenesis among the fifty-one individuals ($p = 0.30$). The twenty-one patients with lung cancer, however, expressed a higher degree of blastogenesis (mean = 32,997 cpm ^{3}H-TdR/Cyt c) than the thirty non-cancer patients (mean = 23,513 cpm ^{3}H-TdR/Cyt c ($p = 0.001$).

For all fifty-one individuals, or for the specific subsets of lung cancer patients and non-lung cancer patients, there was no difference between the levels of mitogen-induced blastogenesis and: a) age of donor, b) smoking history, c) family history of cancer, or d) histologic type of pulmonary location of tumors ($p \geq 0.35$) for all factors. For the lung cancer patients themselves, there was a significantly lower level of blastogenesis observed in those individuals with a family history of cancer, than in those individuals with no family history of cancer ($p = 0.035$; DF = 14 [DF = degrees of freedom]).

The mean AHH/Cyt c activity for those patients with other respiratory disease (but non-cancerous) was 0.47 (see panel B, Fig. 3), which is very similar to the 0.41 units AHH/Cyt c observed in the assay of 161 normal individuals (see panel A, Fig. 3). Thus, the differences between AHH activities in the lung cancer and non-cancer populations were not the result of some selective inhibition of lymphocyte AHH levels by other respiratory diseases. The reasons why seven lung cancer patients expressed AHH/Cyt c levels similar to the non-cancer patients are not known. Analyses of such factors as smoking history, occupation, and cancer-related family history failed to show any particular correlation with these individuals. It is likely that certain other factors must be evaluated which might be related to the expression of lung cancer in humans (e.g., DNA repair capacity, carcinogen detoxification, immunocompetence, etc.).

In the human population, as observed in the murine model system (see Tab. 1 and ref. 2 for details), there are naturally-occurring differences in AHH levels. These differences are probably under a degree of genetic control, and those individuals expressing higher levels of AHH activity are more likely to be associated with the presence of lung cancer. A major difference between the murine system and humans is the fact that it is known that the higher AHH levels in mice are probably the cause of the heightened sensitivity to chemically-induced cancers (see reviews 2,5), while the higher AHH levels in humans could reflect either an inherent susceptibility to lung cancer, or that lung cancers can somehow cause higher levels of lymphocyte-associated AHH activity. This latter hypothesis has merit, for environmental or physiological factors can influence AHH levels in mitogen-activated lymphocytes. These are: seasons of the year (33,59), age (59), smoking habits (54), relative T-cell levels (54), or menstrual cycle (see Fig. 2). Moreover, approximately 70% of the lung cancer patients expressed AHH/Cyt c levels greater than 0.80 units, whereas none of the hospitalized control patients, and less than 10% of the normal volunteers expressed AHH levels this high. Much more work is necessary in order to define the reasons for the higher levels of induced AHH levels in lung cancer patients; however, regardless of the reasons, the fact that about 70% of the lung cancer patients expressed high induced levels of AHH, may justify a more extensive clinical evaluation of suspected lung cancer patients (or normal individuals) who express these high levels.

ACKNOWLEDGEMENT

The authors thank Ms. J. Stinnett for her help in preparing this manuscript.

REFERENCES

1. Mannering, G.H. (1981) Hepatic cytochrome P-450-linked drug-metabolism systems. Concepts in Drug Metabolism, J. Jenner and B. Testa, eds. Marcel-Dekker, Inc., N.Y. pp. 53-165.
2. Kouri, R.E., L.M. Schechtman, and D.W. Nebert (1980) Metabolism of chemical carcinogens. In Genetic Differences in Chemical Carcinogenesis, R.E. Kouri, ed., CRC Press, Boca Raton, Florida.
3. Nebert, D.W., M. Negishi, M.A. Lang, L.M. Hjelmeland, and H.J. Eisen (1982) The Ah locus, a multigene family necessary for survival in a chemically absence environment: Comparison with the immune system. Advanc. Genet. 21:1-52.
4. Thorgeirsson, S.S. and D.W. Nebert (1977) The Ah locus and the metabolism of chemical carcinogens and other foreign compounds. Adv. Cancer Res. 25:149.
5. Kouri, R.E., and D.W. Nebert (1977) Genetic Regulation of Susceptibility to polycyclic hydrocarbon-induced tumors in the mouse. In Origin of Human Cancer, Book A., H. Hiatt, J.D. Watson, and J.A. Winsten, eds. Cold Spring Harbor Laboratory, N.Y., p. 811.
6. Thomas, P.E., R.E. Kouri, and J.J. Hutton (1972) The genetics of aryl hydrocarbon hydroxylase induction in mice: A single gene difference between C57BL/6J and DBA/2J. Biochem. Genet. 6:157.
7. Green, M.C. (1973) Guideline for genetically determined biochemical variants in the house mouse, mus musculus. Biochem. Genet. 9:369.

8. Nebert, D.W., J.R. Robinson, A. Niwa, K. Kumaki, and A.P. Poland (1975) Genetic expression of aryl hydrocarbon hydroxylase activity in the mouse. J. Cell Physiol. 83:393.
9. Poland, A.P., and E. Glover (1975) Genetic expression of aryl hydrocarbon hydroxylase by 2,3,7,8-tetrachlorodibenzo-o-p-dioxin by hepatic cytosol: Evidence for a receptor mutation in genetically non-responsive mice. Mol. Pharmacol. 11:389-398.
10. Kouri, R.E., R.A. Salerno, and C.E. Whitmire (1973) Relationship and sensitivity to chemically-induced subcutaneous sarcomas in various strains of mice. J. Natl. Cancer Inst. 50:363-368.
11. Kouri, R.E. (1976) Relationship between levels of aryl hydrocarbon hydroxylase activity and susceptibility to 3-methylcholanthrene and benzo(a)pyrene-induced cancers in inbred strains of mice. In Carcinogenesis, Vol. 1, Polynuclear Aromatic Hydrocarbons: Chemistry, Metabolism and Carcinogenesis, R. Freudenthal and P.W. Jones, eds. Raven Press, N.Y. pp. 139-151.
12. Legreverand, C., B. Mansour, D.W. Nebert, and J.M. Holland (1980) Genetic differences in benzo(a)pyrene-initiated tumorigenesis in mouse skin. Pharmacol. 20:242-246.
13. Thomas, P.E., J.J. Hutton, and B.A. Taylor (1973) Genetic relationship between aryl hydrocarbon inducibility and chemical carcinogen induced skin ulceration in mice. Genetics 74:655-661.
14. Kouri, R.E., L. Billups, T.H. Rude, C.E. Whitmire, B. Sass, and C.J. Henry (1980) Correlation of inducibility of aryl hydrocarbon hydroxylase with susceptibility to 3-methylcholanthrene-induced lung cancer. Cancer Lett. 9:277-284.
15. Kellermann, G., M. Luyten-Kellermann, M.G. Horning, and M. Stafford (1975) Correlation of aryl hydrocarbon hydroxylase activity of human lymphocyte cultures and plasma elimination rates for antipyrine and phenylbutazone. Drug Metab. Dispos. 3:47-50.
16. Pelkonen, O., E.H. Kaltiala, T.K.I. Larmi, and N.T. Karki (1974) Cytochrome P-450-linked monooxygenase system and drug-linked spectral interations in human liver microsomes. Chem. Biol. Interact. 9:205-216.
17. Harris, C.C., H. Autrup, R. Connor, L.A. Barrett, E.M. McDowell, and B.F. Trump (1976) Interindividual variation in binding of benzo(a)pyrene to DNA in cultured human bronchi. Sci. 194:1067-1069.
18. Cohen, G.M., R. Mehta, and M. Meredith-Brown (1979) Large inter-individual variations in metabolism of benzo(a)pyrene by peripheral lung tissue from lung cancer patients. Inter. J. Cancer 24:129-133.
19. Autrup, H., C.C. Harris, B.F. Trump, and A.M. Jeffrey (1978) Metabolism of benzo(a)pyrene and identification of the major benzo(a)pyrene-DNA adducts in cultured human colon. Cancer Res. 38:3689-3696.
20. Harris, C.C., V.M. Genta, A.L. Frank, D.G. Kaufman, L.A. Barrett, E.M. McDowell, and B.F. Trump (1974) Carcinogenic

polynuclear hydrocarbons bind to macromolecules in cultured human esophagus. Cancer Res. 39:4401-4406.
21. Gough, E.D., M.C. Lowe, and M.R. Juchau (1975) Human placental aryl hydrocarbon hydroxylase: Studies with fluorescence histochemistry. J. Natl. Cancer Inst. 54:819-822.
22. Okuda, T., E.S. Vesell, E. Plotkin, R. Tarone, R.C. Bost, and H.V. Gelboin (1977) Inter-individual and intra-individual variations in aryl hydrocarbon hydroxylase in monocytes from monoozygotic and dizygotic twins. Cancer Res. 37:3904-3911.
23. Paigen, B., E. Ward, K. Steenland, L. Houten, H.L. Gurtoo, and J. Minowada (1978) Aryl hydrocarbon hydroxylase in cultured lymphocytes of twins. Amer. J. Hum. Genet. 30:561-571.
24. Atlas, S.A., E.S. Vesell, and D.W. Nebert (1976) Genetic control of interindividual variations in the inducibility of aryl hydrocarbon hydroxylase in cultured human lymphocytes. Cancer Res. 36:4619-4630.
25. Borresen, A.L., K. Berg, and P. Magnus (1981) A twin study of aryl hydrocarbon hydroxylase (AHH) inducibility in cultured lymphocytes. Clin. Genet. 19:281-289.
26. Kouri, R.E., R.L. Imblum, R.G. Sosnowski, D.J. Slomiany, and C.E. McKinney (1979) Parameters influencing quantitation of 3-methylcholanthrene-induced aryl hydrocarbon hydroxylase activity in cultured human lymphocytes. J. Environ. Path. & Toxicol. 2:1079-1098.
27. Kouri, R.E., J. Oberdorf, D.J. Slomiany, and C.E. McKinney (1981) Aryl hydrocarbon hydroxylase activities in cryopreserved human lymphocytes. Cancer Lett. 14:29-40.
28. Gurtoo, H.L., J. Minowada, B. Paigen, N.B. Parker, and N.T. Hayner (1977) Factors influencing the measurement and the reproducibility of aryl hydrocarbon hydroxylase activity in cultured lymphocytes. J. Natl. Cancer Inst. 59:787-798.
29. Sellmeyer, E., E. Bhettay, A.S. Truswell, O.L. Meyers, and J.D.L. Hansen (1972) Lymphocyte transformation in malnourished children. Arch. Dis. Child. 47:429-435.
30. Levanthal, B.G., D.G. Poplack, G.E. Johnson, R. Simon, C. Bowles, and S. Steinberg (1976) The effect of chemotherapy and immunotherapy on the response to mitogens in acute lymphatic leukemia. In Mitogens in Immunobiology, J.J. Oppenheim and D.L. Rosenstreich, eds. Academic Press, N.Y., pp. 613-623.
31. Fernandez, L.A., J.M. MacSween, and G.R. Langley (1976) Lymphocyte responses to phytohemagglutinin: Age-related effects. Immunology 31:583-587.
32. Kirkpatrick, C.H. (1976) Mitogen and antigen-induced lymphocyte responses in patients with infectious diseases. In Mitogens in Immunobiology, J. Oppenheim and D.L. Rosenstreich, eds. Academic Press, N.Y., pp. 639-656.
33. Paigen, B., E. Ward, A. Reilly, L. Houten, H.L. Gurtoo, J. Minowada, K. Steenland, M.B. Havens, and P. Sartori (1981) Seasonal variations of aryl hydrocarbon hydroxylase activity in human lymphocytes. Cancer Res. 41:2757-2761.
34. Kouri, R.E., R.L. Imblum, and R.A. Prough (1977) Measurement of aryl hydrocarbon hydroxylase and NADH-dependent cytochrome c

reductase activities in mitogen-activated human lymphocytes. Proc. 3rd International Symp. on the Detection and Prevention of Cancer, H. Niebergs, ed. Marcel Dekker, Inc., N.Y., pp. 1659-1676.

35. Prough, R.A., R.L. Imblum, and R.E. Kouri (1976) NADH-cytochrome c reductase activity in cultured human lymphocytes - Similarity to the liver microsomal NADH-cytochromosome b_5 reductase and cytochrome b_5 enzyme system. Arch. Biochem. Biophysics 176:119-126.
36. Selkurt, E., ed. (1971) Physiology, 3rd Edition. Little, Brown and Company, Boston, Mass., p. 802.
37. Arnott, M.A., T. Yamouchi, and D. Johnson (1979) AHH in normal and cancer populations. In Carcinogen: Identification and Mechanisms of Action, A.C. Griffin and C.R. Shaw, eds. Raven Press, N.Y., pp. 147-156.
38. Coomes, M., W. Mason, I. Muijsson, E. Cantrell, D. Anderson, and D. Busbee (1976) Aryl hydrocarbon hydroxylase and 16 - hydroxylase in cultured human lymphocytes. Biochem. Genet. 24:671-685.
39. Emery, A.E.M., N. Danford, R. Anand, W. Duncan, and L. Paton (1978) Aryl hydrocarbon hydroxylase inducibility in patients with lung cancer. Lancet 3:470-471.
40. Gahmberg, C.G., A. Sekkim, T.U. Kosunen, L.R. Holsti, and O. Mekela (1979) Induction of aryl hydrocarbon hydroxylase activity and pulmonary carcinoma. Int. J. Cancer 23:302-305.
41. Guirgis, H., H.T. Lynch, T. Mate, R.E. Harris, I. Wells, L. Caha, J. Anderson, K. Maloney, and L. Rankin (1976) Aryl hydrocarbon hydroxylase activity in lymphocytes from lung cancer patients and normal controls. Oncology 3:105-109.
42. Kellermann, G., C.R. Shaw, and M. Luyten-Kellermann (1973) Aryl hydrocarbon hydroxylase inducibility and bronchogenic carcinoma. N. Eng. J. Med. 298:934-937.
43. Kellermann, G., J.R. Jett, M. Luyten-Kellermann, H.L. Moses, and R.S. Fontana (1980) Variations of microsomal mixed function oxidase(s) and human lung cancer. Cancer 45:1438-1442.
44. McLemore, T.L., R.R. Martin, D.L. Busbee, R.C. Richie, R.R. Springer, K.L. Toppell, and E.T. Cantrell (1977) Aryl hydrocarbon hydroxylase activity in pulmonary macrophages and lymphocytes from lung cancer and non-cancer patients. Cancer Res. 37:1175-1181.
45. McLemore, T.L., R.R. Martin, L.R. Pickard, R.R. Springer, N.P. Wray, K.L. Toppell, K.L. Mattox, G.A. Guinn, E.T. Cantrell, and D.L. Busbee (1978) Analysis of aryl hydrocarbon hydroxylase activity in human lung tissue, pulmonary macrophages and blood monocytes. Cancer 41:2292-2300.
46. McLemore, T.L., R.R. Marti, R.K. Springer, N.P. Wray, E.T. Cantrell, and D.L. Busbee (1979) Aryl hydrocarbon hydroxylase activity in pulmonary alveolar macrophages and lymphocytes from lung cancer and non-cancer patients: A correlation with family histories of cancer. Biochem. Genetics 17:795-806.
47. McLemore, T.L., R.R. Martin, N.P. Wray, E.T. Cantrell, and D.L. Busbee (1978) Dissociation between aryl hydrocarbon hydroxylase

activity in cultured pulmonary macrophages and blood lymphocytes from lung cancer patients. Cancer Res. 38:3805-3811.

48. Snodgrass, D.R., T.L. McLemore, M.S. Arnott, A.C. Griffin, N.P. Wray, and D.L. Busbee (1981) Comparison of aryl hydrocarbon hydroxylase activity in pulmonary alveolar macrophages and lymphocytes from asbestos exposed smoke with and without lung cancer. Chest 80:42S-44S.
49. Bradenburg, J.H., and G. Kellermann (1978) Aryl hydrocarbon hydroxylase inducibility in laryngeal carcinoma. Arch. Otolaryngol. 104:151-152.
50. Trell, E., R. Korsgaard, B. Hood, P. Kitzing, G. Norden, and B.G. Simonsson (1976) Aryl hydrocarbon hydroxylase inducibility and laryngeal carcinomas. Lancet 2:140-146.
51. Trell, E., J. Oldbring, R. Korsgaard, and I. Mattiasson (1977) Aryl hydrocarbon hydroxylase inducibility in carcinoma of renal pelvis and ureter. Lancet 2:612-614.
52. Blumer, J.L., R. Dunn, M.D. Esterhay, T.S. Yamashita, and S. Gross (1981) Lymphocyte aryl hydrocarbon hydroxylase in acute leukemia of childhood. Blood 58:1081-1088.
53. Jett, J.R., E.L. Branum, R.S. Fontana, W.F. Taylor, and H.L. Moses (1979) Macromolecular binding of ^{3}H-benzo(a)pyrene metabolites and lymphocyte transformation in patients with lung cancer and in smoking and non-smoking control subjects. Am. Rev. Resp. Disease 120:369-375.
54. Jett, J.R., H.L. Moses, E.L. Branum, E.F. Taylor, and R.S. Fontana (1978) Benzo(a)pyrene metabolism and blast transformation in peripheral blood mononuclear cells from smoking and nonsmoking and lung cancer patients. Cancer 41:191-200.
55. Paigen, B., H.L. Gurtoo, J. Minowada, L. Houten, R. Vincent, K. Paigen, N.B. Parker, E. Ward, and N.T. Hayner (1977) Questionable relation of aryl hydrocarbon hydroxylase to lung cancer risk. New Eng. J. Med. 297:346-350.
56. Tschantz, C., C.E. Higniti, D.H. Huffman, and D.L. Azarnoff (1977) Metabolic disposition of antipyrine in patients with lung cancer. Cancer Res. 37:3880-3886.
57. Ward, E., B. Paigen, K. Steenland, R. Vincent, J. Minowada, H.L. Gurtoo, P. Sartori, and M.B. Havens (1978) Aryl hydrocarbon hydroxylase in persons with lung or laryngeal cancer. Int. J. Cancer 22:384-389.
58. Kouri, R.E., C.E. McKinney, D.J. Slomiany, D.R. Snodgrass, N.P. Wray, and T.L. McLemore (1982) Positive correlation between high aryl hydrocarbon hydroxylase activity and primary lung cancer by analysis in cryopreserved lymphocytes. Cancer Res. 42:5030-5037.
59. Suolinna, E.M., E. Vanttinen, and A. Aitio (1982) Induction of aryl hydrocarbon hydroxylase activity in cultured human lymphocytes. Toxicol. 24:73-84.

GAMMA-GLUTAMYL TRANSPEPTIDASE EXPRESSION IN NEOPLASIA AND DEVELOPMENT

William L. Richards

McArdle Laboratory
University of Wisconsin
Madison, Wisconsin

INTRODUCTION

Since the discovery (34) that gamma-glutamyl transpeptidase (GGT) can serve as a positive histochemical marker of putative preneoplastic hepatocyte foci and of hepatocellular carcinomas in rats treated with aflatoxin B_1, this enzyme has assumed greater and greater importance as a marker of neoplasia in liver and in several other tissues. The present paper reviews some of the proposed functions of GGT, recent findings on its regulation, characteristics of its expression during carcinogenesis, and recent evidence implicating GGT as a marker of cell differentiation or aging in normal tissues.

Proposed Functions of GGT

GGT, an enzyme that is predominantly membrane-bound in both normal and neoplastic tissues (47,48,62), can catalyze the hydrolysis of glutathione (GSH), S-substituted glutathione derivatives, or other γ-glutamyl compounds (47), and can catalyze the transfer of the γ-glutamyl group from these compounds to neutral amino acids or small peptide acceptors in a transpeptidation reaction (47,48,57, 58). Some of the GGT-catalyzed reactions are shown in Figure 1. Meister (45-47) has proposed that GGT can participate in a 6-enzyme γ-glutamyl cycle (solid lines, Figure 2) that may function in transporting amino acids or peptides across the plasma membrane. A deficiency of GGT (28,47,66,86) or inhibition of this enzyme (26-28) leads to massive urinary excretion of GSH, γ-glutamylcysteine, and cysteine indicating that metabolism and/or transport of these sulfur-containing compounds is a major physiological function of GGT (47). Although amino acids can act as acceptors of the γ-glutamyl group, small peptides (especially glycylglycine) are often much

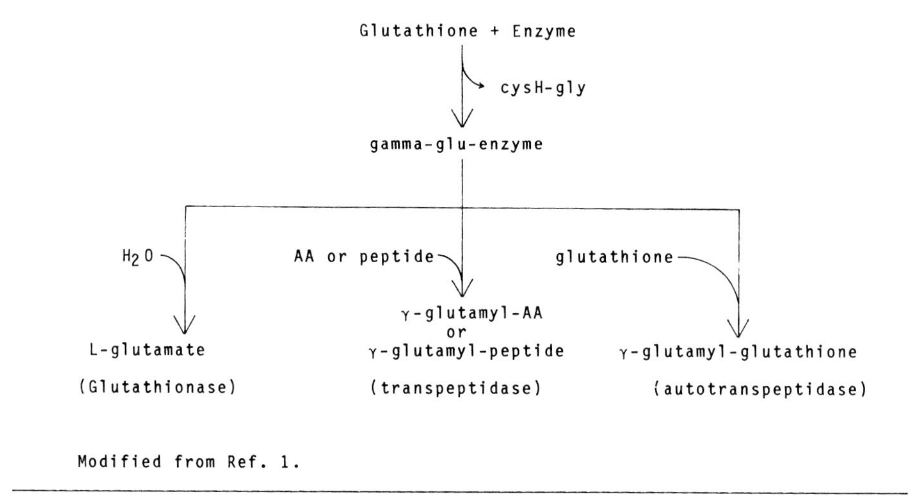

Glutamine

H_2O

(glutaminase)

glutamate + NH_3

Refs. 8 and 78

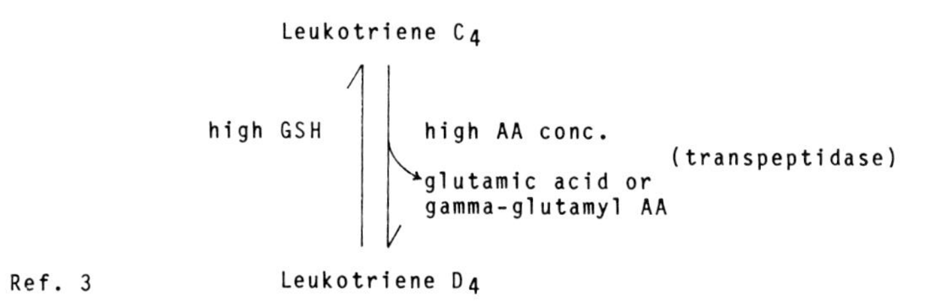

Fig. 1. Reactions catalyzed by GGT.

better acceptors (58,77) and it is possible that the physiologically relevant acceptors in some tissues have not yet been discovered (58). Since the γ-glutamyl cycle is energetically expensive, requiring 3 ATP molecules per turn, Prusiner et al. (57) have proposed an alternative substrate-product reutilization cycle that provides for amino acid and peptide transport without utilizing ATP. Recently, other authors (13,46,84) proposed that GGT and other enzymes of the γ-glutamyl cycle may facilitate the conversion of electrophiles (E^+) to excretable mercapturic acids as shown in Fig. 2 (dashed lines, slanted lettering). Table 1 lists some of the proposed functions of GGT. This diversity of proposed functions suggests

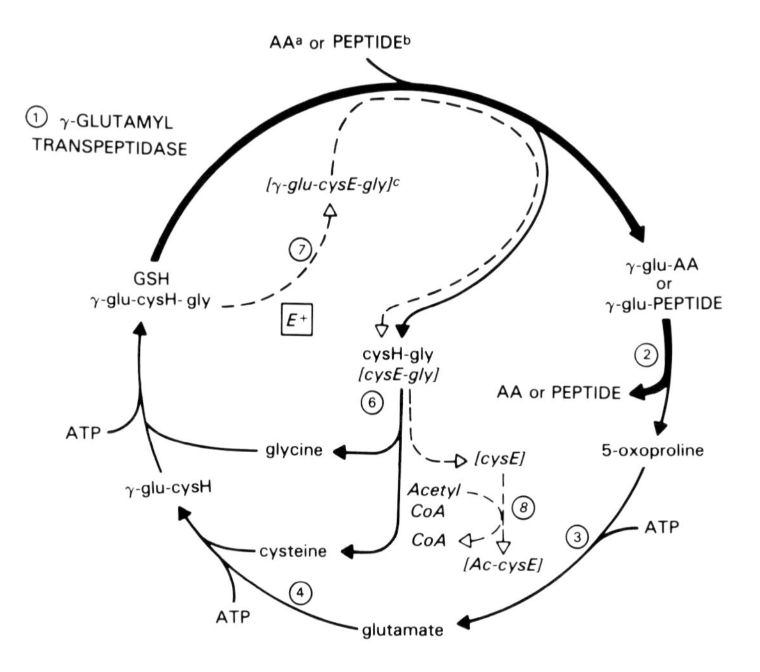

Fig. 2. The γ-glutamyl cycle. [a]Refs. 45 and 53, [b]Ref. 57, [c]Adapted from ref. 84. This cycle is redrawn from ref. 45, with permission.

that GGT may assume different roles in different tissue micro-environments.

Regulation of GGT Activity

Much of the work on regulation of tissue levels of GGT has been done in rat liver. A recent review by Edwards and Lucas (13) cites studies in which liver GGT activity was elevated by administration of phenobarbital (PB), aminopyrine, phenytoin, hexachlorobenzene, or ethanol, by administration of various liver carcinogens, or by portacaval anastomosis. Liver GGT activity can also be increased by feeding of the putative noncarcinogen α-naphthylisothiocyanate (63). Elevation of GGT activity in isolated rat hepatocytes was observed when the cells were exposed to PB *in vitro* or were placed into culture after treatment with a carcinogen *in vivo* (13).

Table 1. Some proposed functions of GGT.

1. Amino acid transport (47)
2. Peptide transport (57)
3. Glutaminase during metabolic acidosis (8,78)
   - pH 6-7 optimum
   - maleate-stimulated
   - P-independent
4. First step in cysteine release from GSH in fetal liver (81)
5. Initiation of GSH oxidation (29,79)
6. Interconversion (3,54) of leukotriene C_4 and D_4, potent smooth muscle contractors and excitors of brain neurons (44)
7. Enhanced resistance to the cytotoxicity of carcinogens and other toxins (13,84)

Two important recent studies (2,12) demonstrate the feasibility of utilizing monolayer cultures of rat hepatocytes to investigate physiologic mechanisms of GGT regulation: (a) Following the demonstration that GGT activity increased gradually with time when rat hepatocytes were cultured on collagen gel/nylon meshes (69), Edwards (12) showed that GGT activity was further increased by addition of dexamethasone to the culture medium (Figure 3). This *in vitro* result is analogous to the results of *in vivo* experiments involving adrenalectomy followed by glucocorticoid replacement therapy (5), and other studies (49). Paradoxically, other authors have reported that dexamethasone suppresses the increase of GGT activity in cultured hepatocytes (7) and it will be interesting to determine whether rat strain, culture medium composition, or cell substratum accounts for this difference in response. Edwards (12) has already demonstrated that the induction of GGT activity by dexamethasone is modulated by pH, glucose concentration, and other hormones. Another study (71) suggests that cortisol may have a biphasic effect on the activity of membrane-bound GGT. Interestingly, the activity of GGT in growth-arrested rat hepatoma cell lines was not altered by any of several hormones, including dexamethasone (62); (b) Althaus et al. (2) showed that $[\text{ADP-ribose}]_n$ metabolism was involved in regulating the expression of GGT activity in primary cultures of adult rat hepatocytes. Inhibitors of ADP-ribosyltransferase (nicotinamide and 3-aminobenzamide) suppressed the expression of GGT activity (Figure 4). Suppression was reversed upon withdrawal of nicotinamide from the culture medium. Moreover, a non-inhibitory analog of one of the ADP-ribosyltransferase inhibitors (3-aminobenzoic acid) had little effect on expression of GGT activity.

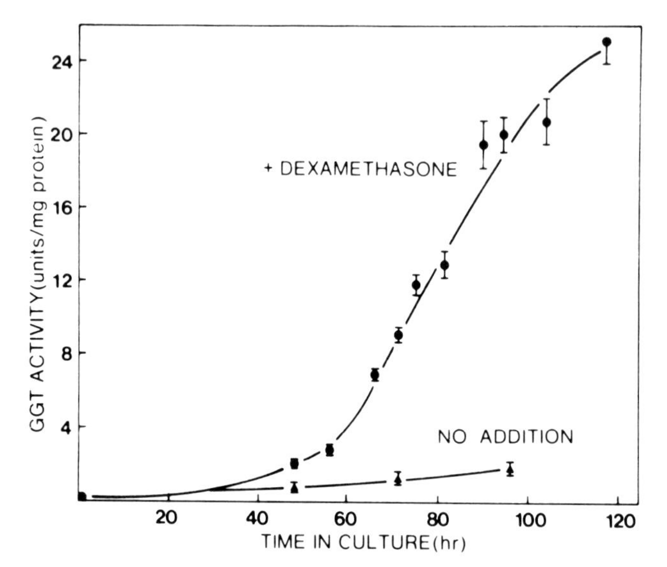

Fig. 3. Induction of GGT activity in adult male Wistar rat hepatocyte monolayer cultures by dexamethasone, 3 μM. The medium was changed daily. Reproduced from ref. 12, with permission.

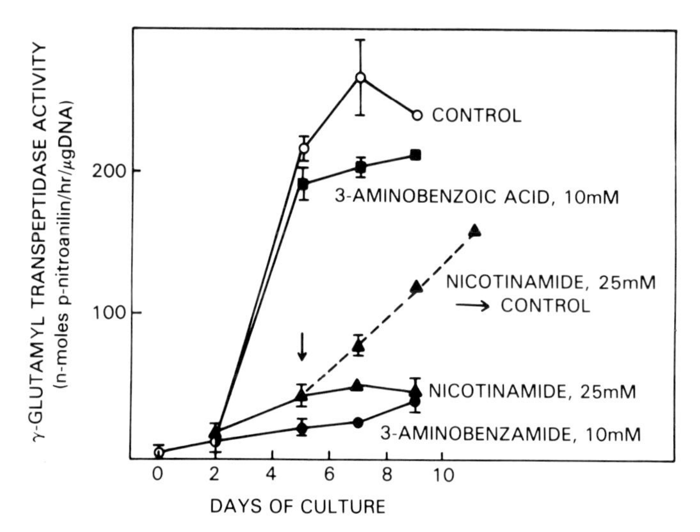

Fig. 4. Effect of nicotinamide, 3-aminobenzamide, and 3-aminobenzoic acid on the expression of GGT activity by adult male Holtzman rat hepatocyte monolayers on collagen gel-nylon meshes. Reprinted by permission from ref. 2. Copyright (C) 1982, Macmillan Journals Limited.

Table 2. Neoplasms demonstrated to have elevated GGT activity.

Mouse skin squamous cell carcinoma (11) [H][a]
Rat and human colon carcinoma (18,50) [B,H]
Rat, human, and mouse mammary carcinomas (9,42,64) [B,H]
Human myeloma>>B cells>T cells (51) [B] (lymphoid cells, mesenchymal origin)
Human lung cancers (30) [B,H]
Rat hyperplastic foci and hepatocellular carcinomas (15,34) [B,H]
Human hepatocellular carcinomas and cholangiocarcinoma (21,62,82) [H]
Endometrium of arachis oil-induced deciduoma, rat uterus (76) [B]
Human epithelial tumors (19) (larynx, urinary bladder, esophagus, breast, colon, prostate, tongue) [H]
Intraepithelial clones in 7,12-DMBA-treated hamster buccal pouch (70) [H]
Rat bladder carcinomas (83) [H]

[a] [ ], indicates histochemical [H] or biochemical [B] study

GGT and Neoplasia

Elevated GGT activity has been demonstrated in neoplasms of several tissues of several animal species (Table 2) and has been used as a diagnostic marker for hepatocellular carcinomas (4,21, 31,73) and other tumors (Tab. 2). However, it is not present in all neoplasms among those considered and hence it cannot be assumed to be a universal marker of the transformed state. Its functional role even in normal tissues is controversial (Tab. 1) and its biochemical significance with respect to tumorigenesis is unclear.

Nevertheless, GGT has become one of the most widely used markers in rat hepatocarcinogenesis studies because it allows the histochemical identification of putative preneoplastic hepatocyte populations (23,34,55,59,68) that would otherwise be undetectable in routine histologic sections. When combinations of histochemical markers were used to reveal enzyme-altered foci in rat liver serial sections, the GGT stain alone revealed a high proportion of foci (Table 3). Iron pigment deficiency will sometimes (32) but not always (56) reveal more foci than GGT.

An important study by Pugh and Goldfarb (59) showed that the ^{3}H-TdR labeling index, used as an index of entry of cells into the S-phase of the cell cycle, was significantly increased above background in GGT-positive rat hepatocyte foci (0.5% in background hepatocytes vs. 2.2% in foci marked by GGT alone) and was further elevated as foci phenotypes deviated further from normalcy (Figure 5), i.e., 2.2% in foci marked by GGT alone; 2.9-3.0% in foci marked by GGT and one other enzyme change; and 4.4% in foci that exhibited

Table 3. GGT incidence during rat liver carcinogenesis protocols.

	Putative Preneoplastic Foci	
Markers Studied[a]	GGT-positive foci as % of all foci detected	References
GGT, ATPase, G6Pase	89	(59)
GGT, ATPase, G6Pase, FE	75	(56)
GGT, ATPase, G6Pase, DTD	90-98	(52)
	Hepatocellular Carcinomas	
Markers Studied	% GGT-positive hepatocellular carcinomas	References
GGT, ATPase, G6Pase	79	(22)

[a] GGT, GGT-positive; ATPase, ATPase-deficient; G6Pase, G6Pase-deficient; FE, iron pigment-deficient; DTD, DT diaphorase-positive.

changes in all three marker enzymes. A similar relationship held (22) when rat hepatocellular carcinomas were studied (Figure 6). Individual cells of GGT-positive foci and hepatomas may exhibit a range of GGT activity from very intense staining to no staining at all. The question of whether the labeling of a foci or hepatoma cell by tritiated thymidine is related to its GGT activity was not directly addressed in the studies of Figs. 5 and 6. Preliminary evidence presented below, however, suggests that focus or hepatoma cells having relatively reduced GGT activity may have a higher probability of being labeled.

Several characteristics of GGT expression in liver carcinogenesis are summarized in Table 4. Items 5 to 7 of this table
Table 2. Neoplasms demonstrated to have elevated GGT activity.
provide evidence of a sequential relationship of GGT-positive foci and hyperplastic nodules to hepatocellular carcinomas. Additional evidence of this sequential relationship comes from studies with other markers of enzyme-altered rat liver foci: (a) Scherer and Emmelot (65) provided evidence that ATPase-deficient island cells induced by a subcarcinogenic dose of diethylnitrosamine in rats serve as targets for further carcinogenic action and thus may be considered as precursor cells in the process of liver tumor formation; (b) Kunz et al. (39) showed that the dose-time dependence for the induction of altered foci (ATPase-deficient) matched exactly that found for the induction of liver tumors; (c) At least two studies (59,60) have shown that rapidly proliferating subpopulations of cells histologically compatible with early hepatocellular carcinomas can arise within enzyme-altered (ATPase-deficient and/or G6Pase-deficient) rat liver foci.

Even though substantial evidence suggests a sequential relationship of GGT-positive and other enzyme-altered rat hepatocyte foci to hepatocellular carcinomas, the existence of hepatocellular carcinomas having negligible GGT activity (Tab. 4) indicates that GGT is not obligatory to maintenance of the fully malignant phenotype. Whether GGT activity contributes to malignant progression in liver remains to be elucidated. Two studies (Tab. 4) have shown that the enzyme-altered foci of mouse liver are GGT-negative unless the mice are administered PB, a promoter of hepatic tumorigenesis. It has not been demonstrated, however, that PB induces GGT activity in enzyme-altered foci of rat liver.

GGT In Proliferating, Developing, Differentiating, and Aging Cells and Tissues

Recent reviews (48,80,84) and research investigations (33,49, 61) describe four different types of changes in GGT expression in developing tissues: (a) Disappearance or reduction of GGT activity with age, e.g, mouse, human and rat liver, and rat brain (see references in 61,80,84); (b) Cyclic appearance of GGT, e.g., reoccurrence of high GGT activity in the outer root sheath of the mouse hair follicle (Figure 7C) during the anagen or growth phase of each successive hair growth cycle (11,61); (c) Programmed GGT appearance in specific cell layers, e.g., GGT changes in the stellate reticulum and in ameloblasts (Figure 7A) in the developing mouse tooth (61); (d) Gradient formation, e.g., the GGT activity gradient observed in the developing mouse intervertebral disc (Figure 7B) (61). Intracellular GGT activity gradients have been observed in kidney brush border cells (48), ameloblasts (Fig. 7A), and outer root sheath cells of the mouse hair follicle (Fig. 7C).

One important question, relevant to understanding the role of GGT in both normal and neoplastic tissues, is whether this enzyme is found predominantly in proliferating cells, in nonproliferating cells, or in both cell categories. A study that examined this question directly (61) found that GGT activity was localized only in mitotically quiescent cellular layers or regions of developing mouse tissues regardless of whether these tissues were derived from epithelium (enamel-producing cells, Fig. 7A; hair follicle cells, Fig. 7C) or from mesenchyme (intervertebral disc cells, Fig. 7B). In Fig. 7A, differentiation of cells participating in tooth enamel formation proceeds in a continuum from left to right. Cell proliferation (cells that incorporate tritiated thymidine into their nuclei have black dots over nuclei) ceased during the presecretory stage (stage 1) prior to the appearance of GGT activity in any of the cell layers. Intense GGT activity was associated with tooth maturation, i.e., with the stellate reticulum (S.R.) during secretion of enamel matrix by secretory ameloblasts (S.A.) and with the ameloblasts during matrix deorganification, mineralization, and maturation (stages 4-6). Fig. 7B shows regions of cell proliferation (cross hatching in upper 3 panels) and of GGT activity

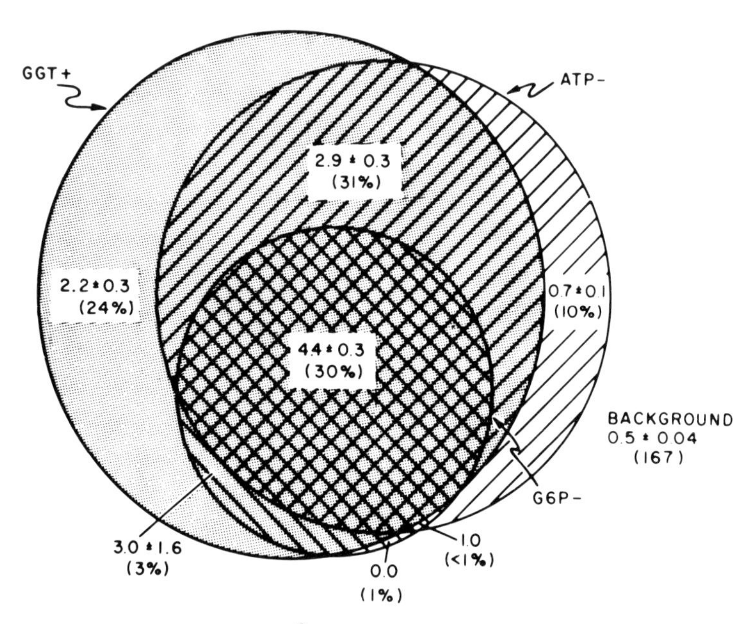

Fig. 5. Labeling indices (^{3}H-thymidine) and enzyme phenotypes of hepatocellular islands of male Buffalo rats. Overlapping circles, percentage distributions (numbers in parentheses) of 195 islands in 5 livers showing 1, 2, or 3 marker changes. The labeling indices (mean ± SE) are indicated for each island phenotype. The labeling index of 0.5 ± 0.04 is indicated for 167 randomly selected background areas. ATP-, ATPase-deficient; G6P-, G6Pase-deficient; GGT+, GGT-positive. Reproduced from ref. 59, with permission.

(shades from light gray to black in lower 3 panels) in the intervertebral discs of the tails of 1-, 10-, and 22-day old mice. The nucleus pulposus (N.P.) at the center of the tail is at the top of each disc diagram. When weak GGT activity (W) first appeared in the disc, adjacent to N.P. in the one-day old mouse, proliferative activity was already confined to the outer two-thirds of the disc and to the connective tissue (C) at the disc periphery. Later, when GGT activity was distributed in a gradient throughout the disc body (D) in the 10-day old, proliferating cells were detected only in the connective tissue outside of the disc. In the 22-day old, GGT activity was associated with the annulus fibrosus or rings of collagen fibers that comprise the supporting framework of the mature disc. Fig. 7C illustrates the location of GGT activity (shades from light gray to black) and proliferative activity (cross hatching) during invagination of the hair follicle from the skin surface and during the stages of initial hair shaft keratinization (Early Anagen) and of maximal hair shaft growth (Mid-Anagen). Following

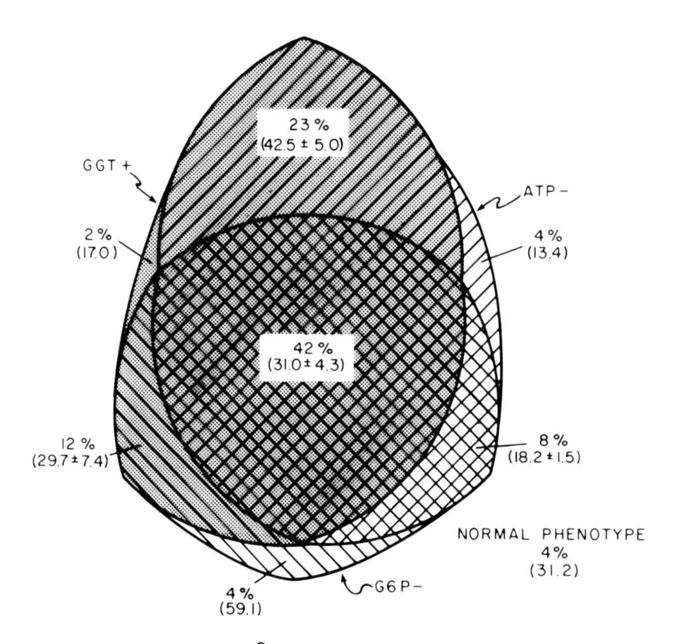

Fig. 6. Labeling indices (^{3}H-thymidine) and enzyme phenotypes of 48 hepatocellular carcinomas of male Sprague-Dawley rats. Overlapping ovals, percentage distribution of carcinomas showing 1, 2, or 3 marker changes. The labeling indices (numbers in parentheses) are indicated for each carcinoma phenotype. Four percent of the carcinomas, found to have normal phenotypes (GGT-, ATP+, G6P+), are indicated outside the Venn diagram. ATP-, ATPase-deficient; G6P-, G6Pase-deficient; GGT+, GGT-positive. Reproduced from ref. 22, with permission.

Mid Anagen, production of the hair shaft ceases and the follicle degenerates to a shortened resting stage configuration, completing one round of a hair growth cycle. A new cycle begins when the hair follicle again begins to elongate and to produce a new hair shaft that pushes out the old or club hair. During invagination, no GGT activity was detected and proliferation was observed throughout the length of the hair follicle. Later, when GGT activity was first detected in a cone adjacent to the region of initial hair shaft keratinization (K, Early-Anagen), proliferation was evident only in a separate region, the lower two-thirds of the hair bulb. During the stage of rapid hair shaft growth (Mid-Anagen), GGT activity was detected in the outer root sheath from the orifice of the sebaceous gland (S) to a point midway up the bulb and mitotic activity was confined to the matrix (M) in the lower half of the bulb.

These data on mouse tissue development show that GGT expression is not necessarily linked to cell proliferation, a conclusion that

Table 4. Characteristics of GGT expression in liver carcinogenesis; possible sequential relationship of GGT-positive foci and hyperplastic nodules to hepatocellular carcinomas.

1. GGT-Positive foci (rat) increased in number and size concomitantly with progressive DNA damage (alkaline elution technique) after partial hepatectomy then DEN (6.6 mg/kg) then > 28 wks .05% PB (72).

2. The ^{3}H-TdR labeling index (LI) was significantly increased in GGT-positive foci (rat) and was further elevated as foci phenotypes deviated further from normalcy (59).

3. The number of GGT-positive foci/cm^2 (rat) increased as a function of DEN dose between 0.3 and 200 mg/kg (20).

4. The number (per cm^3) and size of GGT-positive foci (rat) was proportional to DEN dose between 5 and 50 mg/kg (Richards and Potter, unpublished).

5. Rat liver cells capable of forming GGT-positive foci were transplantable from a donor to a suitably-prepared recipient (40,41).

6. Hepatocytes from AAF-treated rats, when maintained 4 to 12 months in culture, formed proliferative GGT-positive foci that became capable of growth in soft agar and became transplantable (s.c.) to syngeneic neonates, forming hepatocellular carcinomas (36).

7. Persistent hyperplastic nodules (rat) maintained GGT activity and had higher labeling indices than nodules showing a nonuniform loss of GGT activity. A direct transition from benign to malignant cells has been observed in nodules (14).

8. With an initiation/promotion protocol for rat liver tumorigenesis (partial hepatectomy then DEN, 10 mg/kg, then PB), removal of PB from the diet after its administration for 3-4 months resulted in no significant loss in GGT-positive foci/cm^3 (25).

Complications: Some rat and human hepatocellular carcinomas and most mouse hepatomas are GGT-negative (17,22,24, 37,43,59,85).

PB induces GGT activity in mouse liver neoplastic lesions (37,85).

is supported and complemented by evidence of an association of CCT activity with cell differentiation, cell aging, cell senescence, or cell-cell interaction (Table 5). In these investigations, progressive elevation of GGT activity was shown to accompany the differentiation of mouse preadipocytes to adipocytes and the aging of kidney epithelial cells, lung fibroblasts, and periportal rat hepatocytes. Cell-cell interaction also stimulated an increase in the GGT activity of endothelial cells that were cocultured with glial cells. Of special importance is the finding that preadipocytes must arrest their growth at a distinct state in the G_1 phase of the cell cycle, G_D, before differentiation proceeds (67), implying that arrest at this state may also be required for increased GGT expression (Tab. 5) (75).

Table 6 shows that GGT activity remained at control levels during the induction of skin hyperplasia by TPA or of liver hyperplasia following partial hepatectomy. This demonstrates that increases in GGT activity do not occur merely as a function of tissue hyperplasia.

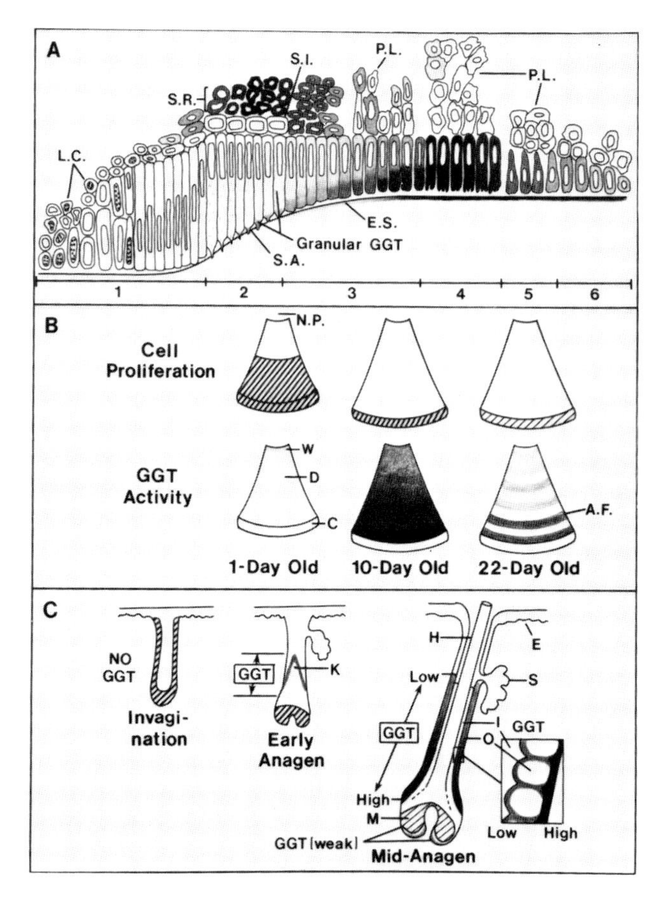

Fig. 7. A, diagrammatic representation of GGT expression and cell proliferation during amelogenesis. L.C., [^{3}H]-thymidine-labeled cell nuclei (black dots over nuclei); S.R., stellate reticulum; S.I., stratum intermedium; P.L., papillary layer; S.A., secretory ameloblast; E.S., enamel surface. Stages of amelogenesis: 1, presecretory; 2, secretory; 3, transition from secretory to early maturation; 4, early to late maturation; 5, late maturation to post-maturation; 6, postmaturation. Shades from light gray to black indicate the intensity of GGT activity.

B, diagrammatic representation of GGT expression and cell proliferation in the developing mouse intervertebral disc. N.P., nucleus pulposus at the center of each disc; D, disc body (fibrocartilage cells); C, connective tissue at disc periphery; W, weak GGT activity first appearing in fibrocartilage cells adjacent to nucleus pulposus. In the 22-day-old mouse, disc GGT was closely associated with the anulus fibrosus (A.F.) or rings of collagen fibers. The density of cross-hatching in the upper 3 drawings indi-

Thus, a growing body of evidence supports the association of GGT with cell differentiation, cell aging, and/or reduced cell proliferation (Tab. 5, Fig. 7) rather than with elevated proliferation. Possible exceptions to this can be pointed out (Table 5), but further investigation of these cells and tissues will be required to determine whether cell proliferation and GGT expression fit a pattern analogous to that of Fig. 7. In this regard, the following data suggest that GGT activity may be associated with regions of reduced cell proliferation in putative preneoplastic rat liver foci and in hepatomas: (a) In foci detected by the GGT histochemical stain, the cells that have incorporated tritiated thymidine into DNA (i.e., are in the S-phase of the cell cycle) often have negligible GGT activity or activity confined to the cell periphery (W.L. Richards and V.R. Potter, unpublished); (b) In a hepatocellular carcinoma that was patchy in GGT staining intensity, areas low in GGT activity had a higher tritiated thymidine labeling index (23 - 36% labeled cells) than areas high in GGT activity (6.5 - 9%) (W.L. Richards and V.R. Potter, unpublished); (c) In the transition of hepatoma cell cultures from log phase (rapid growth rate) to late log or senescence phase (reduced growth rate), GGT activity increased or was expressed only transiently during the late log phase (Table 7) (62). The preliminary nature of (a) and (b) must be stressed. Also Enomoto and Farber (14) found no apparent differences (no data were given) between the labeling indices of GGT-positive and GGT-negative regions of nonpersistent (remodeling) hyperplastic rat liver nodules. Clearly, further experiments of a more quantitative nature will be necessary to define the relationship between GGT expression, tissue maturation, and cell proliferation in neoplasms. Current information on this relationship in normal and neoplastic tissues is shown in Table 8.

While the present review suggests that considerable progress has been made in describing the chemical reactions of GGT and its biological behavior in both normal and neoplastic tissues, much effort will be required to clarify its physiological role in various tissues, to determine the mechanisms that control its expression,

---

cates the frequency of [^{3}H]thymidine-labeled nuclei. In the lower 3 drawings, shades from light gray to black indicate the intensity of GGT activity.

C, diagrammatic representation of GGT expression and cell proliferation in the developing mouse hair follicle. The density of cross-hatching indicates the frequency of cells in metaphase arrest following colchicine administration. Shades from light gray to black indicate the intensity of GGT activity. K, the earliest keratinized material; E, epidermis; H, mature hair shaft; S, sebaceous gland; M, matrix; I, inner root sheath; O, outer root sheath. Reproduced from ref. 61, with permission.

Table 5. Association of elevated GGT activity with cell differentiation, cell aging, cell senescence, or cell-cell interaction.

1. Differentiation of particular cell layers in the developing mouse tooth, hair follicle, and intervertebral disc (61).	Histochemical GGT increase (++ to ++++)
2. Differentiation of mouse preadipocytes to adipocytes <u>in vitro</u> (75).	4-fold GGT increase
3. Aging (?) of cebus monkey kidney epithelial cells (passage 8 vs. 3) (74).	18-fold GGT increase
4. Aging of human embryonic lung diploid fibroblasts (population doubling level 49 vs 22) (75).	8-fold GGT increase
5. Hepatocytes of aging rats (> 30 weeks old) (35).	Histochemical GGT increase in PP cells; Donryu>> Sprague-Dawley; 37-fold GGT increase in Donryu (90 vs. 15 weeks old)
6. Mouse brain capillary endothelial cells cocultured with lethally irradiated rat glioma cells (10).	Histochemical GGT increase (+++)

<u>Possible Exceptions</u>

1. Fetal tissues having high GGT activity and, presumably, high proliferative activity (i.e. human and rat liver; rat intestine and brain) (80,84).
2. Increasing GGT associated with fetalization (α-fetoprotein [AFP], fetal FDP [fructose 1,6-diphosphate] aldolase) of adult rat hepatocytes on collagen gel/nylon meshes (69); note, however, that dexamethasone induces further GGT increases (12).
3. Elevated GGT activity in many neoplasms (Table 2).

Table 6. GGT during induced hyperplasia.

System Studied	Effect on GGT Activity
Hyperplasia of interfollicular epidermis following treatment of mouse skin with TPA (12-0-tetradecanoyl-phorbol-13-acetate) (11).	None
Compensatory liver hyperplasia following partial hepatectomy in rats (6,16,38).	None

Table 7. Transition of hepatoma cell cultures from log phase to late log or senescence phase[a]

Cell Line	Increase in GGT Activity
HTC	7-fold
7777	6-fold
8994	75-fold
H4-II-E-C3	Transient histochemical activity in small cell clusters at late log

[a] Ref. 62

Table 8. Relationship between GGT expression, tissue maturation, and cell proliferation in normal and neoplastic cells and tissues.

Tissue Type	GGT Expression Related to Tissue Maturation?	GGT Found in Regions of Reduced Cell Proliferation
Normal Tissues	Yes (Fig. 7 and Table 5), with possible exceptions (Table 5)	Yes (Fig. 7 and Table 5), with possible exceptions (Table 5)
Neoplasms	?? Aberrant expression (11,62)	Preliminary evidence (Table 7 and text)[a]

[a] Note, however, that Enomoto and Farber (14) found no apparent differences (data not given) between the labeling indices of GGT-positive and GGT-negative regions of nonpersistent (remodeling) hyperplastic rat liver nodules.

and to determine whether the activity of this enzyme contributes in any way to establishing or maintaining the malignant state.

SUMMARY

GGT activity can be elevated in a variety of neoplasms and a cursory examination would suggest that GGT is a marker of cell proliferation. On the contrary, however, recent investigations on a number of different cells and tissues (differentiating mouse tooth, hair follicle, intervertebral disc, preadipocytes; aging human fibroblasts and cebus monkey kidney epithelial cells; Reuber hepatoma cells *in vitro*; compensatory liver hyperplasia following partial hepatectomy in rats; chemically-induced skin hyperplasia in mice; hepatocytes of senescent rat liver; mouse brain capillary endothelial cells cocultivated with rat glioma cells) support the conclusion that GGT is often a marker of cell differentiation, cell aging, critical cell-cell interactions, and/or reduced cell proliferation rather than of elevated cell proliferation. Preliminary studies indicate that GGT may also be a marker of reduced cell proliferation within rat liver carcinomas and preneoplastic foci but these studies require quantitative confirmation.

ACKNOWLEDGEMENTS

The author wishes to thank the several authors who granted permission to redraw or reproduce material from their publications and to thank Frances M. Williams for typing the manuscript.

REFERENCES

1. Allison, R.D., and A. Meister (1981) Evidence that transpeptidation is a significant function of γ-glutamyl transpeptidase. J. Biol. Chem. 256:2988-2992.
2. Althaus, F.R., S.D. Lawrence, Y.-Z. He, G.L. Sattler, Y. Tsukada, and H.C. Pitot (1982) Consequences of altered $[\text{ADP-Ribose}]_n$-metabolism on the expression of fetal functions by adult hepatocytes. Nature (in press).
3. Anderson, M.E., R.D. Allison, and A. Meister (1982) Interconversion of leukotrienes catalyzed by purified γ-glutamyl transpeptidase: Concomitant formation of leukotriene D4 and γ-glutamyl amino acids. Proc. Natl. Acad. Sci., USA 79:1088-1091.
4. Aronsen, K.F., B. Nosslin, and B. Pihl (1970) The value of γ-glutamyl transpeptidase as a screen test for liver tumour. Acta Chir. Scand. 136:17-22.
5. Billon, M.C., G. Dupre, and J. Hanoune (1980) *In vivo* modulation of rat hepatic gamma-glutamyl transferase activity by glucocorticoids. Mol. Cell. Endocrinol. 18:99-108.
6. Cameron, R., J. Kellen, A. Kolin, A. Malkin, and E. Farber (1978) γ-Glutamyltransferase in putative premalignant liver cell populations during hepatocarcinogenesis. Cancer Res. 38:823-829.
7. Coloma, J., M.J. Gomez-Lechon, M.D. Garcia, J.E. Feliu, and J. Baguena (1981) Effect of glucocorticoids on the appearance of gamma-glutamyl transpeptidase activity in primary cultures of adult rat hepatocytes. Experientia 37:941-943.
8. Curthoys, N.P., and T. Kuhlenschmidt (1975) Phosphate-independent glutaminase from rat kidney. Partial purification and identity with γ-glutamyltranspeptidase. J. Biol. Chem. 250:2099-2105.
9. Dawson, J., G.D. Smith, J. Boak, and T.J. Peters (1979) γ-Glutamyltransferase in human and mouse breast tumors. Clin. Chim. Acta 96:37-42.

10. DeBault, L.E., and P.A. Cancilla (1980) γ-Glutamyl transpeptidase in isolated brain endothelial cells: Induction by glial cells in vitro. Science 207:653-655.
11. DeYoung, L.M., W.L. Richards, W. Bonzelet, L.L. Tsai, and R.K. Boutwell (1978) Localization and significance of γ-glutamyl-transpeptidase in normal and neoplastic mouse skin. Cancer Res. 38:3697-3701.
12. Edwards, A.M. (1982) Regulation of γ-glutamyltranspeptidase in rat hepatocyte monolayer cultures. Cancer Res. 42:1107-1115.
13. Edwards, A.M., and C.M. Lucas (1982) γ-Glutamyltranspeptidase as a preneoplastic marker in hepatocarcinogenesis: Expression in hepatocytes isolated from normal and carcinogen-treated rats. In In Vitro Epithelial Cell Differentiation and Neoplasia, Cancer Forum 6, pp. 173-190.
14. Enomoto, K., and E. Farber (1982) Kinetics of phenotypic maturation of remodeling of hyperplastic nodules during liver carcinogenesis. Cancer Res. 42:2330-2335.
15. Fiala, S., and A.E. Fiala (1969) Activation of glutathione in rat liver during carcinogenesis. Naturwissensch. 11:565.
16. Fiala, S., and E.S. Fiala (1973) Activation by chemical carcinogens of γ-glutamyl transpeptidase in rat and mouse liver. J. Natl. Cancer Inst. 51:151-158.
17. Fiala, S., A.E. Fiala, and B. Dixon (1972) γ-Glutamyl transpeptidase in transplantable, chemically-induced rat hepatomas and "spontaneous" mouse hepatomas. J. Natl. Cancer Inst. 48:1393-1401.
18. Fiala, S., A.E. Fiala, R.W. Keller, and E.S. Fiala (1977) Gamma-glutamyl transpeptidase in colon cancer induced by 1,2-dimethylhydrazine. Arch. Geschwulstforsch. 47:117-122.
19. Fiala, S., E.C. Trout, C.A. Teague, and E.S. Fiala (1980) γ-Glutamyltransferase, a common marker of human epithelial tumors? Cancer Detect. Preven. 3:471-485.
20. Ford, J.O., and M.A. Pereira (1980) Short-term in vivo initiation/promotion bioassay for hepatocarcinogens. J. Env. Pathol. Toxicol. 4:39-46.
21. Gerber, M.A., and S.N. Thung (1980) Enzyme patterns in human hepatocellular carcinoma. Am. J. Pathol. 98:395-400.
22. Goldfarb, S., and T.D. Pugh (1981) Enzyme histochemical phenotypes in primary hepatocellular carcinomas. Cancer Res. 41:2092-2095.
23. Goldfarb, S., and T.D. Pugh (1982) The origin and significance of hyperplastic hepatocellular islands and nodules in hepatic carcinogenesis. J. Am. Coll. Toxicol. 1:119-144.
24. Goldfarb, S., T.D. Pugh, and D.J. Cripps (1980) Increased alkaline phosphatase activity -- A positive histochemical marker for griseofulvin-induced mouse hepatocellular nodules. J. Natl. Cancer Inst. 64:1427-1433.
25. Goldsworthy, T.L., and H.A. Campbell (1982) Quantitative evaluation on the stability of enzyme-altered foci (EAF) following phenobarbital (PB) and diethylnitrosamine (DEN) administration. Proc. Am. Assoc. Cancer Res. 23:98.

26. Griffith, O.W., R.J. Bridges, and A. Meister (1978) Evidence that the γ-glutamyl cycle functions *in vivo* using intracellular glutathione: Effects of amino acids and selective inhibition of enzymes. *Proc. Natl. Acad. Sci.*, USA 75:5405-5408.
27. Griffith, O.W., and A. Meister (1979) Translocation of intracellular glutathione to membrane-bound γ-glutamyl transpeptidase as a discrete step in the γ-glutamyl cycle: Glutathionuria after inhibition of transpeptidase. *Proc. Natl. Acad. Sci.*, USA 76:268-272.
28. Griffith, O.W., and A. Meister (1980) Excretion of cysteine and γ-glutamylcysteine moieties in human and experimental animal γ-glutamyl transpeptidase deficiency. *Proc. Natl. Acad. Sci.*, USA 77:3384-3387.
29. Griffith, O.W., and S.S. Tate (1980) The apparent glutathione oxidase activity of γ-glutamyl transpeptidase. *J. Biol. Chem.* 255:5011-5014.
30. Groscurth, P., N. Fleming, and G.S. Kistler (1977) The activity and distribution of gamma-glutamyl transpeptidase (γ-GT) in human lung cancers serially transplanted in nude mice. *Histochemistry* 53:135-142.
31. Hattori, N., N. Sawabu, M. Nakagen, and M. Ishii (1978) Novel gamma-glutamyl transpeptidase isoenzyme occurring in sera of patients with hepatoma. *Scand. J. Immunol. 8* Suppl. 8:507-510.
32. Hirota, N., and G.M. Williams (1979) The sensitivity and heterogeneity of histochemical markers for altered foci involved in liver carcinogenesis. *Am. J. Pathol.* 95:317-328.
33. Igarashi, T., T. Satoh, K. Ueno, and H. Kitagawa (1981) Changes of gamma-glutamyl transpeptidase activity in the rat during development and comparison of the fetal liver, placental, and adult liver enzymes. *Life Sci.* 29:483-491.
34. Kalengayi, M.M.R., G. Ronchi, and V.J. Desmet (1975) Histochemistry of gamma-glutamyl transpeptidase in rat liver during aflatoxin B_1-induced carcinogenesis. *J. Natl. Cancer Inst.* 55:579-588.
35. Kitagawa, T., F. Imai, and K. Sato (1980) Re-elevation of γ-glutamyl transpeptidase activity in periportal hepatocytes of rats with age. *Gann* 71:362-366.
36. Kitagawa, T., R. Watanabe, T. Kayano, and H. Sugano (1980) *In vitro* carcinogenesis of hepatocytes obtained from acetylaminofluorene-treated rat liver and promotion of their growth by phenobarbital. *Gann* 71:747-754.
37. Kitagawa, T., R. Watanabe, and H. Sugano (1980) Induction of γ-glutamyl transpeptidase activity by dietary phenobarbital in "spontaneous" hepatic tumors of C3H mice. *Gann* 71:536-542.
38. Köttgen, E., W. Reutter, and W. Gerok (1978) Induction and superinduction of sialylation of membrane-bound γ-glutamyltransferase during liver regeneration. *Eur. J. Biochem.* 82:279-284.
39. Kunz, W., K.E. Appel, R. Rickart, M. Schwarz, and G. Stöckle (1978) Enhancement and inhibition of carcinogenic effectiveness of nitrosamines. In *Primary Liver Tumors*, H. Remmer, H.M.

Bolt, P. Bannasch, and H. Popper, eds. Baltimore: University Park Press, pp. 261-283.

40. Laishes, B.A., and E. Farber (1978) Transfer of viable putative preneoplastic hepatocytes to the livers of syngeneic host rats. J. Natl. Cancer Inst. 61:507-512.
41. Laishes, B.A., L. Fink, and B.I. Carr (1980) A liver colony assay for a new hepatocyte phenotype as a step towards purifying new cellular phenotypes that arise during hepatocarcinogenesis. Ann. N.Y. Acad. Sci. 349:373-382.
42. Levine, S.E., G. Georgiade, K.S. McCarthy, Jr., and K.S. McCarthy, Sr. (1980) Gamma-glutamyl transpeptidase in lesions of the human mammary gland. Proc. Am. Assoc. Cancer Res. 21:13.
43. Lipsky, M.M., D.E. Hinton, J.E. Klaunig, P.J. Goldblatt, and B.F. Trump (1980) Gamma glutamyl transpeptidase in safrole-induced, presumptive premalignant mouse hepatocytes. Carcinogenesis 1:151-156.
44. Marx, J.L. (1982) The leukotrienes in allergy and inflammation. Science 215:1380-1383.
45. Meister, A. (1973) On the enzymology of amino acid transport. Science 180:33-39.
46. Meister, A. (1981) Metabolism and functions of glutathione. TIBS, September, pp. 231-234.
47. Meister, A. (1981) On the cycles of glutathione metabolism and transport. Curr. Top. Cell. Regul. 18:21-58.
48. Meister, A., S.S. Tate, and L.L. Ross (1975) Membrane-bound γ-glutamyltranspeptidase. In The Enzymes of Biological Membranes, Vol. 3, A. Martonosi, ed., Plenum Press, New York, pp. 315-347.
49. M'enard, D., C. Malo, and R. Calvert (1981) Development of gamma-glutamyltranspeptidase activity in mouse small intestine. Biol. Neonate 40:70-77.
50. Munjal, D.D. (1980) Concurrent measurements of carcinoembryonic antigen, glucose phosphate isomerase, γ-glutamyltransferase, and lactate dehydrogenase in malignant, normal adult, and fetal colon tissues. Clin. Chem. 26:1809-1812.
51. Novogrodsky, A., S.S. Tate, and A. Meister (1976) Gamma-glutamyl transpeptidase, a lymphoid cell-surface marker: Relationship to blastogenesis, differentiation, and neoplasia. Proc. Natl. Acad. Sci., USA 73:2414-2418.
52. Ogawa, K., D.B. Solt, and E. Farber (1980) Phenotypic diversity as an early property of putative preneoplastic hepatocyte populations in liver carcinogenesis. Cancer Res. 40:725-733.
53. Orlowski, M., and A. Meister (1970) The γ-glutamyl cycle: A possible transport system for amino acids. Proc. Natl. Acad. Sci., USA 67:1248-1255.
54. Orning, L., and S. Hammarström (1982) Kinetics of the conversion of leukotriene C by γ-glutamyl transpeptidase. Biochem. Biophys. Res. Commun. 106:1304-1309.
55. Peraino, C., W.L. Richards, and F.J. Stevens. Multistage hepatocarcinogenesis. In Mechanisms of Tumor Promotion. Tumor

Promotion and Carcinogenesis in Internal Organs, T. Slaga, ed. CRC Uniscience, Cleveland (in press).

56. Pitot, H.C. (1980) Characteristics of stages of hepatocarcinogenesis. In Carcinogenesis: Fundamental Mechanisms and Environmental Effects, B. Pullman, P.O.P. T'so, and H. Gelboin, eds. D. Reidel: Hingham, p.p. 219-233.
57. Prusiner, S., C.W. Doak, and G. Kirk (1976) A novel mechanism for group translocation: Substrate-product reutiliation by γ-glutamyl transpeptidase in peptide and amino acid transport. J. Cell Physiol. 89:853-864.
58. Prusiner, P.E., and S.B. Prusiner (1978) Partial purification and kinetics of γ-glutamyl transpeptidase from bovine choroid plexus. J. Neurochem. 30:1253-1259.
59. Pugh, T.D., and S. Goldfarb (1978) Quantitative histochemical and autoradiographic studies of hepatocarcinogenesis in rats fed 2-acetylaminofluorene followed by phenobarbital. Cancer Res. 38:4450-4457.
60. Rabes, H.M., P. Scholze, and B. Jantsch (1972) Growth kinetics of diethylnitrosamine-induced, enzyme-deficient "preneoplastic" liver cell populations in vivo and in vitro. Cancer Res. 32:2577-2586.
61. Richards, W.L., and E.G. Astrup (1982) Expression of γ-glutamyl transpeptidase activity in the developing mouse tooth, intervertebral disc, and hair follicle. Cancer Res. 42:4143-4152.
62. Richards, W.L., Y. Tsukada, and V.R. Potter (1982) Phenotypic diversity of gamma-glutamyl transpeptidase activity and protein secretion in several hepatoma cell lines. Cancer Res. 42:1374-1383.
63. Richards, W.L., Y. Tsukada, and V.R. Potter (1982) γ-Glutamyl transpeptidase and α-fetoprotein expression during α-naphthylisothiocyanate-induced hepatotoxicity in rats. Cancer Res. 42:5133-5138.
64. Sachdev, G.P., G. Wen, B. Martin, G.S. Kishore, and O.F. Fox (1980) Effects of dietary fat and alpha-tocopherol on gamma-glutamyl-transpeptidase activity of 7,12 dimethylbenz-(alpha)anthracene-induced mammary gland adenocarcinomas. Cancer Biochem. Biophys. 5:15-23.
65. Scherer, E., and P. Emmelot (1975) Foci of altered liver cells induced by a single dose of diethylnitrosamine and partial hepatectomy. Their contribution to hepatocarcinogenesis in the rat. Europ. J. Cancer 11:145-154.
66. Schulman, J.D., S.I. Goodman, J.W. Mace, A.D. Patrick, F. Tietze, and E.J. Butler (1975) Glutathionuria: Inborn error of metabolism due to tissue deficiency of gamma-glutamyl transpeptidase. Biochem. Biophys. Res. Commun. 65:68-74.
67. Scott, R.E., D.L. Florine, J.J. Wille, Jr., and K. Yun (1982) Coupling of growth arrest and differentiation at a distinct state in the G_1 phase of the cell cycle: G_D. Proc. Natl. Acad. Sci., USA 79:845-849.
68. Sirica, A.E., and H.C. Pitot (1982) Phenotypic markers of hepatic "pre-neoplasia" and neoplasia in the rat. In

Cancer-Cell Organelles, E. Reid, G.M.W. Cook, and D.J. Morre, eds. John Wiley and Sons, New York, pp. 131-143.

69. Sirica, A.E., W.L. Richards, Y. Tsukada, C.A. Sattler, and H.C. Pitot (1979) Fetal phenotypic expression by adult rat hepatocytes on collagen gel/nylon meshes. Proc. Natl. Acad. Sci., USA 76:283-287.
70. Solt, D.B., and G. Shklar (1982) Rapid induction of γ-glutamyl transpeptidase-rich intraepithelial clones in 7,12-dimethylbenz(a)anthracene-treated hamster buccal pouch. Cancer Res. 42:285-291.
71. Stastny, F., and V. Lisy (1981) Cortisol regulation of gamma-glutamyl transpeptidase in liver, choroid plexus, blood plasma and cerebrospinal fluid of developing chick embryo. Dev. Neurosci. 4:408-415.
72. Stout, D.L., and F.F. Becker (1982) Occurrence of progressive DNA damage coincident with the appearance of foci of altered hepatocytes. Carcinogenesis 3:599-602.
73. Szczeklik, E., M. Orlowski, and A. Szewczuk (1961) Serum γ-glutamyl transpeptidase activity in liver disease. Gastroenterology 41:353-359.
74. Takahashi, S., S. Seifter, and L. Rifas (1978) γ-Glutamyltransferase in human diploid fibroblasts and other mammalian cells. In Vitro 14:282-289.
75. Takahashi, S., and M. Zeydel (1982) γ-Glutamyl transpeptidase and glutathione in aging IMR-90 fibroblasts and in differentiating 3T3 L1 preadipocytes. Arch. Biochem. Biophys. 214:260-267.
76. Tarachand, U., R. Silvabalan, and J. Eapen (1981) Enhanced gamma glutamyl transpeptidase activity in rat uterus following deciduoma induction and implantation. Biochem. Biophys. Res. Commun. 101:1152-1157.
77. Tate, S.S., and A. Meister (1974) Interaction of γ-glutamyl transpeptidase with amino acids, dipeptides, and derivatives and analogs of glutathione. J. Biol. Chem. 249:7593-7602.
78. Tate, S.S., and A. Meister (1974) Stimulation of the hydrolytic activity and decrease of the transpeptidase activity of γ-glutamyl transpeptidase by maleate; identity of a rat kidney maleate-stimulated glutaminase and γ-glutamyl transpeptidase. Proc. Natl. Acad. Sci., USA 71:3329-3333.
79. Tate, S.S., and J. Orlando (1979) Conversion of glutathione to glutathione disulfide, a catalytic function of γ-glutamyl transpeptidase. J. Biol. Chem. 254:5573-5575.
80. Tate, S.S., G.A. Thompson, and A. Meister (1976) Recent studies on γ-glutamyl transpeptidase. In Glutathione: Metabolism and Function, I.M. Arias and W.B. Jakoby, eds. Raven Press, New York, pp. 45-55.
81. Tateishi, N., T. Higashi, K. Nakashima, and Y. Sakamoto (1980) Nutritional significance of increase in γ-glutamyltransferase in mouse liver before birth. J. Nutr. 110:409-415.
82. Uchida, T., H. Miyata, and T. Shikata (1981) Human hepatocellular carcinoma and putative precancerous disorders. Their

enzyme histochemical study. Arch. Pathol. Lab. Med. 105:180-186.

83. Vanderlaan, M., S. Fong, and E.B. King (1982) Histochemistry of NADH diaphorase and γ-glutamyl transpeptidase in rat bladder tumors. Carcinogenesis 3:397-402.

84. Vanderlaan, M., and W. Phares (1981) γ-Glutamyltranspeptidase: A tumor cell marker with a pharmacological function. Histochem. J. 13:865-877.

85. Williams, G.M., T. Ohmori, S. Katayama, and J.M. Rice (1980) Alteration by phenobarbital of membrane-associated enzymes including gamma-glutamyl transpeptidase in mouse liver neoplasms. Carcinogenesis 1:813-818.

86. Wright, E.C., and J. Stern (1979) Glutathionuria: γ-Glutamyl transpeptidase deficiency. J. Inherit. Metab. Dis. 2:3-7.

ORNITHINE DECARBOXYLASE: ALTERATIONS IN CARCINOGENESIS[1]

Mari K. Haddox and Anne R. L. Greenfield

Departments of Internal Medicine
and Pharmacology
University of Texas Medical School
Houston, Texas 77025

The diamine putrescine, the tertiary amine spermidine, and the quaternary amine spermine (collectively referred to as the polyamines) are small, aliphatic, nonprotein nitrogenous bases found in all living cells (1,2). Because of their polycationic nature, they can form tight, noncovalent complexes with negatively charged cellular constituents, such as nucleic acids and proteins. This electrostatic binding is the basis of the known essential roles that polyamines play in nucleic acid and protein synthesis, structure, and function (3,4). In contrast to the inorganic cations, which must be supplied to the cell by the circulation, the polyamines can be synthesized by the cells at times of cellular need. The primary mechanism for a cellular increase in polyamine content is an increase in the activity of ornithine decarboxylase (ODC) which catalyzes the formation of putrescine from ornithine.

ODC: FUNDAMENTAL CONSTITUENT OF NORMAL GROWTH

An increase in ODC activity is a rapid and large (10- to 100-fold) response to growth stimulants, is dependent on protein synthesis and often on RNA synthesis, and is temporally transient, reflecting the apparent short half-life of the enzyme (12-60 min). Increased ODC activity and polyamine biosynthesis and accumulation occur in all growth responses studied to date, including embryonic development, tissue regeneration, cellular hypertrophy, hormonal- and drug-induced tissue growth and neoplastic growth (reviewed in 4-6). Studies employing an irreversible, enzyme-activated inhibitor of ODC activity, difluoromethylornithine (DFMO), have established in

[1] Supported by NIH grant CA32444.

several experimental growth systems, that increased ODC activity and polyamine accumulation are essential for growth to occur (7-10). The growth systems examined have included embryonic development, intestinal maturation and recovery from injury, the growth of cell lines in culture, and the growth of tumor cells in animals.

The induction of ODC is also a general event in the trophic phase of cell cycle progression (reviewed in 6,11). Experiments employing synchronized cells have shown that the enzyme activity begins to increase approximately 2 hours after mitotic exit, and that peaks of activity occur corresponding to G_1 and G_2 phases. The increase during G_1 phase is dependent on transcriptional events and is usually maximal prior to entry into S phase. If the length of G_1 phase is short, the induction phase will overlap with the onset of DNA synthesis. Non-growing cells contain little or no detectable ODC activity, and appear to be arrested at a site prior to the events required for the induction of the enzyme. Certain growth-regulatory arrest signals, such as vitamin A, appear to limit proliferation by acting at this site (12,13). Addition of the ODC inhibitor, DFMO, to cells in culture at the time of cell cycle initiation will immediately arrest progression of some cell lines, while other lines will progress 2 or 3 times through the cycle in the absence of polyamine biosynthesis before the intracellular concentrations fall low enough (via the process of dilution by cell division) to be limiting to cell growth (10,14).

INDUCTION OF ODC DURING TUMOR PROMOTION

One of the earliest and largest intracellular responses to the tumor-promoting phorbol esters is the induction of ODC (15-17). There is a strong correlation between the potency of phorbol esters to promote the formation of epidermal papilloma and their ability to induce ODC. Non-promoting epidermal mitogens are relatively ineffective enzyme inducers (15,18). A more recent study, utilizing a two-stage promotion protocol, has shown that the induction of ODC is associated with stage II of promotion (19). Evidence supporting the suggestion of Boutwell and co-workers (20) that elevations in ODC play a functional role in tumor promotion derive from studies utilizing synthetic and naturally occurring retinoids as inhibitors. There is strong correlation between the potency of the retinoid derivatives to inhibit phorbol ester-stimulated tumor promotion and to inhibit ODC induction (21-23). The inhibition by the retinoids is selective to ODC and is without effect on two other epidermal enzymes increased in activity by the phorbol esters, S-adenosylmethionine decarboxylase, and phosphodiesterase (23). Similar correlations exist for retinoid action to inhibit tumor promoter stimulation of lymphocyte ODC induction and G_0 to S phase transition (24-26), to inhibit carcinogen induction of ODC and tumor formation in the urinary bladder (27), to inhibit ODC induction and G_1 phase progression in synchronized CHO cells (12,13), and to inhibit ODC activity and cell proliferation in neuroblastoma and glioma cells

(28). This combined evidence suggests a role for the induction of ODC in the promotion process.

However, further examination of the effects of other hyperplastic agents on the epidermis have established that, while all tumor-promoting stimulants cause an induction of ODC, not every agent which induces epidermal ODC is a tumor promoter (29,30). Furthermore, inhibitors of tumor promotion, such as the glucocorticoid fluocinolone acetonide, will block promoter-stimulated DNA synthesis without inhibiting the induction of ODC (31,32). Nevertheless, an increase in ODC constitutes the most consistent marker both *in vivo* and *in vitro* of cellular response to tumor promoters. Overall, these studies have led to the conclusion that increased expression of ODC activity constitutes a necessary but not sufficient component of the tumor promotion process.

ODC INDUCTION IN TRANSFORMED CELLS

Many of the early studies linking ODC and cancer focused on identifying an increase in the activity of ODC as a marker of transformation. ODC activity is elevated in spontaneously occurring cancers, such as the mouse L1210 leukemia (33), human neuroblastoma cells (34), and human cutaneous epitheliomas (35). Furthermore, the change in ODC activity correlates with the degree of malignancy of the tumor in comparative studies of fast- and slow-growing Morris hepatomas (36).

In addition to spontaneously occurring cancers, increased ODC activity also occurs in cells transformed in the laboratory by exposure to chemical carcinogens or oncogenic viruses. Several studies have established the relationship between the elevation of ODC activity and tumor induction in the livers of animals maintained on carcinogenic diets (37) or injected with carcinogens (38,39), as well as in carcinogen-treated primary cultures of adult rat liver cells (24). Administration of epidermal, colon, or vesical carcinogens to mice or rats results in large increases in ODC activity in the epidermis (17,41), colon (42), or bladder (27,43), respectively. Similar evidence, correlating increased ODC activity with neoplastic transformation, has been obtained by examining the enzyme activity in normal and virally transformed cells. A comparison between normal mouse 3T3 fibroblasts and a SV40 transformed 3T3 variant, SV101, has established that, under all growth conditions examined, the transformed cells have higher ODC activity (44). Elevated ODC has also been demonstrated in cell lines transformed by several other viruses (45-48).

REGULATION OF ODC IN TRANSFORMED CELLS

While in almost all cases of transformation there is an

observable elevation of ODC activity, the question arises as to whether this increase is actually a reflection of the neoplastic process or rather a marker of stimulation of the proliferative process. Furthermore, changes in ODC expression such as those discussed above could reflect an event inherent to the transformation process, or could simply reflect a secondary consequence of some other carcinogen or virally affected event. These possibilities have been addressed in studies of transformed cell lines as compared to their normal counterparts, particularly those employing thermosensitive mutants of transforming viruses. If the differences in the characteristics of ODC induction are a consequence of the transforming function of the virus, they should be a function of the incubation temperature in cells transformed by the mutant virus in which the transforming function is thermosensitive. Through these types of studies it has become evident that expression of the transformed phenotype involves an inherent alteration in the regulation of ODC expression.

Transformation of chick embryo fibroblasts by Rous sarcoma virus (RSV) leads to an initial 5-fold increase in ODC activity, which precedes morphological alterations by 24 hours (25). By using the temperature-sensitive mutants of RSV, T5, and N468, which multiply at the same rate as the RSV wild-type at both the permissive and non-permissive temperature, but transform only at the lower permissive temperature, the elevated level of ODC has been established to be specifically induced by the onset of the transformation process, independent of cell proliferation or virus infection (45). These findings have been corroborated _in vivo_ when polyamine levels were measured in chorioallantoic membranes infected with RSV and RSV-T5. No elevation in polyamines occurred when RSV-T5 infection was conducted at the non-permissive temperature (49). Similarly, infection of Balb/3T3 cells with the murine sarcoma virus (MSV) caused a marked elevation of ODC activity, while infection with the nontransforming strain RLV caused no alteration in ODC (46). Production of virus is not obligatory for the ODC increase, since an elevation in enzyme activity is also noted in cells transformed by the non-producer Kirsten MSV; nor can the rise in activity be attributed to an increased rate of cellular division, since acutely infected cells fail to divide (46).

Studies of Rat-1 fibroblasts in the non-transformed state and after transformation by the B77 wild-type Rous sarcoma virus (RSV) showed that ODC activity was induced to a greater extent during G_1 phase and was maintained at a higher level during G_1-S phase transition in the transformed cells (50). The induction of ODC in the transformed Rat-1 fibroblast was independent of serum growth factors and could be promoted by the addition of fresh media alone. Feedback repression of the ODC activity in the transformed cell by exogenous polyamine required a 100-fold greater concentration of putrescine than in the normal fibroblast. The Rat-1 cell line transformed by the thermosensitive RSV mutant LA24 displayed characteristics of

ODC regulation similar to those seen in the cell line transformed by the wild-type virus, but only when incubated at the permissive temperature. Furthermore, a shift of this cell line from the non-permissive to the permissive temperature in the absence of any added serum growth factors promoted an induction of the enzyme as a result of the expression of the transforming function of the virus.

A decreased sensitivity to feedback repression by putrescine and the polyamines has also been noted in SV40-transformed 3T3 cells (51) and cytomegalovirus transformed human embryo cells (52). A detailed characterization of a possible intracellular mechanism underlying such a change has been carried out in Syrian hamster fibroblasts (53). Low passage hamster fibroblasts and benzopyrene-transformed fibroblasts, HE68BP, demonstrate under normal culture conditions similar growth parameters, i.e., population doubling time and maximum cell densities. However, the HE68BP line has a neoplastic phenotype, as exhibited by an altered morphology, growth in soft agar, and tumorigenicity. The intracellular polyamine concentration after serum stimulation is greatly elevated in the transformed cells as compared to the control cell line. While the polyamine levels of the normal cells increase only slightly, there is a 13-fold increase in the putrescine level of the HE68BP cells. This difference in polyamine accumulation cannot be attributed to differences in rates of ODC degradation, since under normal conditions the apparent half-life of the enzyme from the two cell types is similar. However, a macromolecular inhibitor of ODC was detected only in normal cells following treatment with high concentrations of extracellular putrescine. Differences in the production of this inhibitor may account for the difference in polyamine accumulation between the normal and transformed cell.

Other alterations in intracellular mechanisms leading to elevated ODC activity may involve stabilization of the enzyme to proteolytic degradation and/or increased rates of enzyme synthesis. While SV40 transformed 3T3 cells possess more ODC activity than the normal parent cells, purification of the enzyme from both cell types revealed no differences in specific activity, Km for substrate, thermostability, sensitivity to antibody, or degradation rate *in vivo* (54). This finding implies that the increased activity in the transformed cell is the result of higher enzyme biosynthesis. However, a difference in half-life has been observed between a hepatoma tissue culture cell line and a variant cell line. The half-life of the enzyme in the variant cell line is 5 to 10 hours compared to 14 minutes in the parental cell line. When ODC was purified from the two cell types, no difference could be found in molecular weight, thermostability, Km for substrate, or inhibition by antibody. However, there was approximately 3 times more soluble enzyme in the variant cells. Since the enzymes are identical, an alteration in the specific deactivation mechanism for the enzyme in the variant cell must be considered (55).

SUMMARY

Pardee has speculated that the restriction points essential to cell cycle progression will represent sites of fundamental difference in regulation between normal and malignant cells (56). In general, the characteristics of ODC expression during the proliferative cycle of transformed cells reflect a diminished response to normal control mechanisms. Whether the altered, apparently less stringent regulatory characteristics of this key event in the growth cycle confer on transformed cells a selective growth advantage, as has been speculated to occur during epidermal transformation (20), remains to be elucidated. However, a large number of studies do suggest that the alteration in ODC regulation may be a universal feature of neoplastic expression.

REFERENCES

1. Tabor, C.W., and H. Tabor (1974) Ann. Rev. Biochem. 45:285-306.
2. Cohen, S.S. (1971) Introduction to the Polyamines, Prentice-Hall, New Jersey.
3. Cohen, S.S. (1978) Nature 274:209-210.
4. Williams-Ashman, H.G., and Z.N. Canellakis (1979) Perspect. Biol. Med. 22:421-453.
5. Russell, D.H., and B.G.M. Durie (1978) Polyamines as Biochemical Markers to Normal and Malignant Growth, Raven Press, New York.
6. Janne, J., H. Poso, and A. Raina (1978) Biochim. Biophys. Acta 473:241-293.
7. Danzin, C., M.J. Jung, J. Grove, and P. Bey (1979) Life Sci. 24:519-524.
8. Fozard, J.R., M.L. Part, N.J. Parkash, J. Grove, P.J. Schechter, A. Sjoerdsma, and J. Koch-Weser (1980) Science 208:505-508.
9. Luk, G.D., L.J. Marton, and S.B. Baylin (1980) Science 210: 1195-1198.
10. Fozard, J.R., and J. Koch-Weser (1982) Trends Pharmacol. Sci. 3:107-110.
11. Russell, D.H., and M.K. Haddox (1978) Adv. Enzyme Regula. 17: 61-87.
12. Haddox, M.K., and D.H. Russell (1979) Cancer Res. 39:2476-2480.
13. Haddox, M.K., K.F.F. Scott, and D.H. Russell (1979) Cancer Res. 39:4930-4938.
14. Janne, J., L. Alhonen-Hongisto, P. Seppanen, and M. Siimes (1981) Med. Biol. 59:443-457.
15. O'Brien, T.G., R.C. Simsiman, and R.K. Boutwell (1975) Cancer Res. 35:1662-1670.
16. Yuspa, S.H., U. Lichti, T. Ben, E. Patterson, H. Hennings, T.J. Slaga, N.H. Colburn, and W. Kelsey (1976) Nature 262:402-404.
17. O'Brien, T.G. (1976) Cancer Res. 36:2644-2653.
18. O'Brien, T.G., R.C. Simsiman, and R.K. Boutwell (1975) Cancer Res. 35:2426-2433.

19. Slaga, T.J., A.J.P. Klein-Szanto, S.M. Fischer, S.M. Weeks, K. Nelson, and S. Major (1980) Proc. Natl. Acad. Sci. USA 77: 2251-2254.
20. Boutwell, R.K. (1977) In Origins of Human Cancer: Proceedings of the Cold Spring Harbor Conferences on Cell Proliferation, Book B, pp. 773-783, (H.H. Hiatt, J.D. Watson, and J.A. Winsten, eds.), Cold Spring Harbor Laboratory, Cold Spring Harbor, N.Y.
21. Verma, A.K., and R.K. Boutwell (1977) Cancer Res. 37:2196-2201.
22. Verma, A.K., H.M. Rice, B.G. Shapas, and R.K. Boutwell (1978) Cancer Res. 38:793-801.
23. Verma, A.K., B.G. Shapas, H.M. Rice, and R.K. Boutwell (1979) Cancer Res. 39:419-425.
24. Kensler, T.W., and G.C. Mueller (1978) Cancer Res. 38:771-775.
25. Kensler, T.W., A.K. Verma, R.K. Boutwell, and G.C. Mueller (1978) Cancer Res. 38:2896-2899.
26. Wertz, P.W., T.W. Kensler, G.C. Mueller, A.K. Verma, and R.K. Boutwell (1977) Nature 277:227-229.
27. Matsushima, M., and G.T. Bryan (1980) Cancer Res. 40:1897-1901.
28. Chapman, S.K. (1980) Life Sci. 26:1359-1366.
29. Marks, F., S. Bertsch, and G. Furstenberger (1979) Cancer Res. 39:4183-4188.
30. Marks, F., G. Furstenberger, and E. Kownatzki (1981) Cancer Res. 41:696-702.
31. Lichti, U., T.J. Slaga, T. Ben, E. Patterson, H. Hennings, and S.H. Yuspa (1977) Proc. Natl. Acad. Sci. USA 74:3908-3912.
32. Yuspa, S.H., U. Lichti, and T. Ben (1980) Proc. Natl. Acad. Sci. USA 77:5312-5316.
33. Russell, D.H., and C.C. Levy (1971) Cancer Res. 31:248-251.
34. Helson, L., C. Helson, A. Majeranowski, S. Hajdi, and S. Das (1976) Proc. AACR and ASCO 17:218.
35. Scalabrino, G., P. Pigatto, M.E. Ferioli, D. Modena, and M. Puerari (1980) J. Invest. Derm. 74:122-124.
36. Williams-Ashman, H.G., G.L. Coppoc, and G. Weber (1972) Cancer Res. 32:1924-1932.
37. Scalabrino, G., H. Poso, G. Holtta, P. Hannonen, A. Kallio, and J. Janne (1978) Int. J. Cancer 21:239-245.
38. Olson, J.W., and D.H. Russell (1979) Cancer Res. 39:3074-3079.
39. Yanagi, S., K. Sasaki, and N. Yamamoto (1981) Cancer Lett. 12:87-91.
40. Wijk, R., H. Louwers, and A. Bisschop (1981) Carcinogenesis 2:27-31.
41. Verma, A.K., E.A. Conrad, and R.K. Boutwell (1980) Carcinogenesis 1:607-611.
42. Takano, S., M. Matsushima, E. Erturk, and G.T. Bryan (1981) Cancer Res. 41:624-628.
43. Ball, W.J., J.S. Salser, and M.E. Balis (1976) Cancer Res. 36:2686-2689.
44. Lembach, K.J. (1974) Biochim. Biophys. Acta. 354:88-100.
45. Don, S., and U. Bachrach (1975) Cancer Res. 35:3618-3622.
46. Gazdar, A.F., H.B. Stull, L.J. Kilton, and U. Bachrach (1976) Nature 262:696-698.

47. Goldstein, D.A., O. Heby, and L.J. Marton (1976) Proc. Natl. Acad. Sci. USA 73:4022-4026.
48. Isom, H.C. (1979) J. Gen. Virol. 42:265-278.
49. Don, S., H. Wiener, and U. Bachrach (1975) Cancer Res. 35: 194-198.
50. Haddox, M.K., B.E. Magun, and D.H. Russell (1980) Cancer Res. 40:604-608.
51. Bethell, D.R., and A.E. Pegg (1979) Biochem. J. 180:87-94.
52. Isom, H.C., and J.T. Backstrom (1979) Cancer Res. 39:864-869.
53. O'Brien, T.G., D. Saladik, and L. Diamond (1980) Biochim. Biophys. Acta 632:270-283.
54. Weiss, J., K. Lembach, and R. Boucek (1981) Biochem. J. 194: 229-239.
55. Pritchard, M.L., A.E. Pegg, and L.S. Jefferson (1982) J. Biol. Chem. 257:5892-5899.
56. Pardee, A.B. (1974) Proc. Natl. Acad. Sci. USA 71:1286-1290.

HUMAN TUMORS: ENZYME ACTIVITIES AND ISOZYME PROFILES

Doris Balinsky

Department of Biochemistry and Biophysics
Iowa State University
Ames, Iowa 50011

INTRODUCTION

My own interest in tumor metabolism was sparked by the unique availability in South Africa, where I was located at the time, of human hepatoma specimens. Human primary hepatoma has a fairly low incidence in Western countries, but an incidence of 1:1,100 in Mozambique. We were able to obtain hepatoma material from migrant workers, mostly from Mozambique, who showed evidence of the disease while working in South Africa. Since I have been in the United States, I have extended these studies to human breast and lung tumors, and found, as I will indicate to you, that they do not show the same enzyme and isozyme profiles. This is an important point to note, since too often generalizations are made about "cancer", as if all cancers were the same, irrespective of cause or tissue of origin.

I will compare our human hepatoma data with those obtained from similar studies made in animals. While taking this approach, I must stress my indebtedness for the fine animal studies which, in many cases, preceded our studies in humans, and gave us the impetus and the guidelines for our own studies. Many similarities between animal and human tumors were found, but also sufficient differences to show that it is not possible simply to extrapolate data obtained from animal studies to the human situation.

ENZYMES OF NUCLEIC ACID METABOLISM

Since cell division and growth occur during tumor development, the activities of enzymes of DNA and RNA synthesis and breakdown have been compared in hepatomas of varying growth rates, in regenerating liver, and during development in rats (1,2,3). In general, in tumors the activities of the biosynthetic enzymes, including those of the so-called "salvage" pathways, increase, while the degradative enzymes decrease, when compared to the activities in normal adult liver (1). Some, but not all, of the activity changes are correlated with tumor growth rate.

Several of the enzymes studied are unique to DNA synthesis, namely DNA polymerase, and enzymes of thymidine metabolism, since thymidine occurs in DNA but not in RNA.

We assayed the activities of DNA polymerase and of three enzymes of DNA synthesis - thymidylate kinase, thymidylate synthetase, and thymidine kinase - in human primary hepatomas and in a rapidly-growing, transplantable 3'-methyldimethylaminoazobenzene (3'-MeDAB)-induced rat hepatoma (4). Figures 1 and 2 show the activities of these 4 enzymes rat and human tumors, normal and host livers, fetal livers and hepatoma cell lines. It should be noted that thymidine kinase levels in the human normal and host livers were considerably higher than in the corresponding rat livers, showing that species differences occur. Also, the levels of thymidylate kinase and DNA polymerase in host livers were, in several cases, higher than in normal livers, suggesting an effect of tumor on the host liver. Several other points emerge. Firstly, it can be seen that all the enzymes were considerably elevated in the rat hepatoma. Three of the enzymes were also elevated in the human hepatomas, but to a lesser extent, while no increase was observed for thymidine kinase. This suggests that human hepatomas may have lower rates of DNA synthesis than rapidly-growing rat hepatomas, and that the synthesis of dTMP is largely effected by the de novo pathway from dUMP and less so by the salvage pathway from thymidine. Furthermore, the enzyme levels in the human hepatomas were considerably lower than in human fetal livers, suggesting lower growth rates. It is of interest that a slow-growing cell line derived from a human hepatoma (AH) showed similar low thymidine kinase activities to solid tumors, whereas a rapidly-dividing cell line derived from another human hepatoma (PH) had 30-fold higher activities. Human hypernephromas (5) and colon carcinomas (6) also do not have increased thymidine kinase activity, while tumors derived from some other tissues do show elevations (5), hence no generalizations about thymidine kinase in human tumors can be made. A very recent study from Japan (6a) showed elevated activities of thymidine kinase in a number of human tumors, including ones originating in lung and liver, stomach, colon, and small intestine. Possibly these were more rapidly- growing tumors than those studied by us (4) or Laszlo et al. (6).

A "fetal" form of thymidine kinase has been demonstrated in human fetal livers and in HeLa and KB cell lines (7), but we did not find this form in human hepatomas (8), so it does not seem to recur generally in human tumors. We did find multiple forms of this enzyme in both liver and hepatoma, but the proportions of these forms were identical in the two tissue types (8). However, both rat and human hepatomas showed the presence of different isozymic forms of DNA polymerase than occurred in normal livers, as evidenced by

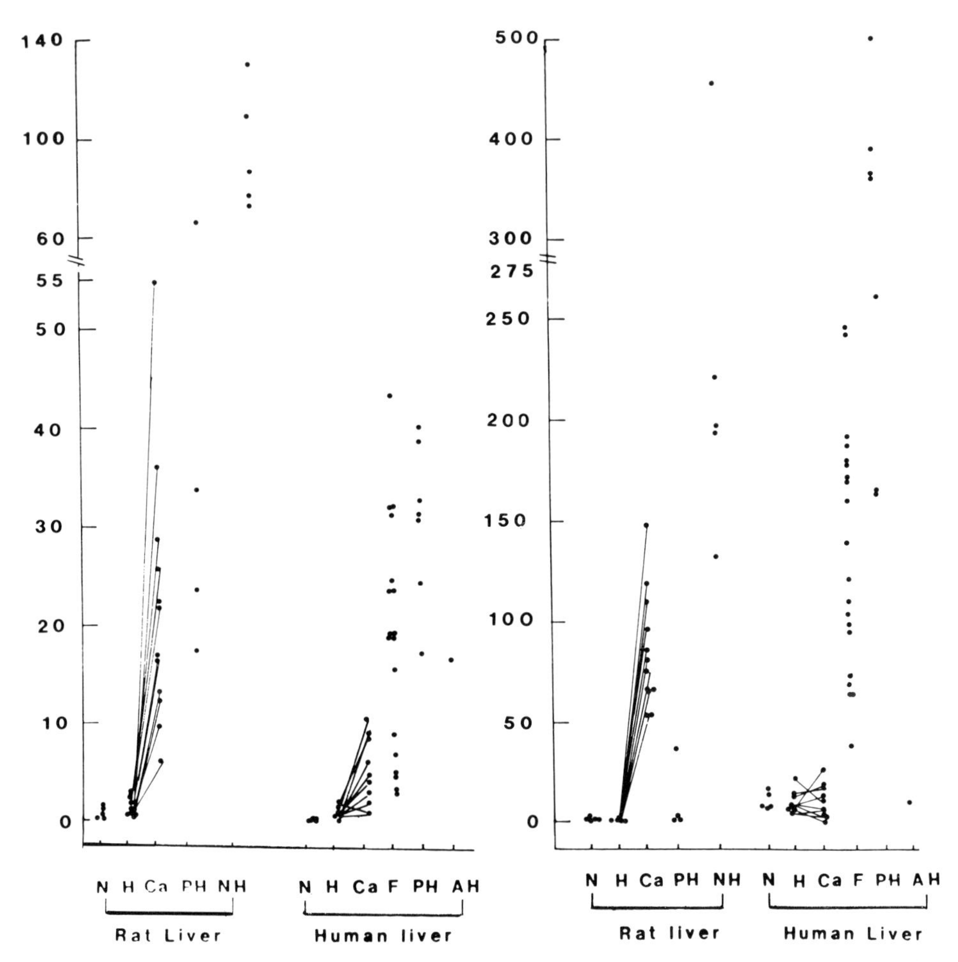

Fig. 1 Activities of DNA polymerase (A) and thymidine kinase (B) in rat and human normal (N) and host (H) livers, hepatomas (Ca) and hepatoma cells in culture (PH, NH, AH).

different template requirements (Table 1). The enzyme from normal and host livers preferentially used native DNA, while that from the tumor preferentially used heat-denatured DNA. The increase in the tumor form correlates with tumor growth rate in Morris hepatomas (9).

ENZYMES OF CARBOHYDRATE METABOLISM

One of the earliest observations in cancer biochemistry, made by Otto Warburg in 1924 (10), was that tumors appeared to have in-increased rates of anaerobic glycolysis, with reduced respiration. This tenet has been examined in detail by several groups of investigators. Many of the studies were made using the series of

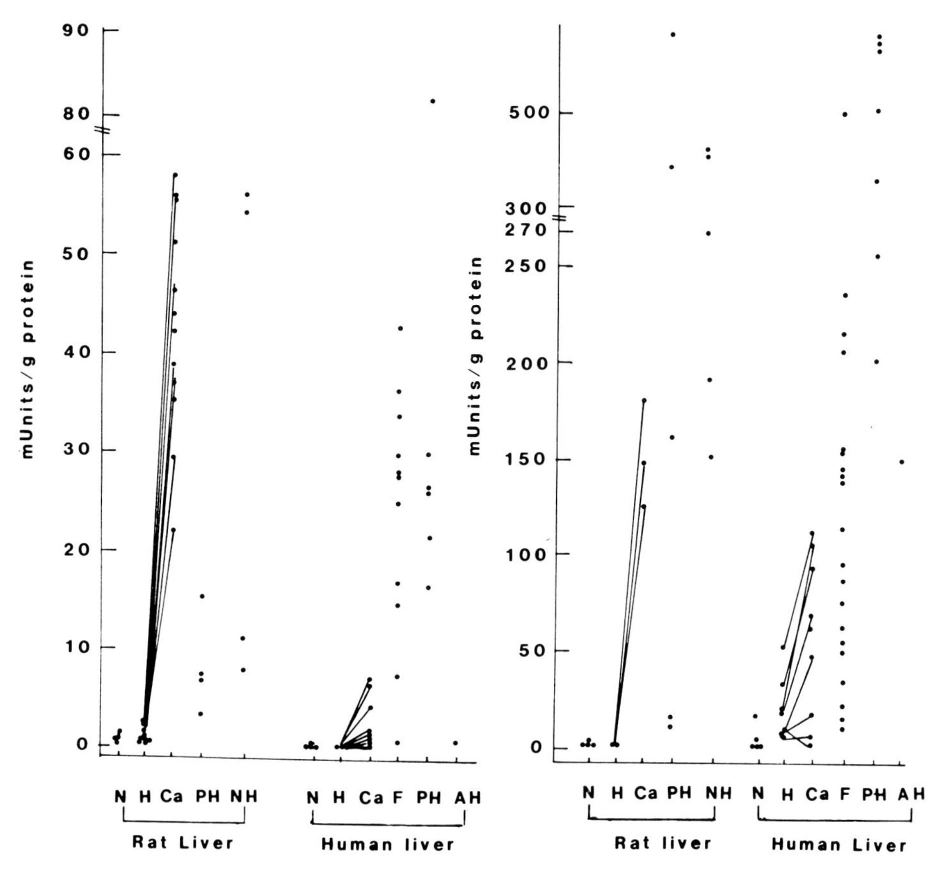

Fig. 2. Activities of thymidylate synthetase (A) and thymidylate kinase (B) in rat and human normal (N) and host (H) livers, hepatomas (Ca) and hepatoma cells in culture (PH, NH, AH).

transplantable, chemically-induced hepatomas developed by Dr. Harold P. Morris, a series of tumors with a wide spectrum of growth rates (11). Warburg's generalization was found not to hold for all tumors, since slowly-growing, well-differentiated hepatomas had very low glycolytic rates (12). Key enzymes of glycolysis (hexokinase, phosphofructokinase, and pyruvate kinase) have been assayed and were found to have increased activities in hepatomas, the increases correlating with tumor growth rate (2). In our own studies with human hepatomas, we assayed 10 glycolytic enzymes, and found only pyruvate kinase to be increased somewhat (13). Table 2 shows the activities of hexokinase, pyruvate kinase, lactate dehydrogenase, and also of malate dehydrogenase in human tumors as a percentage of those in the corresponding host tissues. The activity of lactate dehydrogenase, the terminal enzyme of anaerobic glycolysis, was reduced in the human hepatomas. Lactate dehydrogenase activity also was reduced considerably in all but the most rapidly growing Morris hepatomas

Table 1. DNA polymerase activity in liver and hepatoma.

Source	Activity (milliunits/g protein)		
	DNA Template		
Rat	Native(N)	Denatured(D)	Ratio D/N
Normal liver(6)	2.26 ± 0.38	0.81 ± 0.18	0.38 ± 0.06(3)
Host liver	3.17 ± 0.25(3)	1.31 ± 0.26(15)	0.41 ± 0.02(3)
Hepatoma	14.6 ± 1.7(3)	20.2 ± 3.2 (15)	1.52 ± 0.09
NRH cell line (5)	63.2 ± 14.8	95.9 ± 10.6	1.71 ± 0.22
PRH cell line(5)	30.1 ± 2.6	35.1 ± 10.9	1.12 ± 0.29
Human			
Normal liver(5)	0.33 ± 0.13	0.18 ± 0.11	0.60 ± 0.35
Host liver(11)	1.82 ± 0.41	1.18 ± 0.19	0.78 ± 0.12
1° Hepatoma(12)	4.00 ± 0.77	5.17 ± 0.95	1.43 ± 0.27
Fetal liver (20)	7.06 ± 0.91	17.2 ± 2.6	2.30 ± 0.20
AHH cell liver(1)	9.0	16.6	1.84
PHH cell line(7)	17.0 ± 1.5	30.6 ± 3.0	1.81 ± 0.12

(14). Malate dehydrogenase, which is not directly involved in glycolysis, showed similar activity decreases to lactate dehydrogenase (Table 2).

Quite different results were obtained with tumors originating in other tissues. We recently compared the activities of hexokinase, pyruvate kinase, lactate dehydrogenase, and malate dehydrogenase in human lung (15) and breast (16) tumors. As seen in Table 2, levels of all three enzymes were considerably elevated in lung tumor tissue compared to the uninvolved lung tissue from the same patient. These three glycolytic enzymes were also markedly elevated in malignant human breast tissues compared with uninvolved breast tissues from the same patient. These data show that not all types of human malignant tumors show similar enzyme patterns. Of great interest is the observation that benign human breast tumors showed glycolytic enzyme activities virtually indistinguishable from those of normal breast tissue (16), indicating that malignant tumors can be distinguished from non-malignant ones by enzyme assays. Hilf et al. (17) made similar observations on human breast tissues some years ago. Again, we found that malate dehydrogenase activity closely followed that of lactate dehydrogenase (16).

Table 2. Activities of hexokinase, pyruvate kinase, and lactate dehydrogenases in tumors as a percentage of those in the corresponding host tissues.

	TUMOR SOURCE		
ENZYME	LIVER	BREAST	LUNG
Hexokinase	113.7 ± 17.4(9)	1112 ± 445(5)	183 ± 53(5)
Pyruvate Kinase	186.8 ± 28.4(9)	740 ± 253(7)	946 ± 441(12)
Lactate Dehydrogenase	65.2 ± 14.1(9)	379 ± 92(7)	562 ± 113(12)
Malate Dehydrogenase	56.5 ± 8.3(8)	228 ± 48(7)	530 ± 186(12)

When we examined the enzymes of gluconeogenesis, a very different picture emerged (15,18). Gluconeogenesis is unique to adult liver and kidney, and provides glucose *de novo* from 3-carbon precursors. Figure 3 shows that the activities of pyruvate carboxylase and phosphoenolpyruvate carboxykinase were considerably reduced in all the human hepatomas examined, and also in the rat hepatoma. Glucose 6-phosphatase (18) and fructose 1,6-bisphosphatase (13) were also decreased. Similar data have been observed in animal tumors by others, who showed that the reduced activity correlated with tumor growth rate (1). These data show that the hepatomas have diverged in their enzyme complement from the parent liver tissue. Fetal livers also had low activities, as seen in Fig. 3. This can be explained by the fact that the fetus obtains its glucose from the maternal circulation.

ISOZYMES

Enzyme activity changes have been very helpful in indicating alterations in tumor metabolism. However, isozyme studies can often pinpoint even more clear-cut differences between normal and tumor tissues. Many enzymes occur in multiple forms, or isozymes. Differentiated adult tissues show characteristic isozyme patterns, which frequently become altered in tumor tissues. A few examples will be discussed below.

Lactate Dehydrogenase

Lactate dehydrogenase was the first enzyme shown to occur in isozymic forms, and is probably the most widely studied isozyme system in disease. The enzyme is a tetramer and consists of 2 types

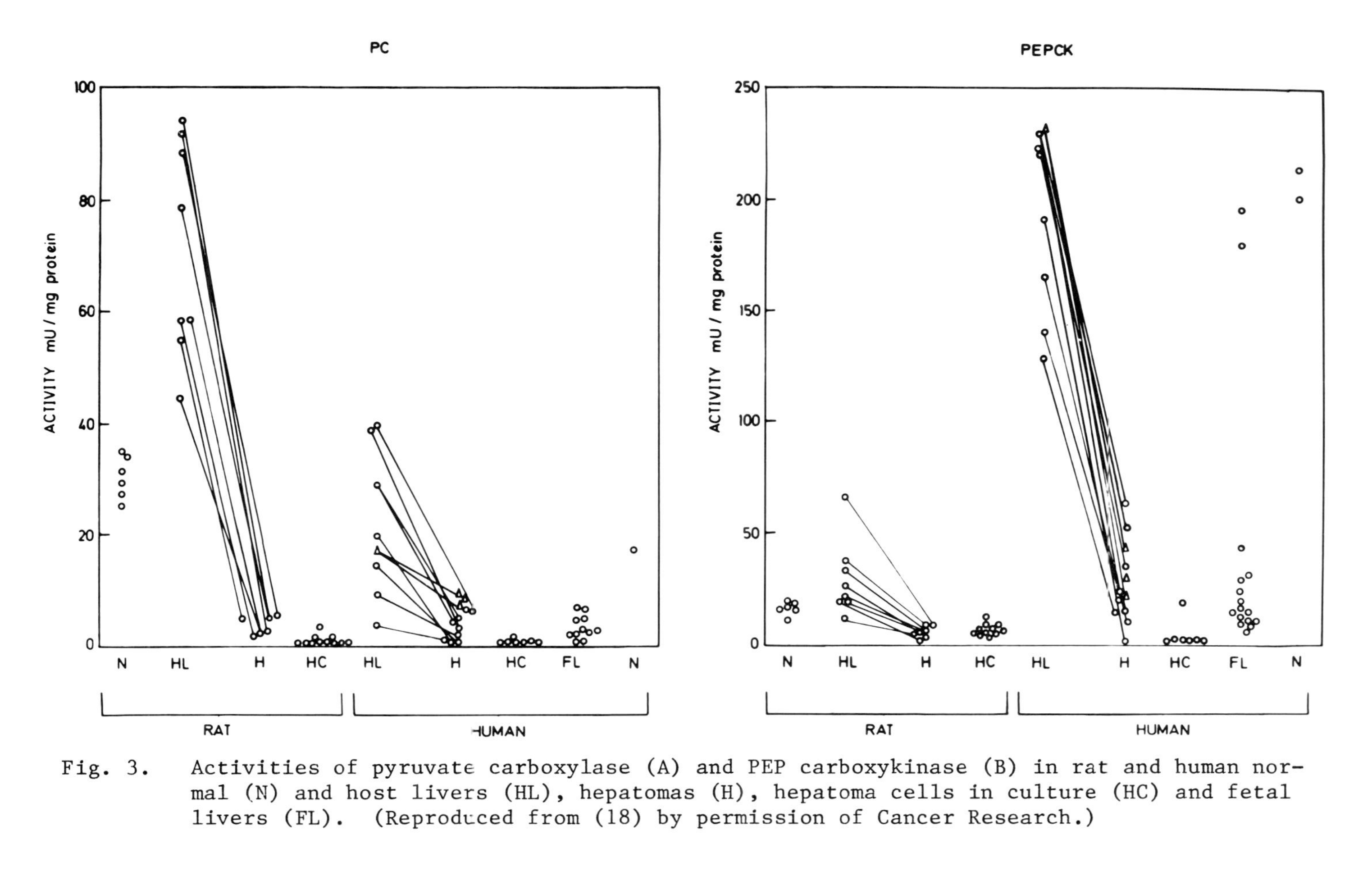

Fig. 3. Activities of pyruvate carboxylase (A) and PEP carboxykinase (B) in rat and human normal (N) and host livers (HL), hepatomas (H), hepatoma cells in culture (HC) and fetal livers (FL). (Reproduced from (18) by permission of Cancer Research.)

of subunits, A (or M) and B (or H), allowing formation of 5 possible isozymes. Early studies with human cancer tissues showed increases of the cathodic forms (19); such studies are still being actively pursued and have, in general, supported these findings. Our own recent studies on human breast tumors showed a shift towards cathodal forms, whereas benign tumors were indistinguishable from normal tissues (16). This is of interest, as it shows that isozyme patterns can be used to distinguish between different types of tumors. Similar data were obtained earlier by Hilf et al. (20). We found, as with enzyme activities, that the most clear-cut changes were observed when comparing tumor tissue with uninvolved breast tissue from the same patient. We also studied human lung tumors, and found similar increases in cathodal forms (15). This is especially noticeable when the percentage of LDH_5 in tumor and uninvolved lung from the same patient are compared. Kaplan has postulated that the cathodic forms are more prominent in more anaerobic tissues (21). Hence, if a cancer is assumed to have a poorer blood supply, or to be actively involved in anaerobic glycolysis, this could explain these increases. On the other hand, our own data in hepatoma tissues showed decreases of cathodic forms in some cases (13). This is understandable when it is appreciated that normal liver has predominantly LDH_5 (M_4); hence, any change from the typical adult tissue would give rise to decreased cathodic isozymes. Some transplantable rat hepatomas similarly showed decreases of LDH_5 (14), as did human rhabdomyosarcomas (22).

A different form of LDH, called LDH-Z, has been found in cultured choriocarcinoma cells (23), though its significance is not clear. Very recently, a new form of LDH has been discovered, LDH_k, which runs very cathodally, and is very sensitive to inhibition by oxygen; it is thus only observed under an atmosphere of nitrogen, or in the presence of cyanide (24). What makes this enzyme of extreme potential interest is that it is produced by Kirsten sarcoma virus, but appears to be derived from the rat genome, since it appears in rat tissues. Very large increases are seen in human tumors compared to uninvolved host tissues, suggesting that this may prove to be a useful tumor marker.

Malate Dehydrogenase

This enzyme occurs in 2 isozymic forms, a cytosolic and a mitochondrial form. The isozyme distribution has not previously been examined in human tumors. We found significantly more mitochondrial malate dehydrogenase in malignant lung and breast tumors than in the corresponding normal tissues (15,16). As with lactate dehydrogenase, we found the increases to be most clear-cut when the percentage of mitochondrial enzyme in the tumor and uninvolved tissue from the same patient were compared. Increases in mitochondrial malate dehydrogenase also occur in experimental mammary tumors in rats (25). Benign human breast tumors showed an isozyme pattern indistinguishable from that of normal tissues (16). This is, thus, another instance in which benign and malignant tumors can be distinguished on the basis of their isozyme patterns.

Pyruvate Kinase

This enzyme occurs in several different isozymic forms (for reviews, see 26,27). Adult differentiated liver has predominantly PK-L_4 with traces of the PK-K_4 isozyme. The K subunits have been considered the prototype of this enzyme, since they predominate in all fetal tissues (28); in addition, they predominate in most adult differentiated tissues, except liver, red cells, brain, and muscle. Muscle has a third form, M_4, which also predominates in adult brain. The M and K subunits show great similarities in amino acid composition and peptide mapping, and are immunologically cross-reactive (26). Recent studies have shown that they are probably coded for by the same gene, but are translated from different messenger RNAs (29). A similar relationship appears to exist between L (liver) and L' (red cell) subunits, which are also closely related and also appear to be coded by the same gene but synthesized by two different types of mRNA (30). The L' form can subsequently be converted to the L form by proteolytic cleavage at the C-terminal end of the molecule (31).

The PK-L_4 isozyme is a highly regulated, allosteric enzyme. In addition to dietary regulation of its level, its activity is regulated by phosphorylation-dephosphorylation, and by allosteric activators and inhibitors. This fine control is advantageous in a tissue where the substrate phosphoenolpyruvate (PEP) may be utilized either in the glycolytic or gluconeogenic pathway, since "futile cycling" between PEP and pyruvate is reduced. The PK-K_4 isozyme shows only slight allosteric properties with respect to PEP, but is strongly inhibited by alanine. This permits it to be distinguished kinetically from the PK-M_4 form, which shows normal hyperbolic kinetics, and is not inhibited by alanine.

Pyruvate kinase in hepatomas. Our studies of pyruvate kinase in human primary hepatomas showed loss of the liver-specific PK-L_4 isozyme, with increases of PK-K_4, which may become the major form (Figure 4). We found that the hepatoma PK-K_4 isozyme has a much lower Michaelis constant for PEP than does the PK-L_4 isozyme in the absence of any activators (32). Hence, at low PEP concentrations in the cell, the activity would be high. Since hepatomas also have considerably reduced levels of gluconeogenic enzymes (13,18), the switch to this less regulated form does not lead to futile cycling between PEP and pyruvate; the tumor is essentially a glycolytic tissue. In rat hepatomas, too, we and others have shown a decrease of PK-L_4 and an increase of PK-K_4 (33,34,35); the changes are more marked in more poorly differentiated carcinomas.

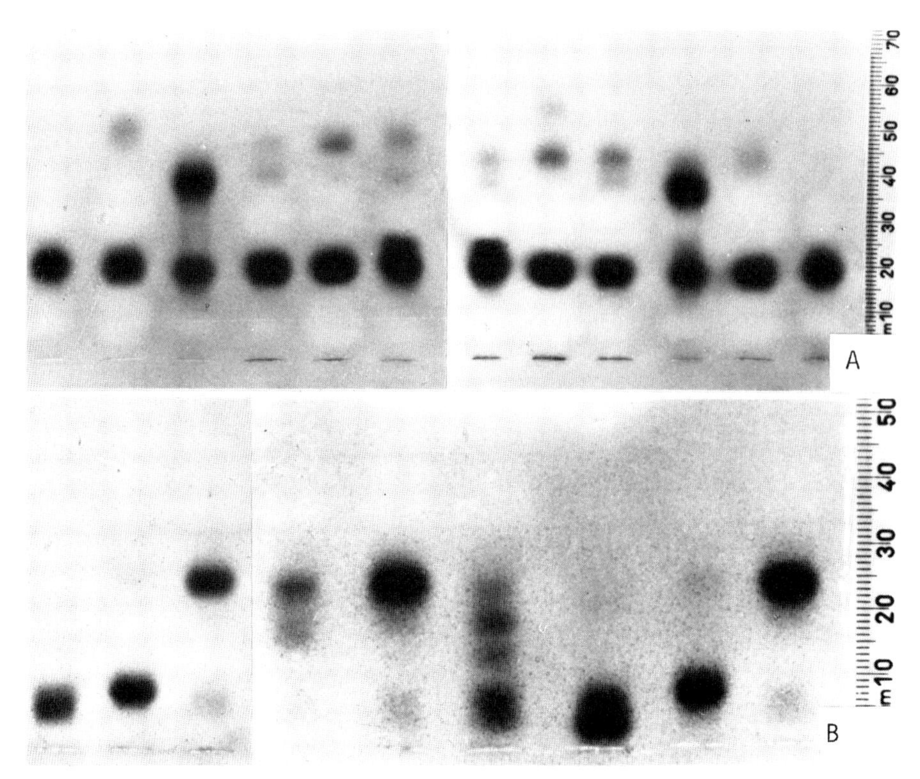

Fig. 4 Isozyme patterns of hexokinase (A) and pyruvate kinase (B) in human livers, hepatomas and hepatoma cells. (A) Left to right: normal muscle, normal liver, hepatoma cells, hepatoma, host liver, fetal liver, fetal liver, hepatoma, hepatoma cells, normal liver, normal muscle. The gel on the left was stained with 0.5 mM glucose; that on the right with 0.1 M glucose to show the fastest band, glucokinase. (B) Left to right: hepatoma cells, normal muscle, normal liver, fetal liver, host liver, hepatoma, hepatoma cells, normal muscle, host liver. (Reproduced from (18) by permission of Cancer Research.)

I should stress the range of data we observed in different human hepatomas. As seen in Figure 5, not all showed the clear-cut PK-L_4 replacement by PK-K_4. Some had isozyme patterns similar to those of normal liver, and some even showed appearance of PK-M_4. Others had hybrids of L and K subunits. This is actually of great interest, as it shows that the L and K subunits must have occurred in the same cell. Adult differentiated hepatocytes have only L subunits, but hepatoma cells can evidently contain both concurrently. Another interesting and important observation was that the apparently uninvolved "host" liver of these patients sometimes showed similar changes. Host livers have been found which had only

$PK\text{-}K_4$, or which had the hybrid forms that are referred to above. The effect of tumors on host liver isozymes has been noted in experimental animal systems, too, and will be discussed in more detail later.

Pyruvate kinase changes in other tumors. As mentioned above, adult brain tissue has predominantly $PK\text{-}M_4$, while fetal brain has mainly K subunits (36). Malignant brain tumors show decreases of $PK\text{-}M_4$ and increases of forms containing K subunits (36). In addition to electrophoretic separation, alanine inhibition, which affects K to a much greater extent than M, has been used to quantitate the degree of malignancy of brain tumors (37). The K_4 form also appears in retinoblastomas (38).

Pyruvate kinase isozymes have been examined also in human breast tumors. ATP inhibition studies (39) had led to the postulate that there was an increase of K subunits in malignant tumors. Our data, using cellulose acetate electrophoresis to separate $PK\text{-}K_4$ from other forms, confirmed these conclusions. The $PK\text{-}K_4$ isozyme was a major band in 28 out of 35 primary malignant breast tumors examined, but in only 8 out of 22 normal breast tissues (16). These data were further strengthened by the observation that MCF-7 human breast cancer cells had virtually only $PK\text{-}K_4$, with faint traces of a more cathodal band. By contrast, organoids obtained by collagenase digestion of normal breast tissue, and consisting purely of normal mammary ductal tissue, had a more cathodal band as the major form; only one specimen had $PK\text{-}K_4$ as a major band Ibsen et al. (40), in studies carried out concurrently with ours, and using only 4 normal breast tissues, found $PK\text{-}K_4$ to predominate, with no other major forms. In malignant breast tumors, he found that $PK\text{-}K_4$ constituted only 35% of total pyruvate kinase activity, but was still the most prominent band.

We found $PK\text{-}K_4$ as a major band in the majority of lung tumors we studied, too, compared to only two-thirds of host lungs and 50% of normal controls (15). Thus, increase of $PK\text{-}K_4$ appears to be a good tumor marker in many different types of tumors.

Hexokinase

Hexokinase consists of three different isozymes, HK I, HK II, and HK III. In addition, there is the so-called "high K_m hexokinase", or glucokinase, which has a much higher Michaelis constant for glucose than does HK. Normal liver from a well-fed person or animal has a high proportion of glucokinase; this disappears on starvation. In addition, liver contains HK III, a form with very low K_m for glucose (high concentrations of glucose are inhibitory to it), HK I, and traces of HK II. In human hepatomas, we found a loss of glucokinase and HK III, and an increase of HK II, as shown in Fig. 4 (13,18). The HK II also increases in many other human tumors (41). In transplantable rat hepatomas, too, we and

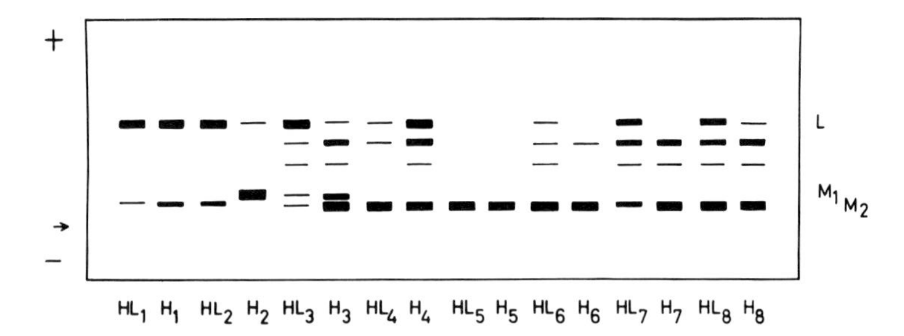

Fig. 5. Diagrammatic representation of pyruvate kinase electrophoretograms for pairs of host livers (HL) and hepatomas (H) from eight patients. (Reproduced from (18) by permission of Cancer Research.)

others have found increased HK II and decreased HK III (33,42), although some workers have found increases of HK III (43).

Aldolase

This is a bifunctional glycolytic enzyme which catalyzes the hydrolysis of fructose 1,6-bisphosphate to two triose phosphates. It is a tetramer consisting of three genetically distinct types of subunits: B, which occurs exclusively in mature hepatocytes; A, which occurs in muscle; and C, which occurs as C_4 and as A-C hybrids in brain tissue. These isozymic forms have been distinguished on the basis of electrophoretic mobility, relative activity with fructose 1,6-bisphosphate and fructose 1-phosphate, and by immunological methods. Schapira's group (44,45) first showed an increase in A and C subunits in hepatomas. We also found alterations of aldolase isozymes in human hepatomas on the basis of differential activity with the two substrates (13). Using immunofluorescent techniques, Schapira's group has shown that adult liver hepatocytes contain only aldolase B, while hepatoma cells contain A, B, and C subunits (45); fetal hepatocytes, too, contain A, B, and C subunits, suggesting that the hepatomas show fetal characteristics.

Alkaline Phosphatase

Another enzyme frequently encountered in clinical studies is alkaline phosphatase. This enzyme is widely distributed in animal tissues, though its function is still obscure. It catalyzes the hydrolysis of a variety of artificial phosphate esters, and is distinguished from the acid phosphatases by having an alkaline pH optimum. A variant of this enzyme, the so-called "Regan" variant, was observed in the tumor and serum of a patient with bronchogenic carcinoma (46). It was found to resemble placental alkaline phosphatase. This variant has since been identified in the sera of patients with a variety of cancers (47). Other cancer-related variants have since been found (48,49,50), some resembling forms found in early placenta or FL-amnion cells. Of interest were findings in a teratocarcinoma of isozymes corresponding to testis, chorion phase 1, FL amnion, and term placenta, showing that genes of various stages of development were being expressed. These data gave rise to the concept of "carcino-placental" isozymes occurring in cancer, in addition to the recurrence of fetal forms previously observed (51).

Esterases

The non-specific esterases use a wide variety of naturally-occurring and artificial substrates. They can be distinguished on the basis of differential staining with different substrates, and by the use of various inhibitors. We compared esterase isozyme patterns in humans in normal and malignant specimens of liver, lung, and breast tissues in a search for additional tumor markers (52). We made an interesting observation on studying breast tumors. Acetylesterases A_4 and A_5 [nomenclature of Coates et al. (53)] were virtually absent from normal tissues, but were a major component of both malignant and benign breast tumors. This was the only enzyme system where benign tumors could be distinguished from normal breast tissue but not from malignant tissue. Butyrylesterase B_3 was a major component of all the tissues examined except some malignant breast tumors. Butyrylesterase B_2 was low in all the breast and lung tissues examined. It was a major component of liver tissue and some hepatomas, but was considerably reduced in many hepatomas, showing again a "switch off" of the gene for a liver-specific isozyme. The study of esterase isozyme patterns thus shows considerable potential for detecting tumor formation in some tissues.

ENZYMES IN FETAL AND REGENERATING LIVER AND FOLLOWING CARCINOGEN FEEDING

Many adult differentiated tissues normally do not divide, i.g., liver. Since tumors represent growing tissue, it is of importance to know whether the changes observed in enzyme activities and isozyme patterns are markers of malignancy or are simply characteristic of dividing tissues. Most of these studies have been made with liver. Comparisons with fetal liver tissue and regenerating liver have been made (34). In both fetal and early regenerating liver, activities of pyruvate kinase and glucokinase were much lower than in normal adult liver. Hexokinase remained unchanged in regenerating liver, but was elevated in fetal liver. Of interest was the observation that, while total pyruvate kinase activity was lower in regenerating liver, PK-K_4 increased considerably. In fetal liver, too, PK-K_4 predominated at first, and was only later replaced by PK-L_4. Aldolase B only appeared in fetal liver at a late state of development (34); fetal hepatocytes contain aldolase A, B, and C subunits (44). One problem in studying fetal tissues is that they contain a large proportion of hemopoietic cells, which contribute their own isozyme patterns. Most studies to date have not specifically examined hepatocyte isozymes. However, the data discussed above show that both regenerating and fetal liver show many of the same isozyme differences as do malignant tumors, hence, due caution must be observed in interpreting whether a biological marker is truly representative of malignancy, or simply represents alteration of adult differentiated tissue enzyme patterns due to growth.

The effects of the carcinogen 3'-MeDAB on glycolytic enzymes have also been examined (34). Glucokinase activity decreased within a few days of carcinogen feeding, while hexokinase increased. Total pyruvate kinase activity decreased, but PK-K_4 increased. These changes occurred before the tissue was committed to becoming malignant, presumably as a result of the toxic effects of the carcinogen.

Recently, detailed studies of aldolase isozyme changes on feeding of toxic substances have been carried out by Schapira's group, using immunofluorescence to localize the A, B, and C subunits. In differentiated adult livers, only B subunits were found in hepatocytes, and A subunits occurred in Kupffer cells (45). Following ingestion of carbon tetrachloride (54), A and C subunits increased, but the increases were limited to the Kupffer cells. Similarly, after ingestion of the carcinogen 3'-MeDAB, A and C subunits appeared in oval and transitional cells, and at a late stage, in sinusoidal cells, while the pre-neoplastic hepatocytes had only aldolase B (55). Only committed hepatoma cells in the hyperplastic nodules showed "switch on" of the genes for aldolase A and C subunits, which occurred there in addition to B subunits. The only other cells in which A, B, and C subunits all occurred were fetal hepatocytes. Thus, the hepatoma aldolase pattern clearly shows similarity to that of fetal hepatocytes. Studies of this kind should be useful for carcinogenesis research, since precancerous cells do not appear to show the isozyme changes observed in the tumor cell.

EFFECTS OF TUMOR ON HOST

The presence of a tumor in an animal or human has been shown to affect the enzyme complement of distant, uninvolved tissues (56,57). Our own studies showed considerably higher levels of thymidylate kinase and DNA polymerase in apparently uninvolved portions of the livers of hepatoma patients than in livers of patients without cancer (4). Also, the hexokinase and pyruvate kinase isozyme patterns of human host livers were sometimes indistinguishable from those of the hepatomas, or were intermediate between those of normal liver and hepatomas (18). Suda (57) first showed increases in hexokinase and pyruvate kinase levels in livers of tumor-bearing animals. Isozyme studies showed that the increases were in HK II and PK-K_4. We have very recently examined this problem using the Dunning mammary carcinoma 5A, transplanted to the backs of Fischer rats. Steady increases of HK II and PK-K_4 activities were observed in the livers from 8 to 15 days after tumor transplantation. The changes were reversible, since removal of the tumor led to decreases in these isozymes. Suda (57) showed that such changes could even occur in livers of animals linked parabiotically to tumor-bearers, showing that contact with the tumor itself was not necessary, but that a humoral factor secreted by the tumor might be responsible for the changes.

SUMMARY AND CONCLUSIONS

In summary, it is evident from the data discussed that altered levels of some enzymes occur in tumors; these alterations depend on the tissue of origin and degree of malignancy of the tumor. Thus, lactate dehydrogenase increases in breast and lung tumors, but decreases in liver tumors. The general pattern is for an increase in enzymes of DNA synthesis and glycolysis, and a decrease in tissue-specific enzymes, including those of gluconeogenesis. More distinctive are isozyme patterns, which frequently differ in tumors from those found in the adult differentiated tissue of origin, sometimes reverting to the fetal pattern. Human tumors in many cases show similar changes to those observed in animal tumors, but there are sufficient differences to warrant study of both systems. Thus, thymidine kinase activity increases in fast-growing rat hepatomas, but is unchanged in human hepatomas. On the other hand, the level of this enzyme in normal human liver is much higher than that in normal rat liver. Due caution must be exercised in interpreting data obtained in studies with carcinogens, since tumor-like changes may occur in tissues during regeneration and in the presence of toxic substances, at a stage where the tissues are not yet committed to cancer. Moreover, similar changes may also occur in uninvolved tissues of the tumor-bearer. This shows that some changes are growth-related (fetal, regenerating), while others may be due to induction of abnormal genetic expression as a result of humoral factors from the tumor. Precise cellular localization of the isozyme changes, e.g., by immunofluorescence, may help to pinpoint the changes more clearly.

ACKNOWLEDGEMENTS

Studies conducted in the author's laboratory were supported by contract N01-CB-84222 and by grants from the South African Medical Research Council, the Council for Scientific and Industrial Research, and the Atomic Energy Board of South Africa.

REFERENCES

1. Weber, G., and M.A. Lea (1966) The molecular correlation

concept of neoplasia. Adv. Enz. Reg. 4:115-145.

2. Weber, G., S.F. Queener, and J.A. Ferdinandus (1971) Control of gene expression in carbohydrate, pyrimidine and DNA metabolism. Adv. Enz. Reg. 9:63-95.

3. Sneider, T.W., V.R. Potter, and H.P. Morris (1969) Enzymes of thymidine triphosphate synthesis in selected Morris hepatomas. Cancer Res. 29:40-54.

4. Cummins, R.R., and D. Balinsky (1980) Activities of some enzymes of pyrimidine and DNA synthesis in a rat transplantable hepatoma and human primary hepatomas, in cell lines derived from these tissues, and in human fetal liver. Cancer Res. 40:1235-1239.

5. Gordon, H.L., T.J. Bardos, Z.F. Chmielewicz, and J.L. Ambrus (1968) Comparative study of the thymidine kinase and thymidylate kinase activities and of the feedback inhibition of thymidine kinase in normal and neoplastic human tissue. Cancer Res. 28:2068-2077.

6. Laszlo, J., P. Ove, W.W. Shingleton, and H.P. Morris (1969) DNA replication and degradation in mammalian tissues. V. A series of slow-growing rat renal tumors and a variety of human cancers. J. Natl. Cancer Inst. 43:1331-1336.

6a. Maehara, Y., H. Nakamura, Y. Nakane, K. Kawai, S. Okamoto, S. Nagayama, T. Shirasaka, and S. Fujii (1982) Activities of various enzymes of pyrimidine nucleotide and DNA synthesis in normal and neoplastic human tissues. Gann 73:289-298.

7. Taylor, A.T., M.A. Stafford, and O.W. Jones (1972) Properties of thymidine kinase partially purified from human fetal and adult tissue. J. Biol. Chem. 247:1930-1935.

8. Cummins, R.R., and D. Balinsky (1979) Thymidine kinases from normal human liver and hepatoma. Isozyme Bull. 12:36.

9. Baril, E., and J. Laszlo (1970) Sub-cellular localization and characterization of DNA polymerases from rat liver and hepatomas. Adv. Enz. Reg. 9:183-204.

10. Warburg, O., K. Posener, and E. Negelein (1924) Über den Stoffwechsel der Carcinomzelle. Biochem. Z. 152:309-344.

11. Morris, H.P., and B.P. Wagner (1968) Induction and transplantation of rat hepatomas with different growth rate (including "minimal deviation" hepatomas). Methods Cancer Res. 4:125-152.

12. Elwood, J.C., Y.C. Lin, V.J. Cristofalo, S. Weinhouse, and H.P. Morris (1963) Glucose utilization in homogenates of the Morris hepatoma 5123 and related tumors. Cancer Res. 23:906-913.

13. Balinsky, D., E. Cayanis, E. Geddes, and I. Bersohn (1973) Activities and isoenzyme patterns of some enzymes of glucose metabolism in human primary malignant hepatoma. Cancer Res. 33:249-255.

14. Rosado, A., H.P. Morris, and S. Weinhouse (1969) Lactate dehydrogenase subunits in normal and neoplastic tissues of the rat. Cancer Res. 29:1673-1680.

15. Balinsky, D., O. Greengard, J.F. Head, E. Cayanis, and S.L. Goldberg (1983) Enzyme activities and isozyme patterns in human lung tumors (submitted for publication).
16. Balinsky, D., C.P. Platz, and J.W. Lewis (1983) Enzyme activities in normal, benign and malignant human breast tissues (submitted for publication).
17. Hilf, R., J.L. Wittliff, W.D. Rector, E.D. Savlov, T.C. Hall, and R.A. Orlando (1973) Studies on certain cytoplasmic enzymes and specific estrogen receptors in human breast cancer and in nonmalignant diseases of the breast. Cancer Res. 33:2054-2062.
18. Hammond, K.D., and D. Balinsky (1978) Activities of key gluconeogenic enzymes and glycogen synthetase in rat and human livers, hepatomas and hepatoma cell cultures. Cancer Res. 38:1317-1322.
19. Goldman, R.D., N.O. Kaplan, and T.C. Hall (1964) Lactic dehydrogenase in human neoplastic tissues. Cancer Res. 24:389-399.
20. Hilf, R., W.D. Rector, and R.A. Orlando (1976) Multiple molecular forms of lactate dehydrogenase and glucose 6-phosphate dehydrogenase in normal and abnormal human breast tissues. Cancer 37:1825-1830.
21. Kaplan, N.O., and T.L. Goodfriend (1964) Role of the two types of lactic dehydrogenase. Adv. Enz. Reg. 2:203-212.
22. Schapira, F., C. Micheau, and C. Junien (1972) Rev. Eur. Etud. Clin. Biol. 17:896.
23. Siciliano, M.J., M.E. Bordelon-Riser, R.S. Freedman, and P.O. Kohler (1980) A human trophoblastic isozyme (lactate dehydrogenase-Z) associated with choriocarcinoma. Cancer Res. 40:283-287.
24. Anderson, G.R., and W.P. Kovacik, Jr. (1981) LDH_k, an unusual oxygen-sensitive lactate dehydrogenase expressed in human cancer. Proc. Natl. Acad. Sci., USA 78:3209-3213.
25. Hershey, F.B., G. Johnston, S.M. Murphy, and M. Schmitt (1966) Pyridine nucleotide-linked dehydrogenases and isozymes of normal rat breast and growing and regressing breast cancers. Cancer Res. 26:265-268.
26. Ibsen, K.H. (1977) Interrelationships and functions of pyruvate kinase isozymes and their variant forms: A review. Cancer Res. 37:341-353.
27. Blair, J.B. (1981) Regulatory properties of hepatic pyruvate kinase. In The Regulation of Carbohydrate Formation and Utilization in Mammals, C.M. Veneziale, ed. University Park Press, Baltimore pp. 121-151.
28. Imamura, K., and T. Tanaka (1972) Multimolecular forms of pyruvate kinase from rat and other mammalian tissues: I. Electrophoretic studies. J. Biochem. (Tokyo) 71:1043-1051.
29. Noguchi, T., and T. Tanaka (1982) The M_1 and M_2 subunits of rat pyruvate kinase are encoded by different messenger RNAs. J. Biol. Chem. 257:1110-1113.

30. Marie, J., M.P. Simon, J.C. Dreyfus, and A. Kahn (1981) One gene, but two messenger RNAs encode liver L and red cell L' pyruvate kinase subunits. Nature 292:70-72.
31. Marie, J., H. Garreau, and A. Kahn (1977) Evidence for a postsynthetic proteolytic transformation of human erythrocyte pyruvate kinase into L-type enzyme. FEBS Lett., 78:91-94.
32. Balinsky, D., E. Cayanis, and I. Bersohn (1973) Comparative kinetic properties of human pyruvate kinases isolated from adult and fetal livers and from hepatoma. Biochemistry 12:863-870.
33. Balinsky, D., E. Cayanis, C.F. Albrecht, and I. Bersohn (1972) Enzymes of carbohydrate metabolism in rat hepatoma induced by 3'-methyl-4-dimethylaminoazobenzene (3'-MeDAB). South African J. Med. Sci. 37:95-99.
34. Walker, P.R. and V.R. Potter (1972) Isozyme studies on adult, regenerating, precancerous and developing liver in relation to findings in hepatomas. Adv. Enz. Reg. 10:339-364.
35. Farina, F.A., J.B. Shatton, H.P. Morris, and S. Weinhouse (1974) Isozymes of pyruvate kinase in liver and hepatomas of the rat. Cancer Res. 34:1439-1446.
36. Van Veelen, C.V.M., H. Verbiest, A.M.C. Vlug, G. Rijksen, and G.E.J. Staal (1978) Isozymes of pyruvate kinase from human brain, meningiomas, and malignant gliomas. Cancer Res. 38:4861-4867.
37. van Veelen, C.W.M., H. Verbiest, K. Zülch, B.A. van Ketel, M.J.M. van der Vist, A.M.C. Vlug, G. Rijksen, and G.E.J. Staal (1979) L-alpha-alanine inhibition of pyruvate kinase from tumors of the human central nervous system. Cancer Res. 39:4263-4269.
38. Beemer, F.A. A.M.C. Vlug, G. Rijksen, and G.E.J. Staal (1982) Hexokinase, pyruvate kinase and aldolase from fetal and adult retina, retino-blastoma cell line, and retinoblastoma. Proc. 4th Intl. Conf. on Isozymes, Austin, Texas.
39. Lopez-Alarcon, L., P. Ruiz, and M. Gosalvez (1981) Determination of the degree of differentiation of mammary tumors by pyruvate kinase analysis. Cancer Res. 41:2019-2020.
40. Ibsen, K.H., R.A. Orlando, K.N. Garratt, A.M. Hernandez, S. Giorlando, and G. Nungaray (1982) Expression of multimolecular forms of pyruvate kinase in normal, benign and malignant human breast tissue. Cancer Res. 42:888-892.
41. Kamel, R., and F. Schwarzfischer (1975) Hexokinase isozymes in human neoplastic and fetal tissues: The existence of hexokinase II in malignant tumors and in placenta. Humangenetik 30:181- 185.
42. Sato, S., T. Matsushima, and T. Sugimura (1969) Hexokinase isozyme patterns of experimental hepatomas of rats. Cancer Res. 29:1437-1446.
43. Weinhouse, S., J.B. Shatton, W.E. Criss, and H.P. Morris (1972) Isozymes in relation to differentiation in transplantable rat hepatomas. Gann Monograph 13:1-17.

44. Hatzfeld, A., A. Weber, and F. Schapira (1975) Biochemical and immunological studies of some carcinofetal enzymes. Ann. N.Y. Acad. Sci. 259:287-297.
45. Hatzfeld, A., G. Feldmann, J. Guesnon, C. Fraysinnet, and F. Schapira (1978) Location of adult and fetal aldolases A, B, and C by immunoperoxidase technique in LF fast-growing rat hepatomas. Cancer Res. 38:16-22.
46. Fishman, W.H., N.R. Inglis, L.L. Stolbach, and M.J. Kraut (1968) A serum alkaline phosphatase isoenzyme of human neoplastic cell origin. Cancer Res. 28:150-154.
47. Fishman, W.H. (1974) Perspectives on alkaline phosphatase isoenzymes. Amer. J. Med. 56:617-650.
48. Nakayama, T., M. Yoshida, and M. Kitamura (1970) L-leucine sensitive, heat-stable alkaline-phosphatase isoenzyme detected in a patient with pleuritis carcinomatosa. Clin. Chim. Acta 30:546-548.
49. Higashino, K., M. Hashinotsume, K.-Y Yang, Y. Takahashi, and Y. Yamamura (1972) Studies on a variant alkaline phosphatase in sera of patients with hepatocellular carcinoma. Clin. Chim. Acta 40:67-81.
50. Fishman, L., H. Miyahama, S.G. Driscoll, and W.H. Fishman (1976) Developmental phase-specific alkaline phosphatase isoenzymes of human placenta and their occurrence in human cancer. Cancer Res. 36:2268-2273.
51. Fishman, W.H., T. Nishiyama, A. Rule, S. Green, N.R. Inglis, and L. Fishman (1976) Onco-developmental alkaline phosphatase isozymes. In Onco-Developmental Gene Expression, W.H. Fishman and S. Sell, eds. Academic Press, New York, pp. 165-176.
52. Balinsky, D., P.S. Jenkins, C.E. Platz, J.W. Lewis, and M.C. Kew (1981) Esterases in human cancer. Proc. Amer. Assoc. Cancer Res. 22:35.
53. Coates, P.M., M.A. Mestriner, and D.A. Hopkinson (1975) A preliminary genetic interpretation of the esterase isozymes of human tissues. Ann. Hum. Genet. 39:1-20.
54. Guillouzo, A., A. Weber, E. LeProvost, M. Rissel, and F. Schapira (1981) Cell-types involved in the expression of foetal aldolases during rat azo-dye hepatocarcinogenesis. J. Cell. Sci. 49:249-260.
55. LeProvost, E., A. Weber, F. Schapira, M. Boisnard-Rissel, and A. Guillouzo (1980) Immunolocalization of fetal aldolase isoenzymes in rat regenerating liver after carbon tetrachloride intoxication. Oncodevelop. Biol. Med. 1:263-272.
56. Herzfeld, A., and O. Greengard (1972) Effect of neoplasms on the content and activity of alkaline phosphatase and gamma-glutamyl transpeptidase in uninvolved host tissues. Cancer Res. 32:1826-1832.
57. Suda, M., T. Tanaka, S. Yanagi, S. Hayashi, K. Imamura, and K. Taniuchi (1972) Dedifferentiation of enzymes in the liver of tumor-bearing animals. Gann Monograph 13:79-93.

ENZYME AND ISOENZYME PATTERNS AS POTENTIAL MARKERS OF NEOPLASIA IN BREAST TISSUE

Russell Hilf

Department of Biochemistry and
University of Rochester Cancer Center
University of Rochester School of Medicine and Dentistry
Rochester, New York 14642

INTRODUCTION

In a recent article, Weinhouse (51) reviewed and discussed some of the concerns relative to the significance of isoenzyme alterations reported in hepatomas, the most widely studied models of neoplasia from the biochemist's viewpoint. He cautions investigators regarding the question of definition of an isoenzyme, i.e., products from different genes vs. post-translational modifications of proteins; the distinction that must be made between physiological alterations and those that are inherent in the lesion; the need to establish the existence of an isoenzyme as a normal component of the tissue at various developmental stages vs. that seen in the transformed tissue; the difficult and unresolved question of tumor heterogeneity; and the extent to which differences in rates of synthesis and degradation, along with other intracellular factors, may alter the actual enzyme activities measured. Despite these reservations, he optimistically concludes that the altered protein composition seen in cancer cells offers an opportunity to elucidate the functional significance of such changes; it offers a promising avenue in the areas of prognosis and management of the disease.

It should also be noted that not all enzymes exhibiting multiple molecular forms are necessarily isoenzymes. The International Union of Biochemists (IUB) Commission on Enzyme Nomenclature has recommended that proper use of the term "isoenzyme" should be restricted to enzymes arising from genetically determined differences in primary structure. However, the term "isoenzyme" is being used in an operational sense when dealing with enzymes demonstrating the same catalytic activities but separable by suitable methods, such as electrophoresis, and where the basis for the multiple forms is lacking. For the purposes here, the operational definition of isoenzyme will be used.

For an isoenzyme to fulfill the ideal role as a diagnostic parameter of neoplasia, it would require that its existence and/or regulation must be elucidated in the non-transformed (normal) tissue. This represents a more complicated situation for studies of breast cancer, since the normal mammary gland can exist in several physiological states. It becomes necessary to establish whether the isoenzyme can be observed in the quiescent gland, in the mammary gland during pregnancy and lactation, and in the gland when it undergoes involution after lactation ceases. Comparison of the mature gland with the fetal gland, akin to those studies done with liver, is difficult, to say the least. The amount of tissue available for study of the mammary gland anlage precludes all but a histochemical approach. For the quiescent gland, even in the mature animal, the predominance of adipose tissue represents an analytical problem, since it is not the adipocyte that becomes neoplastic. Analysis of whole-gland homogenates will yield results that must be interpreted with caution due to cellular heterogeneity. Nevertheless, study of isoenzyme composition of breast cancers could reveal patterns of potential markers and efforts to assign their presence in neoplastic vs. normal tissues are justifiable as a prelude for the more discrete studies relating these enzyme forms to transformation. Hence, initial reservations should be accorded to reports of putative unique isoenzyme markers of breast cancer, in particular for the human disease.

No attempt will be made to summarize all reports dealing with isoenzymes in breast cancer in this chapter. Rather, several selected systems will be discussed from studies of experimental animals as a basis for a summary of selected studies with human tissues. A number of reviews exist on the broader aspects of isoenzymes and cancer (4,16,22,39), but none have specifically dealt with mammary cancers.

Isoenzyme Studies in Rodent Tissues

Hexokinase-glucokinase. Studied more intensively in hepatoma models, a few reports have examined these isoenzymes in mammary glands from pregnant and lactating rats, using starch gel electrophoresis as the separation technique (50). Shatton et al. (40), in an extensive study, reported that two main species (I and II) of hexokinase were found in mammary glands from pregnant and lactating rats, but there were faint bands for isoenzymes III and IV. In contrast, in dimethylbenz(a)anthracene (DMBA)-induced primary mammary tumors, all four isoenzymes were clearly detectable, with strong activity displayed by isoenzyme III. Interestingly, methylcholanthrene-induced mammary tumors, while displaying all four isoenzymes, demonstrated a somewhat different pattern of intensity; isoenzyme IV was weaker than that seen in DMBA-induced tumors.

Farron (7) reported that the mammary gland from pregnant and lactating rats demonstrated type III hexokinase, which was not demonstrable in the gland from virgin rats (note that this latter sample would be primarily adipose tissue since no details were given for preparation of epithelial cells from the gland). Three transplantable tumors all demonstrated considerable amounts of type III hexokinase, results similar to those of Shatton et al. (40) for primary mammary tumors. Thus, it would appear that the mammary tumors display an isoenzyme of hexokinase that is either not expressed or expressed only slightly in normal tissue.

Glutaminase. Knox et al. (23) studied a series of nine transplantable rat mammary tumors, measuring the phosphate-dependent (kidney type) glutaminase activity. The enzyme was absent in the mammary gland from virgin or lactating animals, but was present in all of the tumors. Further, it was reported that a high correlation was observed between glutaminase activity and growth rate of the tumors, and a correlation was also seen between enzyme activity and morphological estimation of undifferentiation. It is surprising that these findings have not, to my knowledge, been extended.

Pyruvate kinase. This has been actively studied in liver and hepatomas where it appears that hepatomas express considerably elevated levels of the kidney (K) isoenzymes, thus demonstrating a shift from the normal adult hepatocyte pattern predominated by the liver (L) form (8,17,45). The K form predominates in most other tissues [except for the muscle (M) form in skeletal muscle and brain], including the mammary gland. It is, therefore, not surprising that studies have not been performed on mammary tumors since they would not be expected to demonstrate any marked differences from the normal gland. However, a recent study by Ibsen et al. (18) on human breast tissues suggests the presence of species that differ based on isoelectric properties (see later explanation).

Aldolase. Electrophoretic separation of aldolase revealed that only the lactating rat mammary gland displayed significant levels of the C isoenzyme. Glands from the pregnant animal, as well as three of four transplantable mammary tumors, showed the presence of the A form of aldolase. The slowest-growing transplantable mammary tumor, however, retained a low level of C type isoenzyme (8).

Lactate dehydrogenase (LDH). The classical demonstrations that neoplastic tissues produce considerable amounts of lactic acid, even under aerobic conditions, has been the subject of numerous studies. The demonstration of Markert and his co-workers (29,30) that LDH was a tetramer consisting of varying proportions of two different monomers was seminal in the development of isoenzymic concepts, including the term "isoenzyme". A mixture of equal concentrations of each subunit, when allowed to recombine, gave five LDH bands by electrophoretic analysis; the species observed occurred in a ratio that would have been predicted for a random association of two different subunits to form tetramers. Kaplan (20) postulated that there were differences in the functional role of each subunit; the muscle (M) isoenzyme functioned best under anaerobic conditions, whereas the heart (H) isoenzyme was more functional under aerobic conditions. Tumors, which are considered to be more anaerobic, would possess higher amounts of the M-type subunit, a general but not universal finding. Indeed, liver, which is considered to be an aerobic tissue, has a preponderance of M-type subunits and in some human hepatomas, there was an increase in the proportion of H-type subunits (3). The functional difference between the isoenzymes was challenged by Wuntch et al. (54), who could not demonstrate any difference between the M or H isoenzyme in their sensitivity for inhibition by pyruvate under conditions that more closely approximated the estimated tissue concentrations of the enzyme.

The development of polyacrylamide gel electrophoretic techniques and optical equipment for scanning the stained gels significantly aided the investigation of isoenzyme composition of normal and neoplastic tissues (6). As with most techniques, caution is necessary when reaching conclusions about changes in enzyme activity after electrophoretic separation and assay. Careful attention to substrate and co-factor requirements, reproducibility, extraction of tissues, and specificity of the reactions is required to standardize the procedures, particularly in situations where a shift in isoenzyme composition may occur compared to the expression of a different isoenzyme species.

One of the first reports of LDH isoenzymes in rodent mammary tumors appeared in 1966 (10). In comparison to normal breast tissue, starch gel electrophoresis of extracts of methylcholanthrene-induced mammary tumors displayed a predominance of the M_4 (LDH-5) species with little or no H subunit-containing species (H_4, H_3M, and H_2M_2 or LDH-1, LDH-2, and LDH-3). Ovariectomy produced no change in the isoenzyme pattern, although the total LDH activity decreased. Richards and Hilf (35) examined the effects of ovariectomy on LDH isoenzymes in dimethylbenzanthracene-induced mammary tumors. Utilizing acrylamide disc gel electrophoresis, followed by staining and spectrophotometric scanning, they reported that the decrease in the total LDH activity was attributable to significant decreases in the M_4 (LDH-5) and M_3H (LDH-4) species. In agreement with Hershey et al. (10), DMBA-induced mammary tumors also displayed very low levels of H-subunit containing species; H_4 (LDH-1) was essentially absent and H_3M (LDH-2) represented approximately 1% of the total LDH activity and was unchanged after ovariectomy-induced tumor regression. Essentially identical results were reported by Lee et al. (24) for the proportion of LDH isoenzyme species in DMBA-induced tumors and for the reduction in M subunits after ovariectomy. They (24) proposed that the consistently undetectable LDH-1 in DMBA-induced tumors could be used as a biological marker.

The pattern of LDH isoenzymes in neoplastic tissues has been compared to that observed in normal rodent mammary glands, examined at various times during pregnancy and lactation, and also after weaning of the young. Reports by Karlson and Carlsson (21) and Simpson and Schmidt (43) indicated that when compared to the gland from virgin animals, the mammary gland during pregnancy and lactation demonstrated a shift in isoenzyme profile towards an increase in LDH-5 (M_4). In the former report, it was stated that LDH-5 was absent in the gland from virgin or early pregnant rats, whereas Simpson and Schmidt (43) observed all five LDH isoenzymes in the quiescent gland. A study from our laboratory (37) demonstrated the presence of all five LDH isoenzymes in the mammary gland, with the major changes occurring in LDH-4 and LDH-5 during late pregnancy and a striking increase in activity seen during mid- to late-lactation. Curiously, although there was a significant decrease in LDH activity after removal of the pups, the decrease did not reach the low levels of activity seen in the glands from virgin rats. Our data agree with the suggestion that the gland, stimulated by the hormonal milieu during pregnancy and lactation, contains more LDH-4 and LDH-5 than the quiescent gland. Lee et al. (24) reported that all five LDH isoenzymes were detectable in the mammary gland from nulliparous rats and that there was a shift in the pattern towards the M-type subunit-containing species during pregnancy and lactation.

Thus, it appears that no unique LDH isoenzyme species was found in mammary carcinomas in rats; rather, neoplastic tissues demonstrated a significant increase in the proportion of M subunit-containing species, LDH-4, and LDH-5. It is interesting that the gland from the lactating animal also possessed a higher proportion of these two species and on this basis, distinction between these two tissues solely by isoenzyme patterns is not likely. The suggestion by Lee et al. (24) that malignant transformation was accompanied by a loss of LDH-1 was not seen by others studying the same or different experimental mammary tumors. Considering the relatively low proportion that LDH-1 represents in normal tissues, and the analytical difficulties that may be encountered, it may not offer a readily identifiable marker of transformation.

<u>Glucose 6-phosphate dehydrogenase (G6PD)</u>. This enzyme exists in genetically independent forms, genetic variants as a result of mutations, multiple enzyme forms resulting from polymers of a single subunit and conformational isomers, and forms resulting from post-translational modifications. An excellent comprehensive review of G6PD appeared in 1979 (25). In the literature cited below, it is most likely that multiple molecular forms of G6PD were being measured by the electrophoretic methods used.

Multiple forms of G6PD have been reported for rat liver, ranging from 3 bands (46) to 7 bands (15); for rat uterus, with 4 bands in endometrial and myometrial preparations (38); and 4 G6PD bands for rat ventral prostate and seminal vesicles (34). Hormonal stimulation of these reproductive organs resulted in increased G6PD activity, primarily reflected as increased G6PD-1 levels (fastest migrating species was called G6PD-1, with successively slower migrating bands labeled 2,3,4, etc.). However, some subtle additional changes were seen depending on the tissue studied (34,38).

For the mammary gland, Hershey et al. (10) reported the presence of two G6PD species, but stated that the faster migrating band was not specific for glucose 6-phosphate or nucleotide adenine diphosphatase (NADP). We examined G6PD species in rat mammary glands during pregnancy and lactation, using acrylamide gel electrophoresis followed by densitometric scanning (37). Three G6PD species were observed; the fastest migrating species (G6PD-1) was essentially absent in the glands from virgin and early pregnant rats, appeared at mid-pregnancy and increased markedly during lactation to become the predominant species. Involution of the gland caused a return to the pattern seen in the quiescent state. The slowest migrating species, G6PD-3, showed little change over the period of study, whereas the intermediate band showed a modest increase during pregnancy and lactation.

In the mouse, however, a somewhat different pattern was observed (11). The slowest migrating species, G6PD-III, was essentially absent during pregnancy but was increased during lactation to comprise about 50% of total G6PD activity. Shreve and Levy (41) reported that the minor form of G6PD from the lactating glands of rats was a molecular weight isomer of the main form, based on electrophoretic mobilities in acrylamide gels at various concentrations; a similar suggestion was made by Hilf et al. (12) for the three major forms of G6PD seen in lactating glands from mice examined by the same techniques. Patterns of G6PD bands may be influenced by oxidation or reduction of cysteine residues; possible relationships between reduced glutathione (GSH) levels and G6PD forms were suggested (12).

In either DMBA-induced or R3230AC transplantable mammary tumors of rats, two G6PD species were observed after electrophoresis on acrylamide gels (35,36). Shreve and Levy (41) also observed two electrophoretic forms of G6PD in R3230AC tumors, and suggested that the minor form in the tumor was a charge isomer, whereas in the normal gland, the minor form was a molecular weight isomer. Only the faster migrating G6PD species displayed hormone-induced changes, either decreasing after ovariectomy or increasing after the administration of estrogen. A somewhat different pattern was seen in mouse mammary tumors and in hyperplastic alveolar nodules (HAN), a morphologically identifiable entity that has an increased neoplastic potential and can be considered to be a pre-neoplastic. In our initial report (13), we observed the presence of significant levels of a rapidly migrating G6PD species in both pre-neoplastic and neoplastic tissues, a form that was virtually absent from the normal gland. In a more extensive study, we essentially confirmed this finding, although low levels of this fastest-migrating species (G6PD-I) were observed in the gland from pregnant mice. Incubation of tissue supernatant preparation with dithiothreitol prior to electrophoresis produced a marked increase in the proportion of G6PD-I in preparations from HAN and adenocarcinomas, with G6PD-I representing about 65% of the total; this was not the case for preparations of normal tissue (11). Subsequently, we demonstrated that the activity of glutathione reductase and the levels of GSH were significantly higher in the abnormal tissues and we proposed that there was a relationship between the distribution of G6PD species and the redox state of the tissue (12). Thus, the experimental turmors of rats resembled the gland of lactation whereas in mice, the tumors more closely resembled the gland during pregnancy. No G6PD species truly unique to transformed cells was found, although the fastest migrating species, and its enhancement by dithiothreitol, offered a potential marker of neoplastic tissue.

Ca++-stimulated ribonuclease. Liu et al. (26) reported on the presence of a Ca++-stimulated ribonuclease in the lactating rat mammary gland, an activity absent from the quiescent (virgin) gland, liver, kidney, lung, uterus, and spleen. It was also absent from the 3924A hepatoma, but enzyme activity was clearly measurable in R3230AC mammary tumors, two spontaneous mammary adenocarcinomas and one spontaneous mammary fibroadenoma. Additional studies conducted by these investigators indicated that this enzyme activity was found in the milk of various species, the highest activity being found in rat, rabbit, and cow (27). It would appear that this ribonuclease could serve as a marker of well-differentiated mammary lesions and further study would seem warranted.

Isoenzyme Studies in Human Breast Cancer

Studies of mammary cancer in animals benefit from the opportunity to examine alterations occurring under a variety of normal physiological conditions, such as pregnancy and lactation. With such data as a baseline, it is possible to identify putative enzyme species unique to neoplasia. Unfortunately, such opportunities do not exist for study of human disease, since the likelihood of obtaining and examining such sequential samples is remote. Because of this, we and others have opted to investigate isoenzymes in breast cancer tissues in comparison with those from samples of non-neoplastic diseases of the breast, as well as in samples classified as "normal". Normal can be defined as the absence of abnormality by light microscopic criteria and usually represents quiescent breast tissue, containing alveolar and ductal epithelium, myoepithelium, connective tissue, and a predominance of adipose tissue. Since it was reasoned that changes in enzymes might reflect proliferative activity, analysis of breast lesions classified as fibroadenoma or fibrocystic disease could offer examples of non-malignant proliferating tissues for comparison with malignant, proliferating tissues. Whether some sub-set of these non-malignant tissues may be pre-neoplastic is an important consideration requiring further examination. Thus, the patterns obtained in the malignant lesions and not shared by benign lesions could offer a potentially useful marker of neoplasia.

Lactate dehydrogenase. One of the first comprehensive reports on LDH in human neoplastic tissues appeared in 1964 (9). The coenzyme analog ratio method was used to obtain the percent LDH-M in the sample, and this was corroborated by starch gel electrophoresis. Malignant human neoplasms displayed an increase in the muscle-type LDH, although no correlation was observed between such changes and histologic grading. For breast tissues, carcinomas displayed from 54% to 96% LDH-M, whereas, benign disease samples, which included fibrocystic disease and fibroadenomas, had 33% to 66% LDH-M. The question arose whether these changes in LDH were representative of a cause or an effect of malignant transformation. Using acrylamide gel electrophoresis and spectrophotometric scanning, we examined LDH isoenzymes in samples of normal, fibrocystic disease, fibroadenoma, and carcinomatous breast tissues (14). Specimens of infiltrating ductal carcinomas demonstrated a 3- to 6-fold increase in the proportion of LDH-5 compared to samples of normal and benign diseases, clearly indicating a shift towards the muscle type LDH component. Normal tissues contained 43% of the muscle-type isoenzyme whereas carcinomas contained about 55% of the muscle-type isoenzyme. Menopausal status had little or no effect on the distribution of isoenzymes in any of the tissues studied. Since the neoplasms displayed a significant increase in total LDH activity, the increase in the proportion of LDH-5 represented an absolute elevation in the amount of muscle-type LDH in the cancers.

Our results confirm and extend those of Goldman et al. (19). However, these results differ somewhat from those of Stanislowski-Birencwajg and Loisillier (44), who reported that the LDH-5/LDH-1 ratio for mastitis and fibroadenomas demonstrated a mean of > 3.0, mammary tissues adjacent to carcinoma had an average ratio of > 6.0, and carcinomas had an average ratio of > 19. Although their data showed a shift towards the muscle-type components, the proportion of M- and H-type LDH components was much higher than in our report (14). Recently, Balinsky et al. (personal communication) also reported that LDH-5 was significantly increased in malignant breast cancers; their data show a higher proportion of the muscle-type isoenzyme in non-malignant tissues than in our study. The reasons for these discrepancies are not known, but it may be due to the different separation medium employed, i.e., acrylamide gels vs. agarose or cellulose acetate.

An interesting recent development is the report of an unusual LDH isoenzyme, originally described in cells transformed by Kirsten murine sarcoma virus but not found in cells transformed by other agents. Activity of this unusual enzyme species, called LDH_k, was markedly elevated in four of five of the human breast carcinomas examined compared to that measured in adjacent, non-tumorous breast tissue (1). This isoenzyme species possesses some unusual properties (compared to LDH species from muscle or heart); LDH_k activity was only expressed under anaerobic atmosphere or in the presence of cyanide. It appears to have a subunit composition comrised of a 35,000 dalton polypeptide plus a 21,000 dalton polypeptide and it demonstrates more basic properties (pI 8.5) than the common LDH species (2). Curiously, the 21,000 dalton polypeptide resembles, in both size and nucleotide binding properties, a polypeptide induced by anaerobic stress. Anderson et al. (2) point out that the subunit structure of cytochrome LDH in yeast is described as having four 57,000 dalton units, each unit composed of a 35,000 and a 22,000 dalton subunit. The yeast enzyme is apparently involved in shifting from aerobic to anaerobic metabolism. While it would be premature to imply a viral etiology of human cancers on the basis of the presence of LDH_k, the unusual activity expressed by LDH_k warrants a more detailed study of its potential as a neoplastic marker.

Glucose 6-phosphate dehydrogenase. A study of multiple forms of G6PD was performed on the same spectrum of human tissues as described above for LDH (14). Three forms of G6PD were noted after acrylamide gel electrophoresis. No differences were observed among samples of normal breast, fibroadenomas, and fibrocystic disease. However, the fastest migrating species, G6PD-I, was significantly elevated in infiltrating ductal carcinomas, concomitant with a significant decrease in the proportion of the slowest migrating species, G6PD-III. These results indicated that, unlike the mouse, the human tissues did not expreses a G6PD species that might distinguish neoplastic from non-neoplastic tissues. Whether the increase in G6PD-I in neoplastic tissue would be seen in normal tissue during pregnancy and lactation is not known.

Pyruvate kinase. Ibsen et al. (18) examined multiple molecular forms of pyruvate kinase in normal, benign, and malignant human breast tissues, using isoelectric focussing and kinetic techniques to characterize the distributions of enzyme species. The predominant form had characteristics consistent with the K_4 isoenzyme. In addition to the K_4 isoenzyme, several other pyruvate kinase forms were separated by these procedures. These forms demonstrated higher activity in neoplastic tissues vs. benign tissues. A question was raised whether the forms with higher isoelectric point (pI) properties were M-type enzyme subunits or post-translational modifications of the K isoenzyme; the latter interpretation is favored by Ibsen et al. (18) based on their kinetic data. This suggestion differs from the explanation proposed by Lopez-Alarcon et al. (28) which was based on inhibition studies of pyruvate kinase activity in human mammary carcinomas of differing degrees of differentiation. Lopez-Alarcon et al. (28) concluded that the greater inhibition of enzyme activity produced by adenosine-S'-triphosphate in the poorly differentiated tumors might be explained by the shift in composition from M- to K- type isoenzyme. Obviously, there will need to be a resolution of these interesting findings.

Aromatase. Osawa et al. (33) recently reported on multiple forms of aromatase activity in placenta, defining activity based on formation of estriol from 16-hydroxylated C-19 androgens (aromatase I) or formation of estrone plus estradiol from androstenedione (aromatase II). In six samples of breast cancer, aromatase II was measurable and confirmed by suppression with antibody reagents made to placental aromatase II. Interest in aromatase activity in breast cancer is based on the potential conversion of androgenic precursors to estrogens by the neoplasm, representing a mechanism whereby tumor growth could be stimulated in the absence of ovarian estrogens. Siiteri (42) however reported aromatase activity in only 40% of human breast cancers, and in those, the activity was quite low; he concluded that significant formation of estrogens from androstenedione by tumors was an unlikely explanation of the lack of estrogen dependence. Whether more than one form of aromatase exists in breast carcinomas remains to be shown.

Acid phosphatase. Serum acid phosphatase activity has been reported to be elevated in the course of breast cancer, and thus may be a potential marker of the disease. Tavassoli et al. (47) recently investigated the pattern of acid phosphatase obtained by acrylamide gel electrophoresis of cytosols prepared from breast cancer tissues in eighteen patients. Only one band of activity, which was inhibited by tartrate, was observed. They noted that the serum acid phosphatase was quite different in mobility and resistance to tartrate, leading them to conclude that the elevated serum acid phosphatase was not derived from the cancer tissue.

Estrogen Receptors in Breast Cancer

In any discussion of biochemical properties of breast cancer, it would be improper not to mention estrogen receptors (ER). While this appears to be out of place in this presentation, since ER are not enzymes, these receptors have prognostic value for endocrine responsiveness and thus represent markers with proven clinical application. Interestingly, they exist in multiple forms and, as such, are worth noting here. Furthermore, isoenzyme patterns seen in human neoplastic tissues may relate to state of differentiation and presence or absence of ER may also relate to such morphological criteria.

Multiple forms of ER can be discerned by centrifugation on low-salt sucrose gradients, the technique originally employed by Toft and Gorski (49) for uteri and adapted by Jensen and his colleagues for breast cancers (19). Under these conditions at low temperatures (4°C), tumors will demonstrate ER sedimenting at 8S and/or 4S; specific binding (competable by unlabeled hormone ligand) is demonstrable and similar for both species. The likely sequence of events in the cell is the initial interaction of estrogens with a cytoplasmic 4S receptor, which is activated to a complex with 5S characteristics (a dimer); this activated form translocates to the nucleus. Differences in affinity of 4S vs. 5S forms have been carefully elucidated by Notides et al. (32). The 8S form seen on sucrose gradients of tumor tissues is most likely an aggregate enhanced by the methods employed, e.g., low temperature and low ionic medium.

An apparent relationship between the presence of 8S vs 4S forms and clinical response to hormonal therapy was noted; patients whose lesions demonstrated only, or mostly, 4S ER species failed to respond favorably, whereas a larger proportion of patients with predominantly 8S ER receptors demonstrated objective improvement to these therapies (52). It was speculated that the 4S species, although capable of binding the hormonal ligand, was not indicative of an intact and functional estrogen axis, i.e., ER were not activatable, or not translocatable to the nucleus (53). If this were borne out, it offers a means to further discriminate among ER-positive patients, only some of whom favorably respond to hormonal therapy. Dao and Nemoto (5) reported that 12 of 22 patients possessing 4S species only (ranging from 11 to >100 fmol/mg cytosol protein) responded to endocrine ablative procedures. However, McCarty et al. (31) reported that only 1 of 14 patients, whose lesions displayed 4S species, responded, results that were comparable to those of Wittliff et al. (52). A caution has been raised to these and related findings, since the presence of proteases in the homogenate preparation may cause an artifactual appearance of smaller molecular forms of receptors (48). Careful evaluation of the possible relationship between molecular species present and response to hormonal therapy should be examined further as a potential marker to distinguish biological behavior.

We have, retrospectively, compared the isoenzyme patterns of LDH and G6PD of several carcinomas according to estrogen receptor status, using >10 fmols/mg cytosol protein as the cut-off for ER positivity. These results, presented in Table 1, demonstrate remarkable similarity in the isoenzyme compositions for ER+ vs.ER- tumors. It would appear from this modest number of samples that no relationship was uncovered between isoenzyme profiles and presence or absence of estrogen receptors.

CONCLUSION

Overall, the usefulness of isoenzymes as a marker for neoplasia has not been fully realized for breast cancer. Several explanations may be offered. First, the more classical findings of isoenzyme shifts have centered around studies of hepatomas in rodents. The relative "uniformity" of the liver as a normal tissue baseline, the multitude of metabolic activities of this organ, and the earlier efforts to define and characterize alterations that occur in a "minimal" transformed state have yielded some generalizations regarding isoenzymes in neoplastic liver. Difficulties encountered in simplistic translation to the mammary gland have probably discouraged many investigators. The need to establish a proper baseline for the mammary gland requires more extensive characterization due to the impact that various physiological states impose on cellular composition and activity of the mammary gland.

Table 1. Multiple molecular forms of lactate dehydrogenase and glucose-6-phosphate dehydrogenase in human breast cancers according to estrogen receptor status.

	ER Positive	ER Negative
LDH - Number of Specimens	49	32
% of Total LDH		
LDH-1	5.5 ± 0.7	5.7 ± 1.0
LDH-2	19.3 ± 1.2	18.6 ± 1.5
LDH-3	35.6 ± 1.2	33.2 ± 1.4
LDH-4	23.7 ± 1.0	26.3 ± 1.5
LDH-5	16.2 ± 1.6	17.0 ± 2.1
G6PD - Number of Specimens	37	23
% of Total G6PD		
G6PD-1	55.6 ± 4.9	58.2 ± 4.0
G6PD-2	39.9 ± 4.9	37.1 ± 4.2
G6PD-3	4.5 ± 0.7	4.8 ± 1.0

Data presented as mean ± standard error, based on analysis by triangulation of the peaks obtained from densitometric tracings.

Second, the variety of experimental models available for study of mammary cancer, each offering its own advantage (and disadvantage) for study of certain aspects of a complex disease, has diluted efforts to explore the question of isoenzyme profiles systematically. Do neoplasms of the mammary glands of mice, rats, dogs, and humans (animals sharing a high propensity for mammary cancer) have common alterations in biochemical parameters? Are mammary tumors induced by chemical carcinogens, X-irradiation, viruses, or hormones (indirectly) similar or dissimilar?

Third, careful longitudinal studies to measure isoenzyme profiles in growing vs. regressing tumors are relatively rare. Since many of the experimental tumors are amenable to such manipulations, i.e., altering the hormonal milieu, investigation of relationships between growth behavior and isoenzyme changes are possible. Some experimental models offer an additional opportunity to examine lesions that do not regress under a similar hormonal milieu, i.e., hormone-independent, DMBA-induced tumors growing after ovariectomy or induction of diabetes. Do such hormone-independent lesions display isoenzyme patterns differing from hormone-dependent lesions?

Fourth, is there a useful model of pre-neoplasia in animals?

The interesting studies of DeOme and his colleagues have gone far to develop the hyperplastic alveolar nodule as such a model. How do isoenzyme profiles in this lesion compare with those in frank carcinomas arising from the nodule?

Answers to most of the above would probably be in hand if a unique, tumor-associated isoenzyme had been uncovered. Unfortunately, most of the studies have reported quantitative, rather than qualitative, changes and such shifts are simply less exciting than the discovery of a "new" species (an exception to this may be the reports of LDH_k, summarized above). But even this has not been exploited to its fullest. In keeping with the title of this conference, and with newer techniques of cell separation, a study carefully examining the time-related changes in isoenzyme patterns in mammary epithelium after administration of a known carcinogen, such as DMBA, would provide important information on isoenzyme changes relative to neoplastic transformation. Do such biochemical changes occur prior to morphologically recognizable alterations? Improved methods for culture of normal mammary epithelium *in vitro* suggest that this experimental system may facilitate investigations of the effect of known carcinogenic agents on isoenzyme patterns at very early times. The likelihood that chemical carcinogens may involve a two-step process for transformation, initiation and promotion, should stimulate examination of each class of chemical for effect on isoenzyme patterns.

Compared to the ability to investigate isoenzymes in animal tumors, restrictions for studies of human lesions lead to less flexibility of experimental protocols. The conduct of a longitudinal study is faced with the problem of serial samples of tissue; skin lesions that might be accessible for surgical removal at various times could offer an opportunity to examine the effects of various interventions on isoenzyme profiles, but the availability of such patients for study is not common. Another possibility would be a longer-term study to compare isoenzyme profiles in primary vs. recurrent lesions. Although interpretation of such results may be compounded by effects of intervening therapy (if employed), it might be possible to observe whether differences in isoenzymes occur and whether such differences may relate to biological behavior. The question of what constitutes a pre-neoplastic lesion, while not yet resolved, needs to be addressed morphologically as well as biochemically; perhaps some biochemical parameters are altered prior to demonstrable changes in structure. Use of immunologically suppressed animals, in whom implants of human tumors can be studied, provides a laboratory model for study of isoenzymes and the influence of host factors, such as hormones. Lastly, with the improving techniques of cell and organ culture, it should be possible to examine the effects of transforming agents on human tissues *in vitro*, perhaps leading to the establishment of an isoenzyme marker of impending neoplasia.

ACKNOWLEDGEMENT

Studies conducted in the author's laboratory were supported by grants CA 12836 and CA 16660, and contracts N01-CB-33880 and N01-CA-74204 from the National Cancer Institute. The continued support by USPHS CA11198, University of Rochester Cancer Center, is gratefully noted.

REFERENCES

1. Anderson, G.R., and W.P. Kovacik, Jr. (1981) LDH_k, an unusual lactate dehydrogenase expressed in human cancer. Proc. Natl. Acad. Sci., USA 78:3209-3213.
2. Anderson, G.R., W.P. Kovacik, Jr., and K.R. Marotti (1981) LDH_k a uniquely regulated cryptic lactate dehydrogenase associated with transformation by the Kirsten Sarcoma Virus. J. Biol. Chem. 256:10583-10591.
3. Balinsky, D., E. Cayanis, E. Geddes, and I. Bersohn (1973) Activities and isoenzyme patterns of some enzymes of glucose metabolism in human primary malignant hepatoma. Cancer Res. 33:249-255.
4. Criss, W.E. (1971) A review of isoenzymes and cancer. Cancer Res. 31:1523-1542.
5. Dao, T.L., and T. Nemoto (1980) Steroid receptors and response to endocrine ablations in women with metastatic cancer of the breast. Cancer 46:2779-2782.
6. Dietz, A.A., and T. Lubrano (1967) Separation and quantitation of lactic dehydrogenase by disc electrophoresis. Anal. Biochim. 20:246-257.
7. Farron, F. (1972) The isoenzymes of hexokinase in normal and neoplastic tissues of the rat. Enzyme 13:233-237.
8. Farron, F.H., H.T. Hsu, and W.E. Knox (1972) Fetal-type isoenzymes in hepatic and non-hepatic rat tumors. Cancer Res. 32:302-308.
9. Goldman, R.D., N.O. Kaplan, and T.C. Hall (1964) Lactic dehydrogenase in human neoplastic tissues. Cancer Res. 24:389-399.
10. Hershey, F.B., G. Johnston, S.M. Murphy, and M. Schmitt (1966) Pyridine nucleotide-linked dehydrogenases and isoenzymes of normal rat breast and growing and regressing breast cancers. Cancer Res. 26:265-268.
11. Hilf, R., R. Ickowicz, J.C. Bartley, and S. Abraham (1975) Multiple molecular forms of glucose-6-phosphate dehydrogenase in normal, preneoplastic and neoplastic mammary tissues of mice. Cancer Res. 35:2109-2116.
12. Hilf, R., R. Ickowicz, J.C. Bartley, and S. Abraham (1978) Relationship between the multiple molecular forms of glucose-6-phosphate dehydrogenase and glutathione in normal, preneoplastic and neoplastic mammary tissues. Cancer Biochem. Biophys. 2:191-197.

13. Hilf, R., W.D. Rector, and S. Abraham (1973) A glucose-6-phosphate dehydrogenase isoenzyme characteristic of preneoplastic and neoplastic mouse mammary tissue. J. Natl Cancer Inst. 50:1395-1398.
14. Hilf, R., W.D. Rector, and R.A. Orlando (1976) Multiple forms of lactate dehydrogenase and glucose-6-phosphate dehydrogenase in normal and abnormal human breast tissues. Cancer 37:1825-1830.
15. Hori, S.H., and S. Matsui (1968) Intracellular distribution of electrophoretically distinct forms of hepatic glucose-6-phosphate dehydrogenase. J. Histochem. Cytochem. 16:62-63.
16. Ibsen, K.H., and W.H. Fishman (1979) Developmental gene expression in cancer. Biochim. Biophys. Acta. 560:243-280.
17. Ibsen, K.H. and E. Krueger (1973) Distribution of pyruvate kinase isoenzymes among rat organs. Arch. Biochem. Biophys. 157:509-513.
18. Ibsen, K.H., R.A. Orlando, K.N. Garratt, A.M. Hernandez, S. Giorlando, and G. Nungaray (1982) Expression of multimolecular forms of pyruvate kinase in normal, benign and malignant human breast tissue. Cancer Res. 42:888-892.
19. Jensen, E.V., G.E. Block, S. Smith, K. Kyser, and E.R. DeSombre (1971) Estrogen receptors and breast cancer response to adrenalectomy. Natl. Cancer Inst. Monograph 34:55-70.
20. Kaplan, N.O. (1964) Lactate dehydrogenase - Structure and function. Brookhaven Symp. Biol., Long Island, New York 17:131-149.
21. Karlsson, B.W., and E.I. Carlsson (1969) Levels of lactic and malic dehydrogenase isoenzymes in mammary gland, milk and blood serum of the rat during pregnancy, lactation and involution. Comp. Biochem. Physiol. 25:949-971.
22. Knox, W.E. (1976) Enzyme patterns in fetal, adult and neoplastic rat tissues. 2nd Edition. Karger: Basel.
23. Knox, W.E., M. Linder, and G.H. Friedell (1970) A series of transplantable rat mammary tumors with graded differentiation, growth rate and glutaminase content. Cancer Res. 30:283-287.
24. Lee, C., O. Oliver, E.L. Coe, and R. Oyasu (1979) Lactate dehydrogenase in normal mammary glands and in 7,12-dimethylbenz(a)anthracene-induced mammary tumors in Sprague-Dawley rats. J. Natl. Cancer Inst. 62:193-199.
25. Levy, H.R. (1979) Glucose-6-phosphate dehydrogenase. Adv. Enzymol. 48:97-192.
26. Liu, D.K., D. Kulick, and G.H. Williams (1979) Ca++-stimulated ribonuclease. A new marker enzyme of differentiated rat mammary tissues. Biochem. J. 178:241-244.
27. Liu, D.K., and G.H. Williams. Species differences in ribonuclease activity of milk and mammary gland. Comp. Biochem. Physiol. (B) 71:535-538.
28. Lopez-Alarcon, L., P. Ruiz, and M. Gosalvez (1981) Quantitative determination of the degree of differentiation of mammary tumors by pyruvate kinase analysis. Cancer Res. 41:2019-2020.

29. Markert, C.L. (1963) Lactate dehydrogenase isoenzymes: dissociation and recombination of subunits. Science 140:1329-1330.
30. Markert, C.L. (1968) The molecular basis for isoenzymes. Ann. N.Y. Acad. Sci. 151:14-40.
31. McCarty, K.S., C. Cox, J.S. Silva, B.H. Woodward, J.A. Mossler, D.E. Haagensen, Jr., T.K. Barton, K.S. McCarty, Sr., and S.A. Wells, Jr. (1980) Comparison of sex steroid receptor analysis and carcinoembryonic antigen with clinical response to hormone therapy. Cancer 46:2846-2850.
32. Notides, A.C., and S. Nielson (1974) The molecular mechanism of the in vitro 4S to 5S transformation of the uterine estrogen receptor. J. Biol. Chem. 249:1866-1873.
33. Osawa, Y., B. Tochigi, T. Higashiyama, C. Yarborough, T. Nakamura, and T. Yamamoto (1982) Multiple forms of aromatase and response of breast cancer aromatase to antiplacental aromatase II antibodies. Cancer Res. (Suppl.) 42:32993-33065.
34. Ozols, R.F., and R. Hilf (1973) Effect of androgen on glucose-6-phosphate dehydrogenase isoenzymes in rat ventral prostate and seminal vesicles. Proc. Soc. Exptl. Biol. Med. 144:73-77.
35. Richards, A.H., and R. Hilf (1971) Glucose-6-phosphate dehydrogenase and lactate dehydrogenase isoenzymes in rodent mammary carcinomas and the effect of oophorectomy. Biochim. Biophys. Acta. 232:753-756.
36. Richards, A.H., and R. Hilf (1971) Effect of estrogen administration of glucose-6-phosphate dehydrogenase and lactate dehydrogenase isoenzymes in rodent mammary tumors and normal mammary glands. Cancer Res. 32:611-615.
37. Richards, A.H., and R. Hilf (1972) Influence of pregnancy, lactation and involution on glucose-6-phosphate dehydrogenase and lactate dehydrogenase isoenzymes in the rat mammary gland. Endocrinol. 91:287-295.
38. Sartini, J., D. Meadows, W.D. Rector, and R. Hilf (1973) Response of endometrial and myometrial glucose-6-phosphate dehydrogenase isoenzymes to estrogen. Endocrinol. 93:990-993.
39. Schapira, F. (1973) Isoenzymes and cancer. Adv. Cancer Res. 18:77-153.
40. Shatton, J.B., H.P. Morris, and S. Weinhouse (1969) Kinetic, electrophoretic and chromatographic studies on glucose-ATP phosphotransferases in rat hepatomas. Cancer Res. 29:1161-1172.
41. Shreve, D.S., and H.R. Levy (1979) Glucose-6-phosphate dehydrogenase from lactating rat mammary gland and R3230AC adenocarcinoma. Enzyme 24:48-53.
42. Siiteri, P.K. (1982) Review of studies on estrogen biosynthesis in the human. Cancer Res. (Suppl.) 41:3269s-3273s.
43. Simpson, A.A., and G.H. Schmidt (1970) Lactate dehydrogenase in the rat mammary gland. Proc. Soc. Exptl. Biol. Med. 133:897-900.
44. Stanislowski-Birencwajg, M., and F. Loisillier (1965) Distribution of LDH isoenzymes in mastitis, fibroadenoma and carcinoma of the human mammary gland. Europ. J. Cancer 1:221-224.

45. Susor, W.A., and W.J. Rutter (1971) A method for the detection of pyruvate kinase, aldolase and other pyridine nucleotide-linked enzyme activities. Anal. Biochem. 43:146-155.
46. Taketa, K., and A. Watanabe (1971) Interconvertible microheterogeneity of glucose-6-phosphate dehydrogenase in rat liver. Biochem. Biophys. Acta. 235:19-26.
47. Tavassoli, M., M. Rizo, and R.B. Johnson, Jr. (1981) Acid phosphatase activity in the cytosol fraction of the breast cancer tissue. Cancer 47:895-898.
48. Tilzer, K.L., R.T. McFarland, F.V. Plapp, J.P. Evans, and M. Chiga (1981) Different ionic forms of estrogen receptor in rat uterus and human breast carcinoma. Cancer Res. 41:1058-1063.
49. Toft, D.D., and J. Gorski (1966) A receptor molecule for estrogens: Isolation from the rat uterus and preliminary characterization. Proc. Natl. Acad. Sci., USA 55:1574-1581.
50. Walters, E., and P. McLean (1967) Multiple forms of glucose-adenosine triphosphate phosphotransferase in rat mammary gland. Biochem. J. 104:778-783.
51. Weinhouse, S. (1982) What are isoenzymes telling us about gene regulation in cancer. J. Natl. Cancer Inst. 68:343-349.
52. Witliff, J.L. (1980) Steroid receptor interactions in human breast carcinoma. Cancer 46:2953-2960.
53. Wittliff, J.L., P.W. Feldhoff, A. Fuchs, and R.D. Wiehle (1981) Polymorphism of estrogen receptors in human breast cancer. In Physiopathology of Endocrine Diseases and Mechanisms of Hormone Action, Alan R. Liss, Inc., New York, pp. 375-396.
54. Wuntch, T., R.F. Chen, and E.S. Vesell (1970) Lactate dehydrogenase isozymes: Further kinetic studies at high enzyme concentration. Science 169:480-481.

CHAIRMEN'S OVERVIEW ON ONCODEVELOPMENTAL MARKERS

Stewart Sell and K. Robert McIntire*

University of Texas Medical School
Department of Pathology and Laboratory Medicine
Houston, Texas 77025

Cancer markers are cell products produced normally during development, or by differentiating normal cells in adult tissues (Figure 1). Such markers include placental products (chorionic gonadotropin), isozymes (alkaline phosphatase, etc.), secreted proteins, (myeloma paraprotein, alphafetoprotein), cell surface gryoproteins (carcinoembryonic antigen), and blood group carbohydrates, as well as lymphocyte surface markers and tumor products detected by monoclonal antibodies.

The first true marker precipitated protein in the urine of a patient with multiple myeloma was reported in 1846 by H. Bence Jones. The urine was given to Dr. Bence Jones by Dr. William McIntire. It took over 100 years to identify this protein as immunoglobulin light chain. The second era of cancer markers began in the 1930s with the identification of hormone production by tumors by Bernhardt Zondek and Harvey Cushing, and the recognition of ectopic hormone production by W.H. Brown, as well as the identification of isozymes by Clement Markert. The third era of cancer markers was ushered in in 1963 by Gari Abelev's discovery of alphafetoprotein and later by Phil Gold's discovery of carcinoembryonic antigen. The fourth era of cancer markers, in the form of monoclonal antibodies, is now upon us, and a fifth era can be expected as recombinant DNA probes are applied to the diagnosis of cancer.

A general principle of most cancer markers so far described is that production of a marker by a cancer is closely related to the tissue that produces that substance normally during development.

*Diagnosis Branch, National Cancer Institute, Bethesda, Maryland 20205

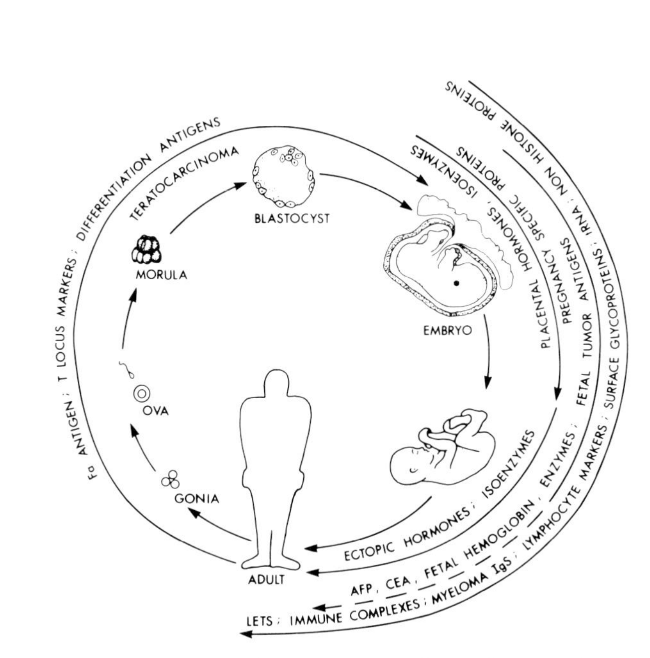

Fig. 1. Some oncodevelopmental markers of cancer.

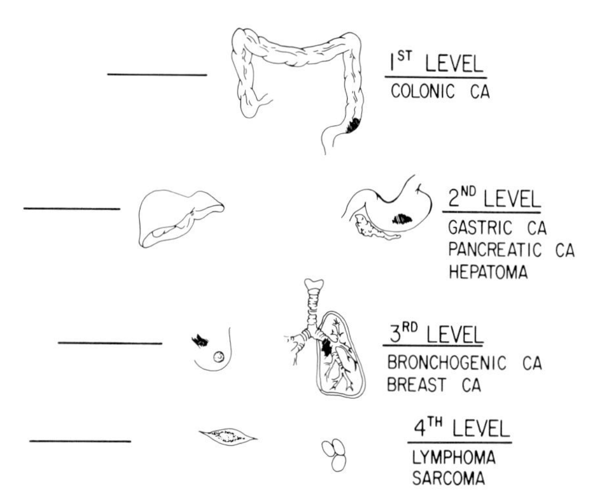

Fig. 2. Carcinoembryonic antigen production (CEA) by human tumors.

Table 1. "Levels" of expression of some "oncodevelopmental" markers.

	LEVELS OF EXPRESSION			
Marker	First	Second	Third	Fourth
CEA	Colon	Stomach Pancreas Liver	Lung	Sarcoma Lymphoma
AFP	Liver Yolk Sac	Colon Stomach Pancreas	Lung	--
Hormone (Serotonin)	Carcinoid	Adrenal Medulla	Lung (Oat Cell)	Lung (Adenoma)
Placental Isozymes	Chorlocar- cinoma	Germ Cell Testis Ovary	Liver	Lung
Immunoglob- ulin	Plasma Cell Myeloma	B Cell Tumor	T Cell Tumor	Monocyte

This principle is illustrated in Figure 2 for carcinoembryonic antigen, and its application to some other markers is listed in Table 1. A given marker is most often found when cancer arises in a tissue that normally produces that marker. The marker is less often produced by tissues that are closely related embryologically, and even less often by cancers that are more distantly related embryologically.

The objective of the diagnostic use of cancer markers is to detect cancers while they are still small enough to be cured. For practical purposes a secreted marker that appears in the blood stream is potentially much more useful as a marker than a cytoplasmic or cell surface marker, with the obvious exception of leukemia markers. The objective of the use of cancer markers to determine carcinogenic potential of chemicals is to identify a marker that is produced by test animals or tissues as a result of carcinogen exposure either during the "precancerous" latent phase or when cancers actually develop. Again, a secreted marker would be more practical, but a definitive tissue marker would be more than acceptable. At the present time the only accepted method to diagnose cancer depends upon morphology. The availability of modern biochemical and immunologic methods provides us with analytic potential that could change this dependency on morphology through the application of marker assays to the study of chemical carcinogenesis.

ALPHAFETOPROTEIN AS A MARKER FOR EARLY EVENTS AND CARCINOMA DEVELOPMENT DURING CHEMICAL HEPATOCARCINOGENESIS

S. Sell,[1] F. Becker,[2] H. Leffert,[3] K. Osborn,[1] J. Salman,[1] B. Lombardi,[4] H. Shinozuka,[4] J. Reddy,[5] E. Ruoslahti,[6] and J. Sala-Trepat[7]

[1] Department of Pathology and Laboratory Medicine University of Texas Medical School, Houston, Texas

INTRODUCTION

Animal bioassays for carcinogens depend on evaluation of morphologic changes in affected organs following carcinogen exposure. In rats and mice this procedure is expensive, and evaluation of the morphologic changes that occur before obvious cancer develops are subject to disagreement among pathologists. In particular, the reversible morphologic changes that occur in the liver upon exposure to chemical hepatocarcinogens have been given different interpretations by different pathologists (1-5). Terms used vary from "preneoplastic," implying a lesion that will eventually develop into a cancer, to "hyperplastic" or "benign proliferative lesion," which implies that it does not develop into cancer.

One of the goals of research in carcinogenesis is to identify other methods of testing animals for carcinogenic effects (6-9). Such methods ideally would include a serum marker for cancer or for "premalignant" or malignant changes that could replace the present need for morphologic or tissue marker analysis (10,11). Since the diagnosis of cancer, given the present state of the art, requires morphologic criteria, demonstration of the usefulness of such a

[2] M.D. Anderson Hospital and Tumor Institute, Houston, Texas; [3] University of California at San Diego, San Diego, California; [4] University of Pittsburgh, Pittsburgh, Pennsylvania; [5] Northwestern University Medical School, Chicago, Illinois; [6] LaJolla Cancer Research Foundation, LaJolla, California; [7] Laboratoire d'Enzymologie du C.N.R.S. Gif. sur-Yvette, France.

marker depends upon correlation with the morphologic changes associated with carcinogen exposure.

The subject of this presentation will be the correlation of evaluations of serum concentrations of alphafetoprotein (AFP) with the morphologic changes in the liver after exposure to a variety of chemical hepatocarcinogens (12). The predictive value of serum AFP in evaluation of carcinogen exposure and the cellular changes induced by carcinogens associated with AFP production will be emphasized.

N-2-ACETYLAMINOFLUORENE (FAA)

FAA is a potent hepatocarcinogen that elicits hepatocellular carcinomas in rats after prolonged exposure (13-16). The cyclic feeding regimen of Teebor and Becker (16) was used in our first experiment. Sprague-Dawley male and female rats were exposed to a 3-week feeding of 0.06% FAA (17). This strain and sex was chosen because male Sprague-Dawley rats are fully susceptible to the carcinogenic effects of FAA, whereas female Sprague-Dawley rats do not produce the active metabolite of FAA because of a deficiency in arylsulfotransferase (Figure 1) (18-20). Although we had planned to give the full four-cycle exposure to FAA, we found that serum AFP

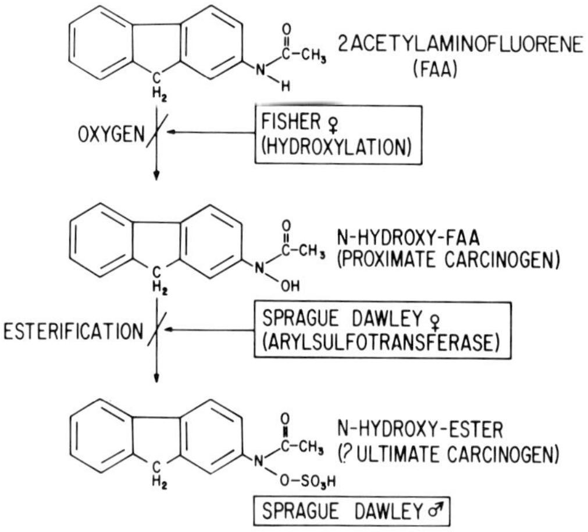

2ACETYLAMINOFLUORENE (FAA)

OXYGEN

FISHER ♀ (HYDROXYLATION)

N-HYDROXY-FAA (PROXIMATE CARCINOGEN)

ESTERIFICATION

SPRAGUE DAWLEY ♀ (ARYLSULFOTRANSFERASE)

N-HYDROXY-ESTER (? ULTIMATE CARCINOGEN)

SPRAGUE DAWLEY ♂

Fig. 1. Metabolism of FAA in rats of different strain and sex. Fischer and Sprague-Dawley male rats are fully susceptible to FAA-induced hepatocarcinogenesis by virtue of the fact that they are able to metabolize the parent carcinogen to the N-hydroxy form. Fischer female rats and Sprague-Dawley female rats are resistant to FAA hepatocarcinogenesis due to deficiencies in enzymes responsible for hydroxalation and esterification, respectively.

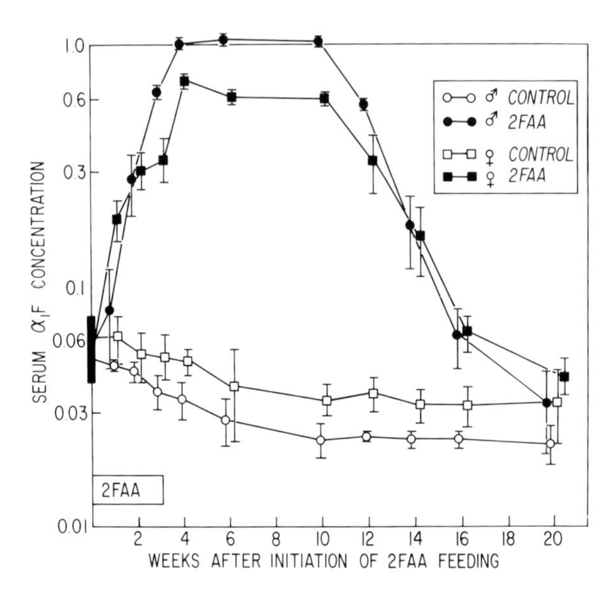

Fig. 2. Serum AFP concentrations in male and female Sprague-Dawley rats fed 0.06% N-2 FAA for three weeks.

concentrations rose above normal within one week and remained elevated for 16 weeks even though carcinogen feeding was discontinued after 3 weeks (Figure 2). This effect is dose related and occurs even with subcarcinogenic doses (Figure 3). Fisher male rats, which are highly susceptible to FAA carcinogenesis, demonstrate higher early elevations of serum AFP than ACI rats (21), which have effects similar to the Sprague-Dawley male rat (Figure 4).

The Fisher male rat was selected for further study since *in vitro* culture methods had been worked out by Hyam Leffert using this strain (22-24). Two-week feeding cycles of FAA were used for Fischer male rats since many rats of this strain are unable to survive an initial 3-week feeding. In contrast to Sprague-Dawley rats and ACI rats, serum AFP levels in Fischer male rats continue to rise during 5 cycles of 2-week feedings of FAA with 1-week rest periods in between (Figure 5) (25). After the 5 cycles, serum AFP did decline, but not to normal levels. Every animal in this experiment developed hepatocellular carcinoma (S. Sell, unpublished data).

This system was used to analyze by immunofluorescence the nature of the cells containing AFP and albumin, and the distribution of fibronectin and laminin in the liver. AFP was found in new cell populations including oval cells, gland-like structures, and atypical hyperplastic areas (Figure 6), but not in normal hepatocytes or nodular areas (over 1,000 nodules per liver were examined) (25). It had been previously noted that the serum concentration of AFP in ACI

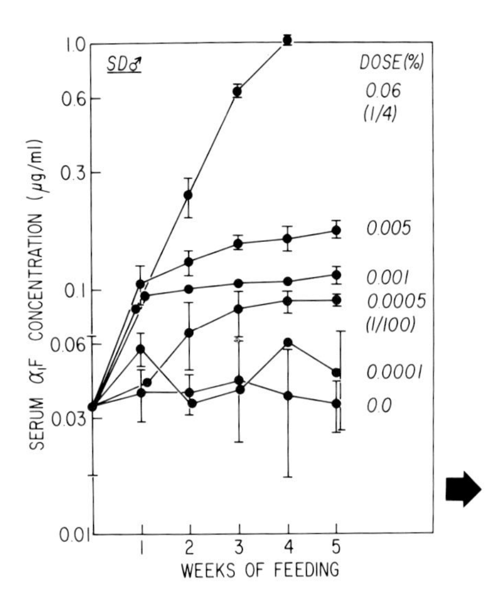

Fig. 3. Serum AFP concentrations in Sprague-Dawley male rats fed different doses of FAA. At the 0.06% the rats have received 1/4th of a carcinogenic dose; at 0.0005% they have recived 1/100th of a carcinogenic dose.

rats acutally fell during the time of maximal nodule formation (Figure 7) (21). Nodules contain less fibronectin and laminin than normal livers, whereas areas of oval cell proliferation contain more laminin and fibronectin than normal livers (31).

To investigate further the development of new cell populations during early carcinogenesis, rats were fed FAA in a choline deficient (CD) diet (26-29). Choline deficiency alone will stimulate a slight AFP elevation associated with proliferation of hepatocytes (26,29). A combination of FAA and choline deficiency produces a much more rapid elevation of serum AFP and a massive increase in small oval cells that first appear in the portal area and rapidly extend throughout the liver lobule (29). After a 2-week feeding of CD-FAA an explosive increase in serum AFP and oval cells occurs, so that by 4 weeks after initial feeding the livers are essentially replaced by oval cells, and the serum AFP concentration is two logs higher than that seen after feeding FAA in a choline-supplemented diet (Figure 8). Immunolfuorescence reveals large numbers of oval cells containing AFP as well as albumin. By 3 weeks, duct-like structures, many containing AFP and albumin, are seen. Autoradiography reveals that small radiolabeled cells appear by 1 day and are located in periportal areas adjacent to the liver lobule (30). Labeling of bile duct cells is not seen until day 3 (Table 1). Oval cell areas stain intensely for fibronectin, and individual groups of oval cells are lightly delineated by a thin layer of laminin (Figure 9) (31). Oval cells also demonstrate infiltrative growth as they appear to invade into the liver lobule.

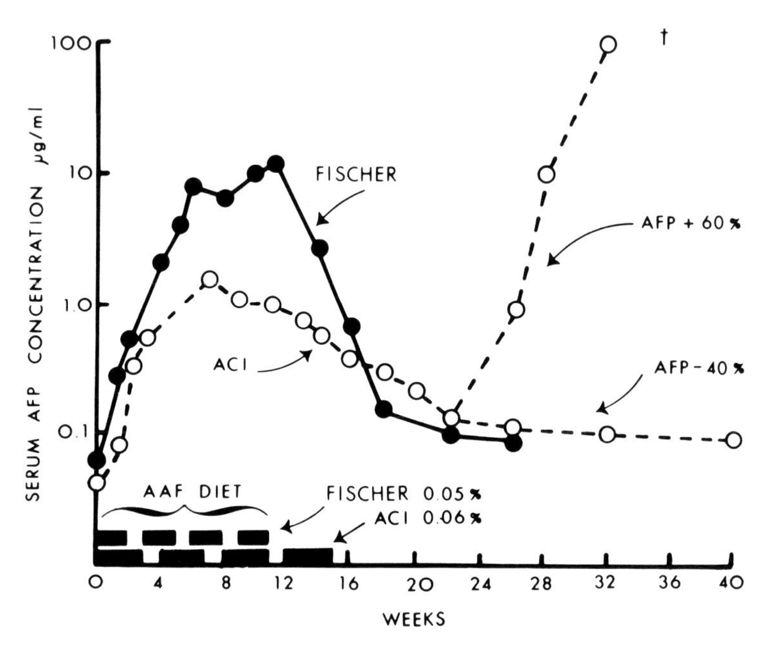

Fig. 4. Comparison of serum AFP concentrations of Fischer and Sprague-Dawley male rats fed four cycles of FAA.

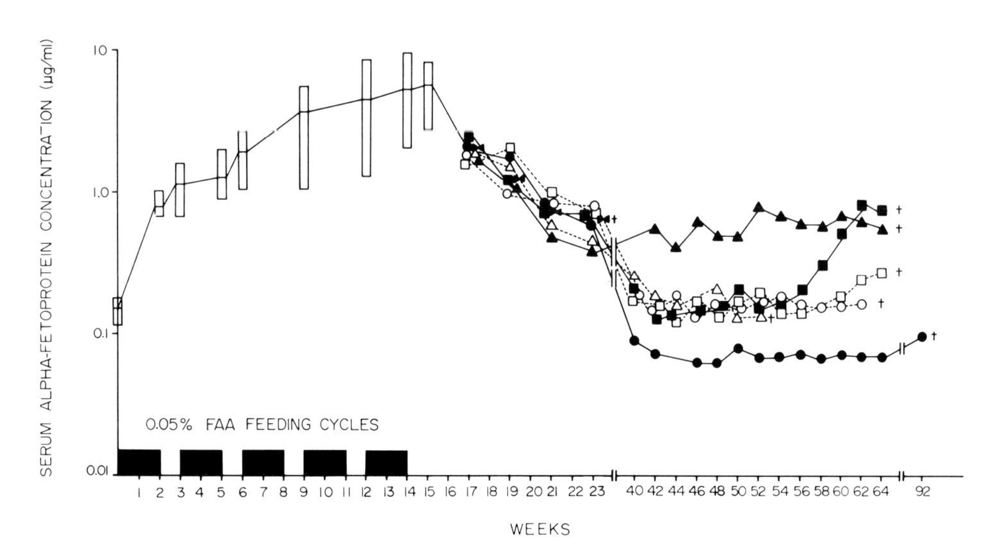

Fig. 5. Serum AFP concentrations of a group of Fischer male rats fed five cycles of FAA and allowed to develop hepatocellular carcinomas.

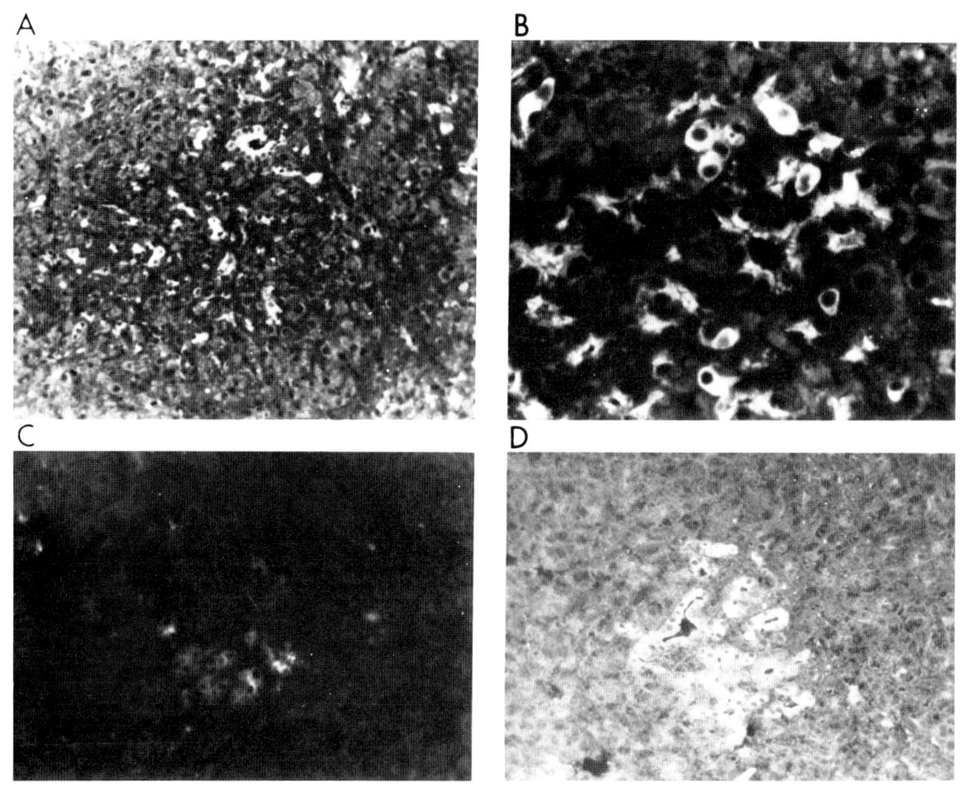

Fig. 6. Localization of AFP in livers of Fischer male rats by immunofluorescence: A) CD-FAA 21 days (X 160); B) CD-FAA 21 days (X 400); C) FAA four cycles (X 160); and D) FAA three cycles (X 160).

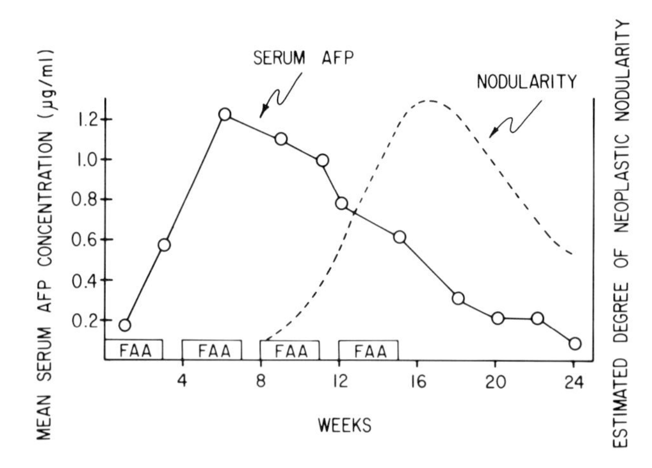

Fig. 7. Lack of correlation of serum AFP concentration and nodule formation in male ACI rats fed four cycles of FAA.

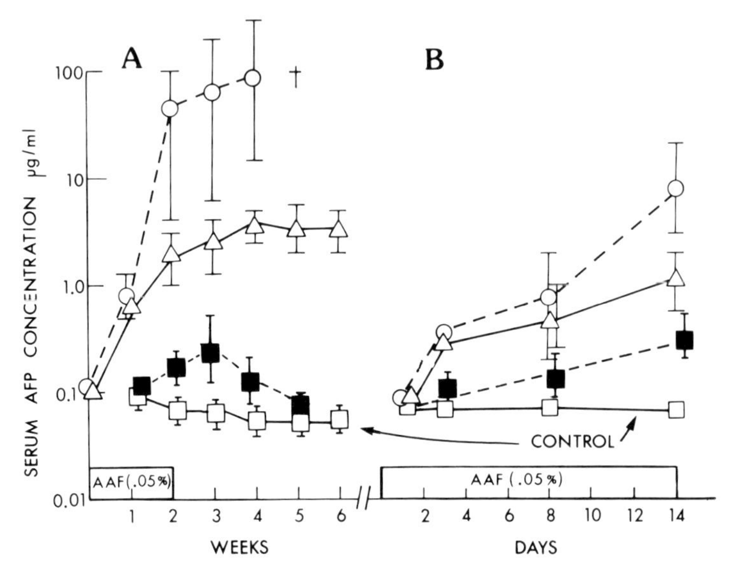

Fig. 8. Serum AFP concentrations in male Fischer rats fed FAA in choline devoid diets: A) Experiment 1; B) Experiment 2. The Serum AFP concentrations are plotted as a function of time after feeding in a two-week cycle of FAA in different diets. The symbol □ – □ represents a choline supplemented (CS) diet; ■ – ■, choline devoid (CD) diet; △ – △, FAA in choline supplemented diet (CSFAA); ○ – ○, FAA in choline deficient diet. The cross (+) indicates time of death of the animals fed CD-FAA. (Reproduced from ref. 29.)

In contrast, nodules contain less fibronectin than normal liver and are sharply delineated from normal liver. Hepatocellular carcinomas have variable fibronectin content ranging from large amounts surrounding each cell to none demonstrable (31).

Table 1. Number of labeled duct cells and number of labeled small nonduct cells in CD-AAF fed rats.

	DUCT CELLS *	NON-DUCT CELLS *
DAY 1	2/300	21/100
DAY 2	1/300	31/100
DAY 3	17/300	67/200

* NO. LABELED/NO. NOT LABELED

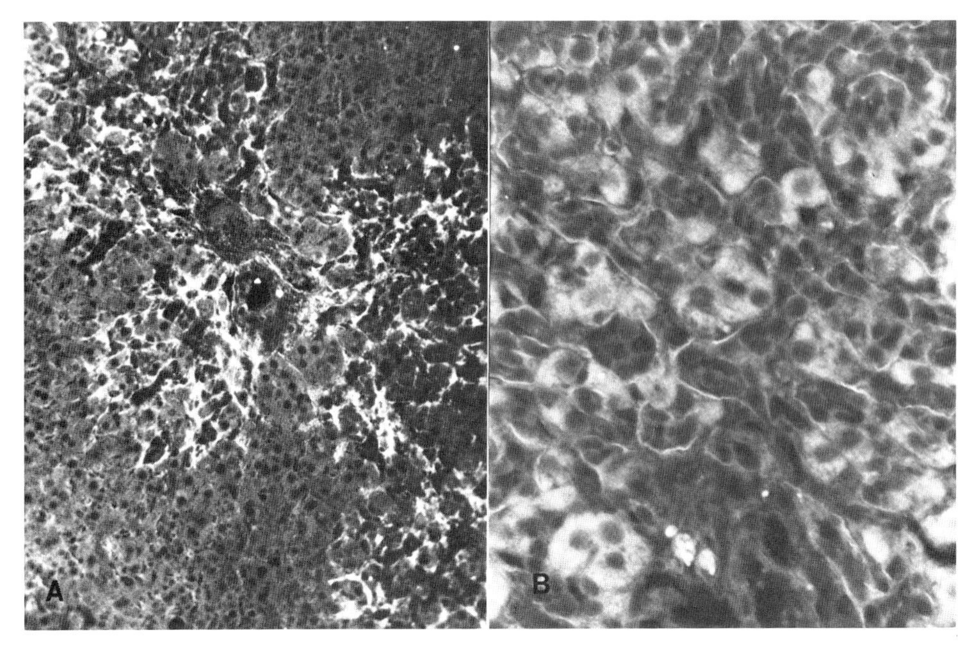

Fig. 9. Fibronectin (A) (X 160) and laminin (B) (X 400) in areas of oval cell proliferation sixteen days after feeding Fischer male rats CD-FAA.

Preliminary electronmicroscopic autoradiography (Figure 10) shows that labeling first appears in small periductular cells that lie on the lobule side of the basement membrane of the true ducts. (S. Sell and J. Salman, unpublished data). They clearly are neither adjacent to the basement membrane, nor do they have villi typical of ductual cells. The nuclei tend to be more oval than duct cell nuclei and there is a broader and more dense area of euchromatin adjacent to the nuclear membrane. The cytoplasm contains few, if any, specific organelles. By day 14, labeled cells can be seen in early duct-like structures (Figure 11). Cells with an intermediate structure between the nondescript oval cells and hepatocytes are also present. A comparison of the properties of oval cells, nodules, and hepatocellular carcinomas is presented in Table 2. It is concluded that oval cells may represent stem cells that are stimulated to proliferation after exposure to a carcinogen and have the capacity to differentiate into duct-like cells and into liver-like cells.

The possibility that acute phase alpha$_2$macroglobulin (α_2M) might also be elevated during carcinogenesis (32) was tested using an RIA for α_2M. No elevation of α_2M was seen during FAA exposure or in rats with transplantable hepatocellular carcinomas, unless necrosis was prominent (33).

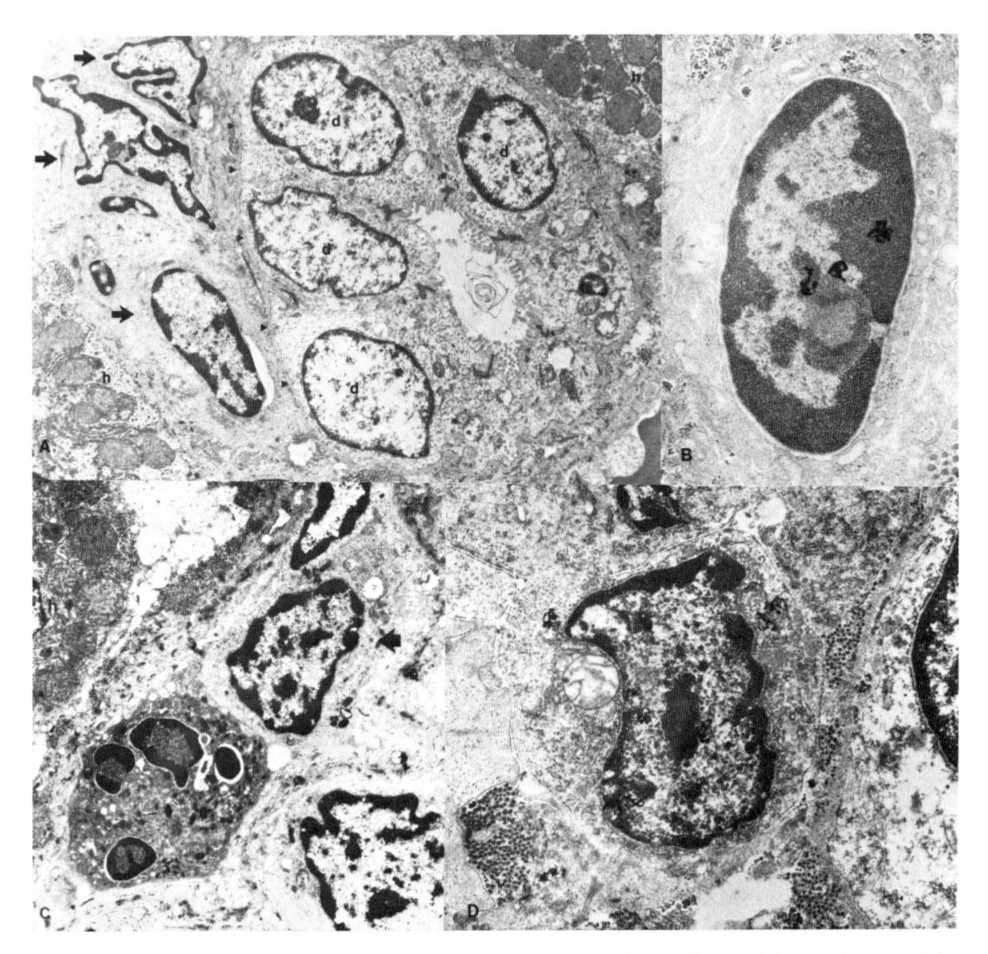

Fig. 10. Electronmicroscopic autoradiographs of periductular cells after feeding CD-FAA for one or two days to Fischer male rats: A) Day 1 (X 8,900); B) Day 2 (X 38,400); C) Day 1 (X 8,900); D) Day 2 (X 14,800). The small triangles in A point to the basement membrane of the bile duct; the arrows to nondescript periductular cells situated between the duct and hepatic parenchymal cells. Radiolabeled periductular cells are shown in B, C, and D. The arrow in C points to a labeled periductular cell. The cell below and to the left is a polymorphonuclear leukocyte.

ETHIONINE (ET)

Ethionine feeding produces an effect similar to that of FAA (Figure 12) (34). In addition, similar early morphologic and AFP enhancing effects are seen when ethionine is fed in a choline deficient diet (35).

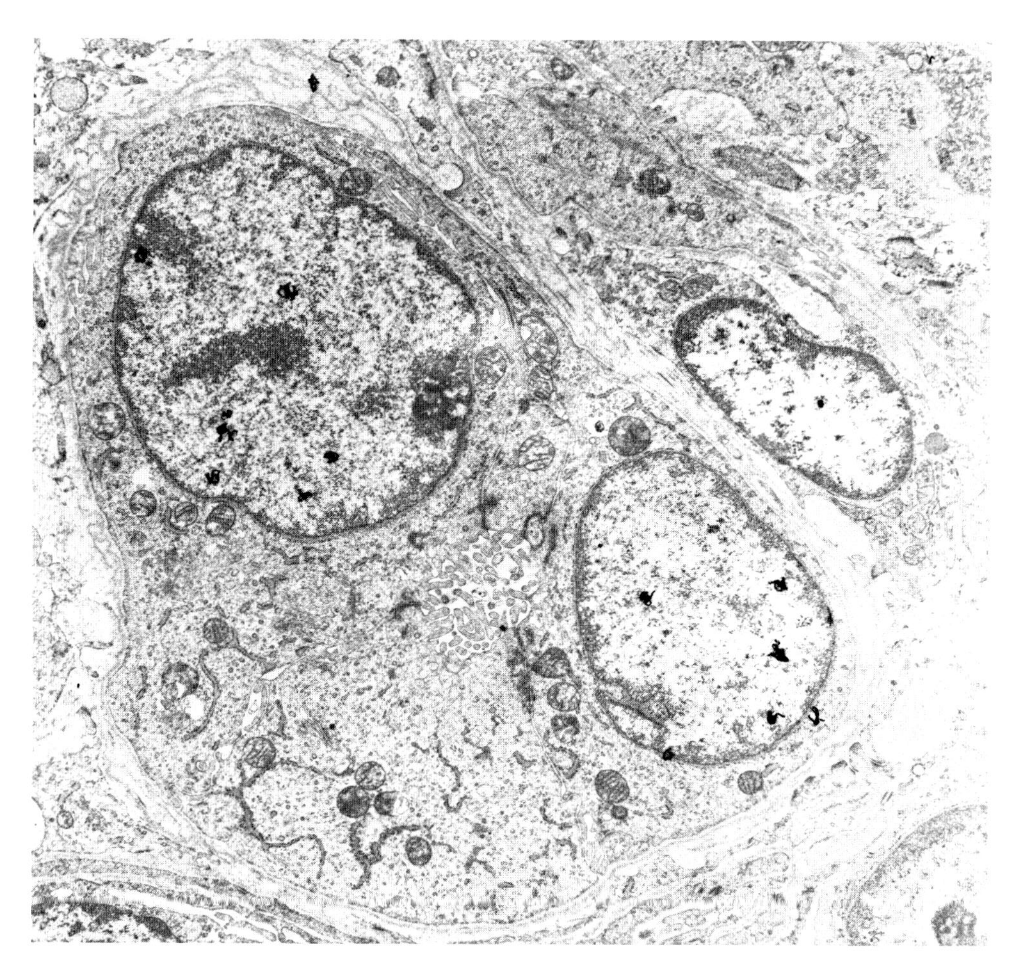

Fig. 11. Electronmicroscopic autoradiograph of midlobular duct-like structure in liver of rat fed CD-FAA for fourteen days. Such structures morphologically resemble ducts but have other properties more like hepatocytes. (See Table 5.)

3'METHYLDIAMINOAZOBENZENE (3'MDAB)

Figure 13 gives the serum concentrations of AFP in rats exposed to 3'MDAB, and to its noncarcinogenic analogues 2-methyl-4-dimethylaminoazobenzene (2MDAB) and p-aminoazobenzene (PAB) given with surgical excision of two lobes of the liver (36). Partial hepatectomy stimulates the expected temporary serum AFP elevation associated with restitutive proliferation. Neither 2-MDAB nor PAB exposure causes a serum AFP elevation whereas 3'MDAB produces a rapid elevation. The serum elevation effect of 3'MDAB is also dose dependent (Figure 14).

Table 2. Comparative properties of oval cells, neoplastic nodules, and hepatocellular carcinomas.

PROPERTY	OVAL CELLS	NODULES	HEPATOMAS
APPEAR ON EXPOSURE TO CARCINOGEN	YES, EARLY	YES, LATER	YES, EVEN LATER
PROLIFERATION	RAPID	RAPID	VARIABLE
INVOLUTION	YES	YES	NO
AFP	YES	NO	YES AND NO
ALDOLASE C	YES	NO	YES AND NO
GGT	YES	YES	YES
ALBUMIN	YES	YES	YES AND NO
INVASIVE	YES	NO	YES
FIBRONECTIN	HIGH	LOW	VARIABLE
TRANSPLANTATION	LIMITED GROWTH	LIMITED GROWTH	PROGRESSIVE GROWTH
IN VITRO	LIMITED GROWTH	SURVIVAL	PROGRESSIVE GROWTH

DIETHYLNITROSAMINE (DEN)

In contrast to the other carcinogens, DEN produces no early serum elevations unless high doses are given (Figure 15) (21,37). At fully carcinogenic doses, AFP elevations occur rapidly and suddenly after little or no preceding increase (21). The phase of normal serum AFP is associated with micronodular formation, with

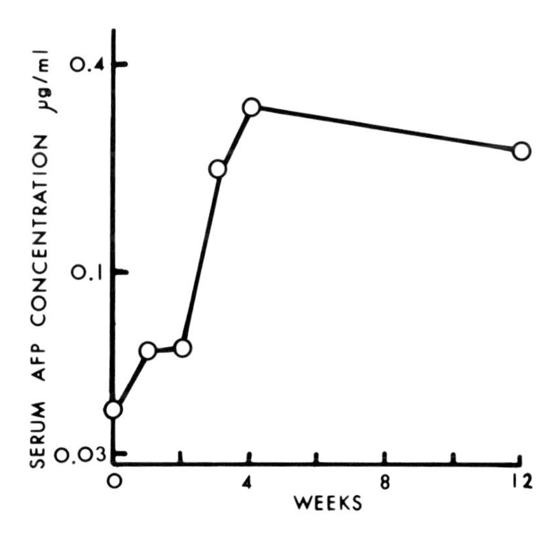

Fig. 12. Serum AFP concentrations in male Sprague-Dawley rats fed 0.05% ethionine.

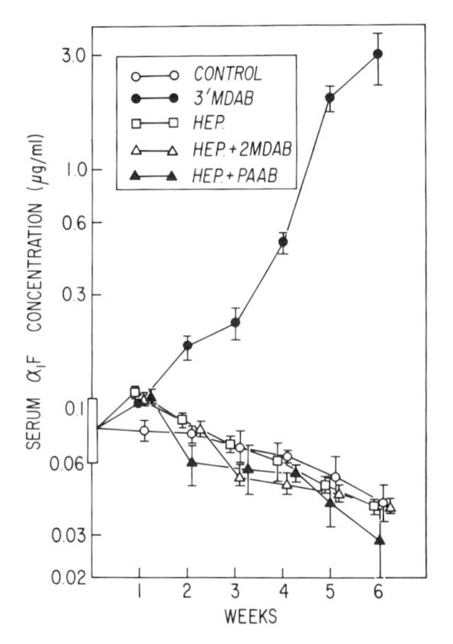

Fig. 13. Serum AFP concentrations in male Sprague-Dawley rats fed 0.06% 3'MDAB, 2'MDAB, and PAAB in a riboflavin-free diet. Rats fed 2'MDAB and PAAB were 2/3rds hepatectomized on Day 0.

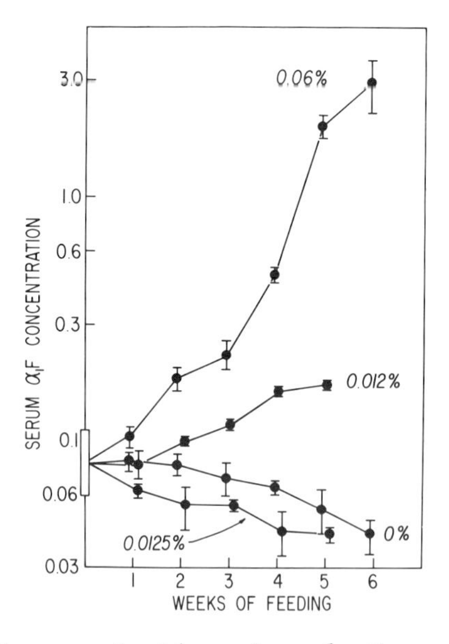

Fig. 14. Serum AFP concentrations in male Sprague-Dawley rats fed different concentrations of 3'MDAB.

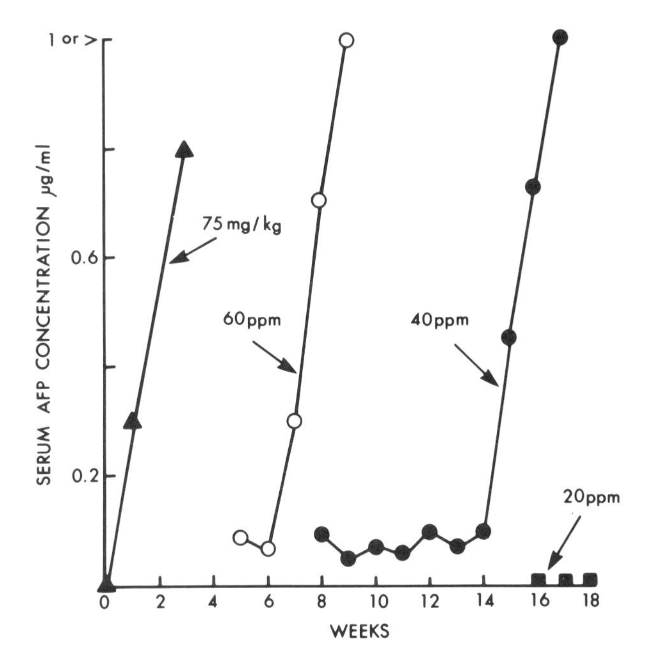

Fig. 15. Serum AFP concentrations of male Sprague-Dawley rats injected IP with 75 mg/kg DEN on Day 0 and in male ACI rats given different doses of DEN in their drinking water.

essentially no oval cell proliferation. At the earliest time of AFP elevation, AFP may be localized in focal adenomatous hyperplastic or glandular areas similar to those seen after FAA exposure (S. Sell and F.F. Becker, unpublished data). The rapid elevations of serum AFP are associated with the development of multiple malignant tumors in the liver. DEN, injected weekly into animals on a choline deficient diet, produces a rapid elevation of serum AFP (Figure 16) and is associated with a massive proliferation of oval cells (37). It should be noted that DEN in a choline supplemented diet also causes an early elevation of AFP, and some oval cell proliferation when high doses are injected IP (intraperitoneally) into Sprague-Dawley male rats (37), in contrast to the absence of an AFP elevation and oval cell proliferation when lower doses of DEN are administered in the drinking water to ACI rats (21).

AZASERINE

Azaserine is a weak hepatocarcinogen noted for its carcinogenic effect on the pancreas. However, administration of azaserine in a choline deficient diet resulted in primary hepatocellular carcinomas (PHC) and in 3/5 rats at 4 months and 5/5 at 6 months. Elevations of serum AFP were seen after 1 and 2 months of feeding in rats

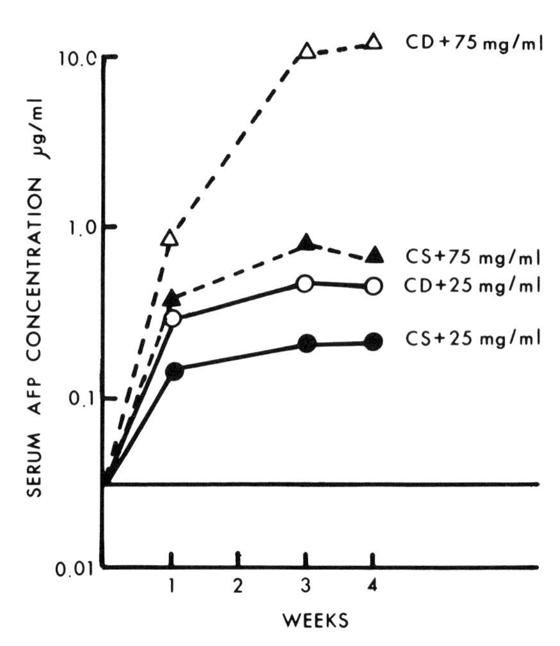

Fig. 16. Serum AFP concentrations in male Sprague-Dawley rats fed a choline deficient or choline supplemented diet and injected with 25 or 75 mg/ml DEN.

exposed to azaserine in either a choline supplemented (1.9 and 2.3 µg/ml) or a choline deficient diet (3.6 and 0.9 µg/ml). After falling to normal (0.06 µg/ml) at 4 months, re-evaluation of serum AFP occurred in rats developing in hepatocellular carcinomas at 6 months.

WY 14643

WY 14643 ([4-chloro-6-(2,3-xylidino)-2-pyrimidinylthio]acetic acid) produces early temporary serum AFP elevations (Figure 17) associated with hepatocyte proliferation (38). After 2 weeks the serum AFP falls to normal and remains low for the long latent period of 60 or more weeks before non-AFP producing tumors become evident. A similar effect is noted with azaserine (Shinozuka et al., unpublished data).

DIAMINODIPHENYMETHANE (DDPM)

Because earlier experiments indicated that oval cells were derived from duct cells (14,39-41), the effect of DDPM, an agent which stimulates bile duct proliferation (42), was tested. The marked bile duct proliferation induced is not associated with AFP production or the appearance of oval cells (43). A comparison of

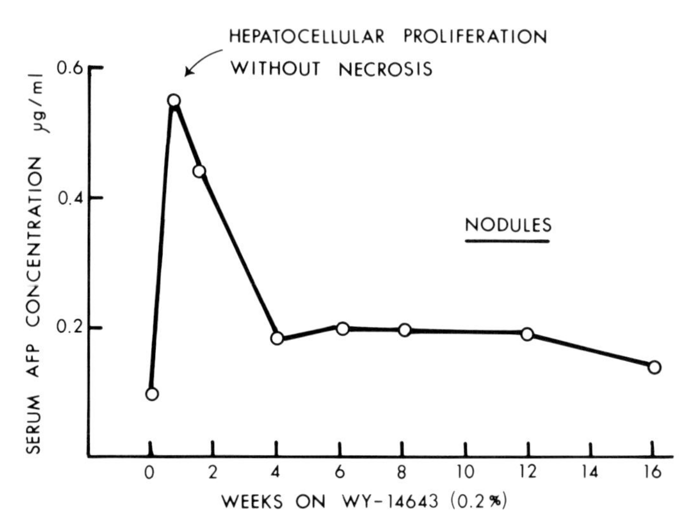

Fig. 17. Serum AFP concentrations in male Fischer rats fed 0.2% WY 14643 for sixteen weeks. (Reproduced from ref. 38.)

some of the properties of duct-like structures appearing after oval cell proliferation induced by CD-FAA, bile ducts stimulated by proliferation by DDPM and normal bile ducts, is given in Table 3. The duct-like structures associated with oval cell proliferation clearly demonstrate different properties than true bile ducts, whether normal or proliferating.

MICE

Serum AFP and histologic changes in mice following exposure to chemical hepatocarcinogens or in strains with a high incidence of spontaneous hepatocellular carcinomas are summarized in Table 4

Table 3. Comparison of duct structures appearing after CD-AAF or DDPM exposure.

	CD-AAF	DDPM	NORMAL DUCTS
AFP	++	0	0
ALB	++	0	0
GGT	++	++	+
LAMININ	+	+++	+++
FIBRONECTIN	+++	+	+
LOCATION	Mid-lobular	Portal	Portal

Table 4. Summary of serum AFP and liver cancer in mice.

AUTHORS		
BECKER, STILLMAN, & SELL *Cancer Res.* 37:870, 1977	(46)	No early elevations in C_3H-A^{vy} fB mice, elevations above 100 ng/ml always predictive of "spontaneous" hepatocellular cancer.
BECKER & SELL *Cancer Res.* 39:3491, 1979	(48)	No early elevations in $C_{57}BL/6$ N mice after exposure to AAF or chlordane although nodular lesions present (BPL). Elevation of AFP > 100 ng/ml predictive of cancer.
JALANKO & RUOSLAHTI *Cancer Res.* 39:3495, 1979	(49)	Early low elevation reported with $C_3H/A/BOM$ mice exposed to O-aminoazotoluene, later higher elevation associated with cancer. In C_3HeB/FeJ mice, spontaneous hepatomas preceeded by slight increase in AFP; high elevations when PHC present. GGT elevations after carcinogen; not with spontaneous hepatocellular cancers.
PRINCER et al. *Eur. J. Can.* 17:1241, 1981	(50)	C_3H-A^{vy}fB mice have elevations of AFP associated with spontaneous PHC. Transient earlier AFP elevation is not associated with later development of PHC. Some mice with high AFP have later decline with evidence of tumor regression.

(46-50). Although there are some differences, perhaps due to different strains or to the use of different chemicals, it may be concluded that serum AFP elevations do not consistently occur in mice early after exposure to chemical carcinogens and that rapid elevations are essentially always associated with development of

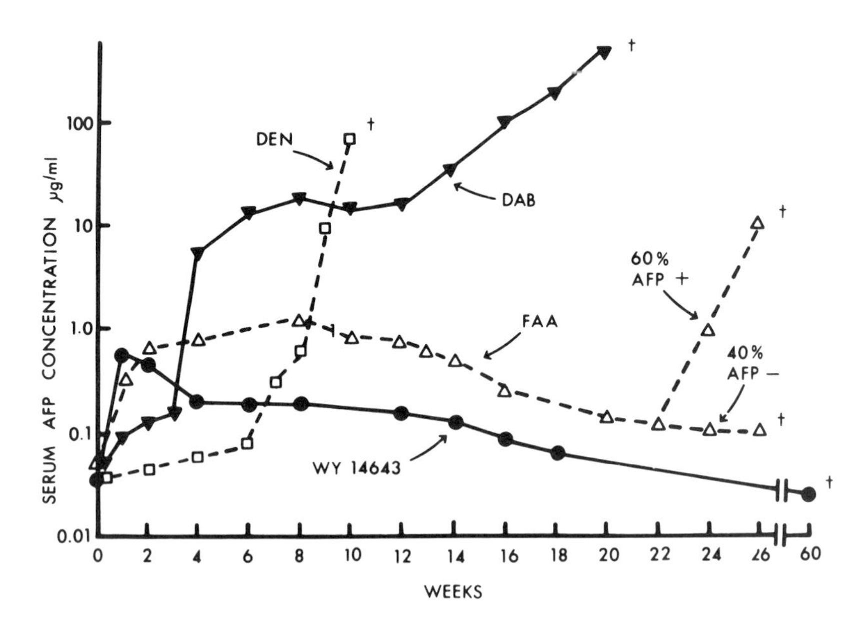

Fig. 18. Composite graph showing serum AFP concentrations of rats fed different carcinogens.

Table 5. AFP production and morphologic changes after exposure of rats to chemical hepatocarcinogens.

	LIVER MORPHOLOGY		AFP PRODUCTION		
CARCINOGEN	EARLY	LATE	EARLY (1-4 wks)	LATE (10-18 wks)	HEPATOMAS
AAF	Foci, oval cells (OC)	Nodules, OC, ADH	++	+	++++ to 0
ETHIONINE	Foci,OC	Nodules, OC, ADH	++	+	+++ to 0
DEN	Foci	Nodules, ADH, ?OC	0	++	+++
DAB	Necrosis, Prolif. OC	Nodules, OC, ADH	++	+++	+++
WY 14643	Hepatocyte Prolif.	Nodules	+	0	0
CHOLINE DEF (CD)	Hepatocyte Prolif.	0	+	0	0
CD + CARCINOGEN (AAF, ETH, DEN)	Massive Oval Cell Prolif.	Nodules, OC	+++	++	+++

hepatocellular carcinomas. It is of interest that oval cell proliferation is not as obvious a feature during chemical carcinogenesis in the mouse as compared to the rat, but that AFP production by PHC is more consistent.

DISCUSSION

A summary of the serum AFP concentrations after exposure of rats to different carcinogens and the associated morphologic changes in the liver is given in Figure 18 and Table 5. Different carcinogens produce different morphologic changes and different kinetics of serum AFP elevations. From these results one must conclude that different hepatocarcinogens produce hepatocellular carcinomas by different mechanisms perhaps involving different "premalignant" cell types.

One must keep an open mind regarding the nature of the so-called premalignant lesion (Figure 19). The oval cell and possibly more differentiated cell types derived from the oval cell as well as nodular cells, must be considered as possible "premalignant" cells.

It seems clear that nodules arise from liver cells with preceding focal changes. The cellular origin of the oval cell is less well defined. There are two possible origins: from duct cells, or from nondescript periductular cells (Figure 10). Both duct cells and periductular cells proliferate after carcinogen exposure, but proliferation of nondescript periductular cells following carcinogen exposure is much more prevalent than proliferation of duct cells. Although oval cells differentiate into structures that resemble ducts, these cells contain AFP and albumin and have other properties

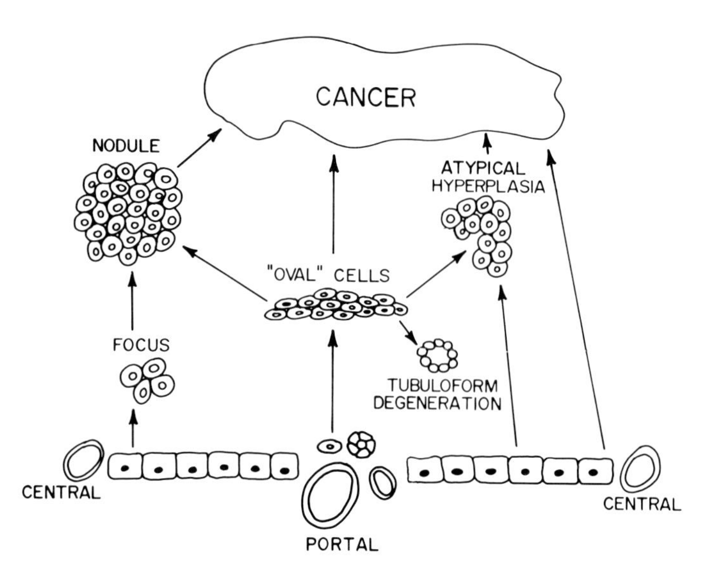

Fig. 19. Possible cellular lineages in experimental hepatocarcinogenesis in rats. Possible relationships between putative premalignant cellular changes in the livers of rats exposed to chemical carcinogens, and the malignant tumor that eventually appears, are depicted. Hepatocellular carcinomas may arise from altered hepatocytes (foci) that progress to nodules and then to cancer. Another putative "premalignant" cell population is the oval cell, which may progress directly to cancer or be a precursor lesion to nodules or areas of atypical hyperplasia that are the ultimate premalignant lesions. It is possible that neither nodules nor oval cells are premalignant and that carcinomas arise from altered hepatocytes either directly or from areas of atypical hyperplasia. (Reproduced from ref. 5.)

different from true bile ducts (Table 3).

In previous studies (39-41) the presence of bile duct-like structures appearing late (1-4 months) after feeding ethionine or DAB, was used as conclusive evidence that oval cells are derived from bile duct proliferation. However, it is likely that the duct-like structures seen as a component of oval cell proliferation represent the differentiation capabilities of oval cells and do not necessarily indicate bile duct origin. The nature of the periductular cells is unclear. Most anatomists have dismissed these cells as fibroblasts as their structural features do not allow precise identification. The presence of fibronectin in areas of oval cell proliferation is consistent with a fibroblast component, however, laminin, which is prominently produced by fibroblasts but is by cells of endothelial or epithelial origin, is also present. The presumptive capacity of oval cells to produce liver proteins and differentiate into duct-like and liver- like cells suggests that oval cells may be pluripotent liver "stem" cells. It is possible that some fibroblast proliferation may also occur as a minor component. Until more precise studies using markers for different cell types are acomplished, the extent of any of the fibroblastic or ductular component to areas of oval cell proliferation will remain unknown.

Regardless of the nature of "preneoplastic" lesions, serum AFP appears to be a reliable marker for determining exposure to hepatocarcinogens in rats. Essentially every carcinogen so far tested produces an elevation of serum AFP at some time during exposure. Difficulties may arise in the interpretation of kinetics of serum AFP elevations with some carcinogens. No problems exist in regard to carcinogens, such as FAA, DAB (4-dimethylaminobenzene), or ethionine as early prolonged elevations are observed. With DEN, elevations occur rapidly when tumors appear, but not during the latent stage. With azaserine (49) and WY 14643 differentiation between hepatocyte proliferative effects and carcinogenic effects is not possible when the carcinogen is in a normal diet. However, as with DEN, the carcinogenic effects may be recognized earlier if azaserine or WY 14643 are given to rats on a choline deficient diet. Shinozuka et al. (49) have shown that a choline deficient diet alters the effect of azaserine so that it becomes a hepatocarcinogen. It is possible that early effects of weak hepatocarcinogens, such as stimulation of AFP production, might be brought out by exposing choline deficient animals, particularly if high doses are used. Serum AFP concentrations have also been reported to be elevated in rats fed purified diets containing high doses (1,000 ppm) of benzo(a)pyrene, suggesting that chemical carcinogens that do not act primarily on the liver may be detected by this means (50).

Study of the molecular mechanisms of AFP production in conjunction with the cellular events during carcinogenesis strongly suggests that expression of the AFP gene is the result of a transcriptional control in a new cell population (51,52). There is no evidence for gene duplication or rearrangement (51), or for changes in mRNA processing (54) in normal cells or hepatocellular carcinomas producing AFP (55). Restrictive endonucleic patterns of the AFP and albumin gene fragments in hepatocellular carcinomas are essentially identical to those of the normal live DNA obtained from the same rat strain as the PHC (J. Boulter and S. Sell, in preparation). There are differences in the restriction endonuclease patterns obtained from liver DNA of different rat strains.

In conclusion, early prolonged elevations of serum AFP or sudden rapid elevations above 1 μg/ml are conclusive evidence of carcinogenic activity. However, "weak" carcinogens may not produce such patterns unless given in high doses or in conjunction with a potentiating system such as a choline deficient diet. From a practical view, short-term determinations of serum AFP concentrations in exposed rats may be used as a definitive marker for carcinogenic doses of FAA, ethionine, DAB, DEN, and perhaps benzo(a)pyrine. For carcinogens such as WY 14643 or azaserine it may be necessary to administer very high doses along with a potentiating diet to stimulate AFP production. Without doubt this system offers distinct advantages over morphologic analysis as the only means of carcinogen evaluation.

SUMMARY

Analysis of serum concentrations of alphafetoprotein (AFP) and cellular location of AFP in tissues by immunofluorescence provide new insights in the events preceding the development of primary hepatocellular carcinomas (PHC) during chemical carcinogenesis. Following exposure of rats to different chemical carcinogens, different patterns of serum AFP elevations are seen. Ethionine (ET) and N-2 fluorenylacetamide (FAA) produce early serum elevations of AFP associated with an increase in AFP containing oval cells. Feeding FAA or ethionine in a choline deficient diet greatly increases the rate of oval cell proliferation and elevation of serum AFP. Oval cells demonstrate invasive properties as they move into the liver lobule, and oval cell areas have a high fibronectin content, whereas, nodules are sharply delineated and contain less fibronectin than normal liver. Light and electron microscopic autoradiographic studies reveal that oval cells arise from a nondescript periductual cell population which is clearly not bile duct in origin. This population differentiates into morphologic intermediates between bile ducts and hepatocytes during later stages of oval proliferation, but typical bile ducts are not produced. Later, during the course of feeding, serum AFP concentrations fall as "neoplastic nodule" development predominates. When PHC develops, approximately 60%-70% of them produce AFP.

Little if any elevation of serum of AFP occurs early after exposure to diethylnitrosamine (DEN), but rapid elevations occur shortly before obvious tumors develop. Essentially all DEN-induced PHC produce AFP. Exposure to DEN in a choline deficient diet results in an immediate rapid elevation of serum AFP associated with striking oval cell proliferation. Early low elevations of AFP follow exposure to 4-dimethylaminobenzene (DAB) and are associated with hepatocellular proliferation. High elevations occur within 2 weeks, well before PHC development. WY 14643 and azaserine produce an early short rise in AFP associated with hepatocellular proliferation, but normal levels are seen over a 60 month time period prior to tumor development. Tumors arising as a result of exposure to WY 14643 do not produce AFP.

All carcinogens tested produce elevations of serum AFP in rats at some time after exposure. This may be associated with hepatocyte proliferation, oval cell proliferation, or PHC proliferation. Stimulation of bile duct proliferation by DDPM is not associated with elevated serum AFP in rats. In mice exposed to chemical hepatocarcinogenesis or which develop spontaneous hepatocellular cancer, early elevations of AFP are not consistently seen, but AFP elevations invariably are found when PHC develop. AFP production by tumors is not associated with changes from normal in AFP gene numbers or in restriction endonuclease patterns. The results suggest that different carcinogens may initiate different cellular pathways to cancer, and that serum AFP determinations may add greatly to the understanding of the early effect of a hepatocarcinogen. Clearly, elevations of serum AFP may be used to determine that a chemical is carcinogenic. However, some carcinogens may not produce serum AFP elevations, so that false negative results are possible.

REFERENCES

1. Farber, E. (1973) Meth. Cancer Res. 7:345.
2. Squire, R.A., and M.H. Levitt (1975) Cancer Res. 35:3214.
3. Schueler, R.L., et al. (1980) Proceedings of a Mouse Liver Tumor Workshop, Silver Spring, Maryland, June 11-13.
4. Becker, F.F. (1982) Cancer Res. 42:3918.
5. Sell, S., and H.L. Leffert (1982) Hepatology 2:77.
6. Farber, E., et. al. (1979) In Carcinogens: Identification and Mechanisms of Action, A.C. Griffin and C.R. Shaw, eds. Raven Press, New York, p. 319.
7. Ames, B.N., W.E. Durston, E. Yamasaki, and F.D. Lee (1973) Proc. Natl. Acad. Sci., USA 70:2281.
8. Heidelberger, C. (1973) Adv. Cancer Res. 18:317.
9. Skehan, P., and S.J. Friedman (1981) In The Transformed Cell, Academic Press, New York, pp. 7-65.
10. Sell, S. (1980) Cancer Markers, Vol. I, Humana Press, Clifton, New Jersey.
11. Sell, S., and B. Wahren, eds. (1982) Cancer Markers, Vol. II, Humana Press, Clifton, New Jersey.
12. Sell, S., F.F. Becker, B. Lombardi, H. Shinozuka, and J. Reddy (1979) In Carcino-Embryonic Proteins, Vol. I, F.-G. Lehmann, ed. Elsevier/North Holland Biomed. Press, p. 129.
13. Kinosita, R. (1937) Trans. Soc. Path. Jap. 27:665.
14. Farber, E. (1956) Cancer Res. 16:142.
15. Reuber, M.D. (1965) J. Natl. Cancer Inst. 34:697.
16. Teebor, G.W., and F.F. Becker (1971) Cancer Res. 31:1.
17. Becker, F.F., and S. Sell (1974) Cancer Res. 34:2489.

18. DeBaun, J.R., E.C. Miller, and J.A. Miller (1970) Cancer Res. 30:577.
19. Gutmann, H.R., D. Malejka-Giganti, E.J. Barry, and R.E. Rydell (1972) Cancer Res. 32:1554.
20. Jackson, C.D., and C.C. Irving (1972) Cancer Res. 32:1590.
21. Becker, F.F., and S. Sell (1979) Cancer Res. 39:1437.
22. Leffert, H.L., and D. Paul (1973) J. Cell. Physiol. 81:113.
23. Sell, S., H.L. Leffert, H. Skelly, V. Muller-Eberhard, and S. Kida (1975) Ann. N.Y. Acad. Sci. 259:45.
24. Koch, K.S., and H.L. Leffert (1980) Ann. N.Y. Acad. Sci. 349:111.
25. Sell, S. (1978) Cancer Res. 38:3107.
26. Lombardi, B., and H. Shinozuka (1979) Int. J. Cancer 23:565.
27. Shinozuka, H., B. Lombardi, S. Sell, and R.M. Iammarino (1978) J. Natl. Cancer. Inst. 61:813.
28. Shinozuka, H., M.A. Sells, S.L. Katyal, S. Sell, and B. Lombardi (1979) Cancer Res. 39:2515.
29. Sell, S., H.L. Leffert, H. Shinozuka, B. Lombardi, and N. Gochman (1981) Gann 72:479.
30. Sell, S., K. Osborn, and H.L. Leffert (1981) Carcinogenesis 2:7.
31. Sell, S., and E. Ruoslahti (1982) J. Natl. Can. Inst. 69:1105.
32. Sarcione, E. (1972) In Embryonic and Fetal Antigens in Cancer, Vol 2, N.G. Anderson, J.N. Coggin, E. Cole, and J.W. Hallerman, eds. Oak Ridge National Laboratory, Oak Ridge, Tennessee, pp. 323-328.
33. Hudig, D., S. Sell, L. Newell, and F.F. Becker (1979) Cancer Res. 39:3715.
34. Shinozuka, H., B. Lombardi, S. Sell, and R.M. Iammarino (1978) Cancer Res. 38:1092.
35. Shinozuka, H., B. Lombardi, S. Sell, R.M. Iammarino (1978) J. Natl. Cancer Inst. 61:813.
36. Becker, F.F., A.A. Horland, A. Shurgin, and S. Sell (1975) Cancer Res. 35:1510.
37. Shinozuka, H., M.A. Sells, S.L. Katyal, S. Sell, and B. Lombardi (1979) Cancer Res. 39:2515.
38. Reddy, J.K., M.S. Rao, D.L. Azarnoff, and S. Sell (1979) Cancer Res. 39:152.
39. Grisham, J.W., and W.S. Hartroft (1961) Lab. Investig. 10:317.
40. Schaffner, F., and H. Popper (1961) Am. J. Pathol. 38:393.
41. Grisham, J.W., and E.A. Porta (1964) Exp. Mal. Path. 3:242.
42. Fukusmima, S., M. Shibata, T. Hibino, T. Yoshimura, M. Hirose, and N. Ito (1979) Toxicol. App. Pharm. 48:145.
43. Sell, S. (in press) Cancer Res.
44. Becker, F.F., D. Stillman, and S. Sell (1977) Cancer Res. 37:870.
45. Jalanko, H., I. Virtauen, E. Engvall, and E. Ruoslahti (1978) Int. J. Cancer 21:453.
46. Becker, F.F. and S. Sell (1979) Cancer Res. 39:3491.
47. Jalanko, H., and E. Ruoslahti (1979) Cancer Res. 39:3495.

48. Princler, G.L., G. Vlahakis, K.H. Kortright, S. Okada, and K.R. McIntire (1981) Eur. J. Cancer Clin. Oncol. 17:1241.
49. Shinozuka, H., S.L. Katyal, and B. Lombardi (1978) Int. J. Cancer 22:36.
50. Boyd, J.N., N. Misslbeck, J.G. Babish, T.C. Campbell, and G.S. Stoewsand (1981) Drug. Chem. Tox. 4:197.
51. Sala-Trepat, J.M., T.D. Sargent, S. Sell, and J. Bonner (1979) Proc. Natl. Acad. Sci., USA 76:695.
52. Sell, S., K. Thomas, M. Michaelsen, J. Sala-Trepat, and J. Bonner (1979) Biochem. Biophys. Acta 564:173.
53. Sala-Trepat, J.M., J. Dever, T.D. Sargent, K. Thomas, S. Sell, and J. Bonner (1979) Biochemistry 18:2167.
54. Nahon, J.-L., A. Gal, M. Franin, S. Sell, and J.M. Sala-Trepat (1982) Nucleic Acids Res. 10:1895.
55. Sell, S., J. Sala-Trepat, T. Sargent, K. Thomas, J.-L. Nahon, T. Goodman, and J. Bonner (1980) Cell Biol. Int. Reports 4:235.

HUMAN CHORIONIC GONADOTROPIN AND ITS SUBUNITS AS TUMOR MARKERS

Judith L. Vaitukaitis

Boston University School of Medicine
Section of Endocrinology and Metabolism
Thorndike Memorial Laboratory
Boston City Hospital
Boston, Massachusetts 02118

INTRODUCTION

Human chorionic gonadotropin (hCG) shares a common quaternary structure of two different subunits with the three pituitary glycoprotein hormones -- thyroid stimulating hormone, luteinizing hormone (LH), and follicle stimulating hormone. Those hormones possess a common alpha subunit but their beta subunits differ, conferring immunologic and biologic specificities (1). The alpha subunit of hCG is comprised of 91 amino acids and two branched oligosaccharide side chains, usually terminating in sialic acid (2). The beta subunit of hCG contains 145 amino acids and six extensively branched oligosaccharide side chains, most terminating with sialic acid (2,3). Human CG is secreted in relatively large quantities by the syncytiotrophoblastic cells of the placenta but may be synthesized at low but significant levels by a wide array of normal tissues (4-6). Human luteinizing hormone (hLH) and hCG share indistinguishable biologic activities. That biologic similarity undoubtedly reflects the presence of a common alpha subunit on those two glycoprotein hormones as well as the extensive structural homology within the beta subunits of those two hormones. The beta subunit of LH contains 115 amino acids and shares at least 80% structural homology with the beta subunit of hCG (7).

The beta subunit of hCG contains a unique 30 amino acid carboxyl terminus with four oligosaccharide side chains; that carboxyl terminus is not present in any other glycoprotein hormone (7,8). The extent of glycosylation of hLH is considerably less than that for hCG and undoubtedly contributes to the markedly different plasma half-lives of those two hormones; the plasma half-life of hCG is 24 to 36 hours whereas the plasma half-life for hLH is approximately 30 minutes. The plasma half-lives of the dissociated subunits of hCG are markedly different from that of the parent hormone. The plasma half-life of hCG alpha is 13 minutes and that for hCG beta is 41 minutes (9). Both subunits are essentially devoid of significant intrinsic biologic activity.

SPECIFIC hCG RADIOIMMUNOASSAYS

Because the biologic effects of both hLH and hCG are similar, no conventional *in vitro* or *in vivo* bioassay reliably differentiates between those two glycoprotein hormones. Available clinical radioimmunoassays for hLH cannot discriminate between hLH and hCG. In fact, most clinical hLH assays incorporate antisera generated to native hCG. Using an antiserum to hCGβ, the first specific hCG assay was developed to selectively measure hCG in plasma or serum samples containing LH, hCG or both hormones (10). During the past decade a variety of other similar hCG radioimmunoassays have been developed and are commonly referred to as "beta subunit assays." That designation is misleading since it implies that only hCGβ is measured when in fact both hCG and hCGβ are detected by those assays. The plasma half-life of hCG is approximately 100 times longer than that for its beta subunit. During pregnancy hCG is the predominant molecular species in peripheral blood with essentially negligible hCGβ but low concentrations of alpha present (11). In the final analysis, the presence of either native hCG or hCGβ in peripheral blood provides important information to the clinician. A wide variety of tumors secrete hCG along with varying concentrations of its subunits (1). Because of the marked difference in plasma half-lives, hCG is the predominant form of hormone in the peripheral blood of patients with tumors secreting hCG and hCGβ.

Essentially, all "specific" hCG assays take advantage of relative rather than absolute immunologic specificity between hCG and hLH. Since hLH is secreted to finite levels in a variety of physiologic settings, relative specificity is a practical approach. Obviously, only a few antisera generated with hCGβ possess sufficient specificity and sensitivity for clinical application.

Antisera have been generated to varying portions of the carboxyl terminus of hCGβ with the hope of identifying specific, sensitive antisera with minimal or no cross-reactivity with hLH. Although the carboxyl terminus antisera are specific for hCG, their sensitivities have been inadequate for determining hCG concentrations in unextracted plasma, serum, or urine samples (12). Moreover, the nonspecific plasma protein effects on carboxyl terminal antisera have been considerably greater than those for antisera generated to hCGβ. To overcome the insensitivity of the carboxyl terminus antisera, some investigators have concentrated urinary samples with Concanavalin A-Sepharose 4B (13). That technique, although cumbersome, can be used to monitor patients with low levels of hCG secreted by their tumors. An alternative immunogen, denatured hCGβ, has also been used with limited success (14). Highly purified hCGβ was initially reduced, S-carbamido-methylated, and coupled to hemocyanin to generate antibody. Relatively specific but not sufficiently sensitive antisera have been generated with that technique.

Although most antisera are generated with in vivo techniques, using rabbits for the most part, in vitro antibody production with hybridomas has been used as well (15). Generation of specific hCG antisera with that technique will be limited as well because of the extensive structural homology between hCGβ and hLHβ. In contrast to in vivo immunization techniques, the antibody produced by a clone, using the hybridoma technique, will be directed to a single antigenic determinant. By contrast, antisera derived from in vivo immunization techniques contain a population of antibodies with varying specificities and sensitivities to many antigenic determinants. The latter may prove to be an advantage in that altered forms of hCG are commonly observed among patients with hCG secreting tumors. Since the beta subunit of hCG confers immunologic specificity, clinically useful antisera may be generated with either technique.

A variety of nonspecific effects may be observed in specific hCG assays. The most commonly observed one is that induced by plasma proteins which interfere with the antigen-antibody interaction. Those effects can usually be overcome by adding "blank" male serum or plasma to nonspecific, maximum binding and standard assay tubes. It is usually imperative that the same concentration of protein be present in all assay tubes. When tissue extracts are assayed, proteases present in those tissue extracts may induce spuriously high concentrations of hCG secondary to degradation of labelled hCG in the assay (16,17). In some cases, protease inhibitors may be added at the time of tissue extraction to overcome those effects.

PHYSIOLOGY OF hCG SECRETION

In peripheral blood and urine hCG is present at readily detectable levels only during pregnancy. Using sensitive, specific assays, hCG first becomes detectable around the time of implantation in the late luteal phase of a menstrual cycle in which conception has occurred (10,18). The hCG levels rise relatively rapidly and attain peak levels of 50-100 IU/ml (Second International Standard hCG) during the mid to late portion of the first trimester of pregnancy. During the last two trimesters of pregnancy, circulating levels of hCG usually range between 1 and 10 IU/ml.

Synthesis of the subunits of hCG is controlled by human chromosomes 10 and 18 (19). In pregnancy there is unbalanced secretion of alpha subunit. Normal placental tissue contains higher concentrations of hCGα mRNA than that for hCGβ. Synthesis of hCG appears to be limited by the concentration of hCGβ mRNA. Low but significant concentrations of the alpha subunit of hCG but negligible concentrations of hCGβ are present in the peripheral blood of normal pregnant subjects. The factors controlling the secretion of hCG by the normal placenta are poorly understood.

In the normal nonpregnant state, low but measurable concentrations of hCG-like activity have been identified in biologic fluids and selected tissues (4-6,20). In general, levels above 5 mIU/ml (Second International Standard hCG) are considered abnormal in the nonpregnant state. The hCG-like immunoreactivity found in tissue extracts, blood and urine of normal individuals has been attributed to incomplete suppression of the genome of normal cells. Fortunately, the level of hCG synthesis and secretion by those tissues is low and does not result in hCG levels in excess of 5 mIU/ml (Second International Standard hCG). Interestingly, hCG-like activity has also been identified in some strains of gram negative bacteria (21). The hCG-like activity isolated from bacteria and normal tissue extracts is indistinguishable from native hCG by biologic, immunologic, and physical criteria.

Human chorionic gonadotropin does not readily cross the blood-brain barrier (22). Less than 1% of hCG crosses the blood into the cerebrospinal fluid. Conversely, hCG does readily cross from the cerebrospinal fluid to peripheral blood. Low but detectable blood levels of hCG sometimes present a diagnostic problem in terms of localizing the tissue site of hCG secretion in a patient. In selected settings, especially in young men with hCG secreting tumors, differential blood-CSF hCG concentrations may help localize the tumor to the brain. In those settings the blood-CSF ratio should be lower than 60:1. Similarly, differential blood:CSF ratios lower than 60:1 have been successfully used to ascertain whether women with gestational trophoblastic disease have CNS metastases (23). However, if a patient has a tumor with the primary site outside the CNS, the ratio may be greater than 60:1 if the peripheral hCG concentrations are in marked excess of those secreted by CNS metastases. Consequently, if an abnormally low ratio is present, it is very helpful. A high ratio does not exclude the presence of CNS metastases, unfortunately.

THE hCG-SECRETING TUMORS

The term "eutopic" secretion connotes the secretion of a hormone by a tissue usually associated with secretion of that hormone under normal physiologic conditions. On the other hand, the phrase "ectopic" secretion usually refers to secretion of a hormone by a tissue not normally associated with synthesis and secretion of that hormone. Obviously, the utilization of those terms may not be relevant in view of evidence that there may be low but significant secretion of an hCG-like substance from essentially all tissues. Secretion of hCG by trophoblastic tissues is termed "eutopic," whereas, those nontrophoblastic tumors secreting hCG are generally considered "ectopic" sources of hCG. Interestingly, nontrophoblastic tumors store hCG in secretory granules whereas hCG is not stored in secretory granules in cells derived from trophoblastic tumors.

Clinical Presentation

Most patients with hCG-secreting tumors have no signs or symptoms of excess circulating hCG. In most cases, circulating hCG levels range between 5 and 20 mIU/ml; however, some patients have hCG levels similar to those encountered during the first trimester of pregnancy and yet have no signs or symptoms of excess hCG secretion. Some patients, however, may present with signs and symptoms directly attributable to secretion of hCG or to concomitant secretion of sex steroid by an hCG-secreting tumor.

Young males with hepatoblastomas, teratomas, or pinealomas secreting hCG may present with signs of precocious puberty (24,25, 26). The signs of precocious puberty are induced from Leydig cell secretion of testosterone induced by tumor hCG secretion. In this setting, affected boys have no or marginal gonadal enlargement since only the Leydig cells are stimulated. The Leydig cells comprise only a small fraction of the gonadal volume. Only with FSH stimulation is marked gonadal enlargement observed. As a result of excess androgen secretion induced by the hCG-secreting tumor, phallic enlargement along with other androgen-induced, secondary sexual characteristics including pubic, axillary and facial hair growth, change in body fat distribution and muscular development may be observed. Girls with extragonadal hCG-secreting tumors rarely present with signs of precocious puberty since both FSH and hCG/LH stimulation of the ovary is needed for granulosa cell proliferation and estradiol secretion.

Men with hCG-secreting tumors may present with gynecomastia as a result of concomitant sex steroid secretion by those tumors (26, 27,28). Similarly, women with hCG-secreting tumors concomitantly secreting estradiol and progesterone, may present with dysfunctional uterine bleeding. An Arias-Stella reaction may be observed in the endometrial tissue of both pre- and postmenopausal women with those tumors.

Trophoblastic Tumors Secreting hCG

Trophoblastic tumors may simply be divided into those derived from the placenta, or gestational trophoblastic neoplasms, and nongestational trophoblastic neoplasms derived from tissues of nonplacental origin. Gestational trophoblastic neoplasms are always associated with secretion of hCG. Gestational trophoblastic neoplasms include hydatidiform mole, chorioadenoma destruens (invasive mole), and choriocarcinoma. Those tumors occur once in every 2,000 pregnancies in well developed areas of the world. In lesser developed areas those tumors may be encountered 10 times more frequently. Molar pregnancy (hydatidiform mole) is the most common form of gestational trophoblastic neoplasm. Eighty percent of affected women require no further therapy other than removal of the molar pregnancy. However, 20% of affected women may develop metastases, transformation of the molar tissue into choriocarcinoma, or persistence of functioning trophoblastic tissue as evidenced by either a constant or increasing level of hCG. Those patients should be treated with chemotherapy.

Patients with choriocarcinoma or with invasive mole should be treated with chemotherapy as soon as that diagnosis is established. Monitoring hCG levels is an invaluable clinical adjunct in monitoring therapy and recurrence of women with gestational trophoblastic neoplasms. If hCG levels remain within normal limits (less than 5 mIU/ml) (Second International Standard hCG) for one year after completion of chemotherapy, the patients are cured.

Nongestational trophoblastic neoplasms arise from the gonads as well as midline structures within the head, chest, and abdomen, most commonly among men. Those tumors are frequently associated with very high circulating levels of hCG. Primary ovarian choriocarcinomas and choriocarcinomas of the testis are examples of nongestational trophoblastic tumors. In contrast to gestational trophoblastic neoplasms, the primary gonadal choriocarcinomas respond less well to chemotherapy.

Nontrophoblastic Tumors Secreting hCG

Tumors of the gonad, stomach, liver, and pancreas are among those tumors most commonly associated with ectopic hCG secretion (1). In fact, patients with hCG-secreting tumors of the stomach frequently have levels of hCG comparable to those found in the first trimester of pregnancy. The range of hCG-secreting neoplasms encompasses leukemias, osteogenic sarcomas, and melanomas, to name but a few. It is not clear why selected tumors are more likely to ectopically secrete hCG than other tumor types. Table 1 lists the relative frequency of the most commonly observed hCG-secreting tumors.

UNBALANCED TUMOR SECRETION OF hCG SUBUNITS

When serum, urine, and aqueous extracts of hCG-secreting tumor tissue are examined by Sephadex G-100 gel chromatography, hCG and negligible concentrations of hCG subunits or altered forms of hCG are observed in the elution profiles of samples obtained from women with gestational trophoblastic neoplasms which are cured with chemotherapy (29).

Table 1. Incidence of immunoreactive hCG (>5mIU/ml) in sera of patients with cancer.

Tumor	No. Examined	No. Positive	Percent
Gastrointestinal	366	68	21
esophagus	12	0	0
stomach	74	17	23
small intestine	23	3	13
pancreas	43	14	33
biliary tract	9	1	11
liver	93	19	21
colon	112	14	12
Gonad			
ovary	47	18	38
testicular	128	80	63
choriocarcinoma	15	15	100
embryonal	62	37	60
seminoma	19	7	37
teratoma/mixed	32	21	66

In contrast, altered forms of hCG and unbalanced secretion of the alpha and beta subunits of hCG are observed in the chromatograms of samples obtained from patients failing to respond to aggressive chemotherapy. Figure 1 depicts the elution profile of an aqueous tumor extract obtained from a woman with gestational trophoblastic disease which failed to completely respond to chemotherapy. Unbalanced secretion of free hCGβ is evident in that chromatogram. Normal placental tissue extracts contain hCG and alpha and little, if any, free hCGβ (11). Figure 2 depicts that relationship. Most nontrophoblastic tumors display unbalanced secretion of hCG and its subunits.

Figure 3 depicts the elution profile of hCG and its subunits in serum and urine obtained from a patient with adenocarcinoma of the stomach. Both hCG and hCGβ are present in the elution profile. In other selected cases, only hCG may be present along with an altered form of hCG, distinct from native hCG and its subunits.

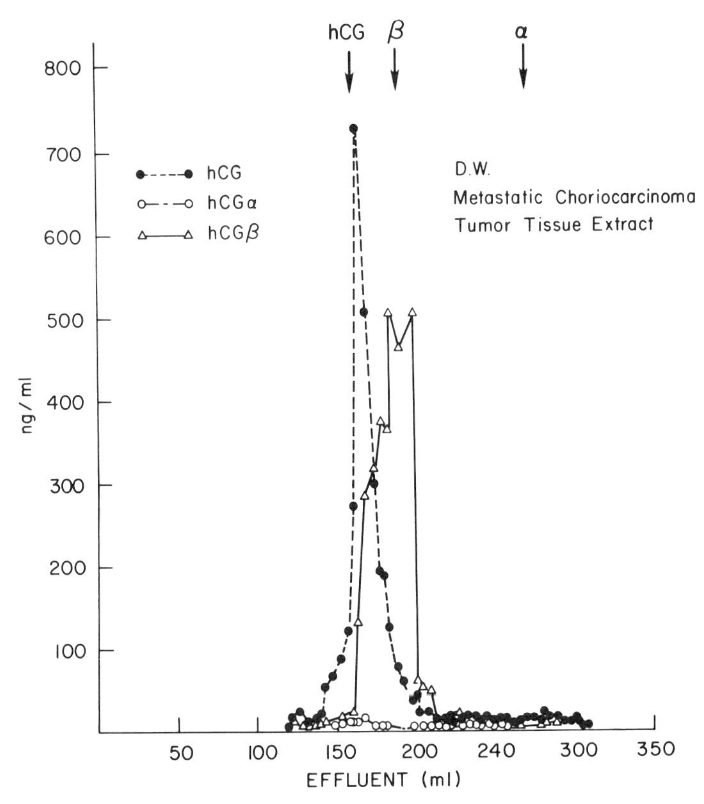

Fig. 1. Elution profile of an aqueous tumor extract obtained from a woman with metastatic gestational choriocarcinoma. After filtration through Sephadex G-100 (2.5 x 90.0 cm), each fraction was assayed in homologous hCG, hCGα, and hCGβ radioimmunoassays. The vertical arrows indicate the points of elution of hCG, hCGβ, and hCGα, respectively. Both native hCG and free hCGβ are present in the extract. (Cited with permission from Vaitukaitis and Ebersole, ref. 29.)

Figure 4 depicts the elution profile after Sephadex G-100 chromatography of a urine sample obtained from a patient with an adenocarcinoma of the stomach. That profile contains a large concentration of an altered immunoreactive form of hCG. Some tumors may secrete only free alpha subunit of hCG or hCGβ subunit rather than native hCG. Malignant carcinoids and functioning malignant islet cell tumors of the pancreas are the most likely candidates for isolated alpha hCG synthesized secretion (30,31). Ectopic production of isolated hCGβ without native hCG is probably even more uncommon (32).

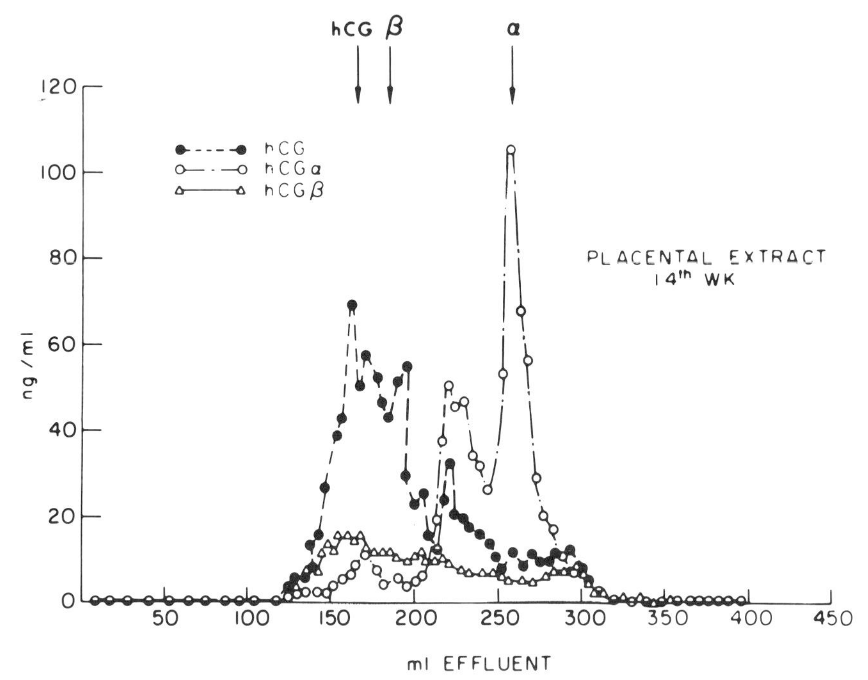

Fig. 2. Elution profile of an aqueous extract of placenta obtained in the 14th week of gestation. After Sephadex G-100 chromatography, each fraction was assayed in homologous hCG, hCGα, and hCGβ radioimmunoassays. The vertical arrows indicate the points of elution of highly purified hCG, hCGβ, and hCGα, respectively.

CLINICAL USEFULNESS OF MONITORING hCG LEVELS

Monitoring blood hCG levels among women with gestational trophoblastic neoplasms and men with hCG-secreting germinal cell tumors of the testis is an invaluable adjunct to conventional diagnostic technique in assisting the clinician's management of the patient's therapy. Among other hCG-secreting tumors, a high incidence of discordance between the tumor's response to chemotherapy and circulating hCG levels is commonly observed. The continued presence of excess circulating hCG is helpful but its absence may not correlate with lack of tumor cell growth. The latter may reflect the fact that chemotherapeutic agents may selectively inhibit cell hCG synthesis or, alternatively, be selectively cytotoxic to one or more cell lines that are synthesizing hCG.

Germinal cell tumors of the testis are the most common neoplasms among men in the 20-34 year old age group. Those tumors

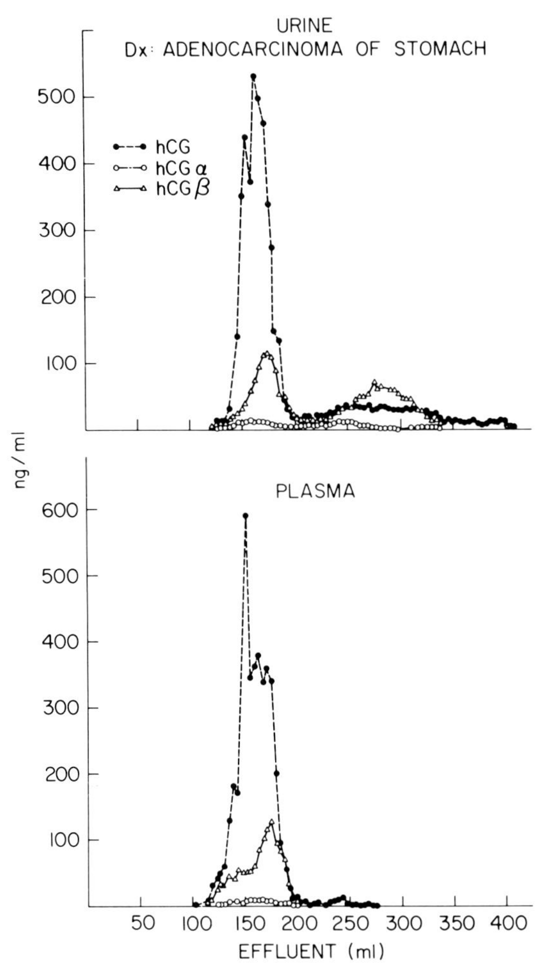

Fig. 3. Elution profile of hCG and its subunits in urine (upper panel) and plasma (lower panel). Urine and plasma were chromatographed through Sephadex G-100 and each fraction assayed in homologous hCG, hCGα, and hCGβ radioimmunoassays. Both native hCG and free beta are evident in the plasma chomatogram. Native hCG, free beta, and an altered immunoreactive form of hCG/hCGβ are evident in the urinary chromatogram. (Published with permission from Vaitukaitis, ref. 26.)

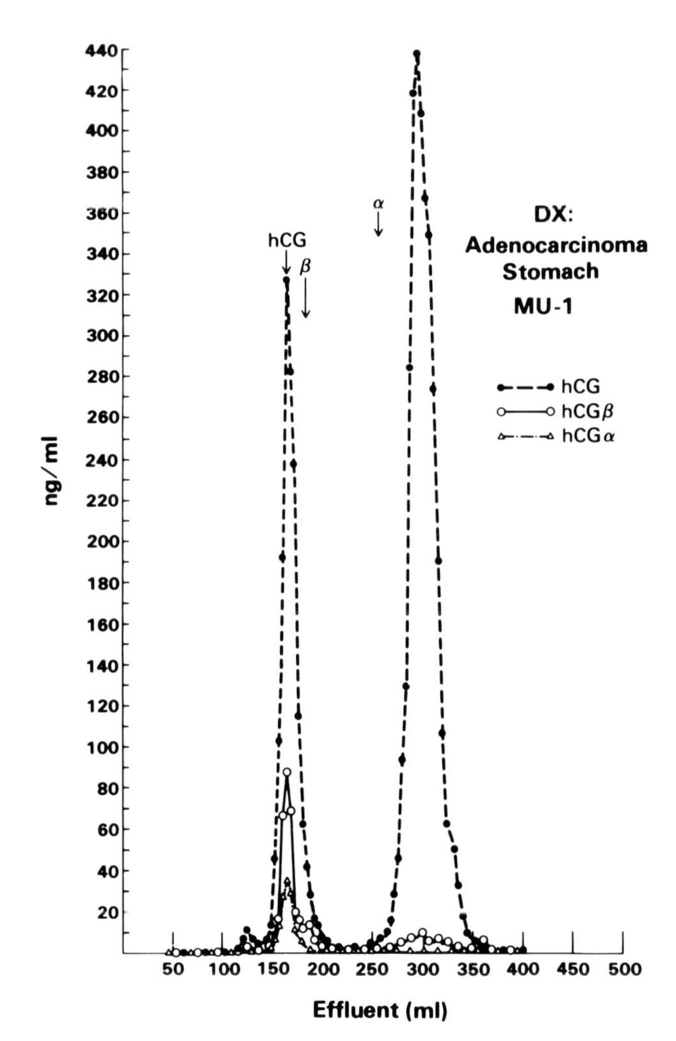

Fig. 4. Elution profile of an aliquot of urine obtained from a patient with adenocarcinoma of the stomach. After gel chromatography through Sephadex G-100, each fraction was assayed in homologous hCG, hCGβ, and hCGα radioimmunoassays. The points of elution of highly purified hCG, its beta and alpha subunits, respectively, are indicated by vertical arrows. A large immunoreactive hCG peak elutes to the right of hCGα. [Reproduced from H.R. Masure, et al. (1981) J. Clin. Endocrinol. Metab. 53:1014]

account for 11% of all cancer deaths in that age group (33). With the advent of more effective combination forms of chemotherapy, the prognosis of affected men has improved considerably (34,35). Approximately 10-30% of men with "pure" seminomas have abnormally high circulating hCG levels. Approximately two thirds of men with embryonal cell carcinomas and all men with choriocarcinomas have abnormally high serum concentrations of hCG. In most cases, the germinal cell tumor is mixed, containing more than one cell type. Conventional diagnostic techniques including lymphangiograms, inferior venacavagrams, and excretory urograms may result in relatively high staging errors of patients with germinal cell tumors (36-39). However, by incorporating the tumor markers, hCG and alpha fetoprotein (AFP) into the clinical staging of men with nonseminomatous germinal cell tumors, the staging error may be markedly reduced (36,39). Approximately 90% of men with nonseminomatous germinal cell tumors will have abnormally high concentrations of either hCG, AFP, or both (36).

The persistence of abnormally high circulating levels of either hCG or AFP usually reflects persistent or recurrent tumor, assuming that abnormal liver function is not the cause of increased circulating AFP levels. Low but significant levels of hCG in men undergoing aggressive chemotherapy for germinal cell tumors portend a poor prognosis provided sufficient time has passed to allow serum levels to become normal. In some cases, the reappearance of immunoreactive hCG in the peripheral blood may be the first sign of recurrent tumor several weeks to months before that tumor becomes evident radiographically.

In some patients presenting with carcinoids and functioning pancreatic islet cell tumors high circulating levels of immunoreactive alpha subunit of hCG may be encountered (30,31). Kahn et al. detected abnormally high circulating levels of immunoreactive alpha hCG in 14 of 27 patients with functioning malignant pancreatic endocrine tumors (31). Those observations have been confirmed in two other studies of functioning pancreatic endocrine tumors in that increased circulating immunoreactive alpha levels reflected malignant degeneration of those adenomas (40,41). In a retrospective study, immunocytochemical localization of alpha subunit of hCG was present in 33 of 48 malignant functioning pancreatic endocrine tumors but in only 1 of 59 apparently benign functioning pancreatic tumors (42). The higher incidence (70%) of immunoreactive alpha hCG in the latter study may simply reflect a difference in level of sensitivity between the immunocytochemical technique and detecting circulating levels of immunoreactive alpha hCG in peripheral blood by radioimmunoassay. As noted previously, the alpha subunit of hCG has a relatively short plasma half-life and consequently, if a tumor secretes low levels of the subunit, peripheral blood levels may be too low to discern from normal levels.

Immunocytochemical localization of alpha hCG in functioning pancreatic tumors may help identify those tumors which are malignant so that appropriate therapy can be initiated before the patient develops widespread metastatic disease.

Since native hCG has a relatively long plasma half-life, monitoring hCG levels across tumor beds does not constitute a useful technique for localizing sites of metastases. On the other hand, if free hCG subunits or altered forms of hCG with markedly shortened plasma half-lives are secreted by the tumor, then differential arteriovenous concentrations may help localize the tumor. As discussed previously, hCG does not readily cross the blood-brain barrier.

In boys presenting with signs of precocious puberty, secondary to a suspected hCG-secreting tumor, differential blood and cerebrospinal fluid concentrations may assist the clinician in localizing

Table 2. Immunocytochemical localization of hCGα in pancreatic endocrine tumors.

Functioning Benign Tumors		No. Positive/No. Examined
Insulinomas		0/23
Glucagonomas		1/5
Gastrinomas		0/8
Vipomas		0/5
Pancreatic polypeptide-omas		0/10
	Total	1/51 (2%)
Functioning Malignant Tumors		
Insulinomas		4/4
Glucagonomas		2/8
Gastrinomas		12/16
Vipomas		13/16
	Total	31/44 (70%)
Nonfunctioning Tumors		
Benign		0/15
Malignant		1/15
	Total	1/30 (3%)

(Modified from Heitz et al., ref. 42.)

the site of the hCG tumor. Intracranial midline teratomas or pinealomas may secrete hCG which in turn stimulates the Leydig cells to secrete testosterone. In some cases, the differential CSF and serum concentrations of hCG may provide the earliest sign of a small intracranial lesion even prior to its becoming detectable by CAT scans or arteriograms.

In summary, monitoring blood levels of hCG is an important clinical adjunct to monitoring patients with gestational trophoblastic tumors and germinal cell tumors of the testis. In selected cases, differential blood:CSF hCG ratios may help localize the site of the primary tumor or ascertain whether CNS metastases are present. Finally, immunohistochemical localization of the alpha subunit of the glycoprotein hormones may identify functioning malignant pancreatic tumors.

ACKNOWLEDGEMENT

Supported by Division of Research Resources, National Institutes of Health, Grant RR-533.

REFERENCES

1. Vaitukaitis, J.L., G.T. Ross, G.D. Braunstein, and P.L. Rayford (1976) Gonadotropins and their subunits: Basic and clinical studies. Recent Prog. Hormone Res. 32:289-331.
2. Bahl, O.P., R.B. Carlsen, R. Bellisario, and N. Swaminathan (1972) Human chorionic gonadotropin: Amino acid sequence of the α and β subunits. Biochem. Biophys. Res. Comm. 48:416-422.
3. Morgan, F.J., S. Birken, and R.E. Canfield (1975) The amino acid sequence of human chorionic gonadotropin: The α subunit and β subunit. J. Biol. Chem. 250:5247-5258.
4. Braunstein, G.D., J. Rasor, and M.E. Wade (1975) Presence in normal human testes of a chorionic gonadotropin-like substance distinct from human luteinizing hormone. New Eng. J. Med. 292:1339-1343.
5. Chen, H.C., G.D. Hodgen, B. Matsuura, J.L. Lin, E. Gross, L.E. Reichert, S. Birken, R.E. Canfield, and G.T. Ross (1976) Evidence for gonadotropin from non-pregnant subjects that has physical, immunologic, and biologic similarities to human chorionic gonadotropin. Proc. Natl. Acad. Sci., USA 73:2885-2889.
6. Yoshimoto, Y., A.R. Wolfsen, and W.D. Odell (1977) Human chorionic gonadotropin-like substance in nonendocrine tissues of normal subjects. Science 197:575-576.
7. Morgan, F.J., S.J. Birken, and R.E. Canfield (1973) Comparison of chorionic gonadotropin and luteinizing hormone: A note on a proposed significant structural difference in the beta subunit. FEBS Lett. 37:101-103.

8. Birken, S., and R.E. Canfield (1977) Isolation and amino acid sequence of COOH-terminal fragments from the β subunit of human choriogonadotropin. J. Biol. Chem. 252:5386-5392.
9. Wehman, R.E. and B.C. Nisula (1979) Metabolic clearance rates of the subunits of human chorionic gonadotropin in man. J. Clin. Endocrinol. Metab. 48:753-759.
10. Vaitukaitis, J.L., G.D. Braunstein, and G.T. Ross (1972) A radioimmunoassay which specifically measures human chorionic gonadotropin in the presence of human luteinizing hormone. Am. J. Obstet. Gynecol. 113:751-758.
11. Vaitukaitis, J.L. (1974) Changing placental concentrations of human chorionic gonadotropin and its subunits during gestation. J. Clin. Endocrinol. Metab. 38:755-760.
12. Chen, H.-C., S. Matsuura, and M. Ohashi (1980) Limitations and problems of hCG-specific antisera. In Chorionic Gonadotropin, S. Segal, ed. Plenum Press, New York, pp. 231-252.
13. Ayala, A.R., B.C. Nisula, H.-C. Chen, G.D. Hodgen, and G.T. Ross (1978) A highly sensitive radioimmunoassay for chorionic gonadotropin in human urine. J. Clin. Endocrinol. Metab. 47:767-773.
14. Pandran, M.R., R. Mitra, and O.P. Bahl (1980) Immunological properties of the β-subunit of human chorionic gonadotropin (hCG). II. Properties of a hCG-specific antibody prepared against a chemical analog of the β-subunit. Endocrinology 107:1564-1571.
15. Khazaeli, M.B., B.G. England, R.C. Dieterle, G.D. Nordblom, G.A. Kabza, and W.H. Bierwaltes (1981) Development and characterization of a monoclonal antibody which distinguishes the β-subunit of human chorionic gonadotropin (βhCG) in the presence of the hCG. Endocrinology 109:1290-1292.
16. Marcio, T., S.J. Segal, and S.S. Korde (1979) Studies on the apparent human chorionic gonadotropin-like factor in the crab ovalipes ocellatus. Endocrinology 104:932-939.
17. Richert, N.D., T.A. Bramley, and R.J. Ryan (1978) Hormone binding, proteases and the regulation of adenylate cyclase activity. In Novel Aspects of Reproductive Physiology. C.H. Spilman, and J.W. Wilks, eds. SP Medical and Scientific Books, pp. 81-106.
18. Braunstein, G.D., J.M. Grodin, J.L. Vaitukaitis, and G.T. Ross (1973) Secretory rates of human chorionic gonadotropin by normal trophoblast. Am. J. Gynecol. 115:447-450.
19. Bordelon-Riser, M.E., M.J. Siciliano, and P.O. Kohler (1979) Necessity for two human chromosomes for human chorionic gonadotropin production in human-mouse hybrids. Somatic Cell Genet. 5:597-613.
20. Borkowski, A., and C. Muquardt (1980) Human chorionic gonadotropin in the plasma of normal, nonpregnant subject. New Eng. J. Med. 303: 298-302.
21. Livingston, V., and A.M. Livingston (1974) Some cultured, immunologic and biochemical properties of Progenitor cryptocides. Trans. N.Y. Acad. Sci. 36:569-582.

22. Bagshawe, K.D., A.H. Orr, and A.G.J. Reishworth (1968) Relationship between concentrations of human chorionic gonadotropin in plasma and cerebrospinal fluid. Nature, London 217:950-951.
23. Bagshawe, K.D., and S. Harland (1976) Detection of intracranial tumors with special reference to immunodiagnosis. Proc. Roy. Soc. Med. 69:51-53.
24. Hung, W., R.M. Blizzard, C.J. Migeon, et al. (1963) Precocious puberty in a boy with hepatoma and circulating gonadotropin. J. Pediatrics 63:895-903.
25. Sklar, C.H., F.A. Conte, S.L. Kaplan, and M.M. Grumbach (1981) Human chorionic gonadotropin-secreting pineal tumor: Relation to pathogenesis and sex limitation of sexual precocity. J. Clin. Endocrinol. Metab. 53:656-660.
26. Vaitukaitis, J.L. (1973) Immunologic and physical characterization of human chorionic gonadotropin (hCG) secreted by tumors. J. Clin. Endocrinol. Metab. 37:505-514.
27. Kirschner, M.A., F.B. Cohen, and D. Jespersen (1974) Estrogen production and its origin in men with gonadotropin-producing neoplasms. J. Clin. Endocrinol. Metab. 39:112-118.
28. Rosen, S., C.E. Becker, and S. Schlaff (1968) Ectopic gonadotropin production before clinical recognition of bronchogenic carcinoma. New Eng. J. Med. 279:640-642.
29. Vaitukaitis, J.L., and E.R. Ebersole (1976) Evidence for altered synthesis of human chorionic gonadotropin in gestational trophoblastic tumors. J. Clin. Endocrinol. Metab. 42:1048-1055.
30. Rosen, S.W., and B.D. Weintraub (1974) Ectopic production of the isolated alpha subunit of the glycoprotein hormones: A quantitative marker in certain cases of cancer. New Eng. J. Med. 290:1441-1447.
31. Kahn, C.R., S.W. Rosen, B.D. Weintraub, S.S. Fajans, and P. Gorden (1977) Ectopic production of chorionic gonadotropin and its subunits by islet cell tumors: Specific marker for malignancy. New Eng. J. Med. 297:565-569.
32. Weintraub, B.D., and S.W. Rosen (1973) Ectopic production of the isolated beta subunit of human chorionic gonadotropin. J. Clin. Invest. 52:3135-3142.
33. MacKay, E.N., and A.H. Sellers (1966) A statistical review of malignant testicular tumors based on the experiences of the Ontario Cancer Foundation Clinics 1938-1961. Can. Med. Assoc. J. 94:889-899.
34. Fraley, E.E., P.H. Lange, and B.J. Kennedy (1979) Germ-cell testicular cancer in adults. New Eng. J. Med. 301:1370-1377 and pp. 1420-1426.
35. Anderson, T., T.A. Waldmann, N. Javadpour, and E. Glatstein (1979) Testicular germ-cell neoplasms: Recent advances in diagnosis and therapy. Ann. Int. Med. 90:373-385.
36. Lange, P.H., K.R. McIntire, T.A. Waldmann, et al. (1976) Serum alpha fetoprotein and human chorionic gonadotropin in the diagnosis and management of nonseminomatous germ-cell testicular cancer. New Eng. J. Med. 295:1237-1240.

37. Scardino, P.T., H.D. Cox, T.A. Waldmann, K.R. McIntire, B. Mittemeyer, and N. Javadpour (1977) The value of serum tumor markers in the staging and prognosis of germ cell tumors of the testis. J. Urol. 118:994-999.
38. Javadpour, N. (1980) Improved staging for testicular cancer markers: A prospective study. J. Urol. 124:58-59.
39. Lange, P.H., L.E. Nochomovitz, J. Rosai, et al. (1980) Serum alpha-fetoprotein and human chorionic gonadotropin in patients with seminoma. J. Urol. 124:472-478.
40. Blackman, M.R., B.D. Weintraub, S.W. Rosen, I.A. Kourides, K. Steinwascher, and M.H. Gail (1980) Human placental and pituitary glycoprotein hormones and their subunits as tumor markers: A quantitative assessment. J. Natl. Cancer Inst. 65:81-93.
41. Stabile, B.E., G.D. Braunstein, and E. Passaro, Jr. (1980) Serum gastrin and human chorionic gonadotropin in the Zollinger-Ellison syndrome. Arch. Surg. 115:1090-1095.
42. Heitz, P.V., M. Kasper, J. Girard, G. Kloppel, J.M. Polak, and J.L. Vaitukaitis (1983) Glycoprotein hormone alpha chain production by pancreatic endocrine tumors - A specific marker for malignancy. Cancer 51:277-282.

PROSTAGLANDINS AND CANCER

Bernard M. Jaffe and M. Gabriella Santoro

Department of Surgery, Downstate Medical Center
State University of New York
Brooklyn, New York 11226

Prostaglandins (PG) are a group of naturally occurring cyclic 20-carbon fatty acids. They are synthesized by virtually all nucleated cells from fatty acids precursors, released from the phospholipid pool of the cell membrane. They are designated by groups, depending on the structure of the cyclopentane ring and are characterized as mono-, bi-, and tri-unsaturated compounds depending on the number of C-C double bonds in the aliphatic side chains. These substances have a broad spectrum of actions and are involved in the control of many physiologic phenomena, including cell growth and differentiation (1,2), immune-function (3), and interferon action (4,5).

Different types of cells produce different amounts and types of prostaglandins. In our first series of studies we demonstrated that tumor tissue produces much larger amounts of PGE than normal tissue. In an experiment in which human colon adenocarcinoma and adjacent normal colonic mucosa were incubated simultaneously in short-term organ culture, carcinoma cells synthesized eight times more PGE than did the normal mucosa (1). An inverse correlation between PGE synthesis and the rate of cell proliferation was reported for four tumor cell lines: HeLa, HEp-2, L, and HT-29 (human adenocarcinoma) (6). Moreover, the amount of prostaglandin produced varies during different phases of cell growth, as has been shown in a mouse neuroblastoma done (N_4), where the rate of PGE production increases abruptly and markedly at confluency (1). These observations suggested that prostaglandins could be involved in the regulation of cell growth, and that manipulation of prostaglandin synthesis by inhibitors (indomethacin, hydrocortisone) or stimulators (dibutyryl cAMP), or by addition of exogenous PGs, could alter cell growth.

The effect of the addition of exogenous PGE_1, dibutyryl cAMP and indomethacin on cell replication in five different tumor cell lines is shown in Table 1. These data show that, in all sytems, inhibition of endogenous PGE synthesis by indomethacin produced an increase in the rate of cell proliferation that, as illustrated by the case of L cells, could be reversed by the addition of physiologic amounts of PGE_1 (10 ng/ml). In B-16 melanoma cells, a second inhibitor of prostaglandin synthesis, hydrocortisone (10^{-7}-10^{-6}M), enhanced by 36.5% the rate of cell replication and this effect was reversed by addition to the medium of subthreshold amounts of PGE_1 (7).

On the other hand, in all systems, stimulation of endogenous PG synthesis by dibutyryl cAMP, and addition of exogenous PGE, resulted in inhibition of cell replication *in vitro* (Tab. 1). PGE_1 and PGE_2 were also active in inhibiting the growth of Friend Erythroleukemia Cells (FLC) derived from a stationary population (Tab. 2) *in vitro*. However, in this system, prostaglandins of the A series were much more effective in inhibiting cell replication measured both by decreases in cell number and by inhibition of ^{3}H-thymidine incorporation. Prostaglandins of the B and F series, thromboxane B_2, and 6-keto $PGF_{1\alpha}$, the metabolite of PGI_2, had no significant effect. Di-M-PGA_2 (a synthetic analog of PGA_2; 16,16 dimethyl PGA_2-methyl ester) and PGD_2 were also effective, but under these conditions, they were slightly toxic for FLC since viability was decreased from 99% to 91% at the end of the experiment (four days of treatment). Prostaglandin D_2 was also found to be a potent inhibitor of B-16 melanoma cell replication *in vitro*. (8). The inhibition was dose-dependent between 3 x 10^{-9}M and 3 x 10^{-6}M (IC_{50} ∿ 0.3μM after six days). At higher concentrations, PGD_2 inhibited DNA, RNA, and protein synthesis in this system. It is interesting to point out that the B-16 melanoma cell line used in this set of experiments (tumor #TBD-811) synthesized arachidonic acid metabolites which comigrated with PGA_2, PGD_2, PGE_2, and $PGF_{2\alpha}$ on a thin layer chromatography system.

The ability of prostaglandins to inhibit tumor cell replication has also been shown *in vivo* in at least two animal models: B-16 melanoma growing in C57Bl/6J mice and Friend Erythroleukemia Cells inoculated subcutaneously in DBA/2J mice (7,9). In the case of the melanoma, treatment with a long-acting synthetic analog of PGE_2 (di-M-PGE_2; 16,16 dimethyl PGE_2-methyl ester, 10μg/mouse/day, subcutaneously) starting on the day of tumor inoculation inhibited the rate of "take" (72% vs. 95%) of tumors, delayed their appearance and increased mouse survival (10). Di-M-PGE_2 treatment also decreased tumor size, as measured by volume, weight, and tumor cell numbers (83.3% decrease), without affecting the animal's body weight (7,10). Treatment with di-M-PGE_2 was also effective in inhibiting tumor growth after the tumor had already developed (Tab. 3). The effect of di-M-PGE_2 was dose-dependent and probably due to a slowing of the

Table 1. Percent changes in viable cell number and in cumulative PGE production after the addition of dibutyryl cAMP, indomethacin, and exogenous PGE_1.

CELL LINE	PGE PRODUCTION (% of Control)		CELL NUMBER (% of Control)			
	1mM dibutyryl cAMP	10^{-9}M Indomethacin	3μM PGE_1	1mM dibutyryl cyclic AMP	10^{-8}M Indomethacin	10^{-8}M Indomethacin + PGE_1 (10ng/ml)
L	+84.5	-79.5	-26.5	-43.3	+29.3	+1.0
HeLa	+26.5	-68.0	-44.3	-22.0	+14.3	-
HEp-2	+56.8	-49.8	-35.8	-73.3	+21.0	-
HT-29	+48.1	-79.0	-55.3	-24.7	+26.0	-
B-16	-	-35.5	-30.0	-	+17.0	-

PGE were dosed by radioimmunoassay (16) in the supernatant of cells growing in monolayer.

Table 2. Effect of prostaglandins and prostaglandin-related compounds on the growth of Friend Erythroleukemia Cells (FLC).

	Number of cells x 10^6/ml
Control	4.25 ± 0.02
PGA_1	1.22 ± 0.14
PGA_2	0.78 ± 0.11
16,16 di-M-PGA_2	0.22 ± 0.04
PGE_1	2.81 ± 0.16
PGE_2	2.35 ± 0.03
16, 16 di-M-PGE_2	1.42 ± 0.57
PGB_2	3.93 ± 0.18
PGD_2	2.19 ± 0.25
$PGF_{2\alpha}$	3.61 ± 0.15
TxB_2	4.20 ± 0.05
6-Keto $PGF_{1\alpha}$	4.73 ± 0.21

*All compounds were tested at a concentration of 1μg/ml. Controls received comparable amounts of ethanol-diluent. Cells were derived from a stationary population. FLC were plated at a density of $2x10^5$ cells/ml and counted four days after plating.

Table 3. Parameters of tumor size after eight days of di-M-PGE_2 or control diluent treatment[a].

	Control	PG-10μg	PG-20μg
Tumor diameters (mm)	11.7±0.7	9.0±1.1	8.6±0.7*
Tumor volumes (mm^3)	763.4±119.7	349.7±89.6*	244.1±74.3*
Tumor weights (mg)	875.8±154.7	458.8±123.4*	271.6±69.9*

* = $p<0.05$

(§) Fourteen days after tumor inoculation ($1x10^6$ viable B-16 melanoma cells) all mice (n=30) developed palpable tumors. After tumor diameters were measured, the mice were weighted, randomized into three equal groups (control, PG-10μg, PG-20μg) and treatment was started. Each mouse received 0.1ml saline containing either 16, 16 d-methyl PGE_2-methyl ester (10 or 20μg) or ethanol diluent, subcutaneously, daily, for eight days.

rate of tumor cell replication, since the degree of inflammation and necrosis of the tumors (determined by histologic examination) were not altered by di-M-PGE_2, while the number of mitoses was decreased in treated tumors (11).

In addition to their effect on cell proliferation, prostaglandins have also been shown to stimulate differentiation in several systems (1,2). In order to study the effect of prostaglandins on cell differentiation, we used Friend Erythroleukemia Cells that can be induced *in vitro* to differentiate into morphologically and functionally mature cells of the erythroid lineage upon stimulation with a wide variety of compounds. Differentiation of FLC is characterized by both morphological and metabolic changes, such as accumulation of RNA for globin synthesis, appearance of new antigens on the cell membrane, increase in iron uptake and heme synthesis, and, finally, hemoglobin production (12,13). We have demonstrated that induction of differentiation by dimethylsulfoxide (DMSO) stimulates the synthesis of endogenous PGE (13), while, as anticipated previously, the addition of exogenous prostaglandins of the E, D, and A series inhibited cell replication, PGA being the most effective. Di-M-PGE_2 was also shown to augment DMSO-induced differentiation, but this prostaglandin had no effect in the absence of DMSO (13).

PGAs, among those tested, were the only prostaglandins that stimulated differentiation in the absence of any inducers in this system in cells coming from either stationary or logarithmically growing populations (Tab. 4). Indomethacin, an inhibitor of the cyclo-oxygenase system involved in PG synthesis, had no effect on DMSO-induced FLC differentiation, while a second inhibitor, hydrocortisone (HC) (that inhibits the liberation of fatty acid precursors from the phospholipid pool of the cell membrane), completely inhibited-induced hemoglobin synthesis without altering cell viability or growth (13) at the same dose at which it inhibited PGE synthesis (10^{-6} - 10^{-7}M). In a study in which several steroid hormones were tested for their effect on DMSO-induced differentiation, only

Table 4. Effect of PGA_1 on FLC differentiation.

Hemoglobin ($\mu g/10^6$ cells) in cells derived from:		
	stationary culture	log culture
Control	0.46 ± 0.06** (n=14)	0.42 ± 0.06 (n=3)
PGA_1 (1μg/ml)	1.29 ± 0.13* (n=12)	1.27 ± 0.10* (n=3)

*p<0.05

**The presence of low amounts of hemoglobin in the control cells is due to spontaneous differentiation which occurs in this cell line.

the 11-17 dihydroxycorticosteroids (dexamethasone, prednisolone, hydrocortisone, and triamcinolone, all inhibitors of PG synthesis) inhibited FLC differentiation, while β-estradiol, cortisone, testosterone, and progesterone, which do not inhibit PG synthesis, had no effect (Fig. 1). However, we were unable to restore the induction of differentiation by adding exogenous prostaglandins. Thus, the inhibitory action of hydrocortisone may be at least partially independent of the effect on prostaglandin synthesis (13).

Finally, we have shown that both PGE and PGA previously found to be usually immunosuppressive in normal animals, partially restore both the humoral and the cellular immunoresponse in immunosuppressed B-16 melanoma-bearing mice (14) and that di-M-PGE_2 treatment augments the effect of chemotherapy in the same model (15).

In conclusion, prostaglandins are involved in many aspects of tumor establishment and growth. The studies described here suggest that prostaglandins of the E, A, and D series can be used to regulate the growth and differentiation of tumor cells.

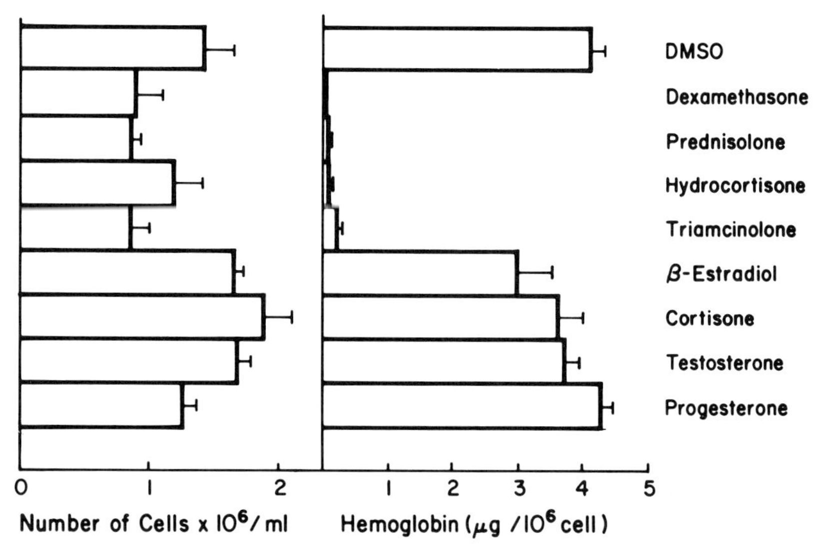

Fig. 1. Effect of steroid hormones on the growth and DMSO-induced differentiation of FLC. The hemoglobin (Hb) content of the cells is expressed as $\mu g/10^6$ cells minus the control value (untreated cells contained 0.43 ± 0.04 μg $Hb/10^6$ cells). Ethanolic solutions of different hormones were dissolved in medium containing 1.5% DMSO. Control media contained the same amount of ethanol (0.005%). Hemoglobin was measured five days after DMSO administration.

REFERENCES

1. Jaffe, B.M., and M.G. Santoro (1977) In The Prostaglandins, Vol. 3, P.W. Ramwell, Ed., Plenum Press, New York, p. 329.
2. Honn, K., R. Bockman, and L.J. Marnett (1981) Prostaglandins 21:833.
3. Goodwin, J.S., and D.R. Webb (1980) Clin. Immun. Immunopath. 15:106.
4. Stringfellow, D.A. (1978) Science 201:376.
5. Pottathil, R., K.A. Chandrobose, P. Cuatrecasas, and D.J. Lang (1980) Proc. Natl. Acad. Sci., USA 77:5437.
6. Thomas, D.R., G.W. Philpott, and B.M. Jaffe (1974) Exp. Cell Res. 84:40.
7. Santoro, M.G., G.W. Philpott, and B.M. Jaffe (1976) Nature 263:777.
8. Simmet, T., and B.M. Jaffe (1983) Prostaglandins 25:47.
9. Santoro, M.G., and B.M. Jaffe (1979) Brit. J. Cancer 39:408.
10. Santoro, M.G., G.W. Philpott, and B.M. Jaffe (1977) Cancer Res. 37:3774.
11. Santoro, M.G., G.W. Philpott, and B.M. Jaffe (1977) Prostaglandins 14:645.
12. Friend, C., W. Scher, J.G. Holland, T. Sato (1971) Proc. Natl. Acad. Sci., USA 68:378.
13. Santoro, M.G., and B.M. Jaffe (1982) In Prostaglandins and Cancer, First International Conference, T.J. Powler, R. Bockman, K. Honn, and P. Ramwell, eds. Alan R. Liss, Inc., New York, p. 425.
14. Favalli, C., E. Garachi, E. Etheredge, M.G. Santoro, and B.M. Jaffe (1980) J. Immunol. 125:897.
15. Hofer, D., A. Dubitsky, J. Marti, M.G. Santoro, P. Reilly, and B.M. Jaffe (1982) J. Surg. Res. 32:555.
16. Jaffe, B.M., H.R. Behrman, and C.W. Parker (1973) J. Clin. Invest. 52:398.

CHAIRMEN'S OVERVIEW ON THE USE OF IMMUNOANALYTICAL METHOD FOR THE CHARACTERIZATION AND QUANTIFICATION OF DNA COMPONENTS STRUCTURALLY MODIFIED BY CARCINOGENS AND MUTAGENS

Ralph A. Reisfeld and Manfred F. Rajewsky*

Scripps Clinic and Research Foundation
LaJolla, California 93037

*Institut für Zellbiologie (Tumorforschung)
Universität Essen, D-4300, Germany

Considerable progress has been made during the past few years with regard to the precise characterization of DNA components structurally modified by chemical carcinogens and mutagens (5,6,14,15). Thus, many of the modifications caused by alkylating N-nitroso compounds have actually been identified (8,15,18). It is quite obvious that highly sensitive methods are required for the detection and quantitation of human DNA that has been structurally modified by chemotherapeutic agents, mutagens, and carcinogens. Such methods are definitely a prerequisite for any analyses of the molecular mechanisms involved in DNA repair, mutagenesis, and malignant transformation.

Radiochromatographic analyses have been largely used for the analysis of chemically modified DNA (1,2); however, the sensitivity of such methods is limited by the specific radioactivity of the agents to be analyzed and thus by relatively large amounts of DNA required for analysis. Consequently, DNA from human cells exposed to nonradioactive environmental agents cannot be analyzed by conventional radiochromatographic methods.

The recently developed immunochemical methods involving polyclonal and especially monoclonal antibodies directed to chemically modified mammalian DNA have offered a particularly attractive alternative since immunoglobulins are well known for their capability to recognize very subtle alterations in molecular structure (11,12,13). The sensitivity and specificity of such immunoanalytical methods utilizing high affinity antibodies are indeed excellent and do not require radioactive labeling of reaction products of chemically modified DNA. Consequently, specific antisera have been developed against naturally occurring modified nucleosides (12), and a number of immunoassays have been described for the detection of DNA components structurally altered by carcinogens and mutagens (3,4,7,9,10, 13,16,17).

It is the major aim of this session to describe and critically evaluate some of the uses of immunoanalytical techniques involving both polyclonal antisera and monoclonal antibodies to analyze structural alterations of DNA in the chromatin of target cells that were caused by chemical carcinogens, mutagens, and environmental agents. Specifically, this includes a review by Dr. B. Erlanger (Columbia University) of methods to make suitable antigens, i.e., altered DNA-methylated BSA conjugates and covalent conjugates between modified bases and carrier proteins with emphasis on the thymine glycol system.

Dr. M. Rajewsky (University of Essen) will describe the characteristics of several high affinity (10% to $>10^{10}$ l/mol) rat monoclonal antibodies against a variety of alkylated DNA components (O^6-ethyldeoxyguanosine, O^6-isopropyldeoxyguanosine, O^6-n-butyldeoxyguanosine, and O^4-ethyldeoxythymidine) and their use in various immunoassays and in the analysis of individual cells by electronically intensified immunofluorescence.

Dr. P.T. Strickland (Frederick Cancer Center) will detail the application of immunochemical methods to detect thymine dimers in single-stranded DNA and polydeoxythymidylic acid following UV irradiation of mouse skin. This approach makes use of specific monoclonal antibodies to localize the binding regions of such reagents directed to thymine dimers in single-stranded poly- or oligonucleotide sequences that are at least four nucleotides long and determines the distribution of these dimers *in situ*.

Finally, Dr. M. Bustin (National Cancer Institute) will describe the use of antibodies to BPDE-modified DNA to determine the influence of chromatin structure on the binding of benzo(a)pyrene metabolites to the genome, particularly carcinogen-modified regions in cellular chromatin, in restriction fragments of SV 40 and in polytene chromosomes. This particular study also involves the application of a method that allows detection of adducts in defined restriction fragments of DNA molecules modified to less than one carcinogen adduct per DNA molecule and the use of polytene chromosomes to directly visualize the carcinogen binding regions.

REFERENCES

1. Baird, W.M. (1979) The use of radioactive carcinogens to detect DNA modifications. In *Chemical Carcinogens and DNA*, P.L. Grover, ed. CRC Press, Boca Raton, p. 59.

2. Beranek, D.T., C.C. Weis, and D.H. Swenson (1980) A comprehensive quantitative analysis of methylated and ethylated DNA using high pressure liquid chromatography. Carcinogenesis 1:595.
3. Briscoe, W.T., J. Spizizen, and E.M. Tan (1978) Immunological detection of O^6-methylguanine in alkylated DNA. Biochemistry 17:1896.
4. Groopman, J.D., A. Haugen, G.R. Goodrich, G.N. Wogan, and C.C. Harris (1982) Quantitation of aflatoxin B_1-modified DNA using monoclonal antibodies. Cancer Res. 42:3120.
5. Grover, P.L., ed. (1979) Chemical Carcinogens and DNA. CRC Press, Boca Raton.
6. Grunberger, D., and I.B. Weinstein (1979) Conformational changes in nucleic acids modified by chemical carcinogens. In Chemical Carcinogens and DNA, P.L. Grover ed. CRC Press, Boca Raton, p. 59.
7. Hsu, I.C., M.C. Poirier, S.H. Yuspa, R.H. Yolken, and C.C. Harris (1980) Ultrasensitive enzymatic radioimmunoassay (USERIA) detects femtomoles of acetylaminofluorene-DNA adducts. Carcinogenesis 1:455.
8. Lawley, P.D. (1976) Carcinogenesis by alkylating agents. In Chemical Carcinogens, C.E. Searle ed. ACS Monograph 173, p. 83. American Chemical Society, Washington, D.C.
9. Leng, M., E. Sage, R.P.P. Fuchs, and M.P. Daune (1978) Antibodies to DNA modified by the carcinogen N-acetoxy-N-2-acetylaminofluorene. FEBS Lett. 92:207.
10. Müller, R., and M.F. Rajewsky (1978) Sensitive radioimunoassay for detection of O^6-ethyldeoxyguanosine in DNA exposed to the carcinogen ethylnitrosourea in vivo or in vitro. Z. Naturforsch. 33c:897.
11. Müller, R., and M.F. Rajewsky (1981) Antibodies specific for DNA components structurally modified by chemical carcinogens. J. Cancer Res. Clin. Oncol. 102:99.
12. Müller, R., J. Adamkiewicz, and M.F. Rajewsky (1982) Immunological detection and quantification of carcinogen-modified DNA components. In Host Factors in Human Carcinogenesis, H. Bartsch and B. Armstrong eds. IARC Scientific Publications No. 39, p. 443. International Agency for Research on Cancer, Lyon, France.
13. Poirier, M.C. (1981) Antibodies to carcinogen-DNA adducts. J. Natl. Cancer Inst. 67:515.
14. Pullman, B., P.O.P. Ts'o, and H. Gelboin, eds. (1980) Carcinogenesis: Fundamental Mechanisms and Environmental Effects. Reidel, Dordrecht, Holland.
15. Rajewsky, M.F. (1980) Specificity of DNA damage in chemical carcinogenesis. In Molecular and Cellular Aspects of Carcinogen Screening Tests, R. Montesano et al., eds. IARC Scientific Publications No. 27, p. 41. International Agency for Research on Cancer, Lyon, France.
16. Rajewsky, M.F., R. Müller, J. Adamkiewicz, and W. Drosdziok (1980) Immunological detection and quantification of DNA

components structurally modified by alkylating carcinogens (ethylnitrosourea). In Carcinogenesis: Fundamental Mechanisms and Environmental Effects, B. Pullman et al., eds. Reidel, Dordrecht, Holland, p. 207.

17. Saffhill, R., P.T. Strickland, and J. Boyle (1982) Sensitive radioimmunoassays O^6-N-butyldeoxyguanosine, O^2-N-butylthymidine and O^4-N-butylthymidine. Carcinogenesis 3:547.
18. Singer, B. (1979) N-nitroso alkylating agents: Formation and persistence of alkyl derivatives in mammalian nucleic acids as contributing factors in carcinogenesis. J. Natl. Cancer Inst. 62:1329.
19. Van der Laken, C.J., A.M. Hagenaars, G. Hermsen, E. Krief, A.J. Kuipers, J. Nagel, E. Scherer, and M. Welling (1982). Measurement of O^6-ethyldeoxyguanosine and N-(deoxyguanosine-8-ul)-N-acetyl-2-aminofluorene in DNA by high-sensitive enzyme immunoassays. Carcinogenesis 3:569.

IMMUNOLOGICAL APPROACHES FOR THE DETECTION

OF DNA MODIFIED BY ENVIRONMENTAL AGENTS

Bernard F. Erlanger,* O.J. Miller,+ R. Rajagopalan,*
and S.S. Wallace**

*Department of Microbiology; +Human Genetics and
Development, and Obstetrics and Gynecology;
Columbia University
New York, New York 10032

INTRODUCTION

This volume deals with a search for biological markers that are expressed as a result of cellular transformation by carcinogenic agents. The rationale for this search is that alteration of DNA by these agents leads to changes in gene expression. It would be useful if we could detect specific changes in DNA structure that accompany the changes in gene expression, and I will discuss how this might be done using immunological procedures.

Immunological methods are sensitive, very specific, and generally speaking, easy to carry out. Their high specificity makes it possible to detect a particular modification in DNA without prior isolation. The techniques that can be applied include, among others, radioimmunassay and enzyme immunoassay, both of which are highly sensitive procedures and simple to perform. In addition, immunocytochemical methods can be employed using fluorescein- or peroxidase-labelled antibodies.

The prime requirement, of course, is the availability of specific antibodies, i.e., antibodies capable of reacting with nucleic acids and their components. Antibodies to proteins and to carbohydrates have a long history as have antibodies to small biologically active or biologically interesting molecules. The work of Landsteiner (1), which goes back to the early 1900s, showed us that, if you want to, you can make antibodies to any low molecular weight

** Department of Microbiology, New York Medical College, Valhalla, New York 10595

compound. All you have to do is attach this molecule to a protein carrier. Immunization with this hapten-protein conjugate elicits antibodies to the protein and to the molecule of interest.

THE PREPARATION OF DNA-REACTIVE ANTIBODIES

The capability of making antibodies to nucleic acids or their components has a relatively short history. The first indication that nucleic acid-reactive antibodies could be elicited came from an experiment of nature, with the discovery that patients with systemic lupus erythematosus (SLE), an autoimmune disease, had circulating antibodies that bound DNA (2,3,4).

The first successful attempt to make antibodies to DNA in the laboratory was reported in 1960 by Levine et al. (5) who immunized rabbits with lysates of T-even bacteriophages. DNA-reactive antibodies were elicited. However, they were specific for denatured DNA of T-even bacteriophages and did not react with denatured DNA from other sources. The reason for this specificity was the presence of a unique pyrimidine base in T-even bacteriophages, namely 5-hydroxymethylcytosine, to which the antibodies were directed.

Despite this limitation, the fact that this experiment could be done encouraged a number of laboratories to increase their efforts to make antibodies that would react with other DNAs. At that time, one of us (BFE) was collaborating with the late S.M. Beiser, an immunologist, and we decided to attack this problem. Our first successful procedure was published in 1962 (6) with Drs. S.W. Tanenbaum and V.P. Butler. Dr. Aaron Bendich of the Sloan-Kettering Institute gave us a reactive purine derivative, 6-trichloromethyl purine, which was capable of reacting with amines. This compound (Figure 1) was covalently linked to the lysine amino groups of bovine serum albumin (BSA) and the conjugate was used to immunize rabbits. Antibodies were raised that were specific for purines and were also able to bind to DNA from many sources if the DNA were denatured. In other words, the antibodies reacted with single-stranded regions of DNA, regions in which there was no base-pairing.

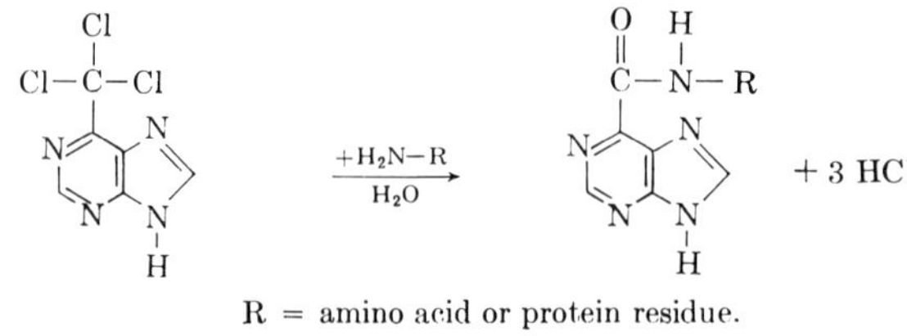

R = amino acid or protein residue.

Fig. 1. Reaction of 6-trichloromethylpurine with amino groups of proteins.

Shortly thereafter, a series of papers appeared including one from our laboratory (7), which really settled the issue of the feasibility of making useful nucleotide-reactive antibodies by relatively simple procedures. One procedure was developed in Parker's laboratory (8). In this procedure, a nucleotide is allowed to react with a protein in the presence of a water soluble carbodiimide, several of which are commercially available. The major reaction is one between the phosphate of the nucleotide and the amino group of lysine in the protein. It is a very simple procedure but a large excess of nucleotide and of carbodiimide are required to make a suitably substituted BSA conjugate. These conjugates are antigenic and antibodies specific for nucleotides are obtained. This is a very important and useful procedure, and, as will be presented later, we make use of it to elicit antibodies to "thymine glycol."

Another procedure (7) is the one developed in our laboratory (Figure 2). It is a simple, effective way to make antibodies to any purine or pyrimidine base and is used widely to make antibodies to carcinogen-modified bases (9). The procedure also makes possible protein conjugates of ribonucleotide sequences by attachment _via_ the three prime end of the molecule (10-13). As many as twenty-five nucleotides can be attached to BSA. The antibodies elicited with these conjugates are highly specific.

An extremely useful procedure was developed in 1964, by Plescia et al. (14) using a different approach. Plescia et al. found that, although DNA alone in solution was not antigenic, a complex of DNA with methylated BSA (MBSA) was a potent immunogen. MBSA is simply

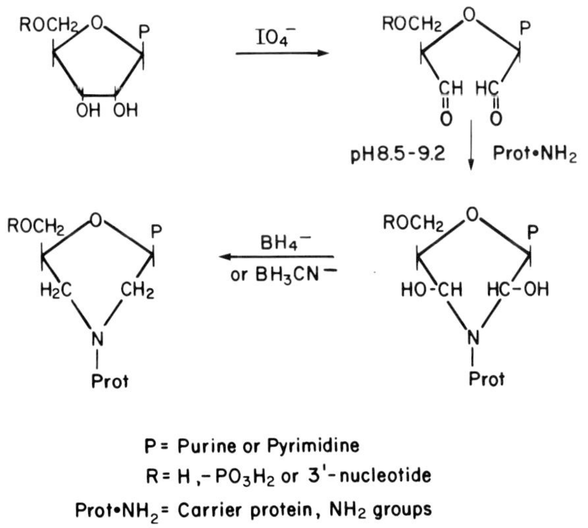

P = Purine or Pyrimidine

R = H, $-PO_3H_2$ or 3'-nucleotide

Prot•NH_2 = Carrier protein, NH_2 groups

Fig. 2. Periodate procedure for the covalent attachment of ribonucleosides and ribonucleotides to proteins.

BSA in which all of the glutamic and aspartic acid carboxyl groups have been esterified to yield a highly basic protein. MBSA forms an immunogenic ionic complex, i.e., a salt, with DNA; DNA-reactive antibodies are stimulated. It is a very useful procedure for making antibodies to DNA, including DNA preparations which have been modified by physical and chemical agents. Antibodies to UV-irradiated DNA have been made this way (15), including the monoclonal antibodies described by Strickland (16) in this volume. It is interesting, however, that native DNA, if mixed with MBSA, still is not antigenic, with one exception: Z or left-handed DNA. Antibodies can be elicited to Z-DNA when it is complexed with the MBSA (17). We have attempted to immunize with Z-DNA in the absence of methylated BSA; it is not antigenic (unpublished data).

PROPERTIES OF DNA-REACTIVE ANTIBODIES

We are now going to describe some experiments that we and other laboratories have carried out with purine- and pyrimidine-specific antisera made by the periodate procedure (Fig. 2) (7). Antisera produced in this way have two prime characteristics. First, they react only with single-stranded regions of nucleic acids in which the purine and pyrimidine bases are not paired. The second property is their high specificity. Both properties make them extremely useful as probes of nucleic acid structure. An example of high specificity is shown in Figure 3, taken from an early paper (18). This is a case in which a rabbit immunized with a conjugate of cytidine-BSA produced antibodies which were essentially

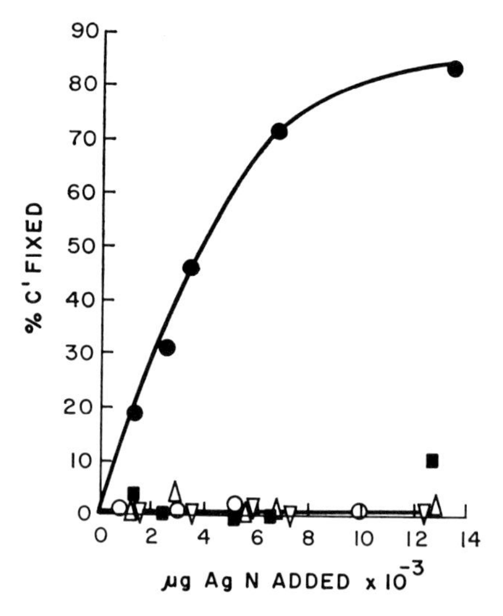

Fig. 3. Reaction of a globulin fraction of antibodies raised to C-BSA with BSA and various conjugates (by complement fixation). ●= C-BSA; ■= T-BSA; △= A-BSA; ○= G-BSA; and ▽= BSA.

monospecific. This is not usually the case but, almost always, highly specific antisera can be obtained by removing the small quantities of cross-reacting antibodies by immunoabsorbent procedures. Another very important example of high specificity is the feasibility of obtaining antibodies to 5-methylcytosine which do not cross-react with cytosine. Shown in Table 1 are some of the antibodies prepared in our laboratory by the periodate procedure. Dr. Rajewsky will describe others in another chapter of this volume (see also, ref. 9).

APPLICATIONS BASED ON SPECIFICITY FOR DENATURED DNA

The two properties described above can be used in ways that are the subject matter of this volume. First, we will deal with specificity for single-stranded regions of DNA. Early work in our laboratory by Klein (19) showed by immunofluorescence that anti-thymidine sera reacted with nuclei of cells only when the cells were in S-phase of the cell cycle. In later work Freeman (20) was able to follow the changes in patterns as the cells went through S-phase.

Table 1. Nucleoside (-tide) specific antibodies prepared in our laboratory.

Specificity	Specificity
Adenosine[19]	UpA (Nahon, unpublished, cited in ref. 36)
Adenosine 5'-phosphate[7]	CpC (Nahon, unpublished, cited in ref. 36)
ATP (Erlanger and Beiser, unpublished)	CpG (Senitzer, unpublished)
Uridine[19]	GpC (Senitzer, unpublished)
Uridine 5'-phosphate[7]	ApG[10]
Guanosine[7]	GpA[10]
Cytidine[7]	AAC[11]
Thymidine (from 5-methyluridine)[7]	AAA[11]
NAD (Erlanger and Beiser, unpublished)	AUG[11]
6-Methyladenosine[35]	AAU[37]
5-Iodouridine[35]	AAUU[37]
5-Bromouridine[35]	AAUUU[37]
ApC (Nahon, unpublished, cited in ref. 36)	AAUUUU[37]
ApA (Nahon, unpublished, cited in ref. 36)	5-Methylcytidine[30]
ApU (Nahon, unpublished, cited in ref. 36)	7-Methylguanosine 5'-phosphate[38, 39]
CpA (Nahon, unpublished, cited in ref. 36)	Poly A[13]

The implication of these experiments is that only cells in S-phase have sufficient single-stranded regions in DNA for a detectable reaction with anti-nucleotide antibodies. This does not necessarily mean that single-stranded regions do not exist in the cell's DNA of G_1 or G_2 of the cell cycle, but if they do, they are either insufficient in number or not accessible to the antibodies. They might be coated with protein, or the single-stranded regions could be too small to yield a signal detectable by light or fluorescence microscopy.

If cells in G_1 are exposed to irradiation, local denaturation (e.g., thymine dimer formation) and subsequent excision prior to repair will make their DNA reactive with antibody. Thus, DNA damage and repair can be measured with anti-nucleotide antibodies (21,22). Any kind of damage that produces single-stranded regions, either directly or because of the repair process, can be detected immunochemically. The results obtained by the rather simple immunochemical procedures parallel those found by isolation of the DNA (21).

The specificity of the antibodies for cells in S-phase can be exploited in other ways. If cells in S-phase can be enumerated, one can determine whether a preparation is multiplying rapidly. For example, biopsies containing cells that are multiplying rapidly will contain more of them in S-phase. Chang et al. (23) have shown that the use of anti-nucleotide sera gave results in agreement with labelling indices found by infusion of [^{3}H]-thymidine. Moreover, the immunological procedure was faster and simpler. Possible applications include better scheduling of chemotherapy or radiation therapy of tumors.

APPLICATIONS BASED ON NUCLEOTIDE SPECIFICITY

Let us now examine some uses of the high specificity of these antibodies. First of all, they can detect DNA modified by such carcinogens as N-acetoxy-N-2-acetylaminofluorene (24), and alkylating agents, in general (9). They can also be used to examine the organization of chromosomes, an application that has the potential of revealing chromosomal abnormalities accompanying cellular transformation. Spreads of metaphase chromosomes prepared by classical methods are treated by procedures that produce selected areas of denaturation. Examples of treatment procedures include partial denaturation (to denature A-T base pairs) (25), photoxidation in the presence of methylene blue (26) [destroys G residues (27)] and ultraviolet irradiation (28-30) (mainly thymine dimer formation). Then the chromosomes are exposed to antisera and evidence of antibody binding is looked for by immunofluorescence of immunoperoxidase procedures.

Shown in Figure 4 is a karyotype of human chromosomes that had been exposed to ultraviolet light and then to rabbit anti-A,

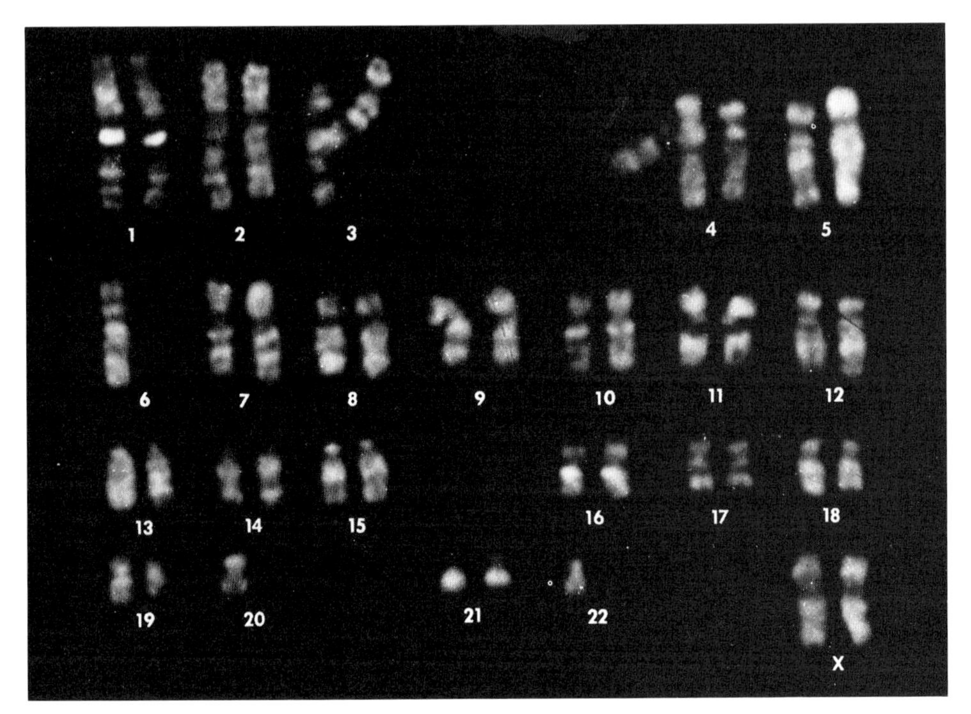

Fig. 4. Human karyotype. Chromosomes exposed to UV-irradiation and then treated with anti-A followed by fluoresceinated sheep anti-rabbit IgG.

followed by fluorescein-labelled anti-rabbit Ig (immunoglobulin) antibody. Banding patterns are obtained which are similar to those seen after quinacrine staining (i.e., Q banding) (28), which is specific for a sub-set of the A-T rich regions of chromomsomal DNA (31). However, in addition, extensive immunofluorescence is seen in the A-T rich heterochromatin near the centromeres of chromosomes 1, 9, and 16, as well as the distal portion of the Y chromosome (not shown).

The same procedure has been carried out with antibodies specific for 5-methylcytosine (32). There are certain highly reactive areas of extensive binding: the heterochromatin of the centromere regions of 1, 9, 16, the short arm of the acrocentric chromosome 15, and the distal portion of the Y chromosome. The "Q-like" banding of anti-A is not seen. In a group of individuals who do not have Down's Syndrome despite the presence of an extra G-group-like chromosome, an abnormal chromosome 15 is seen (Figure 5) (33). The character of this abnormality would not have been apparent without the availability of anti-5-MeC. Thus, nucleotide-specific antibodies have considerable potential for detecting chromosomal abnormalities, which might occur as markers of cellular transformation. Evidence

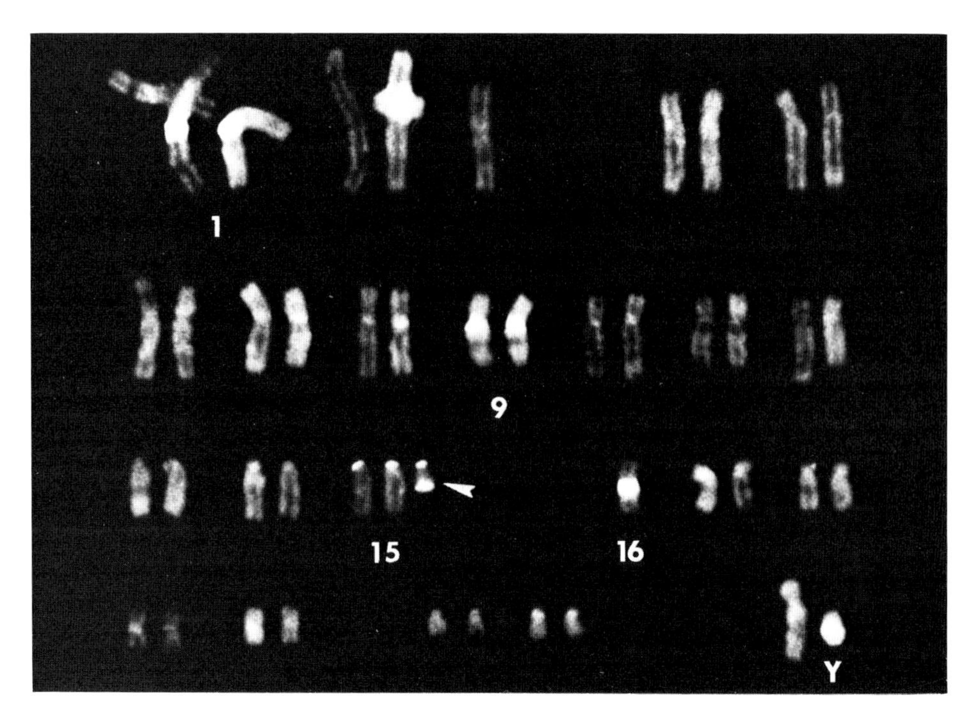

Fig. 5. Karyotype of JF using anti-5-methylcytidine following UV-irradiation. Arrow points to extra chromosome with bipolar fluorescence.

of gene amplification and modified expression can also be seen by immunocytochemical methods (34).

Immunoperoxidase-labelled antibodies can also be used in chromosome studies. They provide greater detail and, moreover, the slides can be stored to provide a permanent record of the experiments. Immunofluorescence fades and the slides are transient. The immunoperoxidase procedure can also be used to provide specimens for electron microscopic examination (30).

Most recently, we have applied immunochemical techniques for the assay of 5,6-dihydroxy-5,6-dihydrothymine (thymine glycol) (Figure 6), a product of the radiolysis of DNA caused by gamma or X-irradiation in aqueous solution. The antibodies were raised by immunization with a BSA (bovine serum albumin) conjugate of 5,6-dihydroxy-5,6-dihydrothymidylic acid (TMP-glycol) prepared by the procedure of Halloran and Parker (8) (using water soluble carbodiimide). The periodate procedure (7) was not applicable in this case because of the susceptibility of thymine glycol to oxidation.

Precipitating titers of antibody specific for thymine glycol

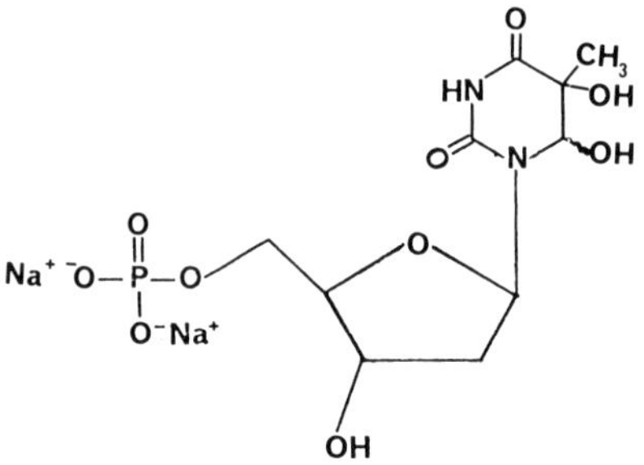

Fig. 6. Sodium salt of the glycol of thymidylic acid (TMP-glycol; see text).

were obtained; no cross-reaction was seen with thymidine, thymidylic acid, 5-methylcytidine, or adenylic acid. The specificity was also demonstrated by enzyme immunoassay using TMP-glycol RSA (rabbit serum albumin), OsO_4- treated DNA, or X-irradiated DNA as the antigen. The specificity was seen both by direct binding studies and be inhibition studies with TMP glycol, thymidine glycol, and thymine glycol. With respect to the sensitivity of the assay, thus far the procedure can detect one thymine glycol in DNA of about 30,000 base pairs in size. There is reason to believe that the sensitivity can be increased even further.

CONCLUSION

We have briefly reviewed the applicability of immunochemical techniques to the detection of DNA or chromatin modified by environmental agents and their potential for the detection of chromosomal abnormalities in transformed cells. The inherent advantages of these techniques, i.e., simplicity, specificity, and sensitivity, should encourage more laboratories to consider their application to problems of carcinogenesis by environmental agents. We hope, of course, that this volume will provide a stimulus. The U.S. Environmental Protection Agency is to be commended for the organization of these proceedings, which have brought together outstanding investigators in diverse fields. It should serve to encourage novel approaches to the problems of carcinogenesis.

ACKNOWLEDGEMENTS

This work was carried out with the support of NIH grants GM 25193, CA 27655 (O.J.M.), AI-06860 (B.F.E.), contract from D.O.E. (S.S.W. and B.F.E.), and the March of Dimes-Birth Defects Foundation (O.J.M.). We wish to acknowledge the invaluable assistance of F. Schneider in the preparation of the manuscript.

REFERENCES

1. Landsteiner, K. (1945) The Specificity of Serological Reactions, Harvard Univ. Press, Cambridge, Massachusetts.
2. Ceppellini, R., E. Polli, and F. Celada (1957) A DNA-reacting factor in serum of a patient with systemic lupus erythematosis diffusus. Proc. Soc. Expt. Biol. Med. 96:572-574.
3. Robbins, W.C., H.R. Holman, H. Diecher, and H.G. Kunkel (1957) Complement fixation with cell nuclei and DNA in lupus erythematosus. Proc. Soc. Expt. Biol. Med. 96:575-579.
4. Seligmann, M. (1957) Evidence for the presence in the serum of patients with lupus erythematosus disseminatus of a substance determining a precipitation reaction with a deoxyribonucleic acids. C.R. Acad. Sci. 245:243-245.
5. Levine, L., W.T. Murakami, H. Van Vunakis, and L. Grossman (1969) Specific antibodies to thermally denatured deoxyribonucleic acid of phage T4. Proc. Nat. Acad. Sci., USA 46:1038-1043.
6. Butler, V.P., S.M. Beiser, B.F. Erlanger, S.W. Tanenbaum, S. Cohen, and A. Bendich (1962) Purine-specific antibodies which react with deoxyribonucleic acids. Proc. Nat. Acad. Sci., USA 48:1597-1602.
7. Erlanger, B.F., and S.M. Beiser (1964) Antibodies specific for ribonucleosides and their reaction with DNA. Proc. Nat. Acad. Sci., USA 52:68-74.
8. Halloran, M.J., and C.W. Parker (1966) The preparation of nucleotide-protein conjugates: Carbodiimides as coupling agents. J. Immunol. 96:373-378.
9. Muller, R., and M.F. Rajewsky (1981) Antibodies specific for DNA components structurally modified by carcinogens. J. Canc. Res. Clin. Oncol. 102:99-113.
10. Wallace, S.S., B.F. Erlanger, and S.M. Beiser (1971) Antibodies to nucleic acids. Immunological studies on dinucleoside phosphate-protein conjugates. Biochemistry 10:679-683.
11. D'Alisa, R., and B.F. Erlanger (1974) Antibodies to the codons ApApA, ApApC and ApUpG. Biochemistry 13:3575-3579.
12. Bonavida, B., S. Fuchs, M. Sela, P.W. Roddy, and H.A. Sober (1972) Specific antibodies to dinucleotides and trinucleotides. Eur. J. Biochem. 31:534-540.
13. Kahana, Z., and B.F. Erlanger (1980) Immunochemical study of the structure of polyadenylic acid. Biochemistry 19:320-324.
14. Plescia, O.J., W. Braun, and N.C. Palczuk (1964) Production of antibodies to denatured deoxyribonucleic acid (DNA). Proc. Nat. Acad. Sci., USA 52:279-285.
15. Levine, L., E. Seaman, E. Hammerschlag, and H. Van Vunakis (1966) Antibodies to photoproducts of deoxyribonucleic acids irradiated with ultraviolet light. Science 153:1666-1667.
16. Strickland, P.T., and J.M. Boyle (1980) Characterization of two monoclonal antibodies specific for dimerized and non-dimerized adjacent thymidines in single-stranded DNA. Photochem. and Photobiol. 34:595-601.

17. Lafer, E.M., A. Moller, A. Nordheim, B.D. Stollar, and A. Rich (1981) Antibodies specific on left-handed Z-DNA. Proc. Nat. Acad. Sci., USA 78:3546-3550.
18. Garro, A.J., B.F. Erlanger, and S.M. Beiser (1971) Pyrimidine-specific antibodies: Reaction with DNAs of differing base compositions. J. Immunol. 106:442-449.
19. Klein, W.J., S.M. Beiser, and B.F. Erlanger (1967) Nuclear fluorescence employing antinucleoside immunoglobulins. J. Exptl. Med. 125:61-70.
20. Freeman, M.V.R., S.M. Beiser, B.F. Erlanger, and O.J. Miller (1971) Reaction of antinucleoside antibodies with human cells in vitro. Exptl. Cell Res. 69:345-355.
21. Liebeskind, D., K.C. Hsu, B.F. Erlanger, and R. Bases (1974) Immunoreactivity of antinucleoside antibodies in X-irradiated cells. Exptl. Cell Res. 83:399-405.
22. Gutter, B., Y. Nishioka, W.T. Speck, H.S. Rosenkranz, B. Lubit, and B.F. Erlanger (1976) Immunofluorescence for the detection of photochemical lesions in intracellular DNA. Exptl. Cell Res. 102:413-416.
23. Chang, T.H., D. Liebeskind, K.C. Hsu, F. Elequin, M. Janis, and R. Bases (1978) Labeling index in clinical specimens estimated by the antinucleoside antibody technique. Cancer Res. 38:1012-1018.
24. Poirier, M.C., S.H. Yuspa, I.B. Weinstein, and S. Blobstein (1977) Detection of carcinogen-DNA adducts by radioimmunoassay. Nature 270:186-188.
25. Miller, O.J., and B.F. Erlanger (1977) Immunological approaches to chromosome banding. ICN/UCLA Symposia Proceedings, Human Cytogenetics VII:87-99.
26. Schreck, R.R., D. Warburton, O.J. Miller, S.M. Beiser, and B.F. Erlanger (1973) Chromosome structure as revealed by a combined chemical and immunochemical procedure. Proc. Nat. Acad. Sci., USA 70:804-807.
27. Simon, M.I., and H. Van Vunakis (1962) The photodynamic reaction of methylene blue (MB) with deoxyribonucleic acid (DNA). J. Mol. Biol. 4:488-499.
28. Schreck, R.R., B.F. Erlanger, and O.J. Miller (1974) The use of antinucleoside antibodies to probe the organization of chromosomes denatured by ultraviolet irradiation. Exptl. Cell Res. 88:31-39.
29. Lubit. B.W., R.R. Schreck, O.J. Miller, and B.F. Erlanger (1974) Human chromosome structure as revealed by an immunoperoxidase staining procedure. Exptl. Cell Res. 89:426-429.
30. Lubit, B.W., T.D. Pham, O.J. Miller, and B.F. Erlanger (1976) Localization of 5-methylcytosine in human metaphase chromosomes by immunoelectronmicroscopy. Cell 9:503-509.
31. Pachmann, U., and R. Rigler (1972) Quantum yield of acridines interacting with DNA of defined base sequence: A basis for the explanation of acridine bands in chromosomes. Exptl. Cell Res. 72:602-608.
32. Miller, O.J., W. Schnedl, J. Allen, and B.F. Erlanger (1974)

5-Methylcytosine localized in mammalian constitutive heterochromatin. Nature 251:636-637.

33. Schreck, R.R., W.R. Breg, B.F. Erlanger, and O.J. Miller (1977) Preferential derivation of abnormal human G-group-like chromosomes from chromosome 15. Human Genetics 36:1-12.
34. Tantravahi, U., R.V. Guntaka, B.F. Erlanger, and O.J. Miller (1981) Amplified ribosomal RNA genes in a rat hepatoma cell line are enriched in 5-methylcytosine. Proc. Nat. Acad. Sci., USA 78:489-493.
35. Sawicki, D.L., B.F. Erlanger, and S.M. Beiser (1971) Immunochemical detection of minor bases in nucleic acids. Science 174:70-72.
36. Beiser, S.M., and B.F. Erlanger (1966) Antibodies which react with DNA. Cancer Res. 26:2012-2017.
37. Wollack, J.B. (1981) Antibodies to oligoribonucleotides of the form A_2U_n: Preparation, properties, and use as probes of oligonucleotide conformation. Ph.D. Dissertation, Columbia University, New York.
38. Meredith, R.D., and B.F. Erlanger (1979) Isolation and characterization of rabbit anti-m^7G-5'-P antibodies of high apparent affinity. Nucl. Acids Res. 6:2179-2191.
39. Castleman, H., R.D. Meredith, and B.F. Erlanger (1980) Fine structure mapping of an avian tumor virus DNA by immunoelectron microscopy. Nucl. Acids Res. 8:4485-4499.

DETECTION OF THYMINE DIMERS IN DNA WITH MONOCLONAL ANTIBODIES

Paul T. Strickland

Laboratory of Immunobiology of Physical
and Chemical Carcinogenesis
LBI - Basic Research Program
NCI - Frederick Cancer Research Facility
Frederick, Maryland 21701

INTRODUCTION

Ultraviolet radiation (UVR) is known to be mutagenic, cytotoxic, and carcinogenic (1,2,3). Cyclobutyl pyrimidine dimers induced in DNA by UVR have been implicated as a major cause of these deleterious biological effects (1,4,5,6). A more complete understanding of the etiological role of pyrimidine dimers in photocarcinogenesis will be facilitated by direct measurement of these lesions in mouse skin (a model) as well as human skin. Several immunological methods for the detection of pyrimidine dimers have been developed that provide alternatives to chromatographic and enzymatic methods previously used to measure these lesions. The initial demonstration of experimentally induced antibodies directed against UV-irradiated DNA (UV-DNA) by Levine et al. (7), prompted a number of laboratories to produce anti-UV-DNA sera by traditional immunization procedures in rabbits (8-16) or by hybridoma methods (17). The present report describes the production, characterization, and use of monoclonal antibodies that are specific for thymidine dimers.

HYBRIDOMA PRODUCTION

Kohler and Milstein (18-20) have reported the development of methods for producing highly specific monoclonal antibodies. These methods utilize the techniques of somatic cell genetics to hybridize lymphoid cells from the spleens of immunized mice with a myeloma cell line that is HGPRT-enzyme deficient. Hybrid clones (hybridomas) are then isolated by selection in HAT (hypoxanthine, aminopterin, thymidine)-supplemented medium. Each hybridoma produces a single species of immunoglobulin (Ig), some of which are specific for the immunizing antigen (21).

In our experiments, BALB/c mice were hyperimmunized with UV-irradiated single-stranded DNA (UVssDNA) and UV-irradiated polydeoxythymidylic acid (UVpolydT) complexed with methylated bovine serum albumin (17) as outlined in Table 1. Mouse spleen cells were collected four days after the final immunization and hybridized with mouse myeloma cells (line MOPC 21 P3-NS1-1-Ag4) by using polyethylene glycol 1,000 (19). After selection of hybrid cells in HAT-supplemented medium (21), supernatants from individual tissue culture wells containing hybrid colonies were screened by a Farr-type precipitation assay for binding to ^{3}H-UVssDNA (22-25). An initial screen of 300 supernatants detected 68 wells with antigen-binding levels significantly above background. Supernatants from positive wells were subsequently screened for binding activity to ^{3}H-UVssDNA, ^{3}H-ssDNA, ^{3}H-UVdsDNA, or ^{3}H-dsDNA. Cells from fourteen of these wells were chosen on the basis of supernatant-binding specificity for cloning in agar. Individual clones were grown in multiwell plates, and supernatant specificities and titers were rechecked. Four hybridomas with high titers and specificity for either UVssDNA or ssDNA were subcloned, and one of each type, designated αUVssDNA-1 and αssDNA-2, was chosen for detailed study (17).

ANTIBODY CHARACTERIZATION

Binding of αUVssDNA-1 was specific for UVssDNA, very low for ssDNA, and undetectable for UVdsDNA (Figure 1), indicating that αUVssDNA-1 will bind photoproduct-containing DNA that has been heat

Table 1. Immunization procedure.

Week #	Treatment	Injection site
1	UVssDNA/mBSA (100μg) + CFA	i.p.
3	UVssDNA/mBSA (200μg) + ICFA	i.p.
5	UVpolydT/mBSA (75μg) + ICFA	i.p.
7	UVpolydT/mBSA (50μg)	i.v.
7 + 4 days	fuse spleen cells	

mBSA = methylated bovine serum albumin

(I)CFA = (in)complete Freund's adjuvant.

i.p. = intraperitoneal

i.v. = intravenous (tail vein).

denatured. This finding was not unexpected since the immunizing antigens were heat-denatured. Competitive inhibition of αUVssDNA-1 binding to ^{3}H-UVssDNA by nonradioactive calf thymus UVssDNA (Figure 2) confirmed that antibodies were directed against UVR-induced alterations in DNA. The percentage inhibition of binding increased in a dose-dependent manner from 10 to 10^4 J/m^2. (The upward curvature of the dose-response relationship is due to the logarithmic dose scale and is not observed on a linear dose scale.)

Antibody specificity was determined by competitive inhibition of αUVssDNA-1 binding to ^{3}H-UVssDNA by various synthetic polynucleotides and oligonucleotides (Figure 3). Several effective inhibitors were found: UVpolydT, UVssDNA, UVp$(dT)_{12-18}$, UVp$(dT)_6$, and UVp$(dT)_4$. No inhibition was observed when up to 10 μg of the following UV-irradiated compounds were tested as inhibitors: UVp$(dT)_2$, UVpdAT, UVpolydAT, UVpolydGC, UVpolyCU, UVpolydA, UVpolydG, UVpoly-dC, UVpolyU, UV-RNA, UV-tRNA, or when nonirradiated compounds were tested. Thus, αUVssDNA-1 is specific for UV-irradiated polynucleotides and oligonucleotides that contain adjacent thymidine residues (Table 2). It is also apparent that inhibition by thymidine oligonucleotides depends on chain length, since inhibitor efficiency decreased in the order UVpolydT > UVp$(dT)_{12-18}$ > UVp$(dT)_6$ > UVp$(dT)_4$ >> UVp$(dT)_2$. Although the quantum yields for dimerization in oligonucleotides are dependent on chain length ($\phi p(dT)_2 = 0.01$ and $\phi p(dt)_{>12} = 0.02$) (26), these differences are not sufficient to explain the preference of antibodies for longer oligonucleotide sequences. This finding was confirmed by increasing the dose of UVR administered to UVp$(dT)_2$. The increase had no effect on inability of this compound to inhibit binding.

Taken together, these data suggest that the antigenic determinants recognized by αUVssDNA-1 are thymidine dimers (T<>T) in single-stranded polynucleotides or oligonucleotide sequences at least four nucleotides long, but not isolated thymidine dimers. Additional experiments (unpublished data) in collaboration with Dr. Paul J. Smith (Chalk River Nuclear Laboratories, Ontario) indicate that the antigenic determinants are photoreversible, a characteristic believed to be unique to pyrimidine dimers. The binding of αUVssDNA-1 to UVpolydT (280 nm UVR) is reduced by exposure of UVpolydT to short wavelength (239 nm) UVR before incubation with antibody. This treatment reduces the saturation level of thymine dimers in UVpolydT (26).

The dependence of antibody binding on oligonucleotide length is an interesting phenomenon. Presumably there are two reasons why longer oligonucleotides (and polynucleotides) containing dimers are preferred antigens. First, the antibody binding site probably encompasses several nucleotides, in which case, binding efficiency would be expected to increase with oligonucleotide length until the correct size of the binding site is attained. The immunoglobulin-combining site has been estimated (27) to be 24 to 36Å in length,

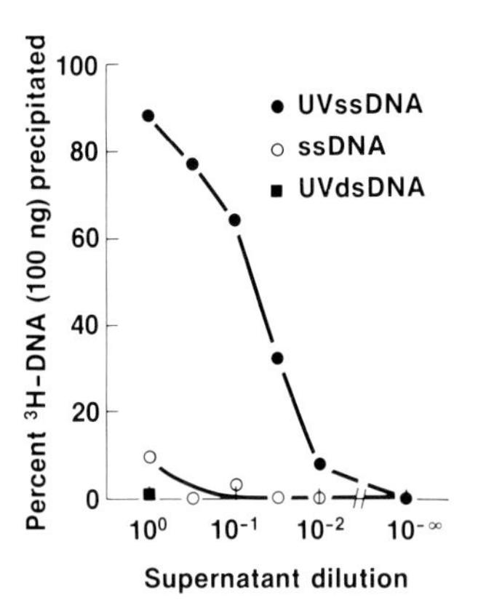

Fig. 1. Mouse monoclonal antibody titration curves. Percent ^{3}H-DNA (100 ng) bound by various dilutions of mouse antibody determined with ^{3}H-ssDNA (○, no UVR; ●, 10^4 J/m^2 UVR) and ^{3}H-dsDNA (■, 10^4 J/m^2 UVR). Reprinted with permission from Ref. 24.

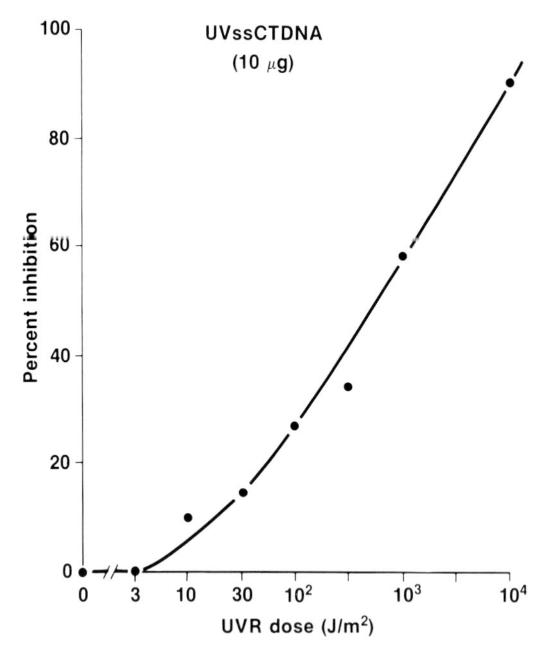

Fig. 2. Dose-dependent inhibition of monoclonal αUVssDNA-1. Competitive inhibition of αUVssDNA-1 binding to ^{3}H-UVssDNA (100 ng) by 10 μg calf thymus ssDNA as a function of UVR exposure administered to calf thymus ssDNA. Reprinted with permission from Ref. 17.

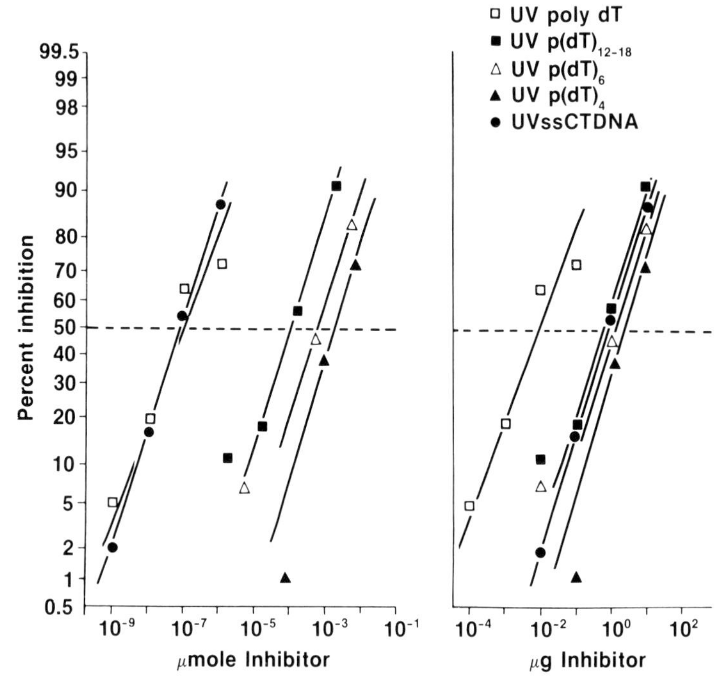

Fig. 3. Competitive inhibition of monoclonal αUVssDNA-1. Inhibition of αUVssDNA-1 binding to ^{3}H-UVssDNA (100 ng) by various amounts of UV-irradiated (10^4 J/m^2) polynucleotides and oligonucleotides. Additional compounds which showed no inhibition are listed in text. Reprinted with permission from Ref. 17.

Table 2. Summary of competitive inhibition of antibody binding.

Inhibitor[a]	Amount required for 50% inhibition (μg)	(n mole)
UVpolydT[c]	0.01	
UVp(dT)$_{12-18}$	0.6	0.12
UVssDNA	0.8	
UVp(dT)$_6$	1.5	0.75
UVp(dT)$_4$	2.5	1.70
UVp(dT)$_2$		>> 14
UVpdAT		>> 15
UVpolydAT	>>10 [b]	
UVpolydGC	>>10 [b]	
UVpolydA	>>10 [b]	
UVpolydG	>>10 [b]	
UVpolydC	>>10 [b]	
UVpolyCU	>>10 [b]	
UVpolyU	>>10 [b]	
UV-RNA	>>10 [b]	
UV-tRNA	>>10 [b]	

[a] No inhibition if compounds not irradiated.

[b] No inhibition detected with 10μg inhibitor.

[c] UV = 1000 J/m^2 (254 nm).

which would accommodate an oligonucleotide not exceeding about 7 to 10 bases. Secondly, as the length of the oligonucleotide approaches ∿ 30 to 50 bases, bivalent immunoglobulin binding may occur (28,29) with a consequent increase in antibody affinity. Several rabbit anti-UV-DNA sera (16,30) with apparent binding sites of between 3 and 6 nucleotides as well as a rabbit anti-DNA serum with binding sites 5 nucleotides long (31) have been reported. The apparent binding site of our monoclonal antibody is about 4 nucleotides long as shown by the fact that UVp$(dT)_6$ and UVp$(dT)_{12-18}$ are only slightly more effective as inhibitors than UVp$(dT)_4$ (Fig.3). The enhanced binding of UVpolydT (and UVssDNA) compared to the oligonucleotides may be due to the presence of bivalent binding.

Two remaining questions regarding the specificity of αUVssDNA-1 are: (i) does the antibody recognize either of the cytosine-containing dimers (C<>T or C<>C)?, and (ii) does the antibody recognize T<>T which is flanked by nonthymidine sequences? Cross reactivity with cytosine-containing dimers was tested by competitive inhibition with defined tetranucleotide sequences pdCCCC and pdCTCT that had been heavily irradiated (Table 3). Neither sequence was an effective inhibitor compared to UVpdTTTT even when exposed to UVR doses that induce numberous cytosine-containing dimers (32).

The influence of flanking bases on antibody binding was investigated by inhibition with UV-irradiated tetranucleotides containing adjacent thymidines (potential T<>T sites) as well as nonthymidine bases. UVpdAATT and UVpdCTTT were bound by αUVssDNA-1, although to a lesser extent than UVpdTTTT (Tab. 3). The decreased level of binding may be caused by: (i) the induction of fewer dimers in short (2 or 3 base) thymidine sequences (33), or (ii) lower affinity of the antibody for T<>T flanked by nonthymidine bases (A or C). High pressure liquid chromatographic analysis of dimer levels in these sequences should provide the information necessary to resolve this question.

APPLICATIONS

In order to understand more fully the role of specific photoproducts in the photocarcinogenic process, it will be necessary to determine their distribution and persistence in experimental photocarcinogenesis models. The antibody described here is being used primarily as a probe for thymidine dimers in irradiated mouse skin.

Two immunological techniques are being employed for this purpose. The first technique is an indirect immunofluorescent assay that detects dimers in fixed cells or frozen mouse skin sections (34). The DNA is denatured *in situ* with mild alkali (0.07 M NaOH in 70% ethanol) according to Cornelis (35), in order to allow dimer-specific antibody binding. Fluorescein isothiocyanate (FITC)-labeled anti-mouse immunoglobulin G is used as the secondary

Table 3. Competitive inhibition by UV-irradiated defined sequence tetranucleotides.

Inhibitor (5 μg)	Percent Inhibition[b]
UVpdTTTT[a]	72
UVpdAATT	50
UVpdCTTT	48
UVpdCCCC	8
UVpdCTCT	0

[a] UV = 2000 J/m^2 (254 nm).

[b] Inhibition of αUVssDNA·1 binding to ^{3}H-UVssDNA (100 ng) by 5 μg inhibitor.

antibody to locate antibodies bound to dimers (8). Thymidine dimers are detected in 10T1/2 cells exposed to either UVC radiation (254 nm) or UVAB radiation (280-400 nm) from an FS40 sunlamp. The UVC radiation is at least ten times more effective than UVAB for dimer induction. In frozen sections of mouse ear and dorsal skin, UVC radiation-induced dimers are detectable in epidermal cells but not in dermal fibroblasts (Figure 4). This procedure is being modified in order to enhance the level of sensitivity.

The second technique, a radioimmunoassay, is used to measure thymine dimers in DNA extracted from mouse skin. This assay is based on competitive inhibition as previously described, using ammonium sulfate precipitation of bound antibody-DNA complexes (see "Antibody Characteriztion"). Purified DNA extracted from mouse skin is used as competitive inhibitor of αUVssDNA-1 binding to ^{3}H-UVssDNA (100 ng). DNA extracted from different anatomical sites (ear, back) and different skin components (epidermal, dermal) is analyzed in this way for dimer content as shown in Table 4. Dimers are detected in DNA extracted from both epidermal and dermal cells from UVC-irradiated dorsal skin and ear skin.

In addition to the two assays discussed here, there are many other immunological methods that could be used with this antibody for determining the distribution and quantity of DNA damage. Other "visualization" methods that have been used include immunoperoxidase (36) and immunoautoradiography (35); the latter provides more quantitative information. Immunofluorescent staining for DNA damage can also be quantitated by means of cytofluorometry (11). Damage in extracted DNA samples can be measured by ELISA (37) and USERIA (38) methods as well as radioimmunoassay. At the molecular level, individually damaged bases can be detected by electron microscopic observation (39,40) of bound antibodies. And finally, antibodies can be used as molecular probes of DNA immobilized on nitrocellulose paper (41).

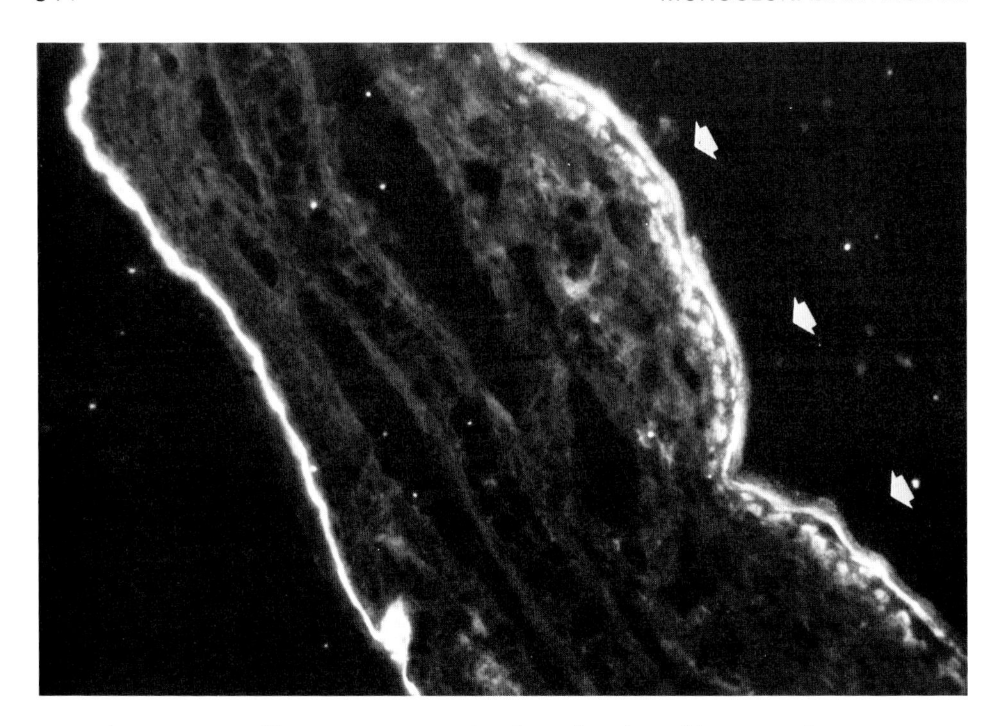

Fig. 4. Immunofluorescent stain for thymine dimers. Cryostat section (6 μm) of ear from mouse exposed to 6.3×10^4 J/m^2 (UVC) stained with monoclonal αUVssDNA-1 and FITC-anti-mouse IgG. Fluorescent nuclei visible just below keratin layer (which stains nonspecifically). Top surface (arrows) of ear directly exposed to UVR; under surface of ear not exposed.

CONCLUSION

Immunological methods for the detection of carcinogen-altered nucleic acids provide a useful tool complementing existing biochemical and physical approaches for studying the process of carcinogenesis. Several immunological procedures (immunofluorescence, immunoperoxidase, immunoautoradiography) allow the observation of DNA damage and repair in cells and tissue sections _in situ_. This is particularly useful when heterogeneous cell populations are being studied and when chemical extraction and subsequent analysis of DNA permits only the average amount of damage sustained by the whole cell population to be measured. Immunological detection of damage in individual cells allows one to determine how DNA damage is distributed within a cell population (or tissue) and whether this distribution changes with time.

A great advantage of immunological assays for DNA damage is that the DNA or chemical adduct under investigation does not have to

Table 4. Competitive inhibition by DNA extracted from UV-irradiated mouse skin.

DNA sample	Percent inhibition[a] No UVR	UVR[b]
Dorsal skin (epidermis)	0	51
(dermis)	0	22
Ear skin (total)	4	24

[a] Inhibition of αUVssDNA·1 binding to ^{3}H-UVssDNA (100 ng) by 3 μg inhibitor (DNA).

[b] UVR = 6.3×10^4 J/m^2.

be radiolabeled. This is particularly useful for analysis of DNA from tissue in which the cells are slowly proliferating and, therefore, cannot be labeled at high specific activity. It also precludes the requirement for a radioactive label on a compound (e.g., carcinogen) in order to detect its binding to DNA. In the former case, the sensitivity (level of detection) of some assays depends on the specific activity of the DNA; in the latter case, the particular compound may be difficult or expensive to label. Thus, the methods discussed here may provide information about carcinogen-damaged DNA that is unobtainable by other approaches.

ACKNOWLEDGEMENTS

Research sponsored by the National Cancer Institute, DHHS, under contract No. NO1-CO-23909 with Litton Bionetics, Inc. The contents of this publication do not necessarily reflect the views or policies of the Department of Health and Human Services, nor does mention of trade names, commercial products, or organizations imply endorsement by the U.S. Government.

REFERENCES

1. Setlow, R.B. (1966) Cyclobutane-type pyrimidine dimers in polynucleotides. Science 153:379-386.
2. Epstein, J.H. (1970) Ultraviolet carcinogenesis. In Photophysiology, A.C. Giese, ed. Vol. V, Academic Press, New York, pp. 235-273.
3. Kripke, M.L., and E.R. Sass, eds. (1978) International Conference on Ultraviolet Carcinogenesis, NCI Monograph, Vol. 50. DHEW Publ. No. (NIH) 78-1532.
4. Cleaver, J.E., and D. Bootsma (1975) Xeroderma pigmentosum: Biochemical and genetic characteristics. Annu. Rev. Genet. 9:19-38.
5. Hart, R.W., R.B. Setlow, and A. Woodhead (1977) Evidence that pyrimidine dimers in DNA can give rise to tumors. Proc. Nat. Acad. Sci., USA 74:5574-5578.

6. Hanawalt, P.C., and P.K. Cooper (1979) DNA repair in bacteria and mammalian cells. Annu. Rev. Biochem. 48:783-836.
7. Levine, L., E. Seaman, E. Hammerschlag, and H. Van Vunakis (1966) Antibodies to photoproducts of deoxyribonucleic acids irradiated with ultraviolet light. Science 153:1666-1667.
8. Tan, E.M. (1968) Antibodies to deoxyribonucleic acid irradiated with ultraviolet light: Detection by precipitins and immunofluorescence. Science 161:1353-1354.
9. Lucas, C.J. (1972) Immunological demonstration of the disappearance of pyrimidine dimers from nuclei of cultured human cells. Exp. Cell Res. 74:480-486.
10. Jarzabek-Chorzelska, M., Z. Zarebska, H. Wolska, and G. Rzesa (1976) Immunological phenomena induced by UV rays. Acta. Derm. 56:15-18.
11. Fukuda, M., N. Nakanishi, T. Mukainaka, A. Shima, and S. Fujita (1976) Combination of Feulgen nuclear reaction with immunofluorescent staining for photoproducts of DNA after UV irradiation. Acta. Histochem. Cytochem. 9:180-192.
12. Saenko, A.S., T.P. Ilyina, V.K. Podgorodnichenko, and A.M. Poverenny (1976) Simple immunochemical method for measuring DNA repair in UV irradiated bacteria. Immunochemistry 13:779-781.
13. Cornelis, J.J., J. Rommelaere, J. Urbain, and M. Errera (1977) A sensitive method for measuring pyrimidine dimers in situ. Photochem. Photobiol. 26:241-246.
14. Fink, A., and G. Hotz (1977) Immunological reaction of UV-induced radiation damage in coliphage DNA. Z. Naturforsch. 32:544-549.
15. Wakizaka, A., and E. Okuhara (1979) Immunologically active lesions induced on double-stranded DNA with ultraviolet. Photochem. Photobiol. 30:573-579.
16. Mitchell, D.L., and J.M. Clarkson (1981) The development of a radioimmunoassay for the detection of photoproducts in mammalian cell DNA. Biochim. Biophys. Acta 655:54-60.
17. Strickland, P.T., and J.M. Boyle (1981) Characterization of two monoclonal antibodies specific for dimerized and nondimerized adjacent thymidines in single-stranded DNA. Photochem. Photobiol. 34:595-601.
18. Köhler, G. and C. Milstein (1975) Continuous cultures of fused cells secreting antibody of predefined specificity. Nature 256:495-497.
19. Köhler, G., S.C. Howe, and C. Milstein (1976) Fusion between immunoglobulin-secreting and non-secreting myeloma cell lines. Eur. J. Immunol. 6:292-295.
20. Galfre, G., S.C. Howe, C. Milstein, G.W. Butcher, and J.C. Howard (1977) Antibodies to major histocompatibility antigens produced by hybrid cell lines. Nature 266:550-552.
21. Herzenberg, L.A., L.A. Herzenberg, and C. Milstein (1978) Cell hybrids of myelomas with antibody-forming cells and T-lymphomas with T cells. In Handbook of Experimental Immunology, D. Weir, ed., Blackwell Scientific Publications, Oxford, Vol. 2 (3rd. Ed.), pp. 251-257.

22. Wold, R.T., F.E. Young, E.M. Tan, and R.S. Farr (1968) Deoxyribonucleic acid antibody: A method to detect its primary interaction with deoxyribonucleic acid. Science 161:806-807.
23. Aarden, L.A., F. Lakmaker, and T.E.W. Feltkamp (1976) Immunology of DNA. I. The influence of reaction conditions on the Farr assay as used for the detection of anti-ds DNA. J. Immunol. Methods 10:27-37.
24. Strickland, P.T., and J.M. Boyle (1981) Application of the Farr assay to the analysis of antibodies specific for UV-irradiated DNA. J. Immunol. Methods 41:115-124.
25. Wakizaka, A., and E. Okuhara (1979) Immunochemical studies on the correlation between conformational changes of DNA caused by ultraviolet irradiation and manifestation of antigenicity. J. Biochem. 86:1469-1478.
26. Deering, R.A., and R.B. Setlow (1963) Effects of ultraviolet light on thymidine dinucleotide and polynucleotide. Biochim. Biophys. Acta 68:526-534.
27. Kabat, E. (1976) Structural Concepts in Immunology and Immunochemistry. Holt, Rinehart, Winston, New York, p. 127.
28. Aarden, L.A., E.R. deGroot, and F. Lakmaker (1976) Immunology of DNA. V. Analysis of DNA/anti-DNA complexes. J. Immunol. Methods 13:241-252.
29. Papalian, M., E. Lafer, R. Wong, and B.D. Stollar (1980) Reaction of systemic lupus erythematosus antinative DNA antibodies with native DNA fragments from 20 to 1200 base pairs. J. Clin. Invest. 65:469-477.
30. Seaman, E., H. vanVunakis, and L. Levine (1972) Serologic estimation of thymine dimers in the deoxyribonucleic acid of bacterial and mammalian cells following irradiation with ultraviolet light and post-replication repair. J. Biol. Chem. 247:5709-5715.
31. Wakizaka, A., and E. Okuhara (1975) Immunological studies on nucleic acids: An investigation of the antigenic determinants of denatured DNA reactive with rabbit anti-DNA antisera by a radioimmunoassay technique. Immunochemistry 12:843-845.
32. Ellison, M.J., and J.D. Childs (1981) Pyrimidine dimers induced in Escherichia coli DNA by ultraviolet radiation present in sunlight. Photochem. Photobiol. 34:465-469.
33. Haseltine, W.A., L.K. Gordon, C.P. Linden, R.H. Grafstrom, N.L. Shaper, and L. Grossman (1980) Cleavage of pyrimidine dimers in specific DNA sequences by a pyrimidine dimer DNA-glycosylase of M. luteus. Nature 285:634-641.
34. Strickland, P.T. (1982) Detection of DNA photoproducts in situ with monoclonal antibodies. Proc. 13th Inter. Cancer Congress (UICC), Seattle, Washington (Abstract #2521).
35. Cornelis, J.J., and M. Errera (1981) Immunocytological detection of pyrimidine dimers in situ. In DNA Repair: A Laboratory Manual of Research Procedures. E.C. Freidberg and P.C. Hanawalt, eds., Marcel Dekker, New York, pp. 31-34.
36. Bases, R., A. Rubinstein, A. Kadish, F. Mendez, D. Wittner, F. Elequin, and D. Liebeskind (1979) Mutagen-induced disturbances

in the DNA of human lymphocytes detected by antinucleoside antibodies. Cancer Res. 39:3524-3530.

37. Hsu, I-C., M.C. Poirier, S.H. Yuspa, D. Grunberger, I.B. Weinstein, R.H. Yolken, and C.C. Harris (1981) Measurement of benzo(a)pyrene-DNA adducts by enzyme immunoassays and radioimmunoassay. Cancer Res. 41:1091-1095.

38. Haugen, A., J.D. Groopman, I-C. Hsu, G.R. Goodrich, G.N. Wagen, and C.C. Harris (1981) Monoclonal antibody to aflatoxin B-modified DNA detected by enzyme immunoassay. Proc. Natl. Acad. Sci., USA 78:4124-4127.

39. deMurcia, G., M-C. Lang, A-M. Freund, R.P. Fuchs, M.P. Daune, E. Sage, and M. Leng (1979) Electron microscopic visualization of N-acetoxy-N-2-acetylaminofluorene binding sites in ColE1 DNA by means of specific antibodies. Proc. Natl. Acad. Sci., USA 76:6076-6080.

40. Slor, H., H. Mizusawa, N. Neihart, T. Kakefuda, R.S. Day, and M. Bustin (1981) Immunological visualization of binding of the chemical carcinogen benzo(a)pyrene diol-epoxide-1 to the genome. Cancer Res. 41:3111-3117.

41. Sano, H., H.D. Royer, and R. Sager (1980) Identification of 5-methylcytosine in DNA fragments immobilized on nitrocellulose paper. Proc. Natl. Acad. Sci., USA 77:3581-3585.

IMMUNOLOGICAL STUDIES ON THE INFLUENCE OF CHROMATIN STRUCTURE ON THE BINDING OF A CHEMICAL CARCINOGEN TO THE GENOME

Michael Bustin, Paul D. Kurth, Hanoch Slor,* and Michael Seidman

Laboratory of Molecular Carcinogenesis
Division of Cancer Cause and Prevention
National Cancer Institute
Bethesda, Maryland 20205

OVERVIEW

The influence of chromatin structure on the binding of benzo-(a)pyrene metabolites to the genome is studied. Carcinogen modified regions, in cellular chromatin, in restriction fragments of SV40, and in polytene chromosomes are detected using antibodies elicited by [r-7, t-8-dihydroxy-t-9, 10-oxy-7,8,9,10-tetrahydrobenzo(a)pyrene] BPDE-modified DNA. A method has been developed which allows detection of adducts in defined restriction fragments derived from DNA molecules modified to less than one carcinogen adduct per DNA molecule. Polytene chromosomes allow direct visualization of carcinogen binding regions. Autoradiography of ^{3}H-BPDE-treated chromosomes followed by immunofluorescence in the presence or absence of RNase allows distinction between the binding to DNA, RNA, and proteins. The results indicate that the nucleosomal structure of chromatin has a very slight effect on the binding of the carcinogen. However, certain loci in the polytene chromosomes display a significant preference for anti-BPDE binding.

INTRODUCTION

The interaction of a chemical carcinogen with the genome of a target cell may lead to transformation. While the mechanism by which the chemical carcinogens cause transformation is not fully

* Present address: Department of Human Genetics, Tel Aviv University, Tel Aviv, Israel.

understood, a variety of studies suggest that the critical step in this process is the formation of covalent carcinogen-DNA adducts. The formation of carcinogen-DNA adducts can be correlated with both mutation and transformation (1-3). Experiments with rodent cells indicated that in some cases the transformation frequency was 10-1,000 times higher than the mutation frequency (3,4,5). The unexpected difference between the frequency of mutation and transformation may be explained in several ways: one of them is the possibility that the interaction of the carcinogens with the DNA in the eukaryotic cell is not completely random.

One of the possible reasons for nonrandom binding of carcinogens to the eukaryotic genome is that the DNA in these cells is complexed with proteins. Since many carcinogens bind not only to DNA but to proteins and RNA as well (6), the distribution of the proteins and RNA along the DNA may affect the interaction of the carcinogen with the various guanosine residues in DNA. Furthermore, the genomic DNA is complexed with histones into discrete subunits called nucleosomes (7). The nucleosome DNA consists of core particle DNA and linker DNA. Several studies have indicated that benzo(a)pyrene binds preferentially to the linker DNA found between two adjacent core particles (8-11). The core particles are positioned in a nonrandom manner upon DNA sequences (7). Transcription and replication of the genomic DNA may affect the structure and position of the nucleosomes along a particular DNA region (7). Higher order packing of the chromatin fiber into solenoids or superbands and ultimately into distinct chromosomes may also affect the manner in which the carcinogen may bind to DNA. Table 1 lists some of the levels of genome structure which may be pertinent to studies on the binding of carcinogens to the eukaryotic genome.

The aim of the studies presented here is to investigate whether the structure of the chromatin fiber affects the manner in which the metabolites of the carcinogen benzo(a)pyrene (BP) bind to chromatin and chromosomes.

Experimental Approach

The SV40 chromosome can be used as a convenient model for the cellular chromatin (12). The DNA of this virus is packed into nucleosomes. The sensitivity of the chromosome to micrococcal nuclease and DNase I is indistinguishable from that of the chromatin found in the nucleated cells. The viral DNA utilizes the cellular enzymes for both replication and transcription. In most preparations of SV40 minichromosomes a fraction of the molecules have a nonnucleosomal origin of replication. Thus, the distribution of BPDE adducts in the core particle DNA, in the linker DNA between core particles, and in nonnucleosomal DNA can be studied. In the present studies the difference between cellular chromatin and SV40 virus has been minimized further by using an encapsidation mutant, tsC219 (13).

Table 1. Levels of genome structure which may affect the binding of a carcinogen.

-DNA sequence
-Specific DNA structures
-Nucleoprotein structures
-Nucleosome positioning
non random organization on the genome
transcriptional events
replication events
-Higher order packing
solenoids, superbeads
chromosomes

Meaningful studies on the distribution of BPDE in the SV40 chromosomes require examination of molecules modified to a very low extent, preferentially at 1.0 BPDE molecule per molecule of SV40. To detect such low levels of modification we are using antisera specific for BPDE-modified DNA (14,15). Purified viral DNA extracted from infected cells which were treated with the ultimate carcinogen BPDE were digested with restriction enzymes. The digest is fractionated on 1% agarose gels and the separated restriction fragments transferred to diazobenzyloxymethyl (DBM) paper (16). The paper is treated with antibody followed by ^{125}I Protein A. The autoradiogram of the treated paper allows identification of the DNA regions which have been modified by the carcinogen. The technique is used to study the interaction of the carcinogen with cellular chromatin and to examine the pattern of modification in both the coding and noncoding strands of SV40.

The effect of higher order packing of chromatin on carcinogen binding was studied using polytene chromosomes obtained from the salivary glands of *Chironomus thummi*. Polytene chromosomes are large chromosomes in which 1,000 or more chromatids are aligned precisely. Tightly coiled chromatin regions appear as bands while more loosely packed chromatin appears as interbands (17). Furthermore, in these chromosomes transcriptional activity can be induced in defined genetic loci. The large size of the chromosomes gives the unique opportunity to study the binding of carcinogens to bands, interbands, and transcriptionally active loci at the resolution obtainable by light microscopy.

Autoradiography of such chromosomes after treatment with ^{3}H-BPDE reveals the binding site of the carcinogen to the proteins, RNA, and DNA in the chromosomes. The binding to RNA and DNA can be separated from the binding to protein by using antibodies to BPDE-DNA in the indirect immunofluorescence technique. These antibodies react with both BP-RNA and BP-DNA; however, treatment of the preparation with RNase prior to antibody addition allows detection of the adduct bound specifically to DNA.

We have found that Chironomus larvae can metabolize the parent hydrocarbon [benzo(a)pyrene]. Therefore, the experiments can also be done with chromosomes that have been modified *in vivo* using the endogenous pathway of BP metabolism.

In both the SV40 system and in the polytene chromosome system the detection of the adduct DNA lesions is based on immunochemical approaches. Therefore, it is of utmost importance to ensure that the antisera are well characterized.

Results and Discussion

Characterization of the antisera. Antisera were obtained from rabbits immunized with BPDE-modified DNA. The specificity was tested by several techniques. Filter binding assay using BPDE-modified and unmodified ^{3}H-Col E 1 DNA revealed that the antibody bound only to the modified DNA. It did not cross react with free BPDE, or proteins treated with BPDE (14). Enzyme linked immunoassays also demonstrate that: a) the binding of the antibody is dependent on the concentration of the adduct; b) that unmodified double-stranded or single-stranded DNA does not bind antibodies, and c) that modified DNA does, but unmodified DNA does not inhibit the binding of antibodies to modified DNA (Figure 1).

When antibodies are added to either modified or unmodified Col E 1 DNA and the mixtures subjected to centrifugation, only the modified DNA binds antibodies creating a complex which is separable from unmodified DNA. Examination of this complex by electron microscopy allows visualization of antibodies bound to the DNA of the plasmid (14). Indirect immunofluorescent studies with this sera also reveal that antibodies bind specifically to the nuclei of cells treated with BPDE (Figure 2). Nonimmune sera did not bind to these cells and the antibody did not bind to cells which were not exposed to PDE. Qualitatively, the intensity of fluorescence could be correlated with the concentration of BPDE to which the cells were exposed. This correlation was observed with solutions containing between 0.25 and 1.0 μg BPDE per ml. Comparison of the intensity of fluorescence among various cells present on a slide suggest that all cells present in a population are susceptible to modification by BPDE (14). The various assays described above indicate that the sera is specific for BPDE-modified DNA.

Distribution of carcinogen in cellular and viral chromatin. To study the distribution of the BPDE adducts in chromatin, we developed a technique which allows immunochemical detection of BPDE-DNA

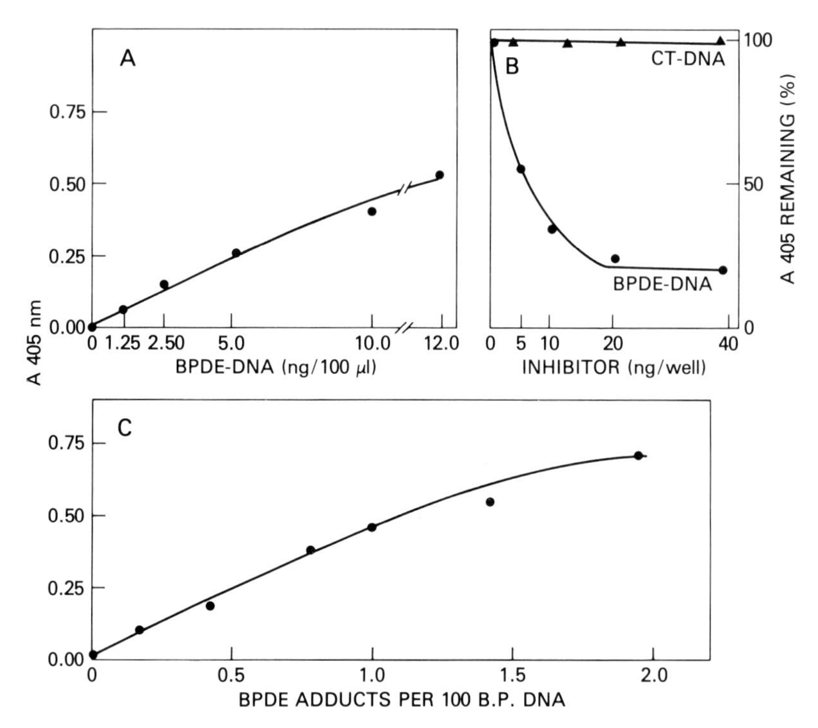

Fig. 1. Enzyme linked immunoassay demonstrating the specificity of anti-BPDE DNA serum. A. Dependence of reaction on antigen concentration. One hundred μl of solutions containing the indicated concentration of BPDE-DNA (modified to 1.0 BPDE per 100 bp DNA) were added to the microtiter plates. B. Inhibition of reaction between BPDE-DNA and anti-BPDE-DNA by preadsorption of the antisera with BPDE-DNA. Microtiter plates were coated with 25 mg BPDE-DNA (1.0 BPDE per 100 bp DNA). Prior to the addition to the plates the antisera were incubated with various amounts (indicated on the abscissa) of either BPDE-DNA or unmodified DNA. C. Antibody binding is dependent on the level of BPDE modification. ^{3}H labeled DNA was modified with ^{14}C-BPDE. The value $^{14}C/^{3}H$ allows calculation of the BPDE/DNA. The wells were coated with 5 mg of DNA containing various amounts of BPDE. In all cases the anti-BPDE was diluted 1:10,000; the second antibody, 1:500. Substrate concentration 1.0 mg/ml. Color developed at 37°C for 30 min.

adducts on DNA fragments separated by gel electrophoresis and transferred from the gel to diazobenzyloxymethyl paper. The technique is outlined in Figure 3. BPDE-modified DNA fragments (either from restriction digestion of plasmid or viral DNA or from micrococcal cvnuclease digestion of chromatin) are resolved by electrophoresis on agarose gels and then transferred to DBM paper. The paper is

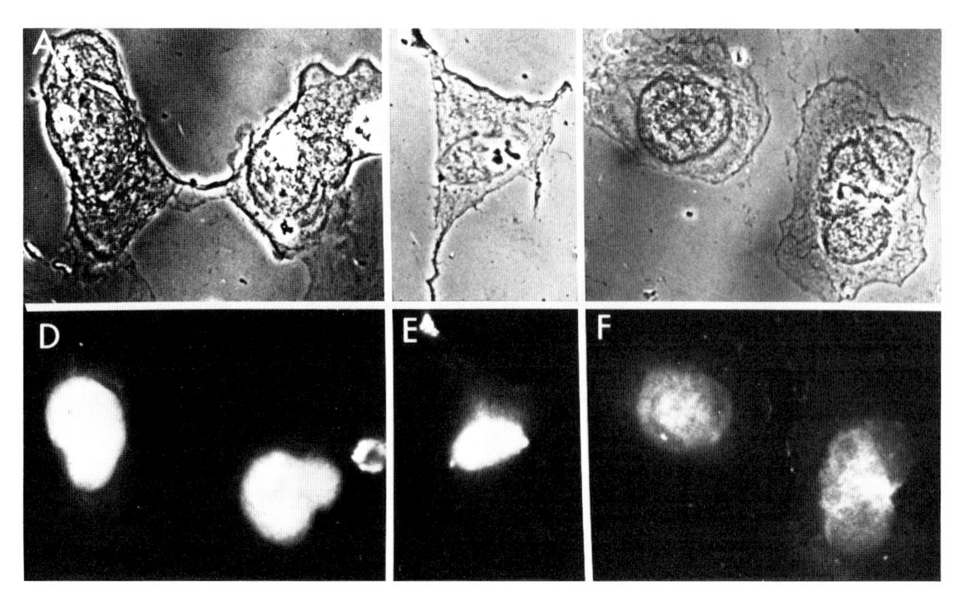

Fig. 2. Specific binding of anti-BPDE-DNA to nuclei of human fibroblasts demonstrated by immunofluorescence. Human KD fibroblasts were treated with A. 0, 5.0 mg/ml; B. E, 1.0 mg/ml; C. F, 0.5 mg/ml BPDE for one hour. After removal of the BPDE, the cells were fixed and processed for immunofluorescence.

incubated with antibody and ^{125}I protein A. Autoradiography reveals the location of the DNA as well as the relative amount of bound carcinogen. The bands can be cut out of the paper and the ^{125}I measured directly. When specific activities measurements were required the DNA was labelled with ^{32}PO4. In experiments with known sequences corrections for G content were made.

The feasibility of using such a technique was tested using a modification series of plasmid pBR322. This plasmid was modified with ^{3}H-BPDE to give molecules with known modification levels ranging from less than one adduct per plasmid to more than 200 adducts per plasmid molecule. After modification to various levels the plasmids were digested with EcoRI yielding the linear, form III, of the plasmid. The DNA was electrophoresed on a 1% agarose gel. Figure 4A shows the Ethidium Bromide stain of the electrophoresed plasmids. It can be seen that all slots contain essentially the same amount of DNA. At modification levels higher than 16 BPDE per plasmid, a band corresponding to the open circular form II DNA appears. Apparently, a fraction of the molecules were modified so that the restriction enzyme can cut only one DNA strand. The DNA was transferred to DBM paper and the paper probed first with antibody and then with ^{125}I Protein A. Figure 4B presents an

1. Modify DNA with BPDE.
   (cells, chromatin, plasmid, viral DNA)
2. Digest with nucleases.
3. Purify DNA fragment by agarose electrophoresis
4. Transfer to DBM paper
5. Incubate with anti BPDE antibody
6. Incubate with ^{125}I-protein A
7. Autoradiography
8. Cut DBM paper, count, correct for G content

Fig. 3. Scheme for localization of a carcinogen adduct on DNA fragments using immunoreplica techniques.

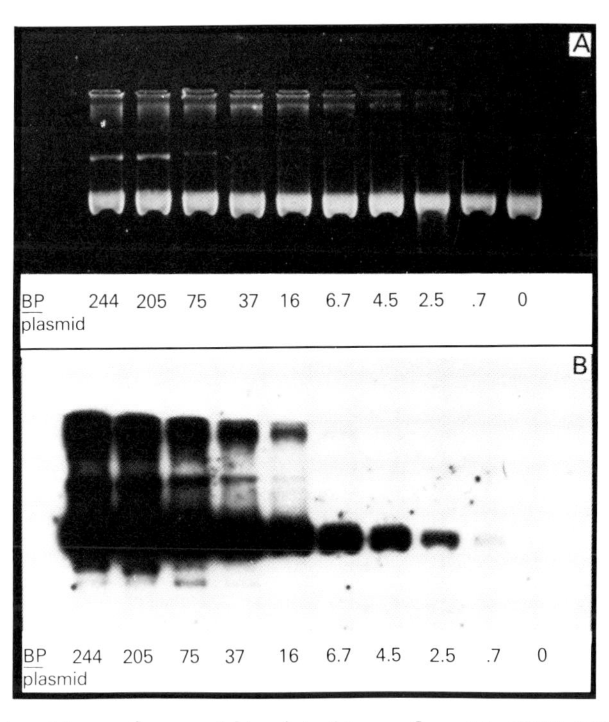

Fig. 4. Detection of specific binding of anti-BPDE-DNA to BPDE-modified DNA by the immunotransfer technique. Each slot contains 1 μg of BPDE treated pBR322 which has been linearized with ECoRI. The molar ratio of BPDE per plasmid is indicated. A. Ethidium Bromide staining of the gels prior to transfer to DBM paper. B. Autoradiogram of the DBM paper after transfer and incubation with antibody and ^{125}I Protein A.

autoradiograph of such a paper. Examination of the autoradiogram reveals that there is a direct correlation between the amount of adduct on the DNA and the amount of IgG-^{125}I Protein A bound.

The autoradiographs were scanned and the height of the various peaks plotted against the amount of BPDE adducts per plasmid. The data presented in Figure 5 indicate a linear relationship up to about 200 BPDE adducts per plasmid. Further experiments revealed that this linear relationship is not affected by the amount of DNA applied to gels. We found that a three-day exposure can detect about 10 ng DNA containing about three BPDE adducts per plasmid. With ^{3}H-BPDE of specific activity of 500 mCi/mmole, this many adducts per plasmid molecule would yield about 0.25 cpm in a scintillation counter. Thus, the transfer technique is significantly more sensitive than direct counting.

Next, we applied this technique to study the binding of BPDE to the cellular chromatin. The BSC1 line of African Green Monkey kidney cells were treated with ^{3}H-BPDE to give a modification level of one BPDE per 50 nucleosomes. The nuclei of these cells were digested with micrococcal nuclease for various times. The DNA purified from the digested nuclei was electrophoresed on agarose gels, transferred to DBM paper, and treated with antibodies and protein A as described before. The autoradiogram shown in Figure 6 indicates that nucleosomes of all fragment sizes contain bound BPDE. Independent measurements indicated that the ratio of BPDE bound per μg DNA from undigested nuclei to that from digested nuclei is 1:7.

These results agree with those reported by others suggesting that the linker DNA between nucleosomes binds more BPDE than the DNA of the core particle. To obtain the correct relative ratio of BPDE per mg DNA in each of the bands, an experiment was done with cells labeled with ^{32}P *in vivo*. After autoradiography the bands were cut from the DBM paper and counted for ^{32}P and ^{125}I. This technique allows direct determination of the BPDE present in each size of polynucleosome (see Table 2). Using the formula suggested by Jack and Brookes (8), it can be calculated that the linker DNA is about 2.5- to 3-fold more accessible to BPDE than nucleosome core DNA.

Because of the apparent enhancement of BPDE binding in linker DNA compared to nucleosomal core DNA, we decided to investigate the BPDE binding in the SV40 chromosome which contains a nonnucleosomal region. To study the distribution of BPDE in SV40 chromosomes, several types of virus preparations were studied. Purified SV40 DNA modified to a level of approximately 20 adducts per viral chromosome was used as a marker for random, *in vitro* modification. It is known that under such conditions the carcinogen binds to all DNA sequences simply as a function of guanine content. The *in vivo* binding was analyzed with viral DNA modified *in vivo* so as to give about 1-3 adducts per viral chromosome. The cells were infected either with wild type virus or with the temperature-sensitive encapsidation

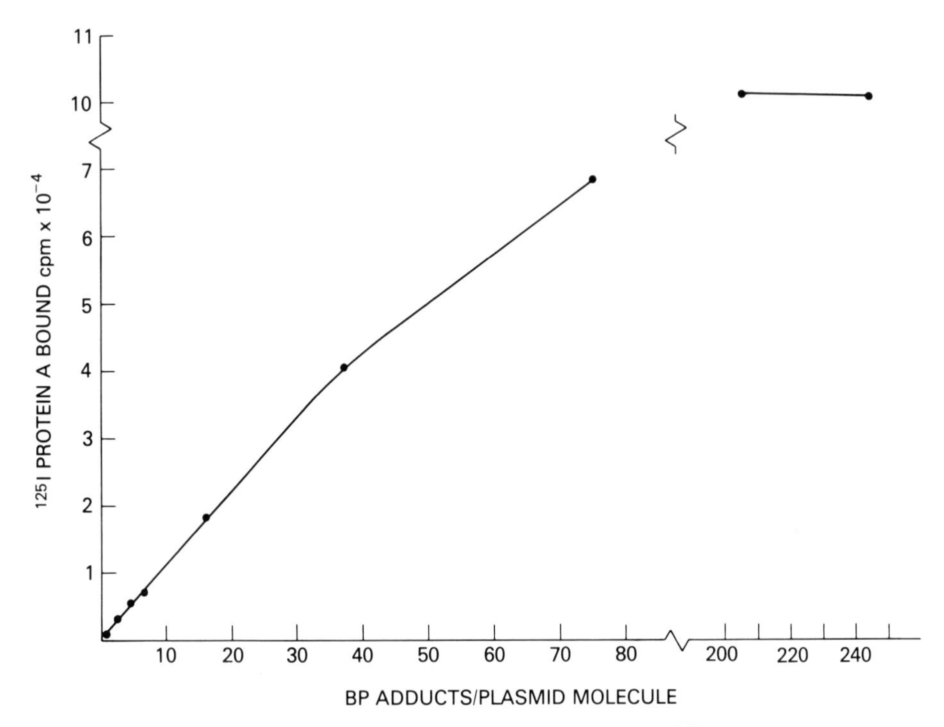

Fig. 5. Antibody binding is proportional to adduct concentration. Data taken from an experiment similar to that presented in Fig. 4. The bands visualized by autoradiography were cut out and the paper was counted.

mutant tsC219. This allows study of the effect of DNA encapsidation on BPDE binding. Both the wild type and the tsC219 mutant DNA bound about twice as much carcinogen as the cellular DNA. Thus, while encapsidation has no effect on adduct binding, the episomal state may have such an effect.

The chromosomes which were of special interest were those with an open origin of replication region. These were selected by treating an SV40 chromosome preparation with Bgl 1, which will cut the chromosomes with an open origin. The linearized form III was separated from the circular form II by electrophoresis in agarose gels. Thus, the distribution of BPDE was studied in four types of SV40 DNA: 1) DNA, modified *in vitro*; 2) DNA, purified from wild type; 3) DNA, purified from tsC219 mutant; and 4) DNA, obtained from tsC219 viruses containing exclusively nonnucleosomal origins. The restriction map of SV40 showing the cutting sites of the enzymes used in this study appears in Figure 7.

The modified viral DNA samples were first digested with Hind III and Hpa II. This combination of enzymes generates DNA fragments ranging in size from 215 to 1,768 bp. The fragment containing the

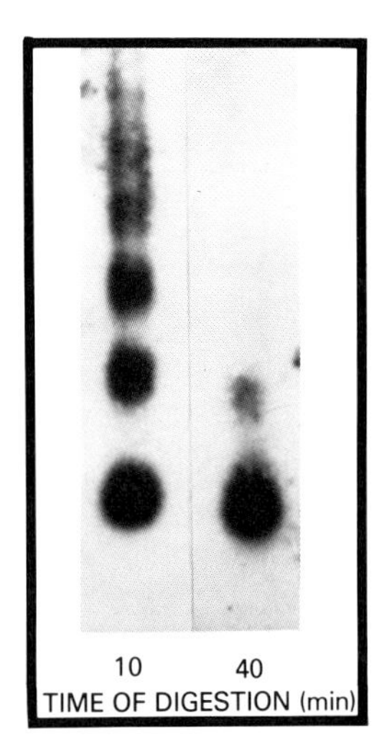

Fig. 6. Detection of BPDE adducts in cellular chromatin. The BSCI cells were incubated with BPDE and then nuclei were prepared and digested with micrococcal nuclease for either 10 min, to give oligonucleosomes of various sizes, or for 40 min, to give primarily core particles.

origin of replication is 415 bp long. The fragments were electrophoresed, transferred to DBM paper, and processed for autoradiography. The results are presented in Figure 8. It can be seen that the patterns of digestion were the same except that the highly modified DNA has additional bands in the 1,200 to 2,000 bp region, probably the consequence of inhibition of the nucleases by the bound carcinogen. The intensity of the bands was quantitated by densitometry and corrections were made for the guanine content of each fragment. These corrections compensated for the differences in antibody binding due to molecular weight of a fraction. A more detailed analysis of the origin promoter regions was done using the three single cut enzymes Bgl 1, Hpa II, and Hae II. The Bgl 1-Hpa II

Table 2. Concentration of BPDE adducts in monomer and oligonucleosomal DNA from cells.

Band Size	10 Min. Digest		40 Min. Digest	
	ng DNA[a]	125I/32P[b]	ng DNA	125I/32P
Monomer	156	1.13	81	1
Dimer	91	1.21		
Trimer	51	1.40		
Tetramer	36	1.52		

a) determined from ^{32}p measurement

b) normalized to the 40 min. monomer

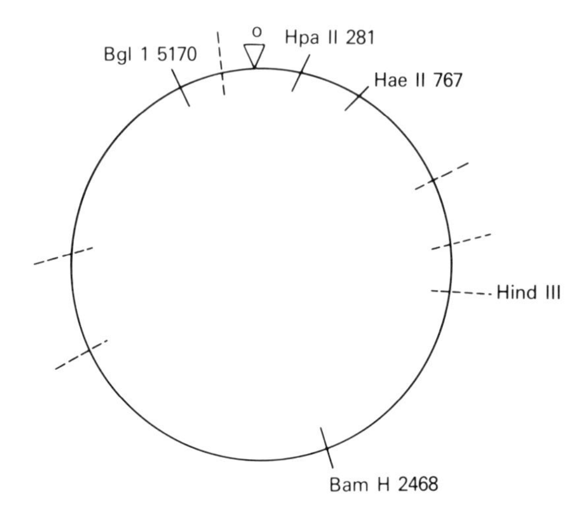

Fig. 7. Restriction map of SV40 indicating the cutting sites of the restriction enzymes used in this study. The number next to the single cut enzymes Bgl 1, Hpa II, Hae II, and Bam H indicate the residue number (starting from position indicated as 0) at which the enzymes cleave the DNA. Hind III cutting sites are indicated by broken lines. Digestion with this enzyme generates fragments of 215, 447, 526, 1,101, 1,169, and 1,768 base pairs.

fragment is a 352 bp fragment which spans the nonnucleosomal origin-promoter region. The 486 bp Hpa II-Hae II fragment is an adjacent, nucleosomal region. As seen from the data in Figure 8C, both fragments bound similar amounts of antibodies.

The 486 bp fragment contains both linker DNA and core particle DNA. It is possible that the differences between a nonnucleosomal and a nucleosomal region would be more apparent if the ori-promoter fragment would be compared to a core particle DNA. For such an experiment SV40 chromosomes isolated from BPDE-treated cells were divided into two parts. The purified DNA from one part was digested with restriction enzymes which give a pattern of well resolved fragments. The chromosomes in the other were digested with the micrococcal nuclease and the DNA purified. The autoradiograms of the DNA fragments (treated as described before) from the two digests is presented in Figure 8D. The amount of antibodies bound by the SV40 restriction fragments was very similar to the amount bound by a random nucleosome dimer obtained by digesting the chromosome with micrococcal nuclease. Because of the efficiency of transfer of a DNA fragment from gels, the paper will affect the results; the correct ratio of antibody per DNA was calculated by using ^{32}P labeled DNA. The spots containing the modified DNA were cut out of the DBM paper and the ratio of antibody bound per DNA determined from the ^{125}I and ^{32}P counts. These ratios, corrected for G content, are listed in Table 3. The $^{125}I/^{32}P$ of the core particle was taken to be one.

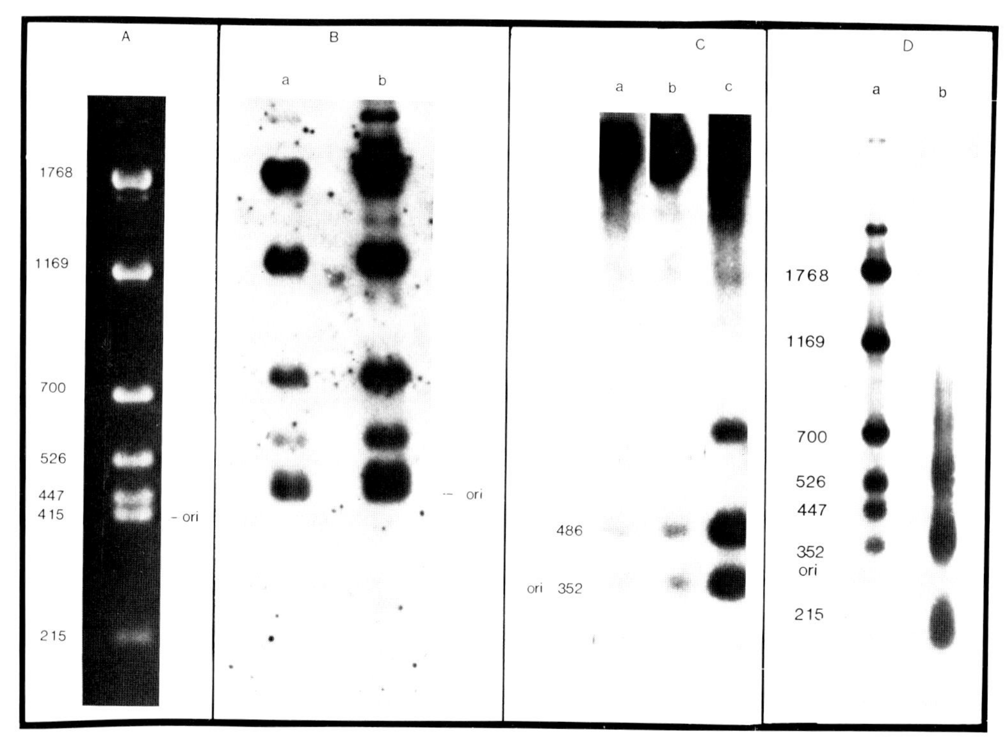

Fig. 8. Analysis of the BPDE binding to various regions of SV40 DNA. A. Ethidium Bromide stain of SV40 digested with Hind III and Hpa II. B. Autoradiogram of the BPDE-modified SV40 DNA digested as in A: a) 2 μg of virus DNA modified *in vivo* to a level of 4 BPDE/DNA, b) 0.5 μg DNA modified *in vitro* to 20 BPDE/DNA. C. The Bgl 1, Hpa II, Hae II digest of DNA: a) modified *in vivo* to 0.8 BPDE/DNA, b) modified *in vivo* to 4 BPDE/DNA, and c) *in vitro* 20 mmole BPDE/DNA. D: a) Bgl 1, Hpa II, and Hind III digest of SV40 DNA modified *in vivo* to about 2 BPDE/DNA and b) micrococcal nuclease digestion of same type of preparation.

It can be seen that the 352 bp DNA fragment containing the ori-promoter region is indeed the most highly modified fragment. However, the $^{125}I/^{32}P$ ratio was only 20% higher than the ratio of any other SV40 DNA fragment or a random cellular dinucleosome fragment. While such relatively minor enhancement in the binding conceivably may be biologically important, our data lead us to conclude that the ori region does not constitute a hot spot for carcinogen binding. It should be noted, however, that others (18, 19) have given different interpretations to results which were very similar to those presented above.

The experiments described above measured the distribution of

Table 3. Distribution BPDE adducts in restriction fragments and nucleosomal DNA of SV40-chromosomes modified *in vivo*.

Treatment	Fragment (Length (bp)	$^{125}I/^{32}P$
micrococcal nuclease	140 (monomer)	1
	400 (dimer)	1.3
Restrict with Hind III, Bgl I and Hpa II	215	1
	352-ori	1.7
	447	1.4
	526	1.1
	700	1.3
	1169	1.4
	1168	1.4

BPDE in duplex DNA found in chromatin structures. Recently, it has been suggested that nucleosomal histones might be more tightly bound to one strand of the DNA duplex than the other (20). Given the apparent distinction by DPDE of linker and nucleosome core DNA, we considered the possibility that the carcinogen would also be sensitive to the putative asymmetry of nucleosome structure on the coding and noncoding strands of a particular gene. For this experiment, we have employed both the wild type SV40 and the encapsidation mutant. The coding sequences of this genome are located on either side of the origin of replication and the coding and noncoding strands have been identified. The viral DNA can be cleaved by restriction enzymes into two fragments which contain the coding sequences on either side of the origin and the strands separated and resolved on agarose gels. The SV40 chromosomes were modified *in vivo* with BPDE and the viral DNA extracted and purified. The BPDE concentration was four moles BPDE/mole SV40 DNA. The DNA was digested with Bam HI and HpAII. The fragments were denatured and electrophoresed on an agarose gel and the standard transfer procedure followed. The results are shown in Figure 9. The coding and noncoding strands from each region of transcription bound essentially the same amount of BPDE. Consequently, we conclude that the coding and noncoding strands of these genes are equally accessible to BPDE.

Distribution of carcinogen adduct in polytene chromosomes. Polytene chromosomes, by virtue of their structure and size, can serve as a convenient experimental system to study and visualize the binding of carcinogens to various regions in chromosomes. Such studies can potentially provide insight into the question of whether a carcinogen preferentially interacts with transcriptionally active regions.

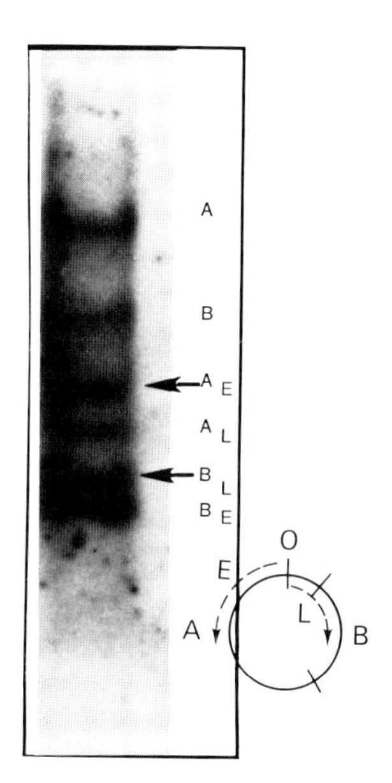

Fig. 9. Both the coding and noncoding strands of SV40 DNA react with BPDE in vivo. The SV40 purified from cells treated with BPDE to give about 4 BPDE/molecule SV40 were digested with Hpa II and Bam H. After denaturation followed by partial renaturation the digest was applied to agarose and processed for antibody binding. The diagram shows, in broken lines, the coding regions of the genome and the cleavage sites of the enzyme. "AE" is the coding strand for early messages while "BL" is the coding strand for late messages.

Initial in vitro studies involved addition of ^{3}H-BPDE to squashes of polytene chromosomes. The location of the bound carcinogen was visualized by autoradiography. The photomicrographs presented in Figure 10 visualize the result of an experiment in which a progressively increasing amount of carcinogen was added to the chromosomes. The progressive increase is reflected in the number of autoradiographic grains visible on the chromosomes. The inset in Figure 10D demonstrates that the BPDE also binds to the glandular material surrounding the chromosomes. This is the consequence of the reactivity of BPDE-1 with various cellular macromolecules (6). The kinetics of the interaction of the carcinogen with the chromosomes were followed by measuring the density of the grains over incremental segments of the chromosomes, semiautomatically, using an Artek counter. The average of all the densities provides a kinetic point for either dose relationship or time relationship studies. Comparison of the density of various increments potentially allows identification of chromosome regions which deviate from the average grain density. Examination of the chromosomes did not reveal a

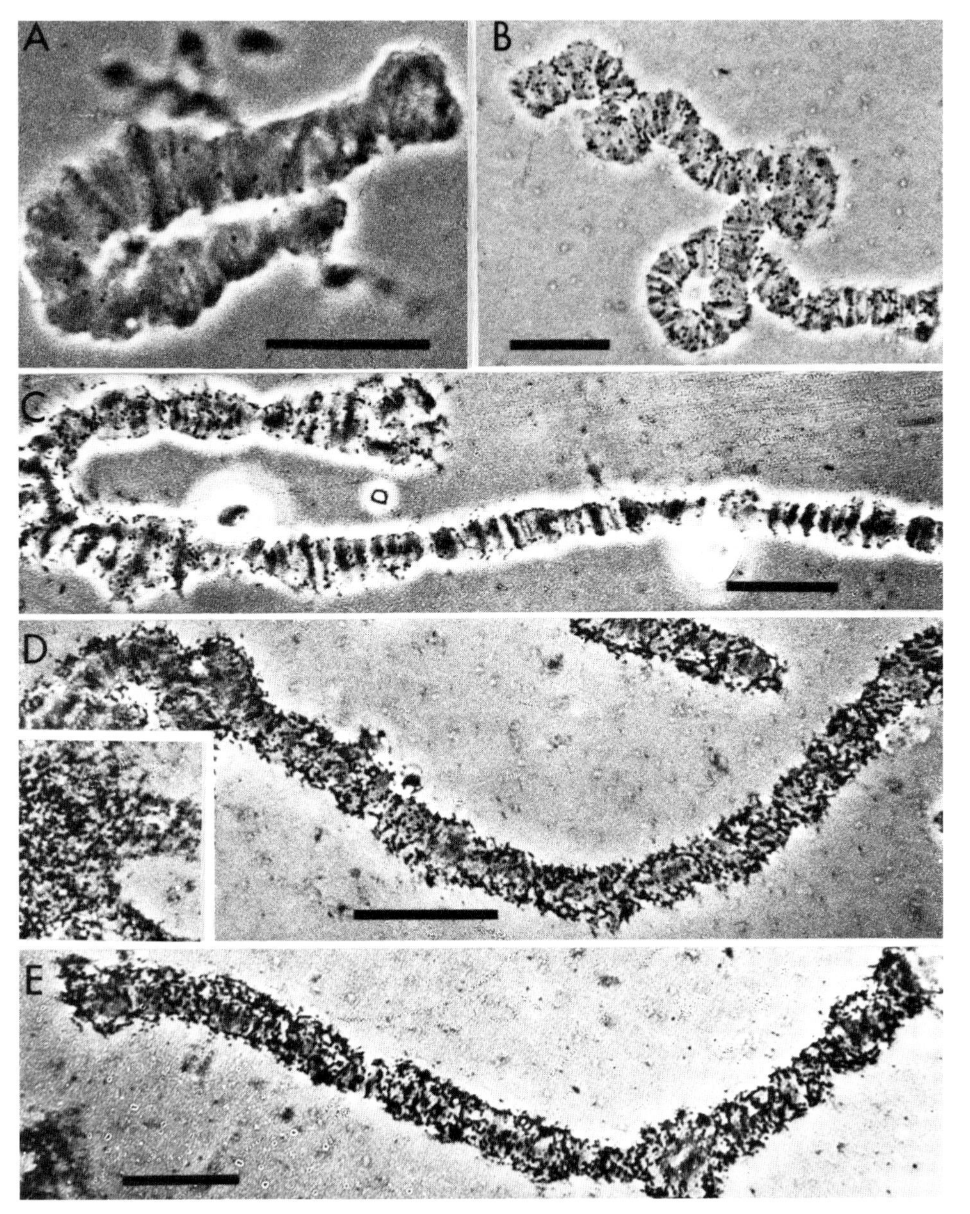

Fig. 10. Visualization of the location of BPDE on polytene chromosomes from Chironomus thummi by autoradiography. Polytene chromosome spreads from salivary glands isolated from the fourth instar of Chironomus thummi were treated with ^{3}H-BPDE and processed for autoradiography. Results in respective sections: A. reacted with 0.10 mM BPDE for 1 min; B. 0.10 mM BPDE for 10 min; C. 0.25 mM BPDE for 10 min; D. 0.1 mM BPDE for 30 min; E. 0.5 mM BPDE for 30 min.

particular region which deviated from the norm. Analysis of the frequency of the grain density over bands and interbands is shown in Figure 11.

The distribution of the grains over the chromosome is bimodal, giving essentially two Poisson distributions, one for bands and one for interbands. Thus, the distribution of grains over bands and interbands is random. A modal count of six grains per band and one for interbands suggests that there is a significantly higher probability that the carcinogen will interact with the bands. Since most of the chromosomal mass resides in the bands (17,21), it can be concluded that the binding is essentially dependent on the chromosomal mass.

The autoradiographic studies visualize the binding of the carcinogen to proteins, RNA, and DNA of the chromosome. The binding to nucleic acids can be separated from that to proteins by using antibodies to BPDE-DNA in the indirect immunofluorescence technique. These antibodies were described in a previous section of the manuscript. To ensure the suitability of these antibodies for such studies their specificity was examined further by: 1) testing them against chromosome preparations which have not been treated with BPDE-I, and 2) testing BPDE-treated chromosomes with antisera which has been preabsorbed with BPDE-DNA (kindly provided by M. Poirier). Positive immunofluorescence was obtained only when BPDE-reacted chromosomes were treated with antiserum specific for BPDE modified DNA. The photomicrographs shown in Figure 12 depict chromosomes treated with ^{3}H-BPDE and examined by both immunofluorescence and autoradiography. After squashing and carcinogen treatment, the chromosomes were processed for immunofluorescence and photographed under phase optics (Figure 12A) and fluorescence optics (Figure 12B). The coverslip was then removed and the slides dipped in Kodak NTB-II emulsion and processed for autoradiography. After development the preparations were photographed again under phase optics (Figure 12C) and fluorescence optics (Figure 12D). The antibody clearly outlines the contour of the chromosome giving a banded pattern. The glandular material is totally devoid of fluorescence. This has been achieved by treating the glands with RNase prior to antibody staining. In the absence of RNase treatment, significant background fluorescence is obtained due to antibody binding to RNA modified by BPDE. Thus, under the condition of these studies the fluorescence visualizes the BPDE adduct bound to DNA. The phase autoradiograph shows significant binding of the ^{3}H-BPDE to the proteins in the glandular material. The grain density on this is the same as that on the chromosome.

Visual inspection of the fluorescence and phase contrast pattern of these chromosomes indicates that, generally, the fluorescence intensity is proportional to the density of the band observed under phase optics. Since the density of the band roughly corresponds with the relative amount of DNA in that region, the immunofluorescence data suggest that the BPDE adduct formation is

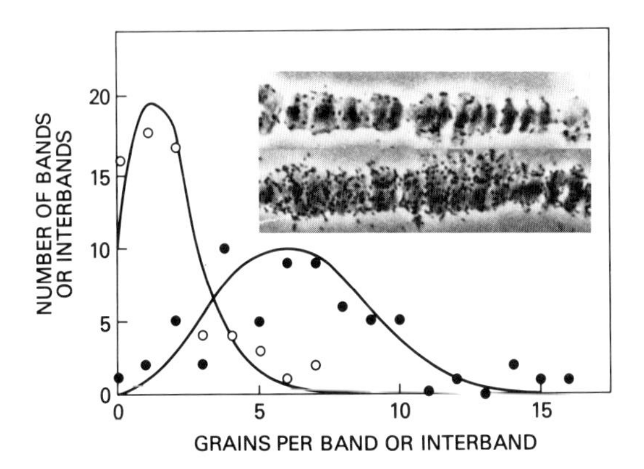

Fig. 11. Distribution of BPDE grains over band and interband regions of polytene chromosomes. The number of grains in 64 bands and 65 interbands of chromosomes displaying a grain density of 0.8 to 1.5 was visually counted. The graph depicts a plot of the number of bands containing a certain number of grains. The insets display representative areas of chromosomes in which the distribution of the grains over the bands and interbands was determined.

proportional to the amount of DNA, i.e., the target size. More detailed observation at lower carcinogen concentrations and higher serum dilutions, however, reveals that some regions fluoresce more intensely than others. A particularly prominent fluorescent band was detected in region B1/c-e of chromosome III (Figure 13).

In this experiment the chromosomes were reacted with decreasing concentrations of BPDE. It can be seen that at high carcinogen concentration the entire chromosome fluoresces. As the concentration of the carcinogen decreases, the bands become more discrete and differences between fluorescent bands are prominent. The band which preferentially binds the carcinogen (marked by arrowhead in Figure 13) is most easily seen at low doses of carcinogen.

The question arises whether the fluorescent bands will display preferential antibody binding when modified under *in vivo* conditions. Homogenates of Chironomus larvae can metabolize BP to give various metabolites resolvable by high pressure liquid chromatography (HPLC). In addition, the homogenate had detectable levels of AHH and P-450 activity. To study the *in vivo* binding of the carcinogen to the chromosomes, larvae were grown in 10 ml media containing 3% DMSO (dimethylsulfoxide). To this media, ^{3}H-BP were added. After 30 min the larvae were removed from this media, and polytene chromosomes were prepared and then processed for autoradiography. The results indicated that detectable levels of ^{3}H-BP derivatives bound to the polytene chromosomes under *in vivo* conditions. Examination of the autoradiograms did indicate that the autoradiographic grains were randomly distributed over the entire length of the

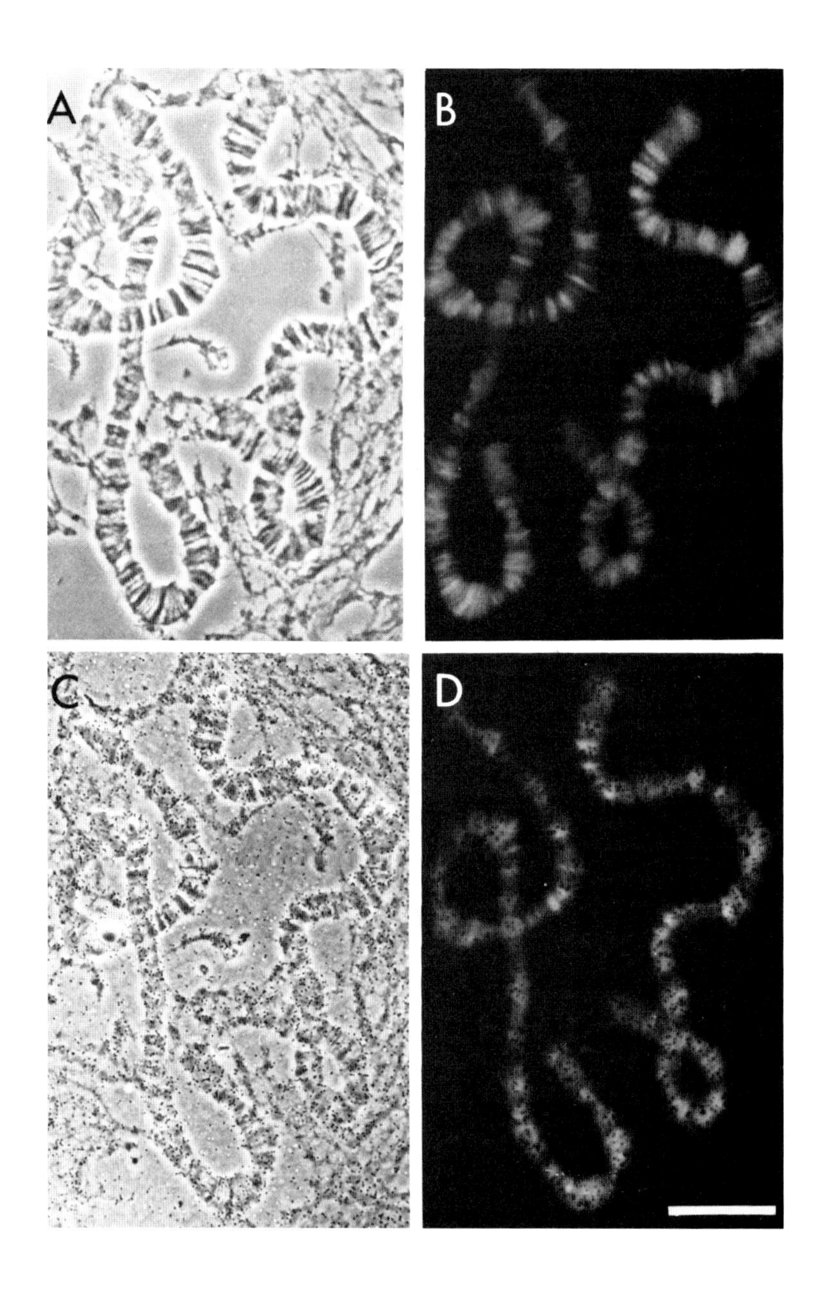
A
B
C
D

chromosome in a pattern indistinguishable from that observed with chromosomes modified *in vitro*. Immunofluorescence studies also revealed that the band which binds the highest amount of antibodies (i.e., fluoresces most intensely) is the same band as that which fluoresces most intensely under *in vitro* conditions. The photomicrograph presented in Figure 14 demonstrates the similarity between the results obtained *in vivo* and those obtained *in vitro*.

The question arises whether this region is a GC satellite which, by virtue of its base composition, binds more BPDE than other chromosomal regions. The chromosomes were treated with quinacrine whose fluorescence is enhanced by the AT-rich regions or with mithramycin which causes enhanced fluorescence in the GC-rich regions of the chromosomes. The results indicate that this region does not contain a disproportionately high density of GC residues. However, we noted an apparent correlation between the intensity of fluorescence with anti-BPDE-DNA antibodies and chromosomal loci which are preferentially active in transcription. Loci active in transcription can be visualized in these chromosomes by autoradiography after incubating either the isolated salivary glands or isolated chromosomes with ^{3}H-uridine. Kinetic measurements indicate that eventually all regions of the chromosomes actively incorporate ^{3}H-uridine, especially at higher salt concentrations. At relatively low salt concentrations (below 30 mM), the autoradiographic grains are localized to a few chromosomal regions. The density of the grains reflects the rate of uridine incorporation.

As shown in the micrograph presented in Figure 15, the regions which stain most intensely with the anti-BPDE-DNA antibodies correspond to the regions which are most active in RNA synthesis. It should be remembered that the immunofluorescence has been done on chromosomes which have been pretreated with RNase. Therefore, the antibody does not bind to modified RNA in the chromosomes. The

Fig. 12. Phase-fluorescence-autoradiographic visualization of the binding of BPDE to polytene chromosomes. Polytene squashes were treated with RNase and with ^{3}H-BPDE and the location of the DNA adducts visualized by indirect immunofluorescence. The coverslips were floated off and the preparations were processed for autoradiography. Thus, the autoradiographic grains visualize binding to DNA, protein, and other macromolecules except RNA while the immunofluorescence visualizes the DNA binding sites. The respective labels indicate: A. phase; B. indirect immunofluorescence; C. phase autoradiogram; and D. fluorescence autoradiogram.

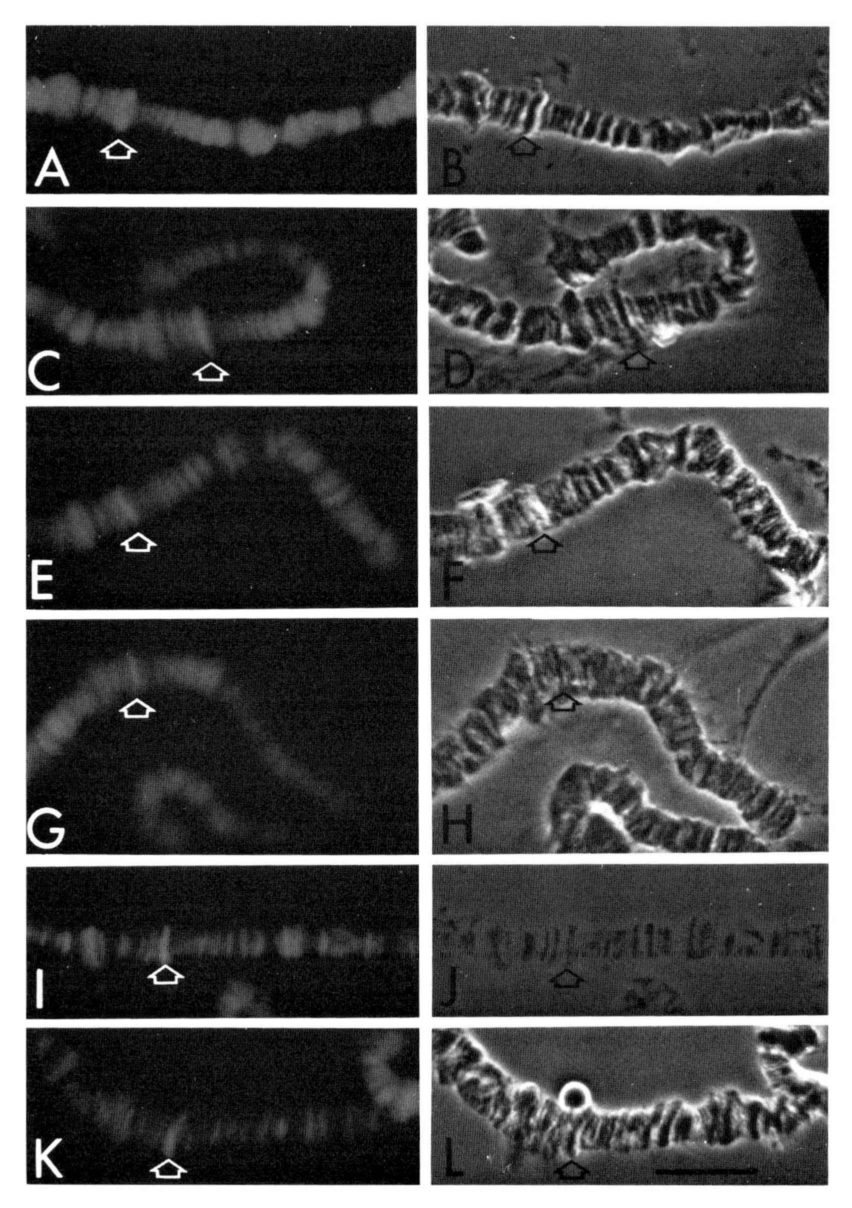

Fig. 13. A preferential anti-BPDE-DNA binding site in polytene chromosomes treated with various levels of BPDE visualized by indirect immunofluorescence. Polytene chromosome squashes were treated with the following concentrations of BPDE: A-B. 200 mM; C-D. 100 mM; E-F. 50 mM; G-H. 25 mM; I-J. 10 mM; K-L. 5 mM; and processed for immunofluorescence. The band which preferentially binds the antibody is indicated by arrowheads. Note that the preferential binding is most easily observed at low doses of BPDE.

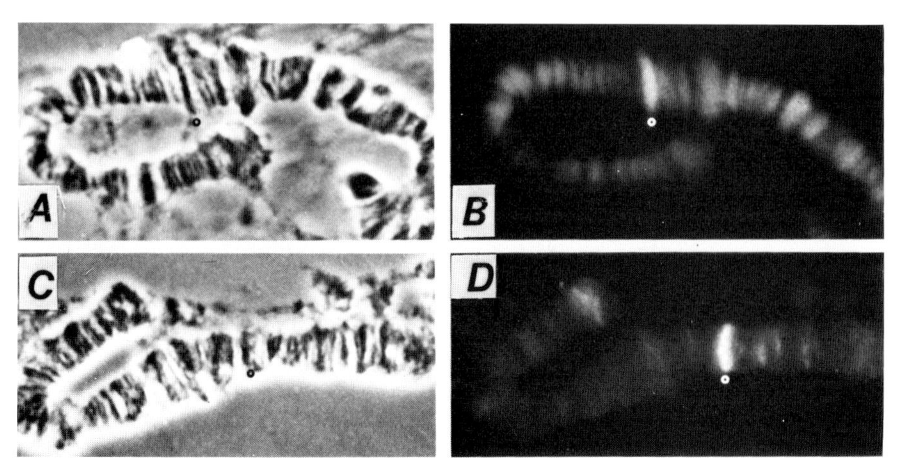

Fig. 14. Immunofluorescence of chromosomes modified *in vivo* and *in vitro*. The band from chromosome III which preferentially binds the antibody is indicated by a circle. The respective labels indicate: A-B. *in vitro* modification; and C-D. *in vivo* modification.

reason for enhanced antibody binding to regions which display augmented transcriptional activity is not known. However, it is known that the antibody binding is dependent on the conformation of the DNA. The BPDE on single-stranded DNA binds antibody more avidly than the same amount of BPDE on double-stranded DNA (unpublished data). It is, therefore, possible that the immunofluorescence results simply indicate augmented antibody binding rather than a local increase in the amount of BPDE bound per nucleotide.

SUMMARY

To the extent that SV40 minichromosomes can be used as a model for functionally differentiated eukaryotic chromatin, the results suggest that the nucleosomal structure of the chromatin fiber does not profoundly affect the binding of BPDE to the genome. The nonnucleosomal origin region in the chromosome showed only a minor enrichment in BPDE binding. While this minor enrichment may be biologically significant, we do not know of any experimental evidence to suggest so. It is possible, however, that in more complex genomes particular regions may display more significant preferential binding. Indeed, in the polytene chromosomes we have detected a region which, by immunofluorescence, displays a local increase in adduct concentration. We wish to emphasize, however, that the immunofluorescence technique visualizes local concentration of antibodies and is not definite proof that the DNA sequences are indeed enriched in adducts. A variety of factors influence the binding of antibodies to chromatin (22). The immunological technique used here

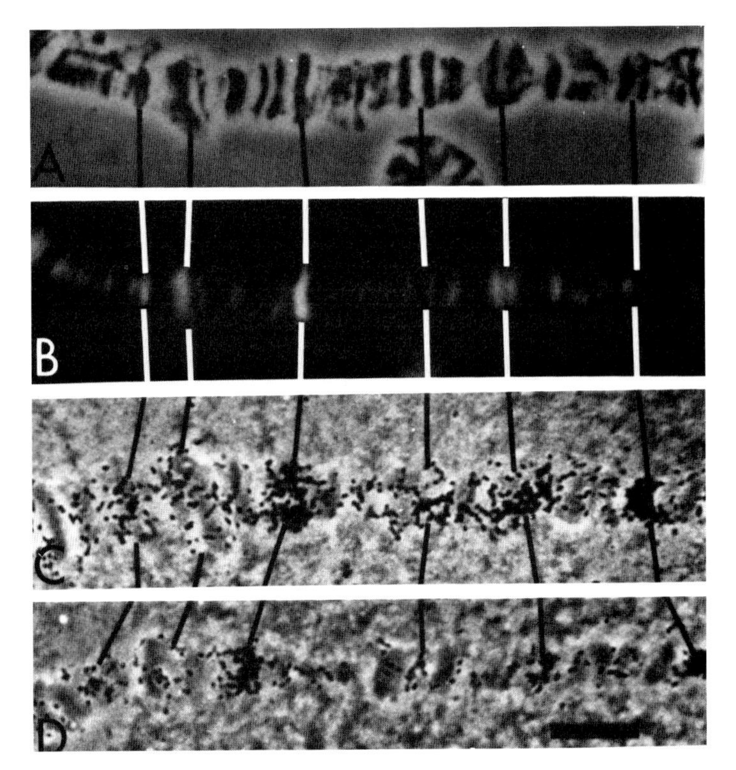

Fig. 15. Immunofluorescence with anti-BPDE-DNA antibodies correlates with transcription. A. Phase contrast of a region of chromosome III which was treated *in vitro* with BPDE. B. Corresponding fluorescence micrograph after treatment with anti-BPDE-DNA. C. Corresponding segment of chromosome III (homologue of that shown in A and B) showing the loci which preferentially incorporate ^{3}H-Uridine. D. Another preparation of chromosomes treated with ^{3}H-Uridine. Note the reproducibility in the loci which preferentially incorporate ^{3}H-Uridine.

to detect a low level of adducts in SV40 restriction fragments may be useful in further studies on the binding of BPDE to the genome. Obviously, this technique is applicable to studies with any DNA adduct against which defined antibodies are available.

ACKNOWLEDGEMENTS

We wish to thank Dr. M. Poirer for many helpful suggestions during the course of this work, and for providing control, comparison sera. We also thank Dr. B. Dunn for proofreading the manuscript and Mrs. H. Jordan for typing the manuscript.

REFERENCES

1. Miller, E.C. (1978) Cancer Research 38:1479-1496.
2. McCann, T., E. Choi, E. Yamasaki, and B.N. Ames (1975) Proc. Nat. Acad. Sci., USA 72:5153-5139.
3. Spanididos, D.A., and L. Simonowitch (1978) Cell 13:651-662.
4. Parodi, S., and B. Brambilla (1977) Mutat. Res. 47:219-229.
5. Lo, K., and T. Kakunaga (1981) Biochem. Biophys. Res. Comm. 99:820-829.
6. Koreeda, M., P.D. Moore, P.G. Wislocki, W. Levin, A.H. Conney, H. Yagi, and D.M. Terina (1978) Science 199:778-781.
7. Igo-Kemenes, T., W. Hortz, and H.G. Zachau (1982) Ann. Rev. Biochem. 51:89-121.
8. Jack, P.L., and P. Brookes (1981) Nucl. Acid Res. 9:5533-5552.
9. Tahn, C.L., and G.W. Litman (1979) Biochem. 18:1442-1469.
10. Koostra, A., T.J. Slaja, and D.E. Olins (1979), Chem. Biol. Interact. 28:225-236.
11. Feldman, G., T. Remsen, K. Shinohara, and P. Cerutti (1978) Nature 274:796-798.
12. DePamphilis, M., and P. Wasserman (1980) Ann. Rev. Biochem. 49:627-666.
13. Chou, T.Y., and R.G. Martin (1974) J. Virol. 13:1101-1108.
14. Slor, H., H. Mizusawa, N. Neihart, T. Kakefuda, R.S. Day, and M. Bustin (1981) Cancer Res. 41:3111-3117.
15. Poirier, M.C., R. Santella, I.B. Weinstein, D. Grunberger, and S. Yuspa (1980) Cancer Res. 40:412-416.
16. Renart, T., T. Reiser, and G.R. Stark (1979) Proc. Nat. Acad. Sci., USA 76:3116-3120.
17. Bostock, C.T., and A.T. Sumner (1978) The Eukaryotic Chromosome. Elsevier, Amsterdam, pp. 233-268.
18. Beard, P., M. Kaneko, and P. Cerutti (1981) Nature 291:84-85.
19. Robinson, G.W., and C.M. Hallick (1982) J. Virol. 41:78-87.
20. Palter, K.B., V.E. Foe, and B.M. Alberts (1978) Cell 18:451-467.
21. Beermann, W. (1972) Results and Problems in Cell Differentiation. W. Beermann, ed., Springer-Verlag, Berlin, Vol. 4, pp. 1-33.
22. Bustin, M. (1979) Current Topics in Microbiology and Immunology 88:105-142.

HIGH-AFFINITY MONOCLONAL ANTIBODIES SPECIFICALLY DIRECTED AGAINST DNA COMPONENTS STRUCTURALLY MODIFIED BY ALKYLATING N-NITROSO COMPOUNDS

Manfred F. Rajewsky, Jürgen Adamkiewicz, Wolfgang Drosdziok, Wilfried Eberhardt, and Ursula Langenberg

Institut für Zellbiologie (Tumorforschung)
Universität Essen (GH)
Hufelandstrasse 55, D-4300 Essen 1
Federal Republic of Germany

INTRODUCTION

Most chemical carcinogens and the majority of cancer chemotherapeutic agents cause structural alterations of DNA in the chromatin of target cells (1-9). In many cases, covalent binding occurs between nucleophilic centers (electron-rich N and O atoms) in cellular DNA and highly reactive, electrophilic derivatives ("ultimate carcinogens") generated from the respective parent compounds either enzymatically or via nonenzymatic decomposition (10-13). While a precise structural characterization is still lacking for the DNA adducts of many chemicals known to react with DNA, the various types of DNA modifications resulting from the exposure of cells to alkylating N-nitroso compounds (alkylnitrosoureas, alkylnitro-nitrosoguanidines, alkylnitrosamines) have, for the most part, been identified (1-2,5-7,14-15).

When cells are exposed to DNA-reactive chemicals *in vivo* or in culture at toxicologically tolerable dose levels, the respective reaction products in DNA are generally formed at very low frequencies. Highly sensitive methods are, therefore, required for the demonstration and quantitation of specific DNA adducts, for their study in relation to DNA repair, mutagenesis, or carcinogenesis, and for the dosimetry of the exposure of cells or individuals to agents affecting the integrity of the genome.

Conventional radiochromatography (16-17) has until recently been the method of choice for the analysis of DNA that is structurally modified by chemical agents. The sensitivity of radiochromatographic techniques is limited by the specific radioactivity of the respective [^{3}H]- or [^{14}C]-labeled agents and by the relatively large amounts of DNA (cells) needed for analysis. Under favorable conditions, one alkylated base can be detected by radiochromatography in $\sim 10^6$ molecules of the corresponding unmodified base. Except for a recently developed ^{32}P "post-labeling" technique (18-19), radiochromatography requires the use of radioactively labeled (i.e., laboratory-synthesized) chemicals for reaction with DNA. Therefore, the detection of specific carcinogen-DNA adducts in (e.g., human) tissues and cells exposed to low doses of nonradioactive (e.g., environmental) agents is not possible. Recently developed immunoanalytical procedures for the detection and quantitation of specific carcinogen-DNA adducts have, however, changed this situation considerably (20-24).

It is well known that immunoglobulins are characterized by an exceptional capability to recognize subtle alterations of molecular structure. Therefore, both the specificity and the sensitivity of immunoanalysis either by antibodies purified from conventional antisera or by monoclonal antibodies are extremely high. Equally important, the antigens, which in the case of naturally occurring (25) or carcinogen-modified nucleosides are haptens against which antibodies are produced, need not be radioactively labeled.

Based on the original work of Erlanger (26-29), Plescia (30-31), and others (25,32-34), immunological assays, in part highly sensitive, have during the past years been established for the quantitation of a number of specific carcinogen-DNA adducts in hydrolyzed DNA, notably for deoxynucleosides modified by N-nitroso compounds (20-21,23-24,35-40), acetylaminofluorene (40-48), benzo-(a)pyrene (49-50), and by aflatoxin B_1 (51-52). In the context of a research program on molecular and cellular mechanisms in chemical carcinogenesis (53-54), we have developed (and are further expanding) a collection of high-affinity monoclonal antibodies directed against alkyldeoxynucleosides produced in the DNA of mammalian cells as a result of exposure to N-nitroso carcinogens (20,23-24). This collection presently includes monoclonal antibodies specific for O^6-ethyl-2'-deoxyguanosine (O^6-EtdGuo), O^6-butyl-2'-deoxyguanosine (O^6-BudGuo), O^6-isopropyl-2'-deoxyguanosine (O^6-iProdGuo), and O^4-ethyl-2'-deoxythymidine (O^4-EtdThd).

EXPERIMENTAL TECHNIQUES

Synthesis of Alkylated Ribo- and Deoxyribonucleosides, and of Radioactive Tracers for Use in Competitive Radioimmunoassay

As alkylribonucleosides, i.e., as the haptens subsequently coupled to carrier proteins for immunization (28,35,38),

O^6-ethylguanosine (O^6-EtGuo) and O^4-ethylthymine riboside (O^4-EtrThd) were synthesized and purified by thin-layer chromatography on silica gel and Sephadex G-10 chromatography as described in refs. 20, 35, and 38. The O^6-isopropylguanosine (O^6-iProGuo) was prepared from 6-chloroguanosine (Pharma Waldhof, Mannheim, Germany) and purified as described in refs. 23 and 55. The O^6-n-butylguanosine (O^6-BuGuo) was synthesized by Dr. R. Saffhill (Paterson Laboratories, Manchester, U.K.). The O^6-EtdGuo, O^6-methyl-2'deoxyguanosine (O^6-MedGuo), O^6-methylguanosine (O^6-MeGuo), O^6-BudGuo, O^4-EtdThd, O^6-ethylguanine (O^6-EtGua), O^6-ethyl-2'-dGMP (O^6-EtdGMP), 7-ethylguanine (7-EtGua), O^6-iProdGuo, and 7-ethyl-2'-deoxyguanosine (7-EtdGuo) were synthesized and purified as described in refs. 20, 23, 35, and 38. High specific [^{3}H]-activity tracers were synthesized according to refs. 20, 23, 35, and 38.

Hapten-Protein Conjugates

For immunization, alkylribonucleosides were coupled to keyhole limpet hemocyanin (KLH; Calbiochem; Marburg, Germany; molecular weight 800,000; ref. 34) as a carrier protein (∿100 hapten molecules bound per molecule KLH; refs. 26, 38).

Immunization

Adult female rats of the inbred BDIX strain (56) were immunized by injection into about 20 intracutaneous sites of an emulgated mixture of 50 μg of hapten-KLH conjugate in 50 μl of phosphate buffered saline (PBS), 50 μl of aluminum hydroxide (Alugel S; Serva, Heidelberg, Germany), and 100 μl of complete Freund's adjuvant. Five weeks later, the animals were boosted by the same procedure. A second intraperitoneal (i.p.) booster injection was administered after 8 weeks, and spleen cells were collected 3-4 days thereafter.

Fusion of Spleen Cells with Myeloma Cells, and Culture and Cloning of Monoclonal Antibody-Secreting Hybridoma Cells

As previously described (20,23), cells of the HAT (hypoxanthine/aminopterine/thymidine)-sensitive rat myeloma line Y3-Ag.1.2.3 (57,58) were fused with spleen cells from immunized BDIX rats, using polyethylene glycol (PEG 4000; Roth, Karlsruhe, Germany). The hybrid cells were cultured in multi-well tissue culture plates (Costar No. 3524; 1-2 x 10^6 cells/well) containing Dulbecco's modified Eagle's-HAT medium, with 1 mM sodium pyruvate, 5 x 10^{-5} M mercaptoethanol, and 20% fetal bovine serum (59). Culture supernatants were tested for the presence of specific antibodies by an enzyme-linked immunosorbent assay (ELISA), and hybridoma cells from positive cultures were cloned, and once more recloned, without aminopterine in the presence of X-irradiated (40 Gy) BDIX rat "feeder" spleen cells. Of the positive clones maintained in culture, aliquots were injected i.p. into Pristan-pretreated, X-irradiated (4 Gy) BDIX rats for growth and antibody production in the ascites. Isotype analysis of monoclonal antibodies was carried out by the Ouchterlony technique (immunoprecipitation in agar), using anti-rat-isotype antisera (Miles GmbH, Frankfurt am Main, Germany).

Concentrations, Affinity Constants, and Isolation of Monoclonal Antibodies

Antibody concentrations in cell culture or ascites fluid, respectively, and the antibody affinity constants for the respective alkylated deoxyribonucleosides, were calculated from the data obtained by competitive RIA according to ref. 38. Antibodies were isolated with the aid of specific hapten-immunosorbents (haptens coupled to epoxyactivated Sepharose 6B, Pharmacia, Uppsala, Sweden) at acid or alkaline pH (38).

Enzymatic Hydrolysis of DNA

DNA from rat tissues was isolated by a modified Kirby method (60) or by hydroxylapatite adsorption (38), and enzymatically hydrolyzed to nucleosides with DNase I (EC 3.1.4.5; Boehringer Mannheim, Mannheim, Germany), snake venom phosphodiesterase (EC 3.1.4.1; Boehringer Mannheim), and grade I alkaline phosphatase (EC 3.1.3.1; Boehringer Mannheim) as described (38). Adenosine deaminase (EC 3.5.4.4; Boehringer Manneheim; 0.3 units/ml) was sometimes used to convert deoxyadenosine (dAdo) to deoxyinosine (dIno) in the DNA hydrolysates prior to analysis. This reaction is complete after 5 min at 20°C and does not lead to measurable dealkylation of O^6-EtdGuo or O^6-BudGuo (in contrast to O^6-MedGuo). The 2'-deoxyguanosine (dGuo) and 2'-deoxythymidine (dThd) concentrations in DNA hydrolysates were determined by peak integration after separation by high pressure liquid chromatography (HPLC). HPLC was also used for the separation of different alkylation products from the same DNA sample, prior to analysis by competitive RIA.

Competitve Enzyme-Linked Immunosorbent Assay (ELISA) and Radioimmunoassay (RIA)

The competitive ELISA was performed in 96-well microtiter plates (type M 129 A; Greiner, Nürtingen, Germany) coated with bovine serum albumin conjugates of the respective haptens (23,38). To test cell culture supernatants for the presence of monoclonal antibodies, they were added to duplicate wells (without or with the respective alkyldeoxynucleoside in a final concentration of 0.15 mM), incubated at 37°C for 90 min, and subsequently washed with PBS containing 0.05% of Triton X-100. Alkaline phosphatase-conjugated goat-anti-rat IgG was then added. After another 90 min incubation at 37°C, the incubation mixture was removed. The wells were thoroughly washed and incubated for 1 hr at 37°C with a 10 mM solution of p-nitrophenyl phosphate (alkaline phosphatase substrate; Sigma-Chemie, Munich, Germany). Thereafter, the plates were either screened for positive wells by eye (wells without added alkyldeoxynucleoside, yellow; wells containing the respective alkyldeoxynucleoside, colorless), or the degree of binding of alkaline phosphatase-conjugated antibody to the solid phase was determined spectrophotometrically (38).

The competitive RIA (61) was carried out as described (34,38). In a total volume of 100 µl of Tris-buffered saline (with 1% bovine serum albumin [w/v] and 0.1% bovine IgG [w/v], each sample contained $\sim 2.5 \times 10^3$ dpm of tracer, an antibody solution diluted to give 50% binding of tracer in the absence of inhibitor, plus varying amounts of hydrolyzed alkylated DNA, or of other natural or modified deoxynucleosides to be tested for cross-reactivity (inhibitor). The samples were incubated for 2 hr at room temperature (equilibrium) and 100 µl of a saturated ammonium sulphate solution (pH 7.0) were then added. After 10 min the samples were centrifuged for 3 min at 10,000 x g, and the [^{3}H]-activity in 150 µl of supernatant was measured by liquid scintillation spectrometry. The degree of inhibition of tracer-antibody binding (ITAB) was calculated as described (38).

RESULTS AND DISCUSSION

Monoclonal antibodies, contrary to antisera raised in animals, are not contaminated with nonspecific antibodies causing reduced specificity. This is, however, not to say that a high degree of antibody purity cannot sometimes also be obtained by extensive affinity purification of specific antibodies from antisera (38). While the detection limit of a competitive RIA is, of course, primarily dependent on the affinity of the antibody for the respective alkyldeoxynucleoside, antibody specificity (i.e., a low degree of cross-reactivity with other DNA constituents) becomes a most important factor when hydrolysates of DNA containing both natural and modified deoxynucleosides are to be analyzed. Even when specific antibodies with very high affinity constants are available, the limit of detection by competitive RIA in a complete DNA hydrolysate is determined by the specific radioactivity of the tracer which cannot be increased above a certain value. It cannot be lowered substantially, unless the level of cross-reactants in the DNA hydrolysate is further reduced or the cross-reactants are completely removed. The latter can, in principle, be achieved by separating the alkylation products in question from a complete DNA hydrolysate with the aid of HPLC prior to the RIA (23). However, in view of the exceedingly low amounts of the products to be measured, this procedure not only requires quantitative recovery of the respective fractions, but also absolutely contamination-free working conditions.

Table 1. Rat x rat hybridoma-secreted, high-affinity monoclonal antibodies specific for various alkyl-deoxynucleosides [examples from a collection of anti-alkyldeoxynucleoside monoclonal antibodies developed at the Institut für Zellbiologie (Tumorforschung), University of Essen, F.R.G.; refs. 20 and 23, and J. Adamkiewicz and M.F. Rajewsky, in preparation].

Alkyl-deoxynucleoside to be detected	Immunogen	Designation of antibody (isotype)	Antibody affinity constant (1/mol)	Radioactive tracer (specific [^{3}H]-activity)	Detection limit of RIA (fmol/100 ul sample)$^{1)}$	Cross-reactivity with hydrolysate of unmodified DNA$^{2)}$
O^6-EtdGuo	O^6-EtGuo-KLH	ER-6 (1gG2b)	2.0×10^{10}	O^6-Et-[8,5'-^{3}H]dGuo (~27 Ci/mmol)	40	2.5×10^6 b)
O^6-BudGuo	O^6-EtGuo-KLH	ER-11 (n.d.)	9.1×10^9	O^6-n-Bu-[8,5'-^{3}H]dGuo (~22 Ci/mmol)	60	2.5×10^6 b)
O^6-iProdGuo	O^6-iProGuo-KLH	ER-05 (1gG2a)	1.2×10^{10}	O^6-iPro-[8,5'^{3}H]dGuo (~17 Ci/mmol)	50	2.4×10^6 b)
O^4-EtdThd	O^4-EtThd-KLH	ER-01 (1gG2a)	1.3×10^9	O^4-Et-[6-^{3}H]dThd (~17 Ci/mmol)	240	1.3×10^6 a)

1) : Determined at 50% ITAB
2) : Value given as the multiple of the 50% ITAB-value for the respective modified DNA. Cross-reactivity determined either for an adenosine deaminase-treated DNA-hydrolysate, or for a mixture of dGuo, dIno, dCyd, and dThd, at the molar ratios for rat DNA.

a-b: Value given for 10% ITAB (a) or 10-15% ITAB (b), respectively, because value for 50% ITAB not reached by the ITAB-curve within the concentration range of the DNA hydrolysate added to the RIA sample.

Et : Ethyl. Bu: Butyl. iPro: Isopropyl. KLH: Keyhole limpet hemocyanin. RIA: Competitive radioimmunoassay.
ITAB: Inhibition of tracer-antibody binding in the RIA.
n.d.: Not determined.

The antibody affinity constants of the best of our presently available anti-alkyldeoxynucleoside monoclonal antibodies are in the range of 1-2 x 10^{10} l/mol (Tab. 1; refs. 20,23-24). When tested by competitive RIA, a number of these monoclonal antibodies cross-react with the constituents of hydrolysates of nonalkylated DNA only to a very low degree; i.e., require a >7 orders of magnitude higher amount of the corresponding unaltered deoxynucleosides to inhibit tracer-antibody binding by 50%. In some cases, 50% inhibition of tracer-antibody binding is not even reached at all by the nonmodified deoxynucleosides at concentrations up to their limit of solubility in the RIA-samples. However, not all monoclonal antibodies in our collection exhibit this extreme specificity. It is, therefore, recommendable to produce and characterize a sufficient number of monoclonal antibodies directed against a given modified deoxynucleoside if maximum specificity and affinity are desired.

At 50% inhibition of tracer-antibody binding in the competitive RIA, the monoclonal antibodies recognizing O^6-alkyl-2'-deoxyguanosines detect, as listed in Tab. 1, 40-60 fmol of the respective alkyldeoxynucleosides in a 100 µl RIA sample. In a probability grid used to transform the sigmoidal inhibition curves into straight lines (38), accurate readings are usually possible at inhibition values <50%. For example, reading at 20% inhibition will reduce the detection limit of monoclonal antibody ER-6 to ∿10 fmol of O^6-EtdGuo in a 100 µl RIA-sample (Tab. 1). Thus, in the competitive RIA, ER-6 detects O^6-EtdGuo at an O^6-EtdGuo/dGuo molar ratio as low as ∿3 x 10^{-7} (i.e., the equivalent of ∿700 O^6-EtdGuo molecules per diploid genome) in a hydrolysate of 100 µg of DNA (equivalent to the DNA of ∿1.6 x 10^7 diploid cells).

Perhaps the most important application of monoclonal antibodies directed against structurally altered DNA components will be the direct detection of specific DNA modifications in individual cells with the aid of immunostaining procedures. Besides other applications this approach will make it possible to demonstrate the presence, and to monitor the level, of specific DNA adducts (e.g., resulting from exposure to carcinogens or chemotherapeutic agents) in very small samples of cells (e.g., human) or biopsy material. Comparative measurements could be made in different types of cells at different stages of differentiation and/or development, with respect to their capacity for enzymatic activation of chemicals to the respective DNA-binding derivatives, the elimination of specific adducts from their DNA, and DNA repair. This might facilitate the detection of individuals with hereditary defects in DNA repair enzymes (risk groups in terms of carcinogen exposure), and the prediction of individual response to cancer chemotherapy.

Although our attempts to use monoclonal antibodies for the direct demonstration of carcinogen-DNA adducts in individual cells by immunofluorescence are still in their beginning, the results so far obtained are promising. Thus, we have recently established a standardized procedure for the quantitation of specific alkyldeoxynucleosides in nuclear DNA by direct immunofluorescence, using tetramethylrhodamine isothiocyanate-labeled monoclonal antibodies and a computer-based image analysis of electronically intensified fluorescence signals (62; J. Adamkiewicz, O. Ahrens, and M.F. Rajewsky, in preparation). Using an anti-O^6-EtdGuo monoclonal antibody, the present detection limit for O^6-EtdGuo in the nuclei of individual cells previously exposed to an ethylating N-nitroso carcinogen (e.g., N-ethyl-N-nitrosourea) is $\sim$700 O^6-EtdGuo molecules per diploid genome, corresponding to an O^6-EtdGuo/dGuo molar ratio in DNA of $\sim 3 \times 10^{-7}$. At the same O^6-EtdGuo/dGuo molar ratio in their DNA, $\sim 1.6 \times 10^7$ diploid cells would be required for the quantification of O^6-EtdGuo by competitive RIA (see above). When used in conjunction with immuno- electron-microscopy and a protein-free DNA spreading technique (63), monoclonal antibodies also permit the direct visualization of specific carcinogen-modified sites in individual DNA molecules (63). Thus, O^6-EtdGuo can be localized in double-stranded DNA by the binding of monoclonal antibody ER-6 (Tab. 1) without requirement for a second antibody carrying an electron-dense label (63; P. Nehls, E. Spiess, and M.F. Rajewsky, in preparation).

ACKNOWLEDGEMENTS

This work was supported by the Deutsche Forschungsgemeinschaft (SFB 102/A9), by the Fritz Thyssen-Stiftung (1980/2/41), and by the Commission of the European Communities (ENV-544-D [B]). The excellent technical assistance of Ch. Knorr, I. Spratte, and U. Buschkamp is gratefully acknowledged.

REFERENCES

1. Lawley, P.D. (1976) Carcinogenesis by alkylating agents. In Chemical Carcinogens, C.E. Searle, ed. ACS Monograph No. 173. American Chemical Society, Washington, D.C., pp. 83-244.
2. Pegg, A.E. (1977) Formation and metabolism of alkylated nucleosides: Possible role in carcinogenesis by nitroso compounds and alkylating agents. Adv. Cancer Res. 25:195-269.
3. Weinstein, I.B. (1977) Types of interaction between carcinogens and nucleic acids. In Mechanisme d'Alteration et de Réparation du ADN, Relations avec la Mutagenèse et la Cancerogenèse Chimique. Colloques Internationaux du CNRS No. 256. Centre National de la Recherche Scientifique, Paris, pp. 2-40.
4. Nagao, M., T. Sugimura, and T. Matsushima (1978) Environmental mutagens and carcinogens. Ann. Rev. Genet. 12:117-159.
5. Grover, P.L., ed. (1979) Chemical Carcinogens and DNA, Vol. I and II. CRC Press, Boca Raton.

6. Kohn, K.W. (1979) DNA as a target in cancer chemotherapy: Measurement of macromolecular DNA damage produced in mammalian cells by anticancer agents and carcinogens. Methods in Cancer Res. 16:291-345.
7. Rajewsky, M.F. (1980) Specificity of DNA damage in chemical carcinogenesis. In Molecular and Cellular Aspects of Carcinogen Screening Tests, R. Montesano, H. Bartsch, and L. Tomatis, eds. IARC Scientific Publications No. 27. International Agency for Research on Cancer, Lyon, pp. 41-54.
8. Waring, M.J. (1981) DNA modification and cancer. Ann. Rev. Biochem. 50:159-192.
9. Singer, B., and J.T. Kusmierek (1982) Chemical mutagenesis. Ann. Rev. Biochem. 52:655-693.
10. Magee, P.N. (1974) Activation and inactivation of chemical carcinogens and mutagens in the mammal. In Essays in Biochemistry, Vol. 10, P.N. Campbell and F. Dickens, eds. Academic Press, London, pp. 105-136.
11. Miller, J.A., and E.C. Miller (1979) Perspectives on the metabolism of chemical carcinogens. In Environmental Carcinogenesis, P. Emmelot and E. Kriek, eds. Elsevier/North-Holland Biomedical Press, Amsterdam, pp. 25-50.
12. Oesch, F. (1982) Chemical Carcinogenesis by polycyclic aromatic hydrocarbons. In Chemical Carcinogenesis, C. Nicolini, ed. Plenum Press, New York, pp. 1-24.
13. Miller, E.C., and J.A. Miller (1981) In search of ultimate chemical carcinogens and their reactions with cellular macromolecules. In Accomplishments in Cancer Research 1980, J.G. Fortner and J.E. Rhoads, eds. J.B. Lippincott, Philadelphia, pp. 63-98.
14. O'Connor, P.J., R. Saffhill, and G.P. Margison (1979) N-nitroso-compounds: Biochemical mechanisms of action. In Environmental Carcinogenesis, P. Emmelot and E. Kriek, eds. Elsevier/North-Holland Biomedical Press, Amsterdam, pp. 73-96.
15. Singer, B. (1979) N-nitroso alkylating agents: Formation and persistence of alkyl derivatives in mammalian nucleic acids as contributing factors in carcinogenesis. J. Natl. Cancer Inst. 62:1329-1339.
16. Baird, W.M. (1979) The use of radioactive carcinogens to detect DNA modification. In Chemical Carcinogens and DNA, P.L. Grover, ed. CRC Press, Boca Raton, pp. 59-83.
17. Beranek, D.T., C.C. Weis, and D.H. Swenson (1980) A comprehensive quantitative analysis of methylated and ethylated DNA using high pressure liquid chromatography. Carcinogenesis 1:595-606.
18. Randerath, K., M.V. Reddy, and R.C. Gupta (1981) ^{32}P-labeling test for DNA damage. Proc. Natl. Acad. Sci., USA 78:6126-6129.
19. Gupta, R.C., M.V. Reddy, and K. Randerath (1982) ^{32}P-postlabeling analysis of non-radioactive aromatic carcinogen-DNA adducts. Carcinogenesis 3:1081-1092.
20. Rajewsky, M.F., R. Müller, J. Adamkiewicz, and W. Drosdziok

(1980) Immunological detection and quantification of DNA components structurally modified by alkylating carcinogens (ethylnitrosourea). In Carcinogenesis: Fundamental Mechanisms and Environmental Effects, B. Pullman, P.O.P. Ts'o, and H. Gelboin, eds. Reidel, Dordrecht-Boston-London, pp. 207-218.

21. Müller, R., and M.F. Rajewsky (1981) Antibodies specific for DNA components structurally modified by chemical carcinogens. J. Cancer Res. Clin. Oncol. 102:99-113.
22. Poirier, M.C. (1981) Antibodies to carcinogen-DNA adducts. J. Natl. Cancer Inst. 67:515-519.
23. Adamkiewicz, J., W. Drosdziok, W. Eberhardt, U. Langerberg, and M.F. Rajewsky (1982) High-affinity monoclonal antibodies specific for DNA components structurally modified by alkylating agents. In Indicators of Genotoxic Exposure, B.A. Bridges, B.E. Butterworth, and I.B. Weinstein, eds. Banbury Report 13. Cold Spring Harbor Laboratory, Cold Spring Harbor, N.Y., pp. 265-276.
24. Müller, R., J. Adamkiewicz, and M.F. Rajewsky (1982) Immunological detection and quantification of carcinogen-modified DNA components. In Host Factors in Human Carcinogenesis, H. Bartsch and B. Armstrong, eds. IARC Scientific Publications No. 39. International Agency for Research on Cancer, Lyon, pp. 463-479.
25. Munns, T.W., and M.K. Liszewski (1980) Antibodies specific for modified nucleosides: An immunological approach for the isolation and characterization of nucleic acids. Prog. Nucleic Acid. Res. Mol. Biol. 24:109-165.
26. Erlanger, B.F., and S.M. Beiser (1964) Antibodies specific for ribonucleosides and ribonucleotides and their reaction with DNA. Proc. Natl. Acad. Sci., USA 52:68-74.
27. Beiser, S.M., S.W. Tanenbaum, and B.F. Erlanger (1968) Purine- and pyrimidine-protein conjugates. Meth. Enzymol. 12:889-893.
28. Erlanger, B.F. (1973) Principles and methods for the preparation of drug-protein conjugates for immunological studies. Pharmacol. Rev. 25:271-280.
29. Meredith, R.D., and B.F. Erlanger (1979) Isolation and characterization of rabbit anti-m^{7}G-5'-P antibodies of high apparent affinity. Nucleic Acids Res. 6:2179-2191.
30. Plescia, O.J., W. Braun, and N.C. Palczuk (1964) Production of antibodies to denatured deoxyribonucleic acid (DNA). Proc. Natl. Acad. Sci., USA 52:279-285.
31. Plescia, O.J. (1968) Preparation and assay of nucleic acids as antigens. Meth. Enzymol. 12:893-899.
32. Halloran, M.J., and C.W. Parker (1966) The preparation of nucleotide-protein conjugates: Carbodiimides as coupling agents. J. Immunol. 96:373-378.
33. Grabar, P., S. Avrameas, B. Taudou, and J.C. Salomon (1968) Formation and isolation of antibodies specific for nucleosides. In Nucleic Acids in Immunology, O.J. Plescia and W. Braun, eds. Springer, Berlin-New York, pp. 79-87.

34. Stollar, B.D., and Y. Borel (1976) Nucleoside specificity in the carrier lgG-dependent induction of tolerance. J. Immunol. 117:1308-1313.
35. Müller R., and M.F. Rajewsky (1978) Sensitive radioimmunoassay for detection of O^6-ethyldeoxyguanosine in DNA exposed to the carcinogen ethylnitrosourea *in vivo* or *in vitro*. Z. Naturforsch. 33c:897-901.
36. Briscoe, W.T., J. Spizizen, and E.M. Tan (1978) Immunological detection of O^6-methylguanine in alkylated DNA. Biochemistry 17:1896-1901.
37. Kyrtopoulos, S.A., and P.F. Swann (1980) The use of radioimmunoassay to study the formation and disappearance of O^6-methylguanine in mouse liver satellite and main-band DNA following dimethylnitrosamine administration. J. Cancer Res. and Clin. Oncol. 98:127-138.
38. Müller, R., and M.F. Rajewsky (1980) Immunological quantification by high affinity antibodies of O^6-ethyldeoxyguanosine in DNA exposed to N-ethyl-N-nitrosourea. Cancer Res. 40:887-896.
39. Saffhill, R., P.T. Strickland, and J. Boyle (1982) Sensitive radioimmunoassays for O^6-N-butyldeoxyguanosine, O^2-N-butylthymidine and O^4-N-butylthymidine. Carcinogenesis 3:547-552.
40. Van der Laken, C.J., A.M. Hagenaars, G. Hermsen, E. Kriek, A.J. Kuipers, J. Nagel, E. Scherer, and M. Welling (1982) Measurement of O^6-ethyldeoxyguanosine and N-(deoxyguanosine-8-YI)-N-acetyl-2-aminofluorene in DNA by high-sensitive enzyme immunoassays. Carcinogenesis 3:569-572.
41. Poirier, M.C., S.H. Yuspa, I.B. Weinstein, and S. Blobstein (1977) Detection of carcinogen-DNA adducts by radioimmunoassay. Nature 270:186-188.
42. Leng, M., E. Sage, R.P.P. Fuchs, and M.P. Daune (1978) Antibodies to DNA modified by the carcinogen N-acetoxy-N-2-acetylaminofluorene. FEBS Lett. 92:207-210.
43. DeMurcia, G., M.C.E. Lang, A.M. Freund, R.P.P. Fuchs, M.P. Daune, E. Sage, and M. Leng (1979) Electron microscopic visualization of N-acetoxy-N-2-acetylaminofluorene binding sites in Col E-1 DNA by means of specific antibodies. Proc. Natl. Acad. Sci., USA 76:6076-6080.
44. Poirier, M.C., M.A. Dubin, and S.H. Yuspa (1979) Formation and removal of specific-acetylaminofluorene-DNA adducts in mouse and human cells measured by radioimmunoassay. Cancer Res. 39:1377-1381.
45. Sage, E., R.P.P. Fuchs, and M. Leng (1979) Reactivity of the antibodies to DNA modified by the carcinogen N-acetoxy-N-aminofluorene. Biochemistry 18:1328-1332.
46. Spodheim-Maurizot, M., G. Saint-Ruf, and M. Leng (1979) Conformational changes induced in DNA by *in vitro* reaction with N-hydroxy-N-2-aminofluorene. Nucleic Acids Res. 6:1683-1694.
47. Hsu, I.C., M.C. Poirier, S.H. Yuspa, R.H. Yolken, and C.C. Harris (1980) Ultrasensitive enzymatic radioimmunoassay

(USERIA) detects femtomoles of acetylaminofluorene-DNA adducts. Carcinogenesis 1:455-458.

48. Kriek, E., C.J. van der Laken, M. Welling, and J. Nagel (1982) Immunological detection and quantification of the reaction products of 2-acetylaminofluorene with guanine in DNA. In Host Factors in Human Carcinogenesis, H. Bartsch and B. Armstrong, eds. IARC Scientific Publications No. 39. International Agency for Research on Cancer, Lyon, France, pp. 541-549.
49. Poirier, M.C., R. Santella, I.B. Weinstein, D. Grunberger, and S.H. Yuspa (1980) Quantitation of benzo(a)pyrene-deoxyguanosine adducts by radioimmunoassay. Cancer Res. 40:412-416.
50. Hsu, I.C., M.C. Poirier, S.H. Yuspa, D. Grunberger, I.B. Weinstein, R.H. Yolken, and C.C. Harris (1981) Measurement of benzo(a)pyrene-DNA adducts by enzyme immunoassays and radioimmunoassay. Cancer Res. 41:1091-1095.
51. Haugen, Å., J.D. Groopman, I.C. Hsu, G.R. Goodrich, G.N. Wogan, and C.C. Harris (1981) Monoclonal antibody to aflatoxin B_1 modified DNA detected by enzyme immunoassay. Proc. Natl. Acad. Sci., USA 78:4124-4127.
52. Groopman, J.D., Å. Haugen, G.R. Goodrich, G.N. Wogan, and C.C. Harris (1982) Quantitation of aflatoxin B_1-modified DNA using monoclonal antibodies. Cancer Res. 42:3120-3124.
53. Rajewsky, M.F., L.H. Augenlicht, H. Biessmann, R. Goth, D.F. Hülser, O.D. Laerum, and L.Ya. Lomakina (1977) Nervous system-specific carcinogenesis by ethylnitrosourea in the rat: Molecular and cellular mechanisms. In Origins of Human Cancer, H.H. Hiatt, J.D. Watson, and J.A. Winsten, eds. Cold Spring Harbor Laboratory, Cold Spring Harbor, New York, pp. 709-726.
54. Rajewsky, M.F. (1982) Pulse-carcinogenesis by ethylnitrosourea in the developing rat nervous system: Molecular and cellular mechanisms. In Chemical Carcinogenesis, C. Nicolini, ed. Plenum Press, New York, pp. 363-379.
55. Gerchman, L.L. J. Dombrowski, and D.B. Ludlum (1972) Synthesis and polymerization of O^6-methylguanosine 5'-diphosphate. Biochim. Biophys. Acta 272:672-675.
56. Druckrey, H. (1971) Genotypes and phenotypes of ten inbred strains of BD-rats. Arzneim.-Forsch. 21:1274-1278.
57. Galfré, G., C. Milstein, and B. Wright (1979) Rat x rat hybrid myelomas and a monoclonal anti-Fd portion of mouse IgG. Nature, London 277:131-133.
58. Köhler, G., and C. Milstein (1975) Continuous cultures of fused cells secreting antibody of predefined specificity. Nature, London 256:495-497.
59. Lemke, H., G.J. Hämmerling, C. Höhmann, and K. Rajewsky (1978) Hybrid cell lines secreting monoclonal antibody specific for major histocompatibility antigens of the mouse. Nature, London 271:249-251.
60. Goth, R., and M.F. Rajewsky (1974) Molecular and cellular mechanisms associated with pulse-carcinogenesis in the rat nervous system by ethylnitrosourea: Ethylation of nucleic acids and

elimination rates of ethylated bases from the DNA of different tissues. *Z. Krebsforsch.* 82:37-64.

61. Farr, R.S. (1958) A quantitative immunological measure of the primary interaction between I BSA and antibody. *J. Infect. Dis.* 103:239-262.
62. Adamkiewicz, J., O. Ahrens, N. Huh, and M.F. Rajewsky (1983) Quantitation of alkyl-deoxynucleosides in the DNA of individual cells by high-affinity monoclonal antibodies and electronically intensified, direct immunofluorescence. *J. Cancer Res. Clin. Oncol.*, 105:A15.
63. Nehls, P., E. Spiess, and M.F. Rajewsky (1983) Visualization of O^6-ethylguanine in ethylnitrosourea-treated DNA by immune electron microscopy. *J. Cancer Res. Clin. Oncol.*, 105:A23.

CHARIMEN'S OVERVIEW ON CARCINOGEN-INDUCED MODIFICATION IN DNA

Arthur C. Upton and George W. Teebor

New York University School of Medicine

New York, New York 10016

Carcinogen-induced modifications in DNA deserve consideration as endpoints in carcinogenicity testing because of the evidence implicating genetic determinants in carcinogenesis; e.g., inherited differences in susceptibility to cancer (6), the common and sometimes specific occurrence of chromosomal abnormalities in cancer cells (8), the frequent correlation in chemicals between genotoxicity and carcinogenicity (7), and the transforming activity of oncogenic nucleotide sequences (2). At the same time, however, the kinetics of carcinogenesis _in vivo_ and of cell transformation _in vitro_ imply that neoplastic transformation usually involves sequential genetic changes, which may vary with the carcinogenic stimulus in question, the stage in the cancer process affected, the target cells at risk, and the capability of affected cells to repair or modify carcinogen-induced alterations in DNA (3,5,10).

There is also evidence that some carcinogens, which do not themselves interact directly with DNA, may alter DNA indirectly through the release of free radicals (e.g., superoxide) (1,4), aberrant methylation (9), or other mechanisms. Thus, although no simple, absolute, or necessary correlation between genotoxicty and carcinogenicity is apparent at this time, elucidation of carcinogen-induced modifications of DNA would appear to be central to an understanding of the mechanisms of carcinogenesis and to the detection and classification of carcinogenic agents.

The papers comprising this section make no attempt to survey existing genotoxicity tests for carcinogens, a subject which has been dealt with elsewhere (7). Instead, other aspects of carcinogen-DNA interactions are analyzed, with a view toward the exploration of new approaches for detecting carcinogens, for characterizing their modes of action, and for identifying and monitoring carcinogen-exposed populations.

Among the subjects to be considered in this section are: 1) the extent to which use of DNA-repair-deficient cell lines may offer special advantages in screening for the detection of carcinogens and in analyzing their mechanisms of action; 2) the characterization of specific nucleotide base modifications induced in DNA by ionizing radiation, as compared with other carcinogenic agents; 3) the occurence of aberrant methylation of DNA as an indirect consequence of cytotoxicity; and 4) the utilization of immunologic probes for specific carcinogen-DNA adducts in the investigation of mechanisms of carcinogenesis and in the surveillance of carcinogen-exposed populations.

REFERENCES

1. Birnboim, H.C. (1982) DNA strand breakage in human leukocytes exposed to a tumor promoter. Science 215:1247-1249.
2. Bishop, J.M. (1982) Oncogenes. Scientific American 246:80-92.
3. Cairns, J. (1981) The origin of human cancers. Nature 289: 353-357.
4. Emerit, I., and P.A. Cerutti (1981) Tumor promoter phorbol-12-myristate-13-induces chromosomal damage via indirect action. Nature 293:144-146.
5. Klein, G. (1981) The role of gene dosage and genetic transpositions in carcinogenesis. Nature 294:313-318.
6. Knudson, A.G. (1981) Genetics and cancer. In Cancer: Achievements, Challenges, and Prospects for the 1980s, J.H. Burchenal and H.F. Oettgen, eds. Grune & Stratton, New York, 1:381-396.
7. Purchase, I.F.H. (1982) An appraisal of predictive tests for carcinogenicity. Mutat. Res. 99:53-71.
8. Sandberg, A.A. (1980) The chromosomes in human cancer and leukemia. Elsevier-North Holland, New York.
9. Shank, R.C., and L.R. Barrows (1981) Toxicity-dependent DNA methylation: Significance to risk assessment. In Health Risk Analysis, C.R. Richmond, P.J. Walsh, E.M.D. Copenhaver, eds. The Franklin Institute Press, pp. 225-233.
10. Upton, A.C. (1982) Role of DNA damage in radiation and chemical carcinogenesis. In Environmental Mutagenesis and Carcinogenesis, T. Sugimura, S. Kondo, and H. Takebe, eds. University of Tokyo Press, Tokyo, pp. 71-80.

USE OF REPAIR-DEFICIENT MAMMALIAN CELLS FOR THE IDENTIFICATION OF CARCINOGENS

John B. Little

Department of Cancer Biology
Harvard School of Public Health
Boston, Massachusetts 02115

INTRODUCTION

The most prominent cellular effects of DNA-damaging agents are cytotoxicity (cell killing), mutagenesis, and the induction of cytogenetic changes including chromosomal aberrations and sister chromatid exchanges. Although bioassays for environmental carcinogens in mammalian cells have employed several of these end points, mutagenesis has generally been considered the one most intimately involved in the process of carcinogenesis. In general terms, the use of hypermutable cells should improve the sensitivity of a bioassay for a carcinogen which is based on its mutagenic potential, particularly if such lines were relatively insensitive to the cytotoxic effects of the carcinogen.

Enhanced sensitivity of mammalian cells to the cytotoxic, cytogenetic, and mutagenic effects of physical and chemical DNA-damaging agents is often associated with the existence of a DNA repair defect. However, such hypersensitivity may result from other molecular mechanisms. Indeed, in the case of fibroblasts isolated from patients with several genetic syndromes characterized by marked hypersensitivity to DNA-damaging agents, only in one case (Xeroderma pigmentosum) has a DNA repair defect been clearly shown to be the basis for the hypersensitivity. Furthermore, established rodent cell lines have been isolated in which hypermutability to ultraviolet (UV) light may or may not be associated with an apparent repair defect.

In this paper I will review some of the available data pertaining to mammalian cell strains or lines which are hypersensitive to DNA-damaging agents, and therefore of potential use as a sensitive bioassay for environmental carcinogens. In terms of such a bioassay, it is not of immediate consequence whether this enhanced sensitivity results from an identifiable DNA repair defect or from another mechanism such as, for example, an enhanced susceptibility to the induction of DNA damage. I will restrict my discussion to cells with an enhanced sensitivity to the induction of mutations. The mutagenic potential of an agent is often associated with its carcinogenic potential. With the exception of cytotoxicity, it is also the cellular end point on which the most data are available with hypersensitive and repair-deficient lines. There is as yet no significant body of data concerning cell lines that are hypersensitive to the induction of malignant transformation.

Following a brief review of current knowledge of the major classes of DNA damage and metabolic repair processes identified in mammalian cells, attention will be first directed to a consideration of human diploid fibroblast strains in which marked hypersensitivity to DNA-damaging agents *in vitro* is associated with a specific genetic disorder. Second, I will describe three systems in which hypersensitive mutants have been isolated from established cell lines of both human and rodent origin.

DNA Damage and Repair

In this section, I will describe very briefly our current knowledge of the general classes of DNA damage induced by physical and chemical agents, as well as the metabolic processes which repair this damage and restore the coding sequence of the DNA molecule. This area has been recently reviewed by Lehmann and Karran (1).

DNA strand breaks. The induction of DNA strand breaks is a prominent effect of exposure to ionizing radiation. Radiation can induce both single-strand and double-strand breaks. Single strand breaks at the phosphodiester bond may be simple and readily ligatable by the action of a polynucleotide ligase. There may be other, more complex single-strand breaks which require several enzymatic activities. Single-strand breaks are rapidly and efficiently rejoined in mammalian cells with a half-time of the order of 10-15 minutes. No cell lines have been identified which appear to be defective in their capacity to rejoin DNA single-strand breaks; this class of damage is, thus, often considered to be of little or no biological significance in itself. Less is known about the biological significance of DNA double-strand breaks. A large fraction of these breaks appear to be rapidly rejoined following exposure. Non-rejoined double-strand breaks are assumed to be lethal lesions.

DNA base adducts. There are two basic types of adducts produced by DNA-damaging agents: bulky adducts, and simple adducts or alkylation products. Bulky adducts are classically produced by chemical carcinogens such as the polycyclic hydrocarbons; pyrimidine dimers induced by ultraviolet light also fall into this category as they produce a distortion of the DNA double helix. Such adducts are generally thought to be repaired by the nucleotide excision repair process. This involves four enzymatic steps: 1) The lesion is recognized and the DNA backbone incised by a specific endonuclease; 2) The region of the DNA strand containing the lesion is excised by the action of an exonuclease (up to 75-100 bases may be excised); 3) The DNA strand is resynthesized under the action of a DNA polymerase using the complementary strand as a template; and 4) The strand is rejoined by a polynucleotide ligase. The adduct is thus removed and the integrity of the DNA restored.

Recent evidence indicates, however, that there is an alternative pathway for the early steps in this process. In some cases, specific glycosylases may recognize the DNA adducts and release the damaged base by cleavage of the N-glycosylic bond, leaving an apurinic or an apyrimidimic (AP) site. This site is recognized by an AP endonuclease which nicks the DNA. The process then continues as in classical nucleotide excision repair. This base excision repair process has several interesting characteristics. First, there appears to be a spectrum of glycosolases with high specificity for specific types of base damage. On the other hand, only a single endonuclease is required; that is, one that will recognize the common abasic site produced by the glycosylic activity. The emerging evidence suggests that bulky DNA adducts in most systems are excised by classical nucleotide excision repair, though the action of a glycosylase may be involved in some cases. For example, the latter pathway appears to be important for the excision of UV light-induced pyrimidine dimers in certain bacterial cells (2).

Alkylation damage involves the methylation or ethylation of various DNA bases by the action of monofunctional alkylating agents. Unlike bulky adducts, these simple products lead to no gross distortion of the DNA helix and, thus, if not repaired, may lead to errors in base pairing during replication. For example, O^6-methylguanine may base pair incorrectly with thymidine. There appear to be two major mechanisms for repairing alkylation products. The first is base excision repair involving the activity of specific glycosylases. The second is an inducible process involving the activity of a methyl-transferase enzyme which removes the methyl group from the base by a process of transmethylation. Evidence has recently been presented which indicates that the latter pathway, originally demonstrated in bacterial cells, is also active in mammalian cells (3).

It appears that other cellular processes deal with DNA base damage which is not excised and remains at the time of replication. However, the existence and nature of these processes in mammalian cells remain poorly understood. "By-pass" repair allows for synthesis past lesions during DNA replication with the random insertion of bases (error-prone repair). "Post-replication" repair is proposed to involve leaving a gap opposite the damaged site during DNA replication; this gap may be filled in by a process of crossing-over or mitotic recombination. Evidence for mitotic recombination in mammalian cells is lacking, and that for error-prone repair processes remains controversial. Evidence for the existence of such repair processes is, thus, derived largely from bacterial models and the empiric observation that large numbers of DNA adducts and pyrimidine dimers are not removed by excision repair processes, and remain in the DNA at the time of replication.

Base hydration products. This type of base damage is classically produced by ionizing radiation and other agents which operate through free radical mechanisms. A classical product is the 5-dihydroxy, 6-dihydrothymine residue resulting from saturation at the C_5-C_6 double bond in thymidine. As in the case of methylation and ethylation products induced by monofunctional alkylating agents, this subtle type of base damage will cause no gross distortion in the DNA helix. Recent evidence suggests that these hydration products may also be recognized by specific glycosylases and repaired by the base excision repair mechanism (2). The extent to which lesion-specific glycosylases may recognize specific hydration products, which may, themselves, be induced by specific classes of DNA-damaging agents, remains to be elucidated.

Crosslinks. DNA-DNA or DNA-protein crosslinks are classically produced by bifunctional alkylating agents. DNA-protein crosslinks can be removed by simple excision repair mechanisms as described for DNA adducts. DNA-DNA crosslinks involve a more complex mechanism which is, as yet, poorly understood in mammalian cells. The best model available is derived from yeast and involves two excision steps leading to a double-strand break; repair of this double-strand break in yeast involves a recombinational mechanism. Somatic cells from certain patients with Fanconi's anemia are the primary model for cell lines specifically hypersensitive to DNA crosslinking agents.

Human Diploid Fibroblasts

Several clinically recognized genetic syndromes have been identified in which the somatic cells are markedly hypersensitive to specific DNA-damaging agents *in vitro* (4). These include Xeroderma pigmentosum (ultraviolet light), ataxia telangiectasia (X-rays), and Fanconi's anemia (DNA crosslinking agents). Xeroderma pigmentosum (XP) was the first such syndrome identified, and represents the prototype for a genetic disorder characterized by hypersensitivity to an environmental agent which is associated with a DNA repair defect. XP is a rare autosomal recessive disease characterized clinically by an extreme hypersensitivity to sunlight. These patients develop multiple skin cancers in areas exposed to sunlight, often within the first few years of life. When skin fibroblasts from XP patients are grown *in vitro*, they are found to be extremely hypersensitive to both the cytotoxic and mutagenic effects of UV light. This is shown graphically in Figures 1 and 2. These cells have been shown to possess a well characterized defect in the early stages of the excision repair process for UV light-induced DNA base damage. Seven different complementation groups have thus far been identified, suggesting that at least seven different genetic loci may be involved in this repair process. A variant XP group also has been found with normal excision repair capacity but an apparent defect in a post-replication repair mechanism. Xeroderma pigmentosum is the only human genetic disorder in which a DNA repair defect has been clearly identified.

The use of XP cell strains as a bioassay for potential carcinogens might have certain advantages over the use of normal human diploid fibroblasts. Mutagenicity would be induced at much lower doses, thus enhancing the sensitivity of the assay. This might be useful, for example, in detecting mutagens in mixtures which were lethal to the cells at higher doses. However, there are several potential problems which arise in respect to such an assay. First, if the increased susceptibility to the induction of mutations is paralleled by an increased susceptibility to cell killing, no advantage might accrue as the assay depends upon scoring mutation frequency per surviving cell. Thus, the same mutation frequency would be observed in normal cells at a similar level of cell killing simply by the use of a higher dose of the carcinogen. Obviously, the most sensitive assay system would be one in which the cell was hypersensitive to the induction of mutations by the carcinogen, but which showed no enhanced cytotoxicity. Mutagenicity could then be studied at a range of doses which produced no cell killing.

Studies have been carried out on the induction of mutations in XP cells by UV light in which the induced mutation frequencies have been normalized for cell survival (5,6). That is, UV-induced mutation frequencies in XP and normal cells have been compared for doses which yielded the same level of survival in both cell types. The results of these studies have not been entirely consistent; one investigation indicates, however, that at least under certain cultural conditions XP cells remain somewhat hypermutable as compared with normal cells even when normalized for the cytotoxic effects of UV light (5). This result would suggest that XP cells could offer an advantage as compared with normal human diploid fibroblasts in terms of their sensitivity to the induction of mutations by certain carcinogens.

The second problem which arises involves mutational specificity. In addition to UV light, XP cells are hypersensitive to certain chemical agents which produce bulky adducts and induce the long patch type of DNA excision repair. However, they show no enhanced sensitivity to many other DNA-damaging agents such as X-rays and alkylating and crosslinking agents which induce different classes of

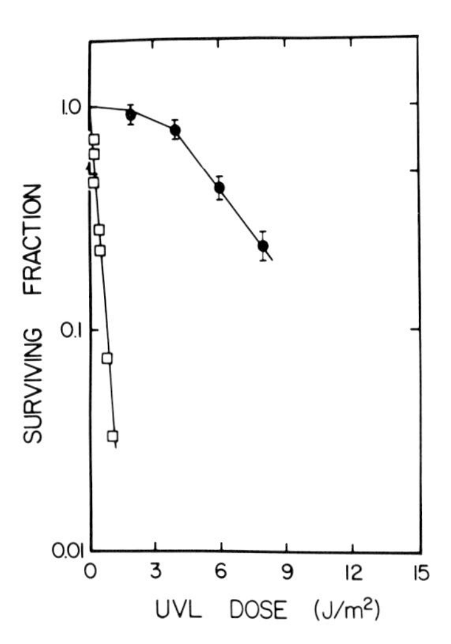

Fig. 1. Ultraviolet light survival curves for a normal human diploid skin fibroblast cell strain (● —— ●), and a strain isolated from a patient with Xeroderma pigmentosum (□ —— □). Note the extreme sensitivity to the cytotoxic effects of UV light of the XP cells. Data from Grosovsky and Little (5).

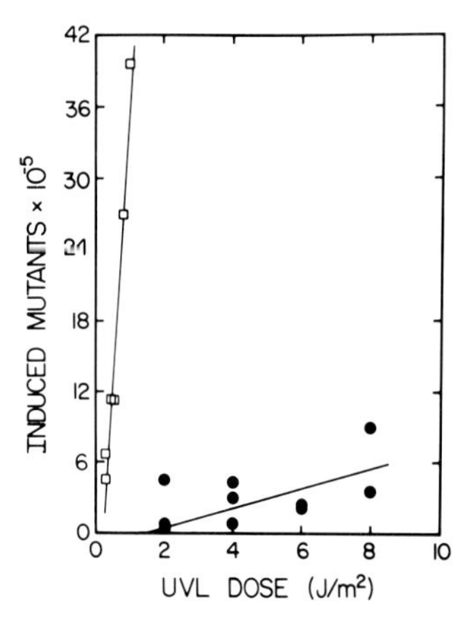

Fig. 2. The induction of mutations to 6-thioguanine resistance by ultraviolet light in a normal human diploid skin fibroblast strain (● —— ●), and in a strain isolated from a patient with Xeroderma pigmentosum (□ —— □). Note the marked hypermutability of the XP cells at very low UV doses. Data from Grosovsky and Little (5).

DNA damage involving different repair mechanisms. Similarly, skin fibroblasts from patients with Ataxia telangiectasia, which are characterized by extreme hypersensitivity to the lethal effects of X-rays, show no enhanced sensitivity to UV light and chemical carcinogens which induce bulky DNA adducts, and only an intermediate response to certain alkylating agents. Cells from patients with Fanconi's anemia, on the other hand, appear fairly specifically hypersensitive to DNA crosslinking agents. Thus, the three known mutant human diploid cell strains with the greatest hypersensitivity to DNA-damaging agents all show significant specificity in terms of the class agents to which they respond.

The third general problem with the use of mutant human diploid fibroblasts as a bioassay system involves the difficulties in working with this type of cell, and the logistics required for each experiment. These mutant cell strains often grow poorly with very low cloning efficiencies as compared with established cell lines. They become contact-inhibited at relatively low cell densities, and have a limited lifespan *in vitro*. Mutagenesis studies with human diploid fibroblasts can, thus, become a heroic undertaking. Newer systems are being developed which may resolve some of these problems. These include the use of established human diploid lymphoblast cell lines (7), and stimulated lymphocytes (8). As yet, however, these techniques have not been applied to the study of induced mutations in repair-deficient cells or cells isolated from patients with hypersensitive genetic syndromes.

In sum, there are at present severe limitations on the use of mutant human diploid cell strains as a sensitive bioassay for environmental carcinogens. There is only one well characterized mutant class available (Xeroderma pigmentosum) in which marked hypermutability, as opposed to enhanced cytotoxicity, has been demonstrated. To a large extent, the hypermutability of XP goes hand in hand with an increased sensitivity to cell killing. Furthermore, XP responds only to one class of carcinogens. Bloom's syndrome fibroblasts are very sensitive to the induction of cytogenetic changes by certain chemical agents, and their spontaneous rate of mutations has recently been reported to be markedly elevated (9). Much better characterization of this phenomenon will be required, however, to determine the potential usefulness of Bloom's syndrome fibroblasts as a sensitive bioassay. Finally, human diploid fibroblasts, in general, are difficult to handle and grow, and the logistics of mutation assays with them are enormous.

With the identification of additional repair-deficient and hypersensitive mutants, especially those with various and well-characterized types of specificity, the use of such a cell system might be a useful adjunct to a bioassay program in which it was important to measure a specific effect in human diploid cells.

Established Cell Lines

Several laboratories have recently developed methods for selecting putative repair-deficient mutants from established rodent and human cell lines. The use of such established cell lines has certain important advantages over human diploid fibroblasts in that they have an indefinite lifespan in culture, can be readily grown in large numbers, and usually have very high cloning efficiencies. These characteristics not only greatly facilitate the carrying out of mutagenesis experiments, but allow the flexibility necessary for the systematic selection of mutant cell lines *in vitro*. I will discuss three such cell systems, two derived from rodent lines and one from a human HeLa line, which illustrates the advantages and problems involved with their use in a sensitive bioassay for carcinogens.

CHO cells. Thompson et al. (10) and Busch et al. (11) described techniques for isolating UV sensitive mutant clones from an established line of Chinese hamster ovary (CHO) cells. These lines resembled human XP mutants in that they showed low or undetectable levels of UV-induced excision repair. Mutants sensitive to alkylating agents also were isolated from CHO cells (10). By the use of hybridization techniques, Thompson, Busch, and their co-workers (12) identified four complementation groups among the forty-four UV-sensitive mutant clones tested, suggesting different classes of mutants representing different defects in one or more steps in the DNA excision repair process. These mutant cell lines were hypersensitive both to cell killing and to the induction of mutagens by UV light. Their characteristics are, thus, very similar to those of Xeroderma pigmentosum. Indeed, the chemical cross-sensitivity observed appears similar to that for XP.

These results suggest that a variety of mutations affecting nucleotide excision repair can be isolated from a cell line that is widely used in mutagenesis studies. The existing UV mutants share some of the same limitations as XP cells in that they are also hypersensitive to the cytotoxic effects of UV light, and appear to be cross-sensitive only to certain classes of potential carcinogens. The fact that other cell lines have been isolated which are specifically sensitive to alkylating agents with cross-sensitivity to gamma-radiation (10) suggests, however, that CHO clones with various mutational specificities can be isolated to provide a spectrum of lines that would recognize many classes of carcinogens.

V-79 cells. Trosko and his co-workers (13,14) have described a selection technique whereby mutant clones of Chinese hamster V-79 cells can be isolated which are hypersensitive to mutagens and/or deficient in UV-induced DNA excision repair. These investigators have studied three of these cell lines in some detail in respect to their sensitivity to the induction by UV light of cell killing,

Table 1. Induction of mutations by ultraviolet light in the UV-7 mutant and parental V79 cell lines.[a]

Selective Agent	Cell Line	UV-dose J/m[b]	Survival %	Total Number Cells Assayed x 10^5	Mutation Frequency Per 10^6 Survivors[c]
Oua^R	V79	20	2.0	0.8	1,093
Oua^R	UV-7	20	1.1	0.3	29,444
$6TG^R$	V79	14	11.2	9.2	221
$6TG^R$	UV-7	14	10.0	8.0	411

[a]Data abstracted from Table 1 of Schulz et al.(14).
[b]Oua^R = Ouabain resistance. $6TG^R$ = 6-thioguanine resistance.
[c]Spontaneous (background) mutation frequencies were 9 and 3 respectively for Oua^R, and 13 and 3 for $6TG^R$.

mutagenesis at three genetic loci, and excision repair as measured by unscheduled DNA synthesis (UDS) (14). The mutant UV-7 line showed markedly decreased levels of UDS as compared with wild type V-79 cells, but only slightly increased sensitivity to the cytotoxic effects of UV light. On the other hand, it was markedly hypermutable by UV light at all three loci tested, although to different degrees.

Experiments illustrating these results are shown in Table 1 for mutations to ouabain resistance and 6-thioguanine resistance. Looking at the data for ouabain resistance, it can be seen that the mutation frequency induced by 20 J/m^2 of UV light was nearly 30-fold higher in the UV-7 line as compared with the parent V-79 line, whereas there was little difference in the cytotoxic effect (2.0 vs. 1.1% survival). About a 2-fold enhancement in 6-thioguanine resistant mutants was induced by 14 J/m^2 with no significant difference in cytotoxicity (Table 1). UV-induced mutability to diphtheria toxin resistance showed an intermediate sensitivity in the UV-7 line (data not shown).

The UV-7 line thus represents a good example of a repair-deficient cell which is significantly hypermutable as compared with the normal, wild-type cells, but only minimally hypersensitive to cell killing. In respect to this characteristic, it represents the prototype cell line for a sensitive bioassay for potential mutagens and carcinogens. On the other hand, the UV-7 line was not sensitive to the chemical carcinogen N-acetoxy-2-acetylaminofluorene (N-Ac-AAF), which produces bulky DNA adducts and usually shows cross-sensitivity to UV light. The other two mutant V-79 cell lines studied by these investigators (14) showed somewhat different characteristics. UV-40 was markedly hypersensitive to the cytotoxic effects of UV light, was somewhat hypermutable at the ouabain resistance locus, but showed only a slight reduction in its DNA excision repair capacity. It was also significantly hypersensitive to the cytotoxic effects of N-Ac-AAF, X-rays, and certain monofunctional alkylating agents. The UV-23 line was abnormally resistant to the cytotoxic effects of UV light, showed an enhanced capacity for excision repair, but no significant difference from the parent V-79 strain in UV-induced mutagenesis.

These results with V-79 cells need to be confirmed, but they are of interest as they suggest that mutant clones may be isolated from established rodent cell lines with varying characteristics in terms of their cytotoxic and mutagenic response to DNA-damaging agents, as well as their repair capacity and the specificity of their response.

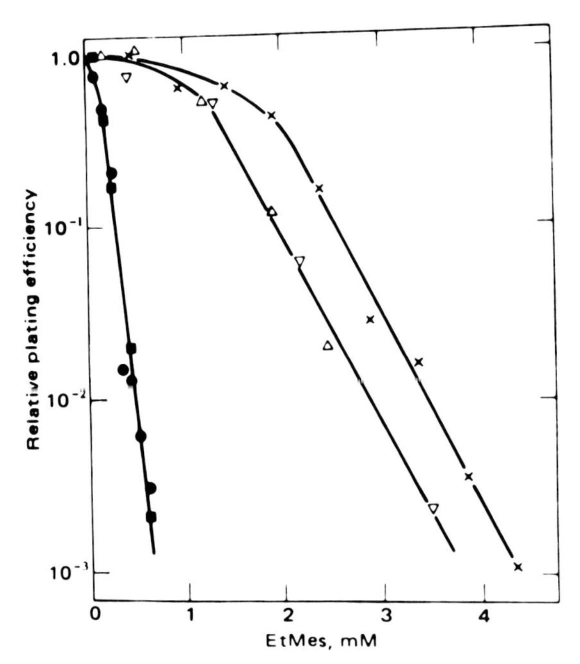

Fig. 3. Ethyl methanesulfonate (Et Mes) survival curves for classical wild type HeLa cells (x), the sensitive S-3 subline (●, ■), and the A6 mutant derived from the S-3 subline (△, ▽). Note the marked resistance to the cytotoxic effects of Et Mes of the A6 mutant as compared with its parental S-3 cell line. Reproduced from Baker et al. (15).

HeLa cells. Baker et al. (15) have studied HeLa cell variants that differ in their sensitivity to monofunctional alkylating agents. These investigators noted that the HeLa S-3 subline was unusually sensitive to the lethal and mutagenic effects of ethyl methane sulfonate (EMS) as compared with the original HeLa line or with other normal human cells. They used the S-3 subline to derive an EMS resistant variant line (designated A-6) which was comparable

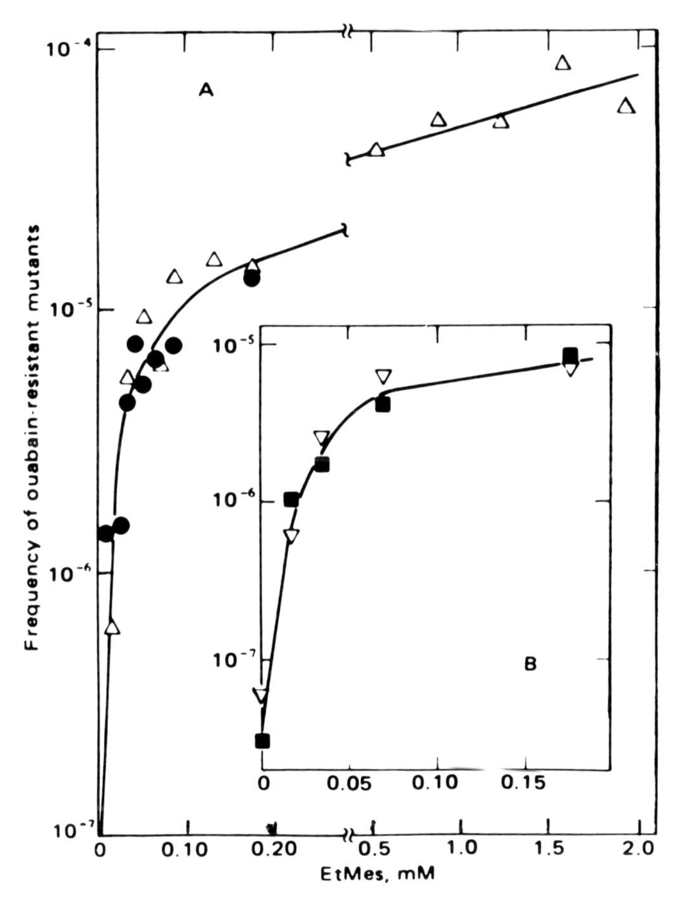

Fig. 4. Induction of mutations to ouabain resistance by ethyl methanesulfonate (Et Mes) in the parental S-3 HeLa cell line (●, ■) and the A6 mutant line (△, ▽). Note that these lines show a similar high sensitivity to the induction of mutations by Et Mes, whereas the A6 line was much more resistant to its cytotoxic effects (Fig. 3); no cell killing occurred with doses up to 0.5 mM. Panels A and B refer to independent experiments. Reproduced from Baker et al. (15).

in its cytotoxic response to the original parent HeLa cell line (Figure 3). Unexpectedly, however, whereas the A-6 subline was resistant to the lethal effects of EMS as compared with the parent S-3 line, it remained similar in its hypermutability. These findings are shown in Figures 3 and 4. As a result, the mutant A-6 line has the characteristic of being resistant to the lethal effects of alkylating agents (wild-type response), but highly sensitive to the induction of mutations by them. Additional studies indicated that the A-6 subline was also resistant to cell killing by several other monofunctional alkylating agents, as opposed to the S-3 line which showed hypersensitivity to all of them. On the other hand, the S-3 line was resistant to the cytotoxic effects of mitomycin-C, ultraviolet light, and X-rays, indicating the specificity of the effect for alkylation damage.

These results suggest that different alkylation products (DNA lesions) may be responsible for the cytotoxic and mutagenic effects of alkylating agents of mammalian cells. The mutant A-6 HeLa cell line may differ from the S-3 parent in a DNA repair process affecting potentially lethal but not potentially mutagenic lesions in DNA. These processes would appear to differ from those involved in cross-link, ionizing, or UV radiation damage. Of particular interest in terms of a sensitive bioassay is the observation that, similar to the UV-7 mutant Chinese hamster V-79 cell line shown in Table 1, the A-6 mutant HeLa cell line is hypermutable by DNA-damaging agents but not hypersensitive to their cytotoxic effects.

CONCLUSIONS

The use of hypersensitive and repair-deficient cell lines represents a potentially interesting approach to the identification of carcinogens based on their mutagenic potential in mammalian cells. As yet, there is no human diploid cell system which is well suited to such an assay. On the other hand, several laboratories have developed techniques for isolating such mutant cell clones from established cell lines in culture. Hypersensitive mutants isolated by these techniques may potentially provide highly sensitive assays for environmental carcinogens and mutagens. However, these mutant lines must have the appropriate specificity for the agent being tested. Most such mutants appear to respond only to specific classes of DNA-damaging agents. This finding suggests that, in order to identify a broad spectrum of potential carcinogens, a variety of mutant tester lines will probably need to be developed as has been done in bacterial assays for mutagenesis.

Clearly, considerable further work is required with respect to the isolation and characterization of mutant cell lines, as well as the determination of their specificity for different classes of carcinogens, in order for this approach to have a clear advantage over existing assays for mutagenesis in normal cells. In particular, this will likely require the development of a battery of cell lines mutant with respect to different types of DNA damage and repair processes which will respond to the broad spectrum of environmental carcinogens.

REFERENCES

1. Demple, B., and S. Linn (1980) DNA N-glycolases and UV repair. Nature 287:203-208.
2. Lehmann, A.R., and P. Karran (1981) DNA repair. Int. Rev. Cytology 72:101-146.
3. Samson, L., and J.L. Schwartz (1980) Evidence for an adaptive DNA repair pathway in Chinese hamster ovary and human skin fibroblast cell lines. Nature 287:861-863.
4. Friedberg, E.C., U.K. Ehmann, and J.I. Williams (1979) Human diseases associated with defective DNA repair. In Advances in Radiation Biology, Vol. 8, J.T. Lett and H.I. Adler, eds. Academic Press, New York, pp. 85-174.
5. Grosovsky, A.G., and J.B. Little. Mutagenesis and cytotoxic effects of ultraviolet light exposure in confluent cultures of Gardner's syndrome, Xeroderma pigmentosum, and normal human fibroblasts (submitted).
6. Maher, V.M., and J.J. McCormick (1976) Effect of DNA repair on the cytotoxicity and mutagenicity of UV irradiation and of chemical carcinogens in normal and xeroderma pigmentosum cells. In Biology of Radiation Carcinogenesis, J.M. Yuhas, R.W. Tennant, and J.B. Regan, eds., Raven Press, New York, pp. 129-145.
7. Thilly, W.G., W.G. DeLuca, E.E. Furth, H. Hoppe IV, D.A. Kaden, J.J. Krolewski, H.L. Liber, T.R. Skopek, S.A. Slapikoff, R.J. Tizard, and B.W. Penman (1980) Gene locus mutation assays in diploid human fibroblast lines. In Chemical Mutagens, F.F. deSerres and A. Hollaender, eds., Vol. 6, Plenum Press, New York, pp. 331-364.
8. Albertini, R.J., K.L. Castle, and W.R. Borcherding (1982) T-cell cloning to detect the mutant 6-thioguanine-resistant lymphocytes present in human peripheral blood. Proc. Natl. Acad. Sci., USA 79:6617-6621.
9. Warren, S.T., R.A. Schultz, C.C. Chang, M.H. Wade, and J.E. Trosko (1981) Elevated spontaneous mutation rate in Bloom's syndrome fibroblasts. Proc. Natl. Acad. Sci., USA 78:3133-3137.
10. Thompson, L.H., J.S. Rubin, J.E. Cleaver, G.F. Whitmore, and K. Brookman (1980) A screening method for isolating DNA repair-deficient mutants of CHO cells. Somatic Cell Genet. 6:391-405.
11. Busch, D.B., J.E. Cleaver, and D.A. Glaser (1980) Large scale isolation of UV-sensitive clones of CHO cells. Somatic Cell Genet. 6:407-418.

12. Thompson, L.H., D.B. Busch, K. Brookman, C.L. Mooney, and D.A. Glaser (1981) Genetic diversity of UV-sensitive DNA repair mutants of Chinese hamster ovary cells. Proc. Natl. Acad. Sci., USA 78:3734-3737.
13. Schultz, R.A., J.E. Trosko, C.C. Chang (1981) Isolation and partial characterization of mutagen sensitive and DNA repair mutants of Chinese hamster fibroblasts. Environ. Mutagenesis 3:53-64.
14. Schultz, R.A., C.C. Chang, and J.E. Trosko (1981) The mutation studies of mutagen-sensitive and DNA repair mutants of Chinese hamster fibroblasts. Environ. Mutagenesis 3:141-150.
15. Baker, R.M., W.C. Van Voorhis, and L.A. Spencer (1979) HeLa cell variants that differ in sensitivity to monofunctional alkylating agents, with independence of cytotoxic and mutagenic responses. Proc. Natl. Acad. Sci., U.S.A. 76:5249-5253.

THYMINE MODIFICATIONS IN DNA AS MARKERS OF RADIATION EXPOSURE

George W. Teebor and Krystyna Frenkel

Department of Pathology
New York University Medical Center
550 First Avenue
New York, New York 10016

INTRODUCTION

Ionizing radiation causes modification of bases in DNA which may contribute to the cytotoxic and mutagenic properties of radiation (1,2). These modifications are thought to result from interinteraction of the DNA bases with hydroxyl radicals derived from the radiolysis of water (1,3-7).

When DNA is irradiated *in vitro*, one of the major stable end products is the ring-saturated thymidine derivative, 5,6-dihydroxy-5,6-dihydrothymidine (thymidine glycol; dTG) (8-10). The irradiated thymine moiety is susceptible to further oxidative modification by hydroxyl radicals leading to ring opening and formation of the relatively unstable N'-formyl-N-pyruvylurea (FPU) (5,9,10). This compound may cyclize to 5-hydroxy-5-methyl hydantoin (HMH) or fragment, leaving a residue such as urea or formaldehyde attached to the deoxyribose moiety (5,9,10). There is evidence that such radiation-induced thymine derivatives are removed from DNA by cellular repair processes (11-14) but neither the specificity nor the extent of the repair has been characterized. Endonucleolytic activities which attack irradiated DNA and DNA which has been chemically oxidized by OsO_4 are present in both bacteria and mammalian cells (11-17). Since oxidation of DNA by OsO_4 causes formation of thymine glycol (TG) as the main oxidation derivative of the thymine (T) moiety (18,19), it is thought that TG is enzymatically repaired when present in DNA. Indeed, it was recently reported that TG was released from oxidized DNA through the action of a N-glycosylase activity present in preparations of *Escherichia coli* (*E. coli*) endonuclease III (20).

Another N-glycosylase activity of *E. coli* effects removal of urea from poly(dA,dT) containing such residues (21). Normal and modified bases are spontaneously released from irradiated DNA, presumably through the attack of hydroxyl radicals on the C'1 position of the deoxyribose moiety rendering the N-glycosyl bond labile (10,22-25). The resulting abasic sites (apurinic/apyrimidinic, AP sites) are subject to attack by AP DNA endonucleases which initiate DNA excision-repair of such sites. Since the modification of basis in DNA by ionizing radiation may constitute a carcinogenic event (2), it is of importance to characterize radiation-induced base derivatives in cellular DNA and to determine whether they are enzymatically repaired in human cells. For this purpose, we developed assays using high pressure liquid chromatography (HPLC), Sephadex LH-20, and micro-derivatization which now makes it possible to measure the content of such derivatives in the DNA of irradiated cells (19,26-29). We also showed that one derivative, 5-hydroxymethyl 2'-deoxyuridine (HMdU), was formed in [^{3}H-methyl]thymidine[dT]-containing DNA as the result of the transmutation of tritium to helium-3 (^{3}H to ^{3}He) in addition to being formed through hydroxyl radical attack on the methyl group of thymine (30).

MATERIALS

[*Methyl*-^{14}C]thymine (49.6 mCi/mmol), [*methyl*-^{14}C]thymidine (48.2 mCi/mmol), [*methyl*-^{3}H]thymidine (6.7 Ci/mmol), and [6-^{3}H]thymidine (16.2 Ci/mmol) were purchased from New England Nuclear, Boston, Massachusetts. Sephadex LH-20 and Sephadex G-50 were obtained from Pharmacia Fine Chemicals, Uppsala, Sweden. *E. coli* W 3100 (*thy*$^{-}$) J. Cairns strain was obtained from the *E. coli* Genetic Stock Center at Yale University School of Medicine, New Haven, Connecticut. Cell and bacteria growth media, calf serum, and L-glutamine were purchased from GIBCO, Grand Island, New York. The 5 hydroxymethyl uracil (HMU), 5-hydroxymethyl 2'-deoxyuridine, and OsO_4 were obtained from Sigma, St. Louis, Missouri.

METHODS

The ^{14}C-containing TG, dTG, HMH, and MH (5-methylene hydantoin) were synthesized according to published methods (10,19,26,27,31-33). *E. coli* were grown in the presence of [^{3}H-methyl]dT, the DNA extracted, purified, then oxidized with OsO_4 or γ-irradiated from a ^{137}Cs source as previously described (19,26). HeLa cells were grown in suspension culture in the presence of either [^{3}H-methyl]dT or [^{3}H-6]dT for one doubling time. Cells were harvested, washed, and exposed to γ-radiation from a ^{137}Cs source (26) and the DNA extracted by the method of Marmur (34). DNA isolated from either *E. coli* or HeLa cells was enzymatically hydrolyzed to 2'-deoxyribonucleosides (19,26), precipitated with acetone, and the supernatant

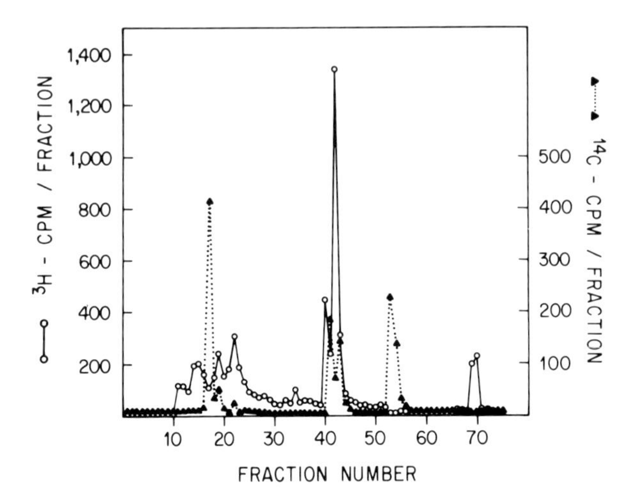

Fig. 1. Chromatography of a mixture of the enzymatic hydrolysate of OsO_4-oxidized [^{3}H-methyl]dT-containing E. coli DNA (o) and [^{14}C]TG, dTG, and T (Δ) on a preparative ODS HPLC column with water as eluent.

analyzed by HPLC. The enzymatic hydrolysates of ^{3}H-DNA were injected into the preparative 5 μ Ultrasphere-ODS (octadecyl silane); C_{18} column with water as eluent, and chromatographed together with UV-absorbing markers including HMU, HMdU, T, and dT, and the ^{14}C-containing thymine derivatives TG, dTG, and HMH. Fractions were collected in test tubes or scintillation vials (26-28).

HPLC fractions which co-chromatographed with dTG, HMH, HMdU, or HMU were collected, evaporated to dryness, acetylated with acetic anhydride in dry pyridine, and analyzed by HPLC with an acetonitrile-water gradient (27-30). Fractions which co-chromatographed with TG were oxidized with sodium periodate to form FPU, then cyclized to HMH by refluxing in water (10), and then acetylated.

RESULTS

Figure 1 shows a HPLC profile of an enzymatic hydrolysate of OsO_4-oxidized [^{3}H]-labelled DNA chromatographed together with [^{14}C]-containing standards. Thymidine was excluded from this and subsequent figures because there were no radioactive peaks eluting between fraction 80 and dT. The small difference in elution time between [^{3}H]-labelled peaks and [^{14}C]-containing marker compounds is due to an isotopic effect (35,36).

Thymidine glycol constituted 85% of the modified thymines and the four early eluting peaks totalled near 15%. Peak 70 amounted to

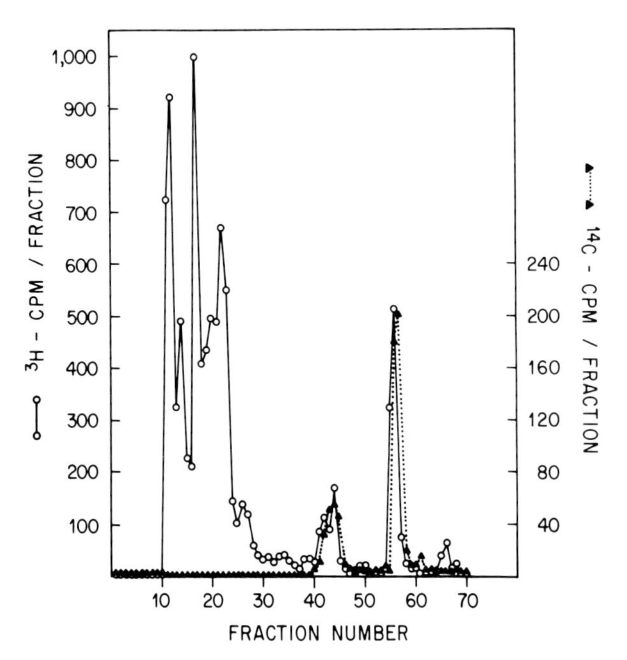

Fig. 2. Chromatography of a mixture of the enzymatic hydrolysate of _in vitro_ gamma-irradiated [^{3}H-methyl]dT-containing _E. coli_ DNA (o) and [^{14}C]TG, dTG, and T (Δ) on a preparative ODS HPLC column with water as eluent.

about 1.5% of the total applied radioactivity and was comparable to that obtained from non-oxidized control DNA. There was no coincidence of [^{3}H]-containing material with [^{14}C]TG indicating that during the overnight enzymatic digestion of the DNA there was no release of TG from DNA. There was also no [^{3}H] present in the [^{14}C]T fractions (ranging from 55-58) showing that oxidation of DNA did not cause spontaneous release of T.

Figure 2 shows the HPLC profile of an enzymatic hydrolysate of single-stranded [^{3}H-methyl]dT-labelled _E. coli_ DNA exposed to 72 krad of γ-radiation. The majority of [^{3}H]-containing products eluted as five peaks between fractions 10 and 30 and contained 15% of the applied radioactivity. There was a peak at fraction 17 which is the elution position of TG. However, the complexity of the pattern precluded its unambiguous identification. Thymidine glycol was identified in fraction 41-45, and T in fraction 55-58. They constituted 1% and 2%, respectively, of the applied [^{3}H]-radioactivity. The presence of T and the probable presence of TG indicated spontaneous release of bases from DNA.

Figures 3 and 4 show the HPLC profiles of enzymatic hydrolysates of [^{3}H-methyl]dT-labelled DNA extracted from unirradiated

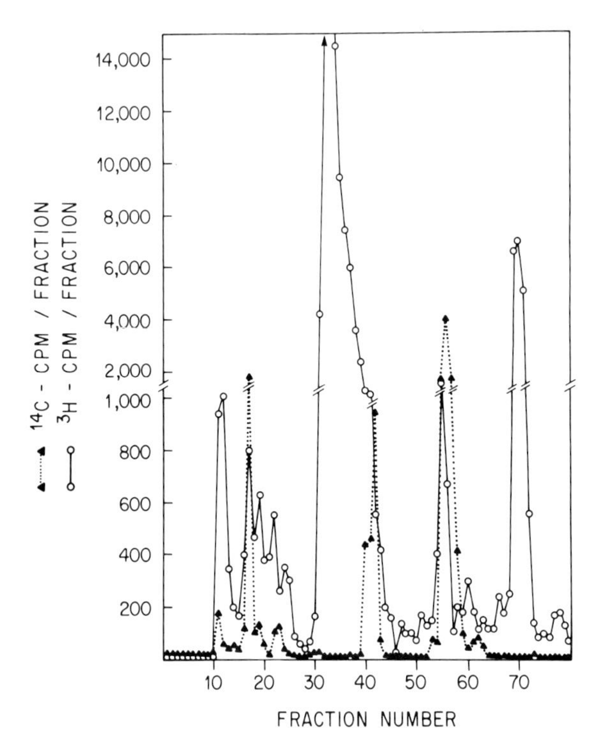

Fig. 3. Chromatography of a mixture of the enzymatic hydrolysate of [^{3}H-methyl]dT-containing DNA (o) isolated from control unirradiated HeLa cells and [^{14}C]TG, dTG, and T (Δ) on a preparative ODS HPLC column with water as eluent.

(Fig. 3) and γ-irradiated (Fig. 4) HeLa cells. A large peak of radioactivity was present in fractions 30 through 40 in the control sample and a much smaller one was present in the irradiated sample. We concluded that this large peak consisted of small oligonucleotides since the control sample was less digested than the irradiated sample. We reasoned that breaks in the DNA of irradiated cells rendered the DNA more susceptible to enzymatic digestion. Even with 99% digestion, an oligonucleotide peak was present (Fig. 4), and if large, this peak may overlap other peaks (Fig. 3). Enzymatic hydrolysates of both unirradiated (Fig. 3) and irradiated (Fig. 4) cellular DNA displayed complex patterns which included five rapidly eluting peaks (fractions 10-30), one of which coincided with [^{14}C]TG, T (fractions 50-60), and a peak at fraction 70.

Total modification of the irradiated DNA was about 0.4% of applied radioactivity. This included the putative oligonucleotide peak which represented 0.1% of the total so that the actual modification level was probably lower. The early eluting peaks constituted 0.07%, T - 0.016%, and peak 70 - 0.2% of the total

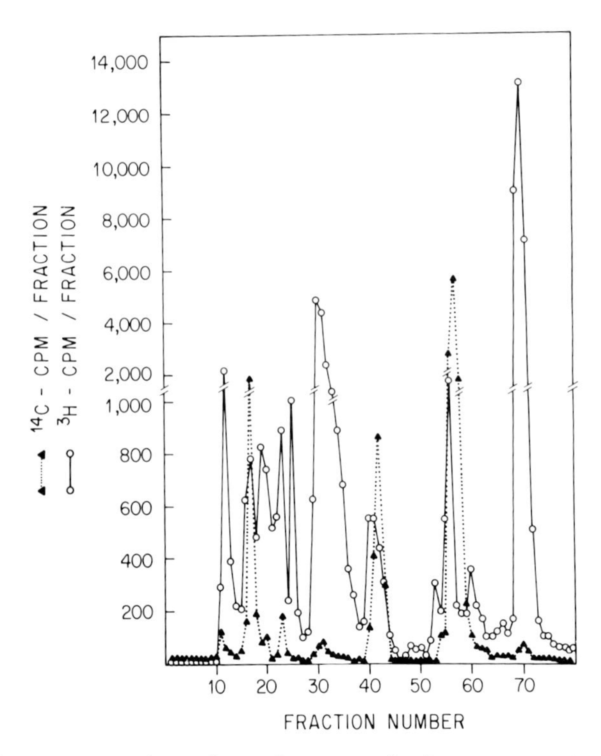

Fig. 4. Chromatography of a mixture of the enzymatic hydrolysate of [^{3}H-methyl]dT-containing DNA (o) isolated from gamma-irradiated HeLa cells and [^{14}C]TG, dTG, and T (Δ) on a preparative ODS HPLC column with water as eluent.

radioactivity. Approximately 0.02% of the [^{3}H]-radioactivity coincided with [^{14}C]dTG and one of the fast-moving peaks eluted together with TG. In unirradiated DNA, the early eluting fractions constituted 0.05%, T - 0.02%, and peak 70 - 0.15% of applied radioactivity. The dTG peak was obscured by the large oligonucleotide fraction. The dTG content of both unirradiated and irradiated DNA was finally determined by chromatography of the appropriate HPLC fractions on Sephadex LH-20 with water and borate buffer, pH 8.6, as eluents (26).

Peak 70 was the major thymine derivative present in cellular DNA which contained [^{3}H-methyl]dT and was apparently increased by 30% by γ-irradiation. This derivative proved to be HMdU as determined by co-chromatography with marker HMdU and subsequent acetylation of both the ^{3}H-containing peak and the UV-absorbing marker (27-30).

5-Hydroxymethyl uracil and HMdU had previously been shown to be minor products when T (4) and dT (5) solutions were exposed to ionizing radiation. Therefore, it was puzzling that HMdU should be the

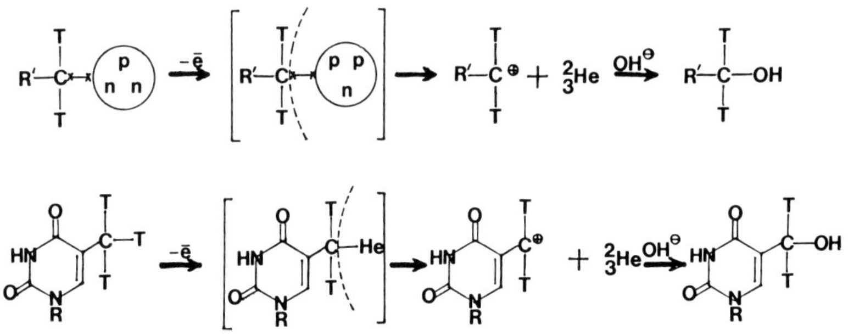

Fig. 5. Proposed mechanism of formation of 5-hydroxymethyl 2'-deoxyuridine from [^{3}H-methyl]thymidine as a result of the transmutation of ^{3}H to ^{3}He. p = proton; n = neutron; $\bar{e}$ = electron; R = H or 2'deoxyribose; R' = alkyl or aryl moiety.

major dT derivative present in such high amounts in the DNA of both unirradiated and irradiated cells. This result is explained by the scheme for formation of HMdU through transmutation of tritium as shown in Figure 5. In compounds in which ^{3}H is linked to carbon, the emission of an electron (beta particle) leads to the formation of an unstable intermediate containing helium-3 (^{3}He) linked to carbon. Such a C-^{3}He bond has a very short half-life (10^{-4} to 10^{-5} sec), and neutral ^{3}He is released by abstracting an electron from carbon, leaving the reactive carbonium ion (37). This carbonium ion interacts with hydroxyl ions derived from water yielding HMdU as the stable end product.

If HMdU is formed as the result of transmutation, then only [^{3}H-methyl]-dT (but not [^{3}H-6]dT) should yield HMdU since transmutation of tritium in position 6 of T could not cause its formation. To prove this hypothesis, HeLa cells were grown overnight in the presence of [^{3}H-6]dT, chased with dT, harvested, and split into two equal populations. One was irradiated with a ^{137}Cs source for a total dose of 18 krad and the other served as control. The effective internal ionizing radiation dose due to ^{3}H decay was the same for both [^{3}H-methyl]dT and [^{3}H-6]dT-containing cells since the specific activities of cellular DNA were the same. Figures 6 and 7 show the HPLC profiles of the enzymatic hydrolysates of [^{3}H-6]dT-containing DNA of unirradiated and γ-irradiated HeLa cells. As can be seen in Fig. 6 (unirradiated cells), there is practically no radioactivity eluting in the region of UV marker HMdU. In contrast, Fig. 7 (irradiated cells) shows a peak coincident with HMdU, the identity of which was confirmed by acetylation and rechromatography on the ODS column.

Figure 8 shows HPLC separation of radiation-induced thymine derivatives, many of which we have synthesized as marker compounds

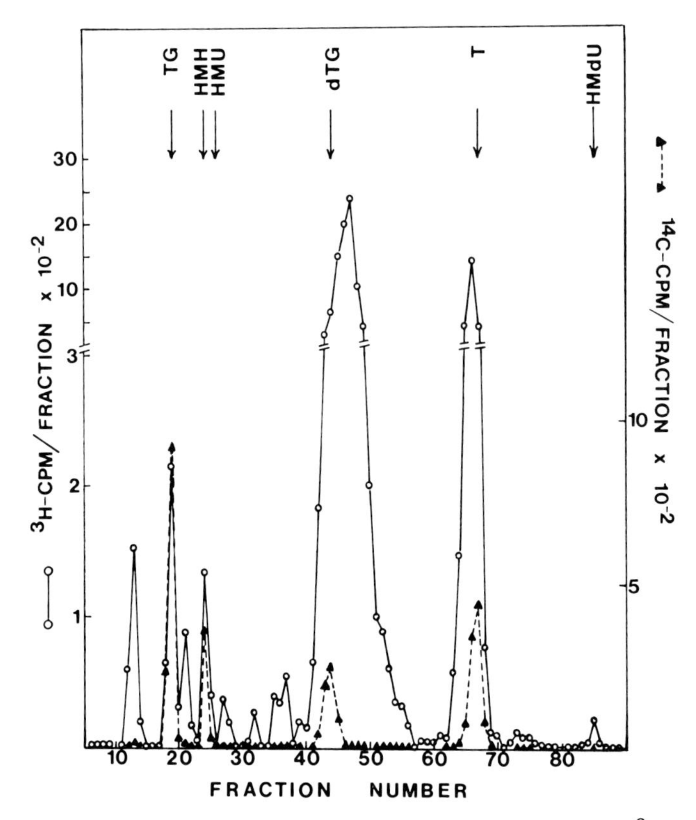

Fig. 6. Chromatography of enzymatic hydrolysate of [^{3}H-6]dT-containing DNA (o) isolated from control unirradiated HeLa cells with marker [^{14}C]TG, HMH, dTG, and T (Δ) and (UV) HMDU on a preparative ODS HPLC column with water as eluent.

in our laboratory. These are used in analyses of irradiated cellular DNA. The compounds synthesized in this laboratory include TG, HMH, FPU, dTG, and MH. HMU and HMdU are commercially available. T and dT are included as internal markers of retention time.

DISCUSSION

The data here presented show that HPLC analysis can be successfully applied to the identification of radiation-induced thymine derivatives in cellular DNA. As shown in Fig. 8, modified bases can be separated from modified nucleosides and from unmodified T and dT. To further corroborate their presence in cellular DNA, we have derivatized them by acetylation and controlled oxidation of TG leading to formation of FPU which is then cyclized to HMH. This last compound can be either acetylated or dehydrated to form MH.

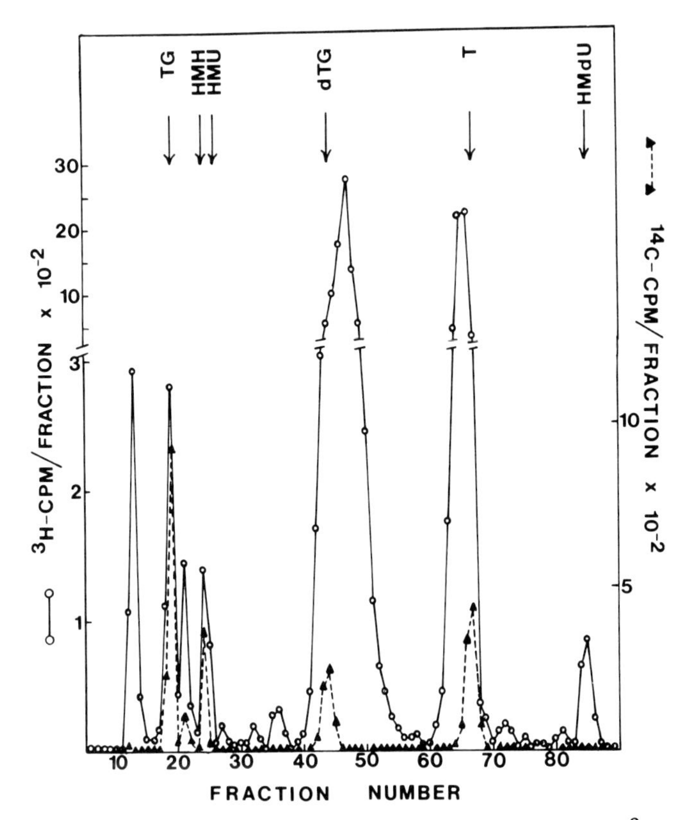

Fig. 7. Chromatography of enzymatic hydrolysate of [^{3}H-6]dT-containing DNA (o) isolated from γ-irradiated HeLa cells with marker [^{14}C]TG, HMH, dTG, and T (Δ) and (UV) HMDU on a preparative ODS HPLC column with water as eluent.

The identification of HMdU in [^{3}H-methyl]dT-containing DNA is the first demonstration of a unique ^{3}H transmutation product formed in cellular DNA. Since the use of [^{3}H-methyl]dT is widespread and HMdU content increases as the rate of beta decay of tritium, analyses of radioactive DNA not freshly prepared may lead to false results.

The demonstration of the formation of HMdU in cellular DNA as the result of ionizing radiation is of considerable interest, since it has been shown that when exogenously administered to mice, HMdU caused diarrhea and leukopenia (38), indicating that it was cytostatic and cytotoxic. HMdU was also cytostatic to several cell lines *in vitro* (38). Although the amount of HMdU formed by ionizing radiation was small in our experiments, it is possible that HMdU contributes to the cytotoxic effects of ionizing radiation. This toxicity may be due to the formation of cross-links between DNA and nuclear protein as the result of HMdU acting as a weak alkylating

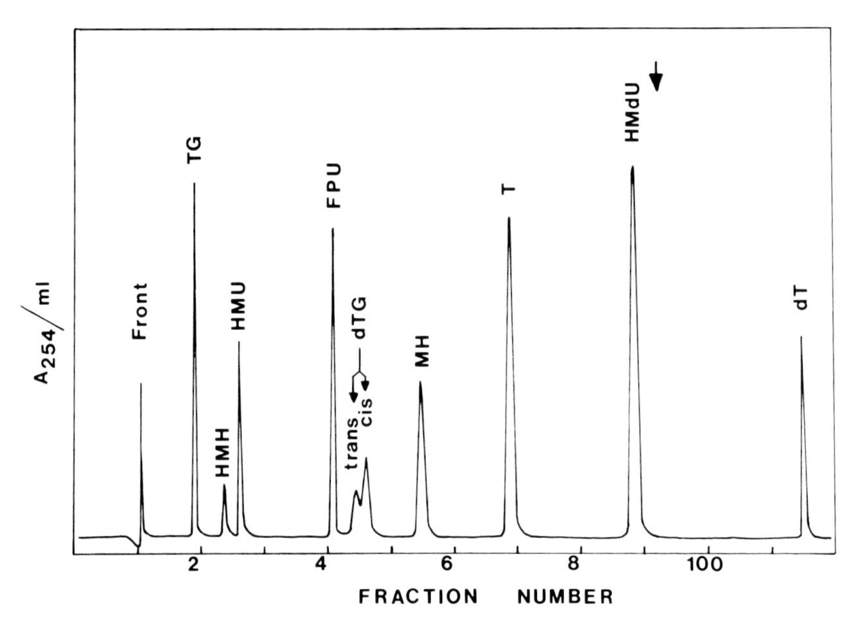

Fig. 8. Chromatography of a mixture of thymine derivatives; thymine glycol (TG) 5-hydroxy-5-methyl hydantoin (HMH), 5-hydroxymethyl uracil (HMU), N'-formyl-N-pyruvylurea (FPU), thymidine glycol (dTG), 5-methylene hydantoin, thymine (T), 5-hydroxymethyl 2'-deoxyuridine (HMdU), and thymidine (dT) on a preparative ODS HPLC column with water as eluent.

agent. The formation of cross-links between DNA and non-histone proteins has been demonstrated in irradiated chromatin (39). The DNA of cells treated with HMdU was reported as being difficult to separate from nucleoprotein and the authors of that report suggested that cross-linking might have occurred (40).

Although the etiologic relationship of radiation-induced DNA base modification to cancer is not yet established, the presence of modified bases in DNA may serve as markers of exposure to radiation. As such, they may aid in identifying populations at risk for the carcinogenic effects of ionizing radiation.

ACKNOWLEDGEMENT

Supported by PHS grants CA16669 and ES02234.

REFERENCES

1. Cerutti, P.A. (1976) Base damage induced by ionizing radiation.

In Photochemistry and Photobiology of Nucleic Acids, S.Y. Wang, ed. Academic Press, New York, Vol. II, pp. 375-401.

2. Little, J.B. (1978) Biological consequences of X-ray induced DNA damage and repair processes in relation to cell killing and carcinogenesis. In DNA Repair Mechanisms, P.C. Hanawalt, E.C. Friedberg, and C.F. Fox, eds. Academic Press, New York and London, pp. 701-711.
3. Scholes, G., J. Weiss, and C.M. Wheeler (1956) Formation of hydroperoxides from nucleic acids by irradiation with X-rays in aqueous systems. Nature 178:157.
4. Myers, Jr., L.S., J.F. Ward, W.T. Tsukamoto, D.E. Holmes, and J.R. Juka (1965) Radiolysis of thymine in aqueous solutions: Change in site of attack with change in pH. Science 148:1234-1235.
5. Cadet, J., and R. Teoule (1975) No. 165 - Radiolyse gamma de la thymidine en solution aqueuse aérée. II. - Caractérisation des produits stables. Bull. Soc. Chim. Fr. 3-4:885-890.
6. Scholes, G. (1976) The radiation chemistry of pyrimidines, purines, and related substances. In Photochemistry and Photobiology of Nucleic Acids, S.Y. Wang, ed. Academic Press, New York, Vol. I, pp. 521-577.
7. Cadet, J., and R. Teoule (1978) Comparative study of oxidation of nucleic acid components by hydroxyl radicals, singlet oxygen and superoxide anion radicals. Photochem. Photobiol., 28:661-667.
8. Hariharan, P.V., and P.A. Cerutti (1972) Formation and repair of γ-ray induced thymine damage in Micrococcus radiodurans. J. Mol. Biol. 66:65-81.
9. Teoule, R., C. Bert, and A. Bonicel (1977) Thymine fragment damage retained in the DNA polynucleotide chain after gamma irradiation in aerated solutions. III. Radiat. Res. 72:190-200.
10. Teoule, R., A. Bonicel, C. Bert, J. Cadet, and M. Polverelli (1974) Identification of radioproducts resulting from the breakage of thymine moiety by gamma irradiation of E. coli DNA in an erated aqueous solution. Radiat. Res. 57:46-58.
11. Hariharan, P.V., and P.A. Cerutti (1971) Repair of γ-ray-induced thymine damage in Micrococcus radiodurans. Nature New Biol. 229:247-249.
12. Painter, R.B., and B.R. Young (1972) Repair replication in mammalian cells after X-irradiation. Mutat. Res. 14:225-235.
13. Mattern, M.R., P.V. Hariharan, B.E. Dunlap, and P.A. Cerutti (1973) DNA degradation and excision repair in γ-irradiated Chinese hamster ovary cells. Nature New Biol. 245:230-232.
14. Mattern, M.R., and G.P. Welch (1979) Production and excision of thymine damage in the DNA of mammalian cells exposed to high-LET radiations. Radiat. Res. 80:874-883.
15. Hariharan, P.V., and P.A. Cerutti (1974) Excision of damaged thymine residues from gamma-irradiated poly (dA-dT) by crude extracts of Escherichia coli. Proc. Natl. Acad. Sci., U.S.A. 71:3532-3536.

16. Gates, F.P., and S. Linn (1977) Endonuclease from Escherichia coli that acts specifically upon duplex DNA damaged by ultraviolet light, osmium tetroxide, acid or X-rays. J. Biol. Chem. 252:2802-2807.
17. Nes, I.F., and J. Nissen-Meyer (1978) Endonuclease activities from a permanently established mouse cell line that act upon DNA damaged by ultraviolet light, acid and osmium tetroxide. Biochim. Biophys. Acta 520:111-121.
18. Kochetkov, N.K., and E.I. Budovskii, eds. (1972) Organic Chemistry of Nucleic Acids, Part B. Plenum Press, London and New York, p. 288.
19. Frenkel, K., M.S. Goldstein, N. Duker, and G.W. Teebor (1981) Identification of the cis-thymine glycol moiety in oxidized deoxyribonucleic acid. Biochem. 20:750-754.
20. Demple, B., and S. Linn (1980) DNA N-glycosylases and UV repair. Nature 287:203-208.
21. Breimer, L., and T. Lindahl (1980) A DNA glycosylase from Escherichia coli that releases free urea from a polydeoxyribonucleotide containing fragments of base residues. Nucl. Acid Res. 8:6199-6211.
22. Ward, J.R., and I. Kuo (1976) Strand breaks, base release, and post-irradiation changes in DNA γ-irradiated in dilute O_2-saturated aqueous solution. Radiat. Res. 66:485-498.
23. Dunlap, B., and P. Cerutti (1975) Apyrimidinic sites in gamma-irradiated DNA. FEBS Lett. 51:188-190.
24. Scholes, G., J.F. Ward, and J. Weiss (1961) Mechanism of the radiation-induced degradation of nucleic acids. J. Mol. Biol. 2:379-391.
25. Rhaese, H.-J., and E. Freese (1968) Chemical analysis of DNA alterations. I. Base liberation and backbone breakage of DNA and oligodeoxyadenylic acid induced by hydrogen peroxide and hydroxylamine. Biochim. Biophys. Acta 155:476-490.
26. Frenkel, K., M.S. Goldstein, and G.W. Teebor (1981) Identification of the cis-thymine glycol moiety in chemically oxidized and gamma-irradiated DNA by HPLC analysis. Biochem. 20:7566-7571.
27. Teebor, G.W., K. Frenkel, and M.S. Goldstein (1982) Characterization of thymine damage in oxidized and gamma-irradiated DNA. Prog. Mutat. Res. 4:301-311.
28. Teebor, G.W., K. Frenkel, and M.S. Goldstein (1982) Identification of radiation-induced thymine derivatives in DNA. Adv. Enz. Reg. 20:39-52.
29. Frenkel, K., M.S. Goldstein, and G. Teebor (1982) Identification of radiation-induced thymine derivatives in cellular DNA. Proc. 73rd Annual Meeting American Assoc. Cancer Res., St. Louis, Missouri 23:6.
30. Teebor, G.W., M.S. Goldstein, and K. Frenkel (1982) Radiation-induced thymine modification in cellular DNA: Differentiation of transmutational and ionization effects of ^{3}H-thymidine. Fed. Proc. Vol. 41 #3, p. 686.

31. Iida, S., and H. Hayatsu (1970) The permanganate oxidation of thymidine. Biochim. Biophys. Acta 213:1-13.
32. Iida, S., and H. Hayatsu (1971) The permanganate oxidation of deoxyribonucleic acid. Biochim. Biophys. Acta 240:370-375.
33. Murahashi, S., H. Yuki, K. Kosai, and F. Doura (1966) Methylenehydantoin and related compounds. I. On the reaction of pyruvic acid and urea: The synthesis of 5-methylenehydantoin. Bull. Chem. Soc. Japan Vol. 39 #7, pp. 1559-1662.
34. Marmur, J. (1961) A procedure for the isolation of deoxyribonucleic acid from microorganisms. J. Mol. Biol. 3:208-218.
35. Klein, P.D. (1966) Isotope fractionation of large molecules. Adv. Chromatogr. 3:3-65.
36. Weinstein, I.B., A.M. Jeffrey, K.W. Jennette, S.H. Blobstein, R.G. Harvey, C. Harris, H. Autrup, H. Kasai, and N. Nakanishi (1976) Benzo(a)pyrene diolepoxides as intermediates in nucleic acid binding in vitro and in vivo. Science 193:592-595.
37. Feinendegen, L.E., and V.P. Bond (1973) Transmutation versus beta irradiation in the pathological effects of tritium decay. In Tritium, A.A. Moghissi and M.W. Carter, eds. Messenger Graphix, Phoenix, pp. 221-231.
38. Waschke, S., J. Reefschlager, D. Barwolff, and P. Langen (1975) 5-Hydroxymethyl-2'-deoxyuridine, a normal DNA constituent in certain Bacillus subtilis phages is cytostatic for mammalian cells. Nature 225:629-630.
39. Olinsky, R., R.C. Briggs, L.S. Hnilica, J. Stein, and G. Stein (1981) Gamma-radiation-induced cross-linking of cell-specific chromosomal non-histone protein-DNA complexes in HeLa chromatin. Radiat. Res. 86:102-114.
40. Matthes, E., D. Barwolff, and P. Langen (1978) Altered DNA-protein interactions induced by 5-hydroxymethyldeoxyuridine in Ehrlich ascites carcinoma cells. Stud. Biophys. 67:115-116.

ABERRANT METHYLATION OF DNA AS AN EFFECT OF CYTOTOXIC AGENTS

Ronald C. Shank

Department of Community and Environmental Medicine
University of California, Irvine, and
University of California Southern Occupational Health Center
Irvine, California 92717

INTRODUCTION

Hydrazine (H_2N-NH_2) is a strong reducing agent, widely used in large amounts as an anticorrosive agent, as an intermediate in chemical syntheses, and as a propellant fuel. The compound at a single, high dose is neurotoxic, producing convulsions and respiratory arrest; at a lower single dose, hydrazine produces marked damage to the liver and kidneys (1,2). In a number of *in vitro* and *in vivo* tests, hydrazine, at relatively low concentration (dose), was positive as a mutagen (3-7). Chronic oral administration of hydrazine, at doses which depressed growth rate, produced pulmonary carcinomas in mice (8) and hepatocellular carcinomas and lung adenocarcinomas in rats and mice (9). When administered by inhalation at four different doses to rats, mice, and hamsters for 6 hr/d, 5 d/wk for 1 yr, hydrazine (only at the maximum tolerated dose) induced microscopic squamous cell carcinomas in the nasal turbinates of rats but no tumors in mice or hamsters (10). Hydrazine is capable of reacting *in vitro* under severe conditions with pyrimidine bases in DNA nucleotides to form the dihydro- or 4-hydrazino-derivatives or to open the pyrimidine rings (11). No one has yet been able to demonstrate a hydrazine-DNA adduct under *in vivo* conditions.

HYDRAZINE AND METHYLATION OF DNA

Recent studies on the metabolism of hydrazine demonstrated that administration of a near LD_{50} of hydrazine to fasted rats resulted in the formation of 7-methylguanine and O^6-methylguanine in liver DNA (12, 13). Fischer 344 rats were given orally 90 mg hydrazine/kg

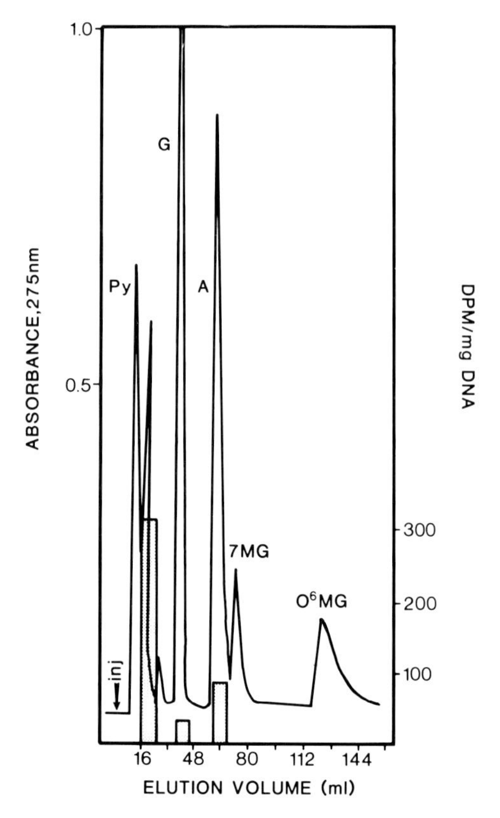

Fig. 1. Elution profile of liver DNA hydrolysate prepared from a control rat given 0.1 M HCl per os and (^{3}H-methyl)methionine ip and killed 6 hr later. DNA was hydrolyzed in 0.1 M HCl; authentic 7-methylguanine and O^6-methylguanine were added to the hydrolysate as carrier, and the hydrolysate was fractionated by high performance liquid chromatography. Elution order: pyrimidine oligonucleotides (Py); guanine (G); adenine (A); 7-methylguanine (7-MG); and O^6-methylguanine (O^6-MG). Amount of radioactivity in each assayed fraction is represented by a solid bar.

body wt in 0.1 M HCl. At the same time the animals were given 20 µCi (^{14}C-methyl)methionine by the intraperitoneal route to label the hepatic pool of S-adenosylmethionine; the radioactive methionine injections were repeated hourly until the animals were decapitated 5 hr after the administration of hydrazine. Liver DNA was isolated, purified, and hydrolyzed for analysis by high performance liquid chromatography. Authentic 7-methylguanine and O^6-methylguanine were added to the hydrolysates as carrier and these fractions were collected after chromatographic separation and assayed for radioactivity. The 7-methylguanine fractions of DNA hydrolysates prepared

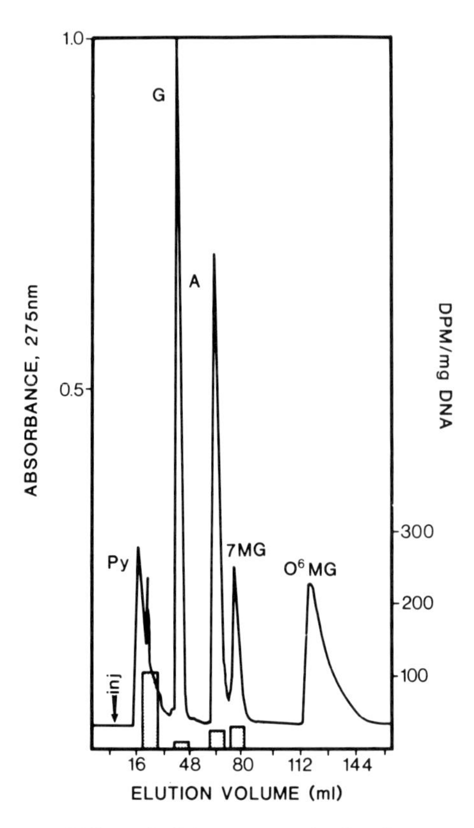

Fig. 2. Elution profile of liver DNA hydrolysate prepared from a treated rat given 60 mg hydrazine/kg body wt in 0.1 M HCl per os and (^{3}H-methyl)methionine ip and killed 6 hr later. Method described in legend to Fig. 1.

from treated, but not from control, rats were radioactive (Figures 1 and 2), suggesting that administration of hydrazine resulted in the aberrant methylation of DNA and that the source of the methyl group was S-adenosylmethionine. Quantitative studies further demonstrated that both 7-methylguanine and O^6-methylguanine formed rapidly and in a ratio of approximately 10 to 1 in liver DNA from poisoned rats, mice, hamsters, and guinea pigs. Kinetic studies in the rat showed that half-maximal levels of the methylation of DNA were achieved 30 min to 45 min after administration of hydrazine and that these bases were removed from DNA at the same rates as were 7-methylguanine and O^6-methylguanine from liver DNA of rats treated with methylating carcinogens, such as dimethylnitrosamine or 1,2-dimethylhydrazine. The O^6-methylguanine was removed much more slowly from liver DNA in hamsters than in rats, which is also consistent with reported slower rates of repair of O^6-methylguanine in hamster liver DNA after treatment with dimethylnitrosamine (14).

The log dose-response curves for methylation of guanine in liver DNA following administration of hydrazine to rats, mice, and hamsters are curvilinear; the relationship becomes linear when the log of the dose is related to the log of the amount of aberrant methylation of DNA, implying that the reaction is dependent upon both the amount of hydrazine reaching the target and the amount or state of the target itself. At the lower doses, where the log dose-response curve is nearly flat, little cytotoxicity is apparent in the liver; at higher doses, where the log dose-response curve becomes exponential, fatty accumulation and periportal and focal necrosis become widespread. This was interpreted as indicating that the methylation response may be related to the cytotoxicity induced by hydrazine, and the experiments have been extended to examine liver DNA from animals treated with a variety of hepatotoxins. To date, the response to carbon tetrachloride, ethanol, thioacetamide, ethionine, dimethylnitrosamine, diethylnitrosamine, nitrosopyrrolidine, aflatoxin B_1, yellow phosphorus, and puromycin has been examined. The S-adenosylmethionine-dependent methylation of guanine in liver DNA has been quantitated reproducibly with dimethylnitrosamine and N-nitrosopyrrolidine; the methylation response has also been seen, but with poor reproducibility, with carbon tetrachloride, ethanol, thioacetamide, aflatoxin B_1, yellow phosphorus, and puromycin, but has never been seen in animals treated with diethylnitrosamine or ethionine. The methylation response appears to be quite sensitive, not only to dose, but also to concentration of the administered hepatotoxin, and the appropriate conditions for reproducible methylation of DNA for these latter compounds have not yet been elucidated.

TOXICITY AND DNA METHYLATION

Methylation of liver DNA in response to a hepatotoxin other than hydrazine can be illustrated by the case of N-nitrosopyrrolidine, a cyclic carcinogen for which no one has yet been able to demonstrate a covalent DNA adduct _in vivo_ until recently. Administration of N-nitrosopyrrolidine results in two DNA adducts, one of which has been identified as 7-methylguanine (15). When rats are treated with (^{14}C-2,5)N-nitrosopyrrolidine, one intensely fluorescent DNA adduct can be isolated and this adduct contains radiolabel; 7-methylguanine can also be isolated and this fraction is not labeled and therefore does not derive from nitrosopyrrolidine. If the experiment is repeated with unlabeled nitrosopyrrolidine and the animals are co-treated with multiple hourly pulses of (^{3}H-methyl)-methionine, the same two DNA adducts are observed in the liver DNA; however, in this case the radiolabel is in the 7-methylguanine fraction and not in the nitrosopyrrolidine-DNA adduct, indicating that the methylation of guanine in the 7-position of DNA in livers of animals treated with nitrosopyrrolidine derives from S-adenosylmethionine and not from the carcinogen itself.

PROPOSED MECHANISMS

This interesting aberrant methylation of liver DNA response to administration of hydrazine and some other hepatotoxins suggests a variety of hypotheses to explain the mechanism of action. All studies completed so far strongly support that S-adenosylmethionine is the source of the methyl group in this response. Labeling the S-adenosylmethionine hepatic pool with radiolabeled methionine labels the resultant 7-methylguanine and O^6-methylguanine in the DNA. Depletion of hepatic levels of S-adenosylmethionine by administration of a high dose of ethionine completely blocks the DNA methylation response to administration of hydrazine; as levels of S-adenosylmethionine return to normal in the recovery from ethionine poisoning, the DNA methylation response to administration of hydrazine also returns. Barrows and Magee (16) and Lindahl (17) have recently shown that S-adenosylmethionine incubated with calf thymus DNA acts as a direct methylating agent, forming non-enzymatically 7-methylguanine, O^6-methylguanine, and 3-methyladenine. This suggested that hydrazine or hydrazine toxicity may induce the non-enzymatic methylation of DNA by S-adenosylmethionine, perhaps by greatly increasing the intracellular concentration of this important biochemical intermediate, or by altering the intracellular distribution of S-adenosylmethionine. Our studies have shown that administration of hydrazine at a high dose causes a slight (approximately 10%) decrease in hepatic concentrations of S-adenosylmethionine, not an increase; studies on the effect of administration of hydrazine on the intracellular distribution of S-adenosylmethionine have not been done. It has been established that the guinea pig is unable to regulate the levels of S-adenosylmethionine as well as most other experimental animals (18). Guinea pigs in our laboratory were given 6.6 mmol methionine/kg body wt and, 4 hr later, 90 mg hydrazine/kg body wt; one hour later the animals were killed and liver S-adenosylmethionine and levels of DNA methylation were measured. Even though administration of methionine resulted in a 20-fold increase in hepatic levels of S-adenosylmethionine, no aberrant methylation of DNA could be detected unless hydrazine had also been administered, and then there was a 40% inhibition of the methylation response (Table 1). These results suggest that S-adenosylmethionine itself may not directly methylate DNA in normal or hydrazine-treated animals.

Perhaps the most obvious mechanism that could explain the aberrant methylation of liver DNA in hydrazine-treated animals is that S-adenosylmethionine could first methylate hydrazine itself, producing monomethylhydrazine (CH_3NH-NH_2). Monomethylhydrazine was shown by Hawks and Magee in 1974 (19) to be metabolized to a methylating agent in the mouse. In studies not yet published, we have shown in the rat that more methylation of DNA results after administration of hydrazine than after administration of an equitoxic dose of monomethylhydrazine. Moreover, administration of equimolar doses of the two compounds to mice results in more

Table 1. S-adenosylmethionine and DNA methylguanine levels in liver after administration of methionine and hydrazine to guinea pigs.

Group	AdoMet* (nmol/g liver)	Methylguanines (μmol/mol DNA guanine 7-methylguanine	O^6-methylguanine
Control	44	N.D.**	N.D.
	24	N.D.	N.D.
	mean = 34		
Hydrazine	14	396	27
	15	328	25
	mean = 15	mean = 362	mean = 26
Methionine	706	N.D.	N.D.
	638	N.D.	N.D.
	mean = 672		
Methionine-Hydrazine	476	280	18
	441	160	13
	mean = 458	mean = 220	mean = 16

*AdoMet = S-adenosylmethionine
**N.D. = none detected

methylation of DNA in the hydrazine-treated animals. In fact, administration of monomethylhydrazine to hamsters at doses which are lethal to this species in 24-48 hr did not result in methylation of liver DNA, while an equimolar dose of hydrazine to hamsters resulted in rapid methylation of almost 0.1% of the guanine of liver DNA.

It could be argued that hydrazine is better able to enter the hepatocyte than is monomethylhydrazine and that administration of the former could result in a higher intracellular concentration of monomethylhydrazine than administration of the parent compound itself. Early studies suggest that both compounds distribute with the body water (1,20), and our laboratory (21) has shown that washed liver slices from rat metabolize (^{14}C)monomethylhydrazine and (^{14}C)glucose to $^{14}CO_2$ at approximately the same rate, suggesting that monomethylhydrazine enters the hepatocyte rapidly. No reliable analytical methods are yet available for the quantitative determination of both hydrazine and monomethylhydrazine together in biological fluids, so the direct experiment on intracellular concentrations of the two compounds cannot yet be done. Also, route of administration has little effect on the toxicity of monomethylhydrazine (1), which suggests that adsorption of these compounds plays little role in the toxicity. Other studies are in progress to test the hypothesis that monomethylhydrazine is an intermediate in the methylation of liver DNA in hydrazine-treated animals, but the evidence available at this time does not support the hypothesis.

Another hypothesis suggested (13) to explain aberrant methylation in hydrazine toxicity points out that there is a normal pathway for methylation of DNA, the normal methylation of cytosine by DNA methyltransferase. This enzyme recognizes a specific sequence of cytosine and guanine in the DNA polymer, and, using S-adenosylmethionine, methylates the 5-position of cytosine; this methylation site is adjacent to the 7-position of guanine, and it may be that in hydrazine toxicity the DNA methyltransferase loses some specificity and methylates this near neighbor, which is the most nucleophilic site in the DNA polymer. Alternatively, hydrazine toxicity could alter the stereochemistry of DNA itself, resulting in methylation of abnormal sites on the polymer. Studies are underway to determine the effect of hydrazine on the specificity of the normal methylating enzyme.

There may be some teleological basis for the observation that the methylated bases found in DNA of poisoned animals are 7-methylguanine and O^6-methylguanine, for these bases, although not known today to be normal constituents of mammalian DNA, are the substrates for specific proteins which are normal constituents of mammalian cells, namely, 7-methylguanine glycosylase (22-24) and O^6-methylguanine demethylase (25,26). It may be, then, that 7-methylguanine and O^6-methylguanine are normal bases in mammalian DNA, present transiently at low concentration and perhaps involved in some regulatory function; the enzymes and proteins could effect the rapid removal of these potentially pro-mutagenic bases once their regulatory function is over and before DNA replication takes place. It has recently been shown that these DNA "repair" proteins have increased activity during S phase in rat liver cells (27,28).

CONCLUSION

These investigations on aberrant methylation of DNA during hepatotoxicity, although in progress for several years now, are still in the preliminary phase. The process for the methylation of DNA during hydrazine toxicity is well established and several studies are in progress in attempt to elucidate the mechanism of action. The studies on hepatotoxins other than hydrazine have progressed slowly and therefore little can be said at this time on the universality of aberrant methylation of DNA in response to cytotoxicity itself. The working hypothesis in the laboratory is this: administration of a hepatotoxin at cytotoxic levels results in: 1) methylation of guanine in DNA, and 2) restorative regeneration of liver cells forcing widespread replication of liver DNA. If the dosing is repeated often, the opportunity for replication of damaged (alkylated and partially apurinic) DNA is increased and such an event may lead to transformation of liver cells and eventually formation of hepatocellular carcinomas. This process would be analogous to that suggested by the Millers (29) for the initiation of the carcinogenic process by chemicals which form highly reactive electrophiles in target cells and covalently bind to the DNA in that organ. The only differences between the two processes would be that one group of chemical carcinogens would be able to react with DNA directly and at relatively low concentrations, while the second group would have to react indirectly, by stimulating S-adenosylmethionine-dependent methylation of DNA, which presumably requires doses of the agent sufficiently high to cause cytotoxicity.

The proposed hypothesis, then, would serve to unify the covalent DNA adduct theory in chemical carcinogenesis and would offer a potentially important warning in studies on early biological markers in cancer studies: the inability of a chemical to combine covalently with DNA may not be sufficient evidence to conclude that DNA adducts, and DNA damage, do not occur; the liver cells, at least, appear capable of using an endogenous agent to achieve that end, given sufficient provocation.

REFERENCES

1. Witkin, L.B. (1956) Acute toxicity of hydrazine and some of its methylated derivatives. AMA Arch. Ind. Health 13:34-36.
2. Weir, F.W., J.H. Nemenzo, S. Bennett, and F.H. Meyers (1964) A study on the mechanism of acute toxic effects of hydrazine, UDMH, MMH, and SDMH. Aerospace Medical Research Laboratories Technical Report TDR-64-26, Wright-Patterson Air Force Base, Ohio.
3. Jain, H.K., and P.T. Shukla (1972) Locus specificity of mutations in Drosophila. Mut. Res. 14:440-442.
4. Rohrborn, G., P. Propping, and W. Buselmaier (1972) Mutagenic activity of isoniazid and hydrazine in mammalian test systems. Mut. Res. 16:189-194.
5. Kimball, R.F., and B.F. Hirsch (1975) Test for the mutagenic action of a number of chemicals on Haemophilus influenzae with special emphasis on hydrazine. Mut. Res. 30:9-20.
6. Herbold, B., and W. Buselmaier (1976) Induction of point mutations by different chemical mechanisms in the liver microsomal assay. Mut. Res. 40:73 84.
7. MacRae, W.D., and H.F. Stich (1979) Induction of sister-chromatid exchanges in Chinese hamster ovary cells by thiol and hydrazine compounds. Mut. Res. 68:351-365.
8. Biancifiori, C., and R. Ribacchi (1962) Pulmonary tumours induced by oral isoniazid and its metabolites. Nature 194:488-489.
9. Severi, L., and C. Biancifiori (1968) Hepatic carcinogenesis in CBA/Cb/Se mice and Cb/Se rats by isonicotinic acid hydrazide and hydrazine sulfate. J. Natl. Cancer Inst. 41:331-349.
10. MacEwen, J.D., E.H. Vernot, C.C. Haun, E.R. Kinkaed, and A. Hall (1981) Chronic inhalation toxicity of hydrazine: Oncogenic effects. Proc. 10th Conf. Environ. Toxicol. AFAMRL-TR-81-56, Air Force Aerospace Medical Research Laboratory, Wright-Patterson Air Force Base, Ohio, pp. 1-63.
11. Lingens, F., and H. Schneider-Bernloehr (1965) Reaction of naturally-occurring pyrimidine bases with hydrazine and methyl-substituted hydrazines. Annalen der Chemie 686:134-144.

12. Barrows, L.R., and R.C. Shank (1981) Aberrant methylation of liver DNA in rats during hepatotoxicity. Toxicol. and Appl. Pharmacol. 60:334-345.
13. Becker, R.A., L.R. Barrows, and R.C. Shank (1981) Methylation of liver DNA guanine in hydrazine hepatotoxicity: Dose-response and kinetic characteristics of 7-methylguanine and O^6-methylguanine formation and persistence in rats. Carcinogenesis 2:1181-1188.
14. Stumpf, R., G.P. Margison, R. Montesano, and A.E. Pegg (1979) Formation and loss of alkylated purines from DNA of hamster liver after administration of dimethylnitrosamine. Cancer Res. 39:50-54.
15. Hunt, E.J., and R.C. Shank (1982) Evidence for DNA adducts in rat liver after administration of N-nitrosopyrrolidine. Biochem. Biophys. Res. Comm. 104:1343-1348.
16. Barrows, L.R., and P.N. Magee (1982) Nonenzymatic methylation of DNA by S-adenosylmethionine in vitro. Carcinogenesis 3:349-351.
17. Rydberg, B., and T. Lindahl (1982) Nonenzymatic methlation of DNA by the intracellular methyl group donor S-adenosyl-l-methionine is a potentially mutagenic reaction. EMBO J. 1:211-216.
18. Hardwick, D.F., D.A. Applegarth, D.M. Cockcroft, P.A. Ross, and R.J. Calder (1970) Pathogenesis of methionine-induced toxicity. Metabolism 19:318-391.
19. Hawks, A., and P.N. Magee (1974) The alkylation of nucleic acids of rat and mouse in vivo by the carcinogen 1,2-dimethylhydrazine. Br. J. Cancer 30:440-447.
20. Dost, F.N., D.L. Springer, B.M. Krivak, and D.J. Reed (1979) Metabolism of hydrazine. Aerospace Med. Res. Lab. Tech. Report, AMRL-TR-79-43, Wright-Patterson Air Force Base, Ohio.
21. Shank, R.C. (1979) Comparative metabolism of propellant hydrazine. Aerospace Med. Res. Lab. Tech. Report, AMRL-TR-79-57, Wright-Patterson Air Force Base, Ohio.
22. Laval, J., J. Piere, and F. Laval (1981) Release of 7-methylguanine residues from alkylated DNA by extracts of Micrococcus leutes and Escherichia coli. Proc. Natl. Acad. Sci., USA 78:852-855.
23. Singer, B., and T.P. Brent (1981) Human lymphoblasts contain DNA glycosylase activity excising N-3 and N-7 methyl and ethyl purines but not O^6-alkylguanines or 1-alkyladenines. Proc. Natl. Acad. Sci., USA 78:856-860.
24. Margison, G.P., and A.E. Pegg (1981) Enzymatic release of 7-methylguanine from methylated DNA by rodent liver extracts. Proc. Natl. Acad. Sci., USA 78:861-865.
25. Pegg, A.E. (1978) Enzymatic removal of O^6-methylguanine from DNA by mammalian cell extracts. Biochem. Biophys. Res. Comm. 84:166-173.
26. Craddock, V.M., A.R. Henderson, and S. Gash (1982) Nature of the constitutive and induced mammalian O^6-methylguanine DNA repair enzyme. Biochem. Biophys. Res. Comm. 107:546-553.
27. Rabes, H.M., R. Kerler, R. Wilhelm, G. Rode, and H. Riess (1979) Alkylation of DNA and RNA by ^{14}C-dimethylnitrosamine in

hydroxyurea-synchronized regenerating rat liver. Cancer Res. 39:4228-4236.

28. Pegg, A.E., W. Perry, and R.A. Bennett (1981) Effect of partial hepatectomy on removal of O^6-methylguanine from alkylated DNA by rat liver extracts. Biochem. J. 197:195-201.
29. Miller, J.A. (1970) Carcinogenesis by chemicals: An overview - G.H.A. Clowes memorial lecture. Cancer Res. 30:559-576.

IDENTIFICATION OF CARCINOGEN-DNA ADDUCTS BY IMMUNOASSAYS

Miriam C. Poirier, Juichiro Nakayama,
Frederica P. Perera*, I. Bernard Weinstein*, and
Stuart H. Yuspa

Laboratory of Cellular Carcinogenesis
and Tumor Promotion
National Cancer Institute
National Institutes of Health
Bethesda, Maryland 20205

INTRODUCTION

In recent years several laboratories have elicited highly avid rabbit and monoclonal antisera specific for particular three-dimensional structures of DNA damaged by physical and chemical agents (1,2). These antisera have been utilized as probes for the presence of carcinogens bound to DNA in various animal and cell-culture model systems in order to investigate mechanisms of carcinogenesis (3-6). Immunoassays established with such antisera are among the most highly sensitive quantitative procedures known for the determination of carcinogen-DNA adducts (1,2) and have been adapted also for morphological localization studies (7,8,9). In addition, since the techniques permit measurement of unlabeled DNA-bound products, the possibility of searching for evidence of chemical exposure in DNA from human tissues has become exceedingly attractive. Such human studies are of recent inception, however, and the data are sparse. This presentation will focus on quantitation by immunoassay of DNA adducts (Figure 1) of 2-acetylaminofluorene (2-AAF) and benzo(a)-pyrene (BP) in cultured cells and intact animals. In addition, preliminary evidence will be presented indicating that BP-DNA adducts may be measured in lung and lung-tumor DNA obtained from lung cancer patients.

---

* Division of Environmental Sciences, School of Public Health, Columbia University College of Physicians and Surgeons, New York, New York, 10032.

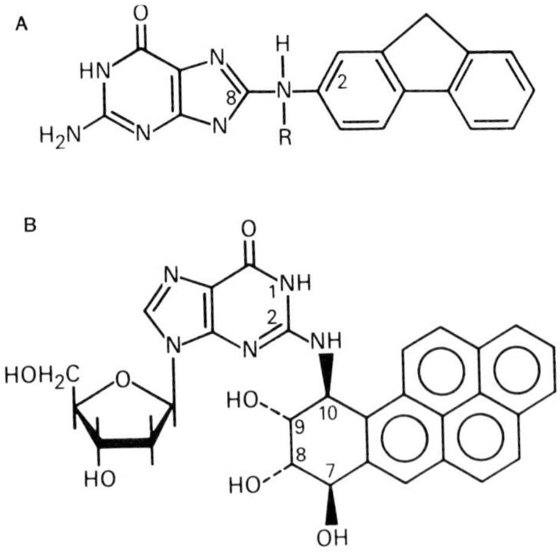

Fig. 1. The major *in vivo* adducts formed upon interaction of (A) 2-acetyl-aminofluorene or (B) benzo(a)pyrene with DNA. When R = H, the (A) structure is guanosin-(8-yl)-aminofluorene (G-8-AF). When R = acetyl, the (A) structure is guanosin-(8-yl)-acetylaminofluorene, (G-8-AAF). The (B) structure is trans-(7R)-N^2- [10-(7β,8α,9α)-trihydroxy-7,8,9,10-tetrahydrobenzo(a)pyrene)-yl]-deoxyguanosine.

MATERIALS AND METHODS

Anti-guanosin-(8-yl)-acetylaminofluorene

Preparation of the immunogen, immunization of rabbits, antibody characteristics, and details of the radioimmunoassay (RIA) procedure have been described previously (4,10,11). Procedures for the culture of primary epidermal cells, exposure to N-acetoxy-acetylaminofluorene (N-Ac-AAF) and the preparation of DNA on cesium chloride gradients are in references 12, 13, and 14. DNA preparations were the same for cells and mouse and human tissues used in all of these studies.

Anti-guanosin-(8-yl)-aminofluorene

This antiserum was elicited against guanosin-(8-yl)-aminofluorene (a gift of F. A. Beland) coupled to bovine serum albumin in a fashion analogous to that previously described for the anti-G-8-AAF (10). Rabbits were injected intramuscularly with 70.5 mg of adduct on four weekly occasions and bled at weekly intervals between two and four months after the first injection. The enzyme-linked immunosorbent assay (ELISA) was performed essentially as reported for the BP-DNA antibody (15,16), with specifics shown in the legend to Figure 2. The RIA was performed with a ^3H-G-8-AF tracer prepared by deacetylation of ^3H-G-8-AAF using conditions previously described (17).

Anti-BP-DNA

Preparation of the BP immunogen, immunization, and characteristics of the antiserum have been published previously (18). Detailed procedures for the BP-DNA ELISA have been described (15,16). The arrangements for procurement of human tissue samples, and details concerning source, maintenance, and intraperitoneal injection procedure for BALB/c mice are in ref. 16. For experiments with mouse skin, BALB/c mice were maintained similarly and BP was applied topically during a resting phase of the hair cycle in 200 μl of acetone to a shaved dorsal of 6 cm^2. DNA was prepared from the heat-separated epidermis (19).

RESULTS AND DISCUSSION

Antibody Characteristics and Immunoassay Procedures

Highly avid antisera elicited against either individual nucleoside adducts coupled covalently to carrier proteins or chemically modified DNAs electrostatically attracted to methylated proteins have affinity constants in the range of 10^8-10^{10} L/mole (1,2). Antibody characterization indicates antigenic specificity directed towards the combination of carcinogen bound to base or carcinogen bound to DNA, with little or no affinity for either the carcinogen or nucleic acid alone. Based on a three-dimensional recognition of adduct structure, specificity towards adducts of one carcinogen which differ slightly (by a hydroxyl, methyl, or acetyl group) is frequently similar, so that competition occurs in the same concentration range as the immunogen. On the other hand, adducts of one carcinogen which are structurally very different from each other will not usually be recognized by the same antiserum. Thus, these antisera are specific probes for both differences in carcinogen structure and differences in adduct structure.

The most useful quantitative assay techniques in which these antibodies have been employed are competitive radioimmunoassays (RIA) and enzyme-linked immunosorbent assays (ELISA), both of which can be performed with many variations (1,2). In competitive immunoassays, two chemically identical haptens compete in a concentration-dependent fashion for an antibody binding site. One of these haptens can be identified by a secondary variable (radioactivity in the case of RIA, and binding to a microtiter well for ELISA), and is always employed in a constant quantity. Increasing concentrations of the other hapten (called the inhibitor) will competitively inhibit binding between the antibody and the constant component. This

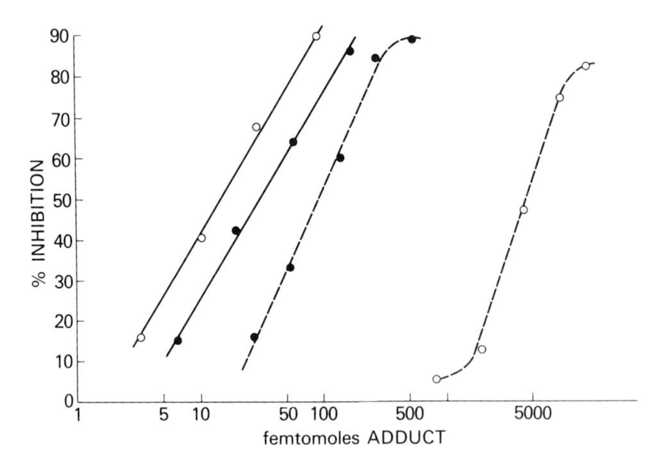

Fig. 2. Standard curves for RIA (---) and ELISA (——) utilizing anti-G-8-AF ( ) and anti BP-DNA (0). RIA conditions were as follows: for anti-BP-DNA, serum diluted 1:200, [^{3}H]-BPdG tracer as previously described (18), inhibitor was BP-DNA; for anti-G-8-AF, serum diluted 1:2,000, [^{3}H]G-8-AF tracer (27,000 cpm, 25 Ci/mmole) inhibitor was dG-8-AF; both assays using non-equilibrium conditions and goat-anti-rabbit-IgG. BP-DNA ELISA conditions were essentially the same as previously described (16) with 1.5 ng of 1.4% modified BP-DNA coated per well, antiserum diluted 1:69,000 and BP-DNA inhibitor. The G-8-AF ELISA conditions utilized 3.0 ng of AF-DNA (7.4% modified) coated per well, antiserum diluted 1:100,000 and AF-DNA inhibitor. Both ELISAs were performed with goat-anti-rabbit-IgG conjugated to alkaline phosphatase and P-nitro-phenyl phosphate as substrate.

concentration-dependent increasing inhibition plotted as a function of inhibitor concentration becomes the standard curve with the immunogen as inhibitor. The percent of inhibition with an unknown sample can be translated into quantity of adduct by comparison with the standard curve. In the case of RIA, which is performed in tubes, a second reagent is used to precipitate the antigen-antibody complex. The ELISA, performed in microtiter wells, relies upon a color-generating enzymatic end point in which the amount of substrate hydrolyzed is directly proportional to the amount of bound antibody. Procedures for both of these assays utilizing carcinogen-DNA adduct antisera have been published in detail elsewhere (14,15, 17,20). Figure 2 shows representative standard curves for both RIA and ELISA, and it is apparent from these data that ELISA is the more sensitive of the two immunoassays with these antisera. Since 50 µg of carrier DNA per microtiter well does not alter the standard curve, a 50 μg sample of unknown DNA can be tolerated so that the lower limit of sensitivity for the most sensitive ELISA is in the range of one adduct per 10^7 nucleotides of DNA.

Adduct Determination in Cultured Cells, Intact Animals, and Human Tissue

The data presented here will focus on the quantitation of carcinogen-DNA adducts in cells and animal and human lung tissue with emphasis on validation of the immunoassays by comparison with other techniques. The antisera used in these studies are specific for the major *in vivo* adducts formed upon interaction of DNA with 2-AAF or BP, and the structures of these adducts are shown in Fig. 1. Extensive characterization of the antisera have been published previously (8,10,11,18).

Exposure of mouse keratinocytes to N-acetoxy-acetylaminofluorene. The two antisera employed in these studies were elicited against guanosin-(8-yl)-acetylaminofluorene(G-8-AAF), 10) and its deacetylated derivative, G-8-AF. The first antiserum has been shown to be specific for both the acetylated and deacetylated C-8 adducts of 2-AAF with DNA (4,17). The second antiserum is specific primarily for the original immunogen, G-8-AF, with an insignificant degree of cross-reactivity for the acetylated C-8 adduct, particularly if assayed by ELISA. Figure 2 shows that both the RIA and ELISA standard curves are very sensitive.

Early attempts to validate the quantitation of carcinogen-DNA adducts by immunoassays focused on a dose response in primary BALB/c epidermal cells exposed for 1 hr to increasing concentrations of the activated carcinogen N-acetoxy-acetylaminofluorene (N-Ac-AAF). With doses in the range of 10^{-6} M to 10^{-4} M, the adduct levels observed by RIA were 5±5 μmole adduct/mole DNA-P to 90±5 μmole/mole DNA-P respectively, with the most rapid increase at concentrations greater than 5×10^{-5} M (17). In the BALB/c epidermal cells after exposure to 10^{-5} M of N-Ac-AAF, a comparison of binding detectable by radioactivity and RIA indicated that the RIA was consistently recognizing $\geq$75% of the total DNA-bound products (4). Further analysis of DNA profiles by RIA indicated that in these cells, 97% of the C-8 adduct formed was dG-8-AF, and 3% was dG-8-AAF (11). These ratios were subsequently confirmed by HPLC analysis after exposure of the primary mouse keratinocytes to a radioactive N-Ac-AAF (M. Poirier and F. Beland, unpublished observations). The HPLC analysis also indicated that radioactivity associated with the two C-8 adducts comprised 83% of the total DNA-bound products, and very little (2-3%) radioactivity eluted in positions similar to markers for the N^2 adduct (21) or the ring-opened deacetylated adduct (22). Thus, in primary mouse keratinocytes exposed to 10^{-5} M N-Ac-AAF, with binding levels in the range of 100-200 fmoles/μg DNA, the deacetylated C-8 adduct was the predominant C-8 adduct species.

More recent studies have focused upon attempts to alter the total binding levels and proportions of acetylated and deacetylated C-8 adducts with the ultimate intention of assessing the effects of these manipulations on epidermal cell transformation. Preincubation with 10^{-5} M paraoxon has been shown to effectively block the formation of the deacetylated C-8 adduct in BALB/c epidermal cells exposed to 10^{-5} M of N-Ac-AAF, resulting in adduct levels of approximately 1-2 fmoles/μg DNA (11). This concentration of paraoxon is non-toxic to cultured cells and specifically inhibitory for a microsomal deacetylase enzyme required for the formation of the dG-8-AF adduct (23). If the dosage of N-Ac-AAF is increased several fold in the presence of paraoxon, the DNA adduct levels can be increased substantially, and most of the C-8 adduct bound to the DNA in this instance is acetylated. Thus, varying the conditions of N-Ac-AAF exposure allows manipulation of the quantity and proportion of C-8 adducts on the DNA of mouse keratinocytes, and ongoing studies are attempting to discern the biological consequences of these alterations.

Adduct formation in skin and lung DNA of mice exposed to BP. An antiserum elicited against DNA modified with the major deoxyguanosine adduct of BP has been utilized for the detection of BP-DNA adducts in skin and lung of intact mice exposed to initiating doses of BP. This antiserum does not distinguish between the 7R and 7S

Table 1. Dose response for BP-DNA adduct formation in mouse skin after topical application.

	fmol BP-DNA adduct/μg DNA	
BP Exposure	ELISA	$[^3H]$ BP
μmoles/mouse	(mean ± S.E.)[a]	
0.050	1.2 ± 0.25	
0.100	1.8 ± 0.45	
0.250	3.4 ± 0.63	12.9[b]
0.500	3.8 ± 0.67	
1.000	3.6 ± 0.67	
1.500	2.2 ± 0.2[c]	

[a]Native DNA from the skins of two-four animals combined for each experiment. Data are pooled values from five separate experiments.

[b]Native DNA from the skins of four animals combined for RNAse and proteinase K digestion and subsequent liquid scintillation spectrometry.

[c]Mean ± range for two experiments.

Table 2. Dose response for BP-DNA adduct formation in mouse lung after intraperitoneal injection.

BP Exposure	fmol BP-DNA Adduct/μg DNA	
	ELISA	[^{3}H] BP
μmoles/mouse	(mean ± S.E.)[a]	
0.40	0.14 ± 0.02	
1.00	0.19 ± 0.03	
2.00	0.18 ± 0.01	
4.00	0.47 ± 0.05	1.2 ± 0.28[b]
8.00	0.37 ± 0.03	
12.00	0.36 ± 0.04	

[a]Native DNA from the lungs of two-three animals per dose; data combined for six separate experiments.
[b]Duplicate animals.

anti and syn deoxyguanosine adducts (18), and has been shown to cross-react with both the deoxyadenosine adduct and BP-RNA (8).

Therefore, this antiserum exhibits a rather broad specificity for the most predominant BP-nucleic acid adducts formed *in vivo*. A very sensitive ELISA has been developed for use with this antibody (Fig. 2, o——o), and investigations of adduct formation in lung and skin from mice exposed to BP have compared binding levels obtained with this assay as well as with radioactive BP. The rationale for these studies is two-fold: to investigate the formation and persistence of BP-DNA adducts in mouse skin upon topical exposure to initiating doses of BP, and to evaluate the limit of sensitivity of the ELISA in order to search for evidence of BP-DNA adducts in human lung tissue.

Dose response data for formation of BP-DNA adduct in mouse skin and lung are shown in Tables 1 and 2. Exposure to BP was topical on a defined area of skin, or by intraperitoneal injection for the lung. The skin is remarkable in that a topical dose of BP, 16- to 20-fold lower than that injected intraperitoneally, gave binding levels approximately 10-fold higher than those observed in lung. In both tissues a plateau in the levels of antibody-recognizable adducts was reached at or above doses initiating for tumors (100 mg/kg for lung and 200-300 nmoles for skin) (24,25). A radioactive BP was employed in experiments designed to validate the binding levels determined by ELISA. In both lung and skin the adduct levels determined by ^{3}H-BP were 2- to 4-fold higher than values observed by ELISA. The biological samples assayed by ELISA (Tabs. 1 and 2) were native DNAs prepared on CsCl buoyant density gradients.

The same DNAs assayed after denaturation gave binding levels two-fold higher because of greater antibody specificity for single-stranded BP-DNA (8,18). However, since occasional loss of adducts was observed upon DNA denaturation, and since the ELISA results were less variable with more uniform samples, the decision was made to assay the biological samples as native, knowing that the adduct levels were really about two-fold higher than the values observed. Some of this difference can be attributed to the decrease in antibody recognition of BP-DNA in the native state and some is undoubtedly due to DNA-bound or intercalated products which the antibody is unable to detect. From the date presented in Tab. 2, the lower limit of sensitivity established for the BP-DNA ELISA, with 50 μg of carrier DNA added per microtiter well, is 0.1 fmole adduct/μg DNA (16). This is approximately one carcinogen molecule in 10^7 nucleotides of DNA.

BP-DNA ELISA of DNA from human lung and lung tumor. In carefully determining the lower limit of sensitivity for the BP-DNA ELISA, our intention was the application of this technology to quantitate carcinogen damage in human DNA. The results obtained in this study were difficult to assess because the levels of adducts observed were near the lower limit of ELISA sensitivity and have not yet been confirmed by other procedures. Nevertheless, the antigen activity measured in the biological samples was removed when the specific antiserum was absorbed with the original immunogen, BP-DNA. Obtaining valid control tissue (devoid of BP-DNA adducts) proved to be one of the most difficult aspects of this investigation, second only to attempts to reduce and evaluate the variability inherent in the ELISA (16).

Lung tissue from tumors and surrounding normal areas was obtained as both surgery and autopsy specimens from patients hospitalized due to lung cancer or other diseases. Approximately fifty samples of lung and/or lung tumor DNA from a total of twenty-seven individuals were assayed by ELISA. Control DNA samples were prepared on CsCl gradients, with each new group of experimentals, from three large autopsy specimens of human lung obtained from individuals who died of causes other than lung cancer and who either never smoked or quit smoking at least five years previously. These control DNAs consistently gave values similar to each other and to unmodified calf thymus DNA in assays performed on different days. Thus, they provided a stable baseline from which to evaluate the other patient samples. Table 3 lists particulars for those samples, from five individuals found to be consistently positive in quadruplicate wells in two or more separate ELISAs. All of these individuals had primary carcinoma of the lung and were active, passive, or former smokers.

Table 3. Assay of human DNA from lung and lung tumors by BP-DNA ELISA.

Patient	Tissue (Diagnosis)	fmoles BP-DNA/ μg DNA	Smoking History
A. Patients with Lung Tumors			
1. With Primary Carcinoma of the Lung:			
1	Lung (PDC/SC)[a]	0.14[b]	Active
2	Lung	NS[c]	
	Lung Tumor (AC)	0.17	Former
3	Lung	0.18	Former
	Lung Tumor (AC)	0.18	
4	Lung	0.17	Passive
	Lung Tumor (AC)	0.14	
5	Lung	NS	Former
	Lung Tumor (AC)	0.17	
	Bronchus	0.14	
6-14	Lung and/or Lung Tumor (PDC, SC, AC)	NS	Includes active, passive, former and non-smokers
2. With Other Lung Tumors:			
15-21	Lung and/or Lung Tumor (MC, CT, L, G)	NS or 0[d]	
B. Patients Without Lung Tumors			
22	Lung (cardiomyopathy)	0	Never smoked
23	Lung (myocardial infarction)	0	Former, stopped 1976 No passive smoking
24	Lung (retinoblastoma)	0	Never smoked
25-27	Lung (MM, CNS, AB)	NS or 0	Smoking history unknown

[a]PDC = poorly differentiated carcinoma; SC = squamous cell carcinoma; AC = adenocarcinoma; MC = metastatic carcinoma; CT = carcinoid tumor; L = lipoma; G = granuloma; MM = multiple myeloma; CNS = CNS disease; AB = asthma/bronchiectasis.
[b]All DNAs were not denatured but were utilized after dialysis of CsCI peaks against water.
[c]NS = not significant.
[d]0 = below lower limit of detectability (0.08 fmole/μg DNA).

Among the remaining nine patients with primary carcinoma of the lung and thirteen other patients giving less than significant values in the ELISA, some were active or former smokers, or exposed passively to cigarette smoke. Smoking history and other epidemiologic data were obtained for only fifteen of the patients, including the three individuals whose lung tissue served as controls. These data (Tab. 3) did not show a correlation between ELISA values and cigarette exposure, possibly because of the small sample size. However, the observation that five patients with primary carcinoma of the lung were exposed to cigarette smoke appeared to have positive values suggests that screening of large numbers of smokers and/or individuals occupationally exposed to BP would be a reasonable approach in searching for a dosimeter for the biologically effective dose of BP.

CONCLUSIONS

The studies presented here constitute prototypes for the use of immunoassays to detect evidence of biological exposure to chemical carcinogens. In model systems, in both primary epidermal cell culture and intact mice, validation of results obtained with RIA and ELISA has been achieved through the use of radioactively labeled carcinogens and by comparison with literature values for the type and quantity of adduct formed. For example, in the cultured cells, binding values for 10^{-5} M of N-Ac-AAF exposure are consistent with those previously reported by Amacher et al. (26), and the nature of the predominating C-8 adduct, dG-8-AF, has been confirmed by HPLC and is similar to that found in normal human fibroblasts by Maher et al. (27). In mouse skin, values for BPdG adduct determined by ELISA correspond to literature values for similar doses of BP applied topically in different strains of mice (28,29).

Validation of the data obtained with human lung samples is somewhat more difficult and may require repetition of similar experiments in several laboratories. At the present time, Shamsuddin et al., using our antibody in a different immunoassay, have reported detecting positive antigenicity in DNA from white blood cells of roofers and foundry workers (30). Fourteen out of forty-eight individuals gave values ranging between 0.04 fmole/mg DNA and 2.48 fmole/μg DNA with a mean of 0.51 fmole/μg DNA. Since, in our studies, large doses of BP injected intraperitoneally in mice (1-3 mg/20-25 gm mouse) were required to yield maximum binding levels of 0.4 fmoles/μg DNA in lung tissue, it is surprising that the cumulative human exposure through occupation or smoking would reach such comparatively high levels. However, the extent of human exposure is unknown, and individual variations in hydrocarbon metabolism are known to be considerable (31). In any case, the human data presented here are promising, although they require additional validation.

To further extend this immunotechnology in assaying for human exposure we have elicited an antibody against a DNA modified *in vitro* with the chemotherapeutic agent, *cis*-diamminedichloroplatinum (II) (*cis*-DDP) (32). This antiserum is specific for adducts of *cis*-DDP formed both *in vitro* and *in vivo* and is currently being utilized in an investigation designed to determine adduct levels (by ELISA) in nucleated blood cells of cancer patients. This study avoids some of the difficulties of the human lung study, since exposure of patients can be readily documented and may be a useful adjunct to clinical treatment by correlating drug dosage with biological effectiveness (33).

In summary, antisera specific for carcinogen-DNA adducts have become useful probes for investigating mechanisms of carcinogen-DNA interactions in experimental model systems. They may also become useful markers for evidence of chemically induced DNA damage in the human populations.

ACKNOWLEDGEMENTS

Thanks are extended for ongoing discussion and collaborations to Drs. R. Santella, D. Grunberger, and F.A. Beland. The technical assistance of E. Patterson, C. Thill, M.A. Dubin, and R. Shores are gratefully acknowledged.

REFERENCES

1. Poirier, M.C. (1981) Antibodies to carcinogen-DNA adducts. *J. Natl. Cancer Inst.* 67:515-519.
2. Muller, R., and M.F. Rajewsky (1981) Antibodies specific for DNA components structurally modified by chemical carcinogens. *J. Cancer Res. Clin. Oncol.* 102:99-113.
3. Muller, R., and M.F. Rajewsky (1980) Immunological quantification by high-affinity antibodies of O^6-ethyldeoxyguanosine in DNA exposed to N-ethyl-N-nitrosourea. *Cancer Res.* 40:887-896.
4. Poirier, M.C., M.A. Dubin, and S.H. Yuspa (1979) Formation and removal of specific acetylaminofluorene-DNA adducts in mouse and human cells measured by radioimmunoassay. *Cancer Res.* 39:1377-1381.
5. Poirier, M.C., B'Ann True, and B.A. Laishes (1982) Formation and removal of (Guan-8-yl)-DNA-2-acetylaminofluorene adducts in liver and kidney of male rats given dietary 2-acetylaminofluorene. *Cancer Res.* 42:1317-1321.
6. Kyrtopoulos, S.A., and P.F. Swann (1980) The use of radioimmunoassay to study the formation and disappearance of O^6-methylguanine in mouse liver satellite and main-band DNA following dimethylnitrosamine administration. *J. Cancer Res. Clin. Oncol.* 98:127-138.

7. deMurcia, G., M.E. Lang, A. Freund, R.P.P. Fuchs, M.P. Daune, E. Sage, and M. Leng (1979) Electron microscopic visualization of N-acetoxy-N-2-acetylaminofluorene binding sites in ColE1 DNA by means of specific antibodies. Proc. Natl. Acad. Sci., USA 76:6076-6080.
8. Poirier, M.C., J.R. Stanley, J.B. Beckwith, I.B. Weinstein, and S.H. Yuspa (1982) Indirect immunofluorescent localization of benzo(a)pyrene adducted to nucleic acids in cultured mouse keratinocyte nuclei. Carcinogenesis 3:345-348.
9. Slor, H., H. Mizusawa, N. Neihart, T. Kakefuda, R.S. Day, and M. Bustin (1981) Immunochemical visualization of binding of the chemical carcinogen benzo(a)pyrene diol-epoxide 1 to the genome. Cancer Res. 41:3111-3117.
10. Poirier, M.C., S.H. Yuspa, I.B. Weinstein, and S. Blobstein (1977) Detection of carcinogen-DNA adducts by radioimmunoassay. Nature 270:186-188.
11. Poirier, M.C., G.M. Williams, and S.H. Yuspa (1980) Effect of culture conditions, cell type, and species of origin on the distribution of acetylated and deacetylated deoxyguanosine C-8 adducts of N-acetyoxy-2-acetylaminofluorene. Molec. Pharm. 18:581-587.
12. Yuspa, S.H., and C.C. Harris (1974) Altered differentiation of mouse epidermal cells treated with retinyl acetate in vitro. Exp. Cell Res. 86:95-100.
13. Lieberman, M.W., and M.C. Poirier (1973) Deoxyribonucleoside incorporation during DNA repair of carcinogen-induced damage in human diploid fibroblasts. Cancer Res. 33:2097-2103.
14. Poirier, M.C. (1980) Measurement of formation and removal of adducts of N-acetoxy-2-acetylaminofluorene. In DNA Repair, E.C. Friedberg and P.C. Hanawalt, eds. Marcel Dekker, Inc., New York, pp.143-153.
15. Hsu, I-C., M.C. Poirier, S.H. Yuspa, D. Grunberger, I.B. Weinstein, R.H. Yolken, and C.C. Harris (1981) Measurement of benzo(a)pyrene-DNA adducts by enzyme immunoassays and radioimmunoassay. Cancer Res. 41:1091-1095.
16. Perera, F.P., M.C. Poirier, S.H. Yuspa, J. Nakayama, A. J. Aretzski, M.M. Curnen, D.M. Knowles, and I.B. Weinstein (1982). A pilot project in molecular cancer epidemiology: Determination of benzo(a)pyrene-DNA adducts in animal and human tissues by immunoassays. Carcinogenesis 3:1405-1410.
17. Poirier, M.C., and R.J. Connor (1982) Radioimmunoassay for 2-acetylaminofluorene-DNA adducts. In Methods in Enzmology, H. van Vunakis and J. Langone, eds. Academic Press, New York, Vol. 84:607-618.
18. Poirier, M.C., R. Santella, I.B. Weinstein, D. Grunberger, and S.H. Yuspa (1980) Quantitation of benzo(a)pyrene-deoxyguanosine adducts by radioimmunoassay. Cancer Res. 40:412-416.
19. Marrs, J.M., and J.J. Voorhees (1971) A method for bioassay of an epidermal chalone-like inhibitor. J. Invest. Dermatol. 56:174-181.

20. Hsu, I.C., M.C. Poirier, S.H. Yuspa, R.H. Yolken, and C.C. Harris (1980) Ultrasensitive enzymatic radioimmunoassay (USERIA) detects femtomoles of acetylaminofluorene-DNA adducts. Carcinogenesis 1:455-458.
21. Westra, J.G., E. Kriek, and H. Hittenhausen (1976) Identification of the persistently bound form of the carcinogen N-acetyl-2-aminofluorene to rat liver DNA in vivo. Chem. Biol. Interact. 15:149-164.
22. Kriek, E., and J.G. Westra (1980) Structural identification of the pyrimidine derivatives formed from N-deoxyguanosin-(8-yl)-2-aminofluorene in aqueous solution at alkaline pH. Carcinogenesis 1:459-468.
23. Irving, C.C. (1966) Enzymatic deacetylation of N-hydroxy-2-acetylaminofluorene by liver microsomes. Cancer Res. 26:1390-1396.
24. Shimkin, M.B., and G.D. Stoner (1975) Lung tumors in mice: Application to carcinogenesis bioassay. Adv. Cancer Res. 21:1-58.
25. Brookes, P., and P.D. Lawley (1964) Evidence for the binding of polynuclear aromatic hydrocarbons to the nucleic acids of mouse skin: Relation between carcinogenic power of hydrocarbons and their binding to DNA. Nature 202:781-784.
26. Amacher, D.E., J.A. Elliott, and M.W. Lieberman (1977) Differences in removal of acetylaminofluorene and pyrimidine dimmers from the DNA of cultured mammalian cells. Proc. Natl. Acad. Sci., USA 74:1553-1557.
27. Maher, V.M., R.M. Hazard, F.A. Beland, R. Corner, A.L. Mendrala, J.W. Levinson, R.H. Heflick, and J.J. McCormick (1980) Excision of the deacetylated C-8-guanine DNA adduct by human fibroblasts correlates with decreased cytotoxicity and mutagenicity. Proc. American Assoc. for Cancer Res. 21:71 (abstract 286).
28. Pereira, M.A., F.J. Burns, and R.E. Albert (1979) Dose response for benzo(a)pyrene adducts in mouse epidermal DNA. Cancer Res. 39:2556-2559.
29. Ashurst, S.W., and G.M. Cohen (1982) The formation of benzo(a)-pyrene-deoxyribonucleoside adducts in vivo and in vitro. Carcinogenesis 3:267-273.
30. Shamsuddin, A.M., N.T. Sinopoli, K. Hemminki, R.R. Boesch, and C.C. Harris. Detection of benzo(a)pyrene-DNA antigenicity in white blood cells of roofers and foundry workers (submitted for publication).
31. Harris, C.C., B.F. Trump, R. Grafstrom, and H. Autrup (1982) Differences in metabolism of chemical carcinogens in cultured human epithelial tissues and cells. J. Cell Biochem. 18:285-294.
32. Poirier, M.C., S.J. Lippard, L.A. Zwelling, H.M. Ushay, D. Kerrigan, C.C. Thill, R.M. Santella, D. Grunberger, and S.H. Yuspa (1982) Antibodies elicited against cis-diamminedichloroplatinum (II)-DNA adducts formed in vivo and in vitro. Proc. Natl. Acad. Sci., USA 79:6443-6447.

33. Perera, F., and I.B. Weinstein. Molecular epidemiology and carcinogen-DNA adduct detection: New approaches to studies of human cancer causation. J. Clin. Res. (in press).

CHAIRMAN'S OVERVIEW ON INTEGRATION AND EXPRESSION OF THE POLYOMA VIRUS ONCOGENES IN TRANSFORMED CELLS

Claudio Basilico with Lisa Dailey, Sandra Pellegrini, Robert G. Fenton, and Franca La Bella

Department of Pathology
New York University School of Medicine
New York, New York 10016

INTRODUCTION

Polyoma is a small DNA tumor virus which can cause tumors in a variety of rodents, particularly the mouse, hamster, and rat. *In vitro*, the virus exhibits two types of interactions with susceptible cells: a productive or lytic interaction in which the virus multiplies and kills the infected cells, and a transforming interaction, in which a fraction of the infected cells becomes phenotypically "transformed" and tumorigenic, without substantial virus production or cell death. The type of interaction depends mainly on the species of the infected cells. Permissive mouse cells, although transformable, give rise almost exclusively to a lytic interaction, while in nonpermissive cells, such as rat and hamster, transformation is easily detected (1).

The genome of Polyoma virus is a double-stranded, closed circular DNA molecule of 5.3 kb. It has been completely sequenced and its organization well studied. Specifically the viral DNA molecule is divided into early and late transcription units (Figure 1). The early region comprises nucleotide sequences mapping from 73 to 26 m.u., while transcription of the late region proceeds counterclockwise on the opposite DNA strand from 67 to 26 m.u. (1). Nucleotides comprised between 67 and 73 m.u. contain noncoding regulatory sequences, including the origin of replication, early and late promoters, and the so-called "enhancer" sequences, required for *in vivo* expression of the early genes (2,3). Transcription of the early region generates a single mRNA precursor which is spliced differentially to produce the three mRNAs encoding the early viral proteins [large, middle, and small (tumor) T-Antigen] (see Map, Fig. 1). Efficient transcription of the late region requires the initiation of viral DNA replication, and differential splicing of the mRNA precursor(s) yields the three mRNAs for the viral structural proteins VP1, VP2, and VP3 (1).

About 50% of the total coding of polyoma DNA is dedicated to the synthesis of proteins which are involved in transformation. These are the three early proteins, and it has been conclusively shown that polyoma transformation requires the integration of the viral genome into the DNA of the host cells, and the expression of the early viral genes. Integration is a necessary, although not sufficient condition for the expression of viral genes in a transformed cells as it ensures the regulated transmission of these DNA sequences to the cells' progeny (1).

With the purpose of understanding the mechanisms regulating the association of the genome of an oncogenic virus and that of transformed cells, and the effects of such an association on the expression of viral functions, we have studied the process of integration of Polyoma viral DNA into the DNA of transformed rat cells, the rearrangements of viral and host DNA which accompany or follow integration, and the modality of transcription of integrated viral genomes. The results of some of these studies will be summarized in this paper.

Integration

Polyoma DNA encodes three early proteins: small (20k), middle (56k), and large (100k) T-Antigen (Fig. 1). While middle T-Antigen is the viral gene-product mainly responsible for conferring to cells a transformed phenotype (4-7), it has long been known that large T-Ag plays a role in the establishment of transformation (8-10). Polyoma temperature sensitive mutants of the A complementation group (ts-A, producing a thermolabile large T-Ag) are greatly impaired in the ability to induce transformation at the non-permissive temperature. However, the phenotype of the cells remains transformed when ts-A transformed cells are shifted from 33°C to 39.5°C (8-11). These findings suggested the possibility that large T-Ag promoted the integration of polyoma DNA into the genome of the infected cells, and we undertook experiments aimed at elucidating the role of this protein in transformation.

We determined first the efficiency of transformation of the DNA of ts-A mutants of Polyoma virus at 39°C or 33°C, and compared it with that of wt viral DNA or DNA fragments, some of which could not encode an intact large T-Antigen (12). The results of all these experiments showed that the efficiency of transformation was about 20-fold higher in the presence of a functional viral A gene-product than in its absence (Table 1) but, as expected, the phenotype of cells transformed under different conditions (all expressing at least small and middle T-Antigen) was indistinguishable (12).

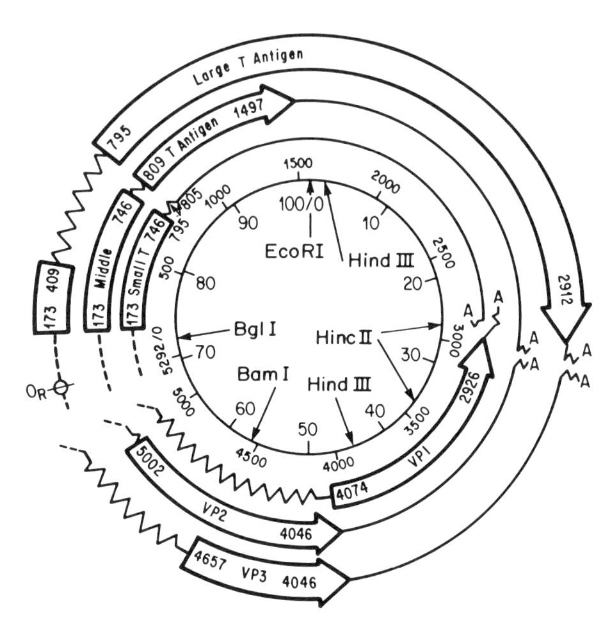

Fig. 1. Physical and functional map of Polyoma virus DNA (A2 strain). The physical map is divided into map units (starting clockwise from 0 at the EcoRl site) or nucleotide number (starting clockwise from 0 at the origin of replication). The mRNAs for the three T-Antigens and the late proteins VP1, VP2, and VP3 are shown with their coding regions (▭); 3' noncoding regions (—); and intervening sequences (ʌʌʌ). The numbers within the coding regions represent initiation and termination codons and splice junctions. OR = origin of viral DNA replication. Adapted from (1).

We then turned our attention to the mechanism by which large T-Antigen could have promoted a high efficiency of transformation. We examined the pattern of integration of Polyoma viral DNA into the DNA of transformed cells (13) by Southern blot analysis of restriction enzyme digests of high m.w. DNA. Ours and other laboratories had already established that the arrangement of Polyoma viral DNA integration in cells transformed under standard conditions was almost invariably that of a tandem head-to-tail insertion of more than one viral DNA molecule (14-16). We found, however, that cells transformed in the absence of a functional A protein mostly contained nontandem insertions of viral DNA segments shorter than the infecting molecules (12).

Polyoma large T-Ag is known to be necessary for the initiation of viral DNA replication (1), and if the main role of Polyoma

Table 1. Transforming ability and type of integration of the DNA of various Polyoma mutants or of viral molecules cloned into bacterial plasmids.*

DNA	TRANSFORMATION		MOST COMMON TYPE of INTEGRATION	
	39°	33°	39°	33°
wt, native	100	100	Tandem Δ	Tandem
ts A, native	4.8	100	S. Copy Δ	Tandem
ts A-pBR322	5	10	N.D.	N.D.
ts A pML	5	20	N.D.	N.D.
ORI⁻ tsA ∇	5	7	N.D.	S. Copy
ORI⁻ wt ∇	7	7	N.D.	S. Copy

* The results are the average of several independent experiments. Rat F2408 cells were transfected with the Polyoma DNA molecules indicated and transformation was measured from the frequency of cells forming colonies in agar medium at 20-25 days after plating. Data are expressed in % of wild-type values.

Δ This notation does not refer to the number of independent viral insertions into the host DNA, but to the type of arrangement of each viral insertion (tandem vs less than single copy).

∇ These molecules were excised from plasmid vectors before transfection.

N.D. = Not Done

large T in the establishment of transformation is allowing the formation of tandem head-to-tail integration, this integrated arrangement may result from a replication process. We therefore tested the efficiency of transformation of viral DNA molecules which were deficient in replication for reasons other than a lack of T-Antigen function. We first tested the transforming ability of Py molecules cloned at the Bam I site (in the late region) in the plasmid pBR322. Such molecules are defective in replication because of the interference of not well defined "poison" sequences of pBR (17). When ts-A DNA was cloned in pBR322, it became no longer ts for transformation, and transformed at 33°C with an efficiency similar to that of native ts-A DNA at 39°C. Cloning into pML, a poisonless derivative of pBR322 (17), increased the efficiency of transformation at 33°C, which was still, however, lower than that of native ts-A DNA (Tab. 1)

We then tested the efficiency of transformation of polyoma deletion mutants, which have a cis-acting defect in viral DNA replication, because of the deletion of large T-Antigen binding sequences near the origin of replication (3; L. Dailey and C. Basilico, in preparation). One of these origin-defective mutants is in a ts-A background, and produces a ts large T-Antigen. It was found that this mutant was longer ts for the ability to transform but, like the ts-A pBR clone, transformed at low efficiency both at 33°C or 39°C. Furthermore, even at 33° the pattern of integration was predominantly single-copy (or less) (S. Pellegrini and C. Basilico, in preparation). Accordingly, the origin-defective mutant producing wt large T-Ag (3) transforms at low efficiency at any temperature (Tab. 1).

These findings clearly show that viral DNA replication is required for a high efficiency of transformation, as well as for the formation of tandem insertions. A model accounting for these results is described in the "Discussion" that follows.

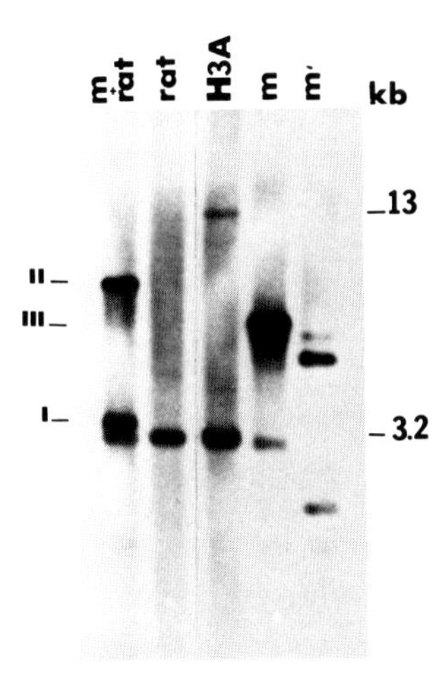

Fig. 2. Arrangement in normal rat cell DNA of DNA sequences homologous to those flanking the viral insertion in the Polyoma transformed line H3A. DNAs were digested with the restriction enzyme Bgl II, fractionated on agarose gels, and transferred to nitrocellulose filters. The filters were hybridized with a ^{32}P-labeled pBR322 clone containing a small portion of viral DNA (approximately 0.7 Kb) and 1.1 Kb of host DNA sequences 3' to the viral insertion of H3A. Rat is normal rat cell DNA, m+rat is a mixture of Polyoma DNA form I and II and rat cell DNA, m and m' are molecular weight markers.

We also wanted to study possible rearrangements of the host DNA which had occurred as a result of integration. It had been suggested that integration of SV40 or Polyoma DNA led to deletions of host DNA at the site of insertion (18-19). To verify whether this was also true in our cell lines, we cloned in Lambda Charon 30 (20) the entire viral insertion of the ts-A H3A line, which had been the object of several studies in our laboratory (21,22). The DNA fragment cloned contained sequences between the cleavage sites of the restriction enzyme Bgl II (does not cleave Py DNA) proximal to the viral insertion, and thus comprised the viral sequences and about 7 kb of flanking host DNA. By subcloning appropriate restriction fragments into pBR322 we obtained two plasmids containing a limited amount of viral DNA sequences and 0.7 and 1.1 kb of host DNA corresponding to the left or right flanking host sequences, respectively. With these probes we examined the arrangement of these host sequences in the parental untransformed rat F2408 cells. It appears that the two host sequences are not adjacent in the parental cells, but are separated by the undetermined length of DNA sequences (more than 5 kb). Figure 2 shows the results obtained with Bgl II digestion. In H3A, the "right" probe recognized the viral insertion ($\sim$ 13 kb) and a fragment of 3.2 kb. The latter fragment is also present in normal rat cells. If this fragment were joined to the left flanking sequences, its m.w. should be about 7 kb. Similar conclusions were reached by analysis with other restriction enzymes. Therefore, also in the case of H3A, viral integration has led to a deletion of host DNA. Furthermore, the viral insertion is haploid, since the restriction fragments recognized by the "right" and "left" probes in the parental rat cells are also present in the transformed H3A line, indicating that viral DNA has inserted in only one of the two identical sites existing on the homologous chromosomes.

Expression of Integrated Viral Genomes

Once the polyoma genome is integrated into the host DNA, transformation is maintained by the transcription and translation of specific viral genes. It therefore became important to define the modalities of viral transcription in transformed cells, and to determine whether they differ from those manifested during productive infection. We particularly wanted to study whether the physical state of integration modified in any way the control of expression of the early genes.

The modality of transcription of integrated DNA sequences, when an intact early region is present, shows the following characteristics: transcription invariably starts from viral promoters; no evidence of host initiated transcription can be detected. Transcription is limited to the early region of polyoma DNA. Studies of polyadenylated cytoplasmic mRNAs reveal the presence of three types of RNAs, indistinguishable in their 3' and 5' ends and splice junctions from those detected early in lytic infection, coding for small, middle, and large T-Ag (Fig. 1). In the absence of free

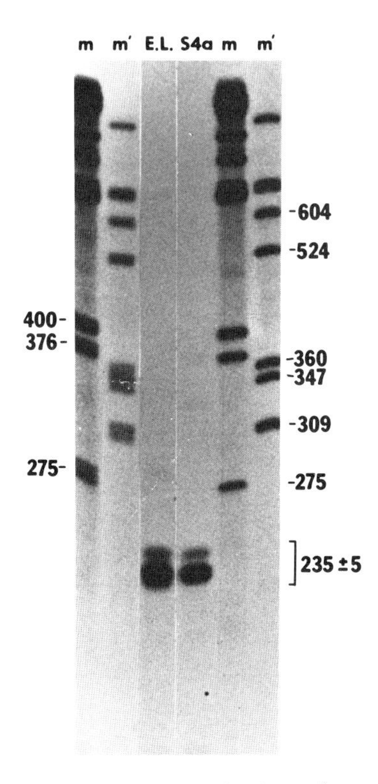

Fig. 3. High resolution mapping of the 5' ends of Polyoma early mRNAs present during lytic infection and in cell line S4a. 5 μg of early lytic (E.L.) or 10 μg of S4a poly(A)+ cytoplasmic RNAs were hybridized to the E-strand of the HinfI-4 fragment of polyoma DNA (66.4-77.8 m.u.; nucleotides 5073 to 385) which had been ^{32}P-labeled at its 5' end. Nuclease S1-resistant DNA was size fractionated on an 8% polyacrylamide/50% urea thin gel and bands visualized by autoradiography. Markers are 5' end labeled Py DNA cleaved with HpaII (m) or HinfI (m'). The protected DNA fragments of 232-237 nucleotides in length map the major 5' ends to positions 148-153 on the genomic sequences near 73.3 m.u.

viral DNA production, no evidence of stable transcripts hybridizing to polyoma late sequences can be found (23).

Figure 3 shows an experiment in which the 5' ends of the early mRNAs produced by a transformed cell line are compared with those observed during the lytic cycle. It can be seen that the major 5' ends of these RNA species are identical, supporting the idea that transcription initiates at identical promoters.

Similarly, splice junctions and 3' ends of mRNAs transcribed from complete early region sequences are the same (23). These findings support the idea that viral transcription is regulated mainly by viral controlling elements, and to substantiate this hypothesis we determined the effects of large T-Antigen on the rate of transcription of early mRNAs. During lytic infection, large T-Antigen binds near the origin of viral DNA replication, thus decreasing early transcription in the late phase of the infectious cycle. Inactivation of large T-function results in a 5- to 10-fold increased rate of transcription (1). A similar effect was detected in polyoma-transformed cells (24). Shift of Py ts-A transformed cells to the temperature that is nonpermissive for large T-function (39°C) results in a 5- to 10 fold higher rate of early transcription. Thus, integrated viral DNA molecules are still subject to repression of transcription by this viral protein, and transcription of integrated viral DNA molecules seems to be mainly regulated by viral controlling elements.

On the other hand, several forms of new transcripts are observed in transformed cells when the attention is focused on the 3' ends of the messengers. Deletion of the major early region poly(A) attachment site at 25.8 m.u. on the polyoma map results in the generation of readthrough transcripts and increased utilization of an alternative poly(A) addition signal at 99 m.u. (25). This can be produced by the joining of viral sequences upstream from 25 m.u. to host DNA, or by deletions which remove this region from the viral genome (23). In the first case, transcripts extend variable distances into host DNA and are polyadenylated at a host site. When defective viral DNA molecules are tandemly arrayed, multimeric viral transcripts are produced, and these are homologous both to early and anti-late viral sequences (23).

Transcripts polyadenylated at 99 m.u. have been precisely mapped at their 3' ends. Polyadenylation occurs at 10 and 30 nucleotides downstream from the AAUAAA sequence at nucleotides 1,475-1,480. Utilization of the second polyadenylation site produces a mRNA which has sufficient information to code for a complete middle T-protein. However, polyadenylation at the first site produces a mRNA species lacking the termination code for middle T-Ag (R.G. Fenton and C. Basilico, unpublished results).

DISCUSSION

Integration

The results presented in this paper show that polyoma virus DNA can integrate into the genome of transformed cells in tandem or single copy arrangement. The efficiency of transformation is greatly dependent on the presence of a functional large T-Antigen which is, however, not necessary for the expression of the transformed phenotype. Restriction enzyme analysis of a number of transformed lines showed a strong correlation between the presence of a functional large T-Antigen and the integration pattern of viral genomes in "head-to-tail" tandem repeats (12). Cells transformed in the absence of a functional large T-Antigen generally contained one or more single copy insertions (12).

Polyoma large T-Antigen is known to be involved in viral DNA replication (1), and therefore the association of its function with tandem integration suggests that this type of integration may result from a replication process. The involvement of replication rather than recombination in the formation of tandem insertions is also suggested by the experiments with molecules which are defective in viral DNA replication for reasons other than absence of large T-function. We were able to show that these molecules transform at low efficiency, and that this efficiency was not altered by the presence or absence of an active large T-protein. In addition, these molecules were found to be preferentially integrated in less than single-copy arrangement. It appears, therefore, that inhibition of replication (no matter how obtained) results in suppression of tandem formation. Thus, even if recombination were involved in this process, it would have to be postulated that a replication step is a necessary condition for recombination to follow.

In conclusion, the model best supported by our results postulates that polymers of viral DNA arranged in "head-to-tail" tandems are the precursors of the integrated form (26). Such polymers are likely to be produced by replication of the viral DNA by mechanisms that allow DNA synthesis without separation of the newly formed DNA molecules. A rolling circle mechanism of viral DNA synthesis would meet many of the requirements of this hypothesis. The rolling circle is both a replicative intermediate and an oligomer containing viral genomes in "head-to-tail" sequence. Such a structure could integrate by a double crossing-over event, leading to insertions of multiple viral genomes. Models including a double crossing-over event would predict the deletion of the host DNA between the two cross-over sites. This prediction is borne out by our data and that of other investigators (18,19), showing that SV40 or Polyoma integration generally leads to a deletion of host DNA sequences at the site of insertion. The model represented in Fig. 4 is therefore consistent with all the presently available data.

Expression of Integrated Genomes

The regulation of transcription of integrated polyoma genomes depends mainly on viral controlling elements. No evidence of transcription initiating from host DNA has been detected, and the mRNAs produced from integrated intact early regions are indistinguishable from those seen early in the infectious cycle. In addition, negative regulation of early transcription by large viral T-Antigen operates in transformed cells at approximately the same level as in

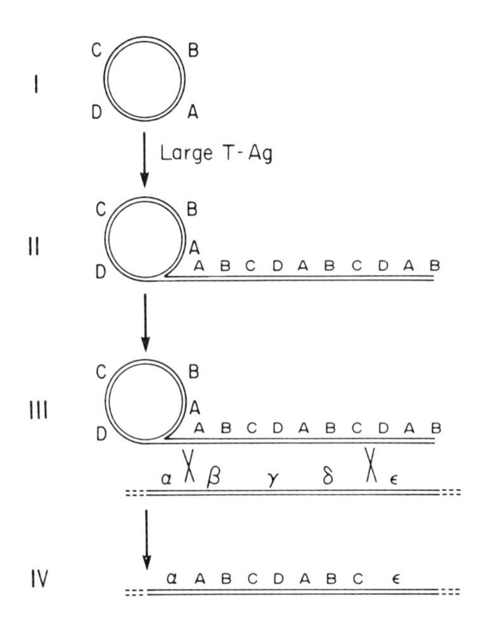

Fig. 4. A model of the mechanism by which Polyoma DNA integrates in a tandem arrangement. The circular viral DNA molecule (I) initiates replication (requiring large T-Antigen) through a rolling circle mechanism. This produces tandem repeats of viral DNA sequences in a head-to-tail arrangement (II). A double-crossing over event (III) between these oligomers and cellular DNA leads to tandem integration with concomitant loss of the host DNA sequences comprised between the cross-over sites (IV).

productive infection (24). The major alterations of transcripts observed result from deletion of 3' early region sequences, resulting from the joining of viral to host DNA at the site of integration. When deletions of 3' early region sequences including the major polyadenylation signal at 25.8 m.u. occur, novel viral transcripts are produced, consisting of fused transcripts of viral and host DNA (23). In addition, the poly(A) addition at 99 m.u. is used much more frequently (23). It is possible that sequences around 26 m.u. signal transcription termination and when these are deleted, transcription continues into host or viral sequences normally not transcribed. However, this does not explain the shift observed to the usage of the polyadenylation signal at 99 m.u. It seems likely that the relative efficiency of poly(A) addition depends on other sequences in the vicinity of the AAUAAA signal, and work is in progress to investigate this possibility.

In conclusion, while integration could have a profound effect on the expression of viral functions, such as the physical interruption of early coding sequences, the virus seems to have developed two strategies to overcome these problems. The first is that of tandem insertion, which allows the preservation of intact DNA sequences coding for transforming proteins irrespective of the region of the viral DNA where integration occurs. The second is the utilization of an otherwise rarely used polyadenylation signal at 99 m.u., which allows the formation of mRNA molecules coding for the main transforming protein (i.e., middle T-Antigen) even when sequences which are part of the "early" transcriptional unit are deleted as a result of integration.

One final point regards the question of whether the host cell controls in any way the expression of viral genes, once they become integrated into the host cell genome. Although the results discussed in this paper clearly indicate that the fine regulation of viral transcription is mediated by viral elements, a higher order of control at the level of chromatin structure, site of integration, etc., cannot be excluded and is, in fact, suggested by some data obtained on transcription of integrated SV40 DNA molecules.

Taking advantage of host mutations, which somehow interfere with the expression of the transformed phenotype, it has been possible to show that in a few SV40 transformed mouse cell lines, transcription of integrated SV40 molecules is strongly cell-cycle regulated, such that G1-arrested cells do not transcribe integrated SV40 DNA and do not express viral antigens (27). Transcription of other cellular genes proceeds unimpaired (28). Whether this cell cycle control results from the availability of host factors necessary for SV40 transcription, or from the site of viral DNA integration is not known. In the latter case, it would have to be postulated that SV40 preferentially integrates in regions of the host DNA that are not accessible to transcription in G1 because of chromatin conformation or other undefined factors. Clearly further work is needed to resolve this issue and this work may be useful also for understanding the factors regulating the expression of animal cell genes.

ACKNOWLEDGMENT

This work was supported by PHS Grants CA16239, CA11893, and CA26070 from the National Cancer Institute.

REFERENCES

1. Tooze, J., ed. (1980) DNA Tumor Viruses, Cold Spring Harbor Press, Cold Spring Harbor, New York.
2. Soeda, E., J.R. Arrand, N. Smolar, J.E. Walsh, and B.E. Griffin (1980) Nature 283:445-453.

3. Tyndall, C., G. La Mantia, C.M. Thacker, and R. Kamen (1981) Nuc. Acids Res. 9:6231-6250.
4. Griffin, B.E., and E. Maddock (1979) J. Virol. 31:645-656.
5. Hassell, J.A., W.C. Topp, D.B. Rifkin, and P.E. Moreau (1980) Proc. Natl. Acad. Sci., USA 77:3978-3982.
6. Treisman, R., U. Novak, J. Favaloro, and R. Kamen (1981) Nature 292:595-600.
7. Templeton, D., and W.J. Eckhart (1982) J. Virol. 41:1014-1024.
8. Fried, M. (1965) Proc. Natl. Acad. Sci., USA 53:486-491.
9. DiMayorca, G., J. Callender, G. Marin, and R. Giordano (1969) Virology 38:126-133.
10. Eckart, W. (1969) Virology 38:120-125.
11. Fluck, M.M., and T.L. Benjamin (1979) Virology 96:205-228.
12. Della Valle, G., R. Fenton, and C. Basilico (1981) Cell 23:347-355.
13. Southern, E. (1975) J. Mol. Biol. 98:503-515.
14. Birg, F., R. Dulbecco, M. Fried, and R. Kamen (1979) J. Virol. 29:633-648.
15. Basilico, C., S. Gattoni, D. Zouzias, and G. Della Valle (1979) Cell 17:645-659.
16. Basilico, C., D. Zouzias, G. Della Valle, S. Gattoni, V. Colantuoni, R. Fenton, and L. Dailey (1980) Cold Spring Harbor Symp. Quant. Biol., Cold Spring Harbor Press, Cold Spring Harbor, New York, 44:611-619.
17. Lusky, M., and M. Botchan (1981) Nature 293:79-81.
18. Stringer, J.R. (1982) Nature 296:363-366.
19. Hayday, A., E.H. Ruley, and M. Fried (1982) J. Virol. 44:67-77.
20. Rimm, D.L., D. Horness, J. Kucera, and F.R. Blattner (1980) Gene 12:301-312.
21. Colantuoni, V., L. Dailey, and C. Basilico (1980) Proc. Natl. Acad. Sci., USA 77:3850-3854.
22. Dailey, L., V. Colantuoni, R.G. Fenton, F. La Bella, D. Zouzias, S. Gattoni, and C. Basilico (1982) Virology 116:207-220.
23. Fenton, R.G., and C. Basilico (1981) J. Virol. 40:150-163.
24. Fenton, R.G., and C. Basilico (1982) Virology 121:384-392.
25. Kamen, R., J. Favaloro, J. Parker, R. Treisman, L. Lania, M. Fried, and M. Mellor (1980) Cold Spring Harbor Symp. Quant. Biol., Cold Spring Harbor Press, Cold Spring Harbor, New York, 44:63-75.
26. Chia, W., and P.W. Rigby (1981) Proc. Natl. Acad. Sci., USA 78:6638-6642.
27. Basilico, C., and D. Zouzias (1976) Proc. Natl. Acad. Sci., USA 73:1931-1935.
28. La Bella, F., E.H. Brown, and C. Basilico (submitted for publication).

ACTIVATION OF CELLULAR ONC (C-ONC) GENES:

A COMMON PATHWAY FOR ONCOGENESIS

W.S. Hayward, B.G. Neel, S.C. Jhanwar, and R.S.K. Chaganti

Memorial Sloan-Kettering Cancer Center

New York, New York 10021

INTRODUCTION

Retroviruses can be classified into two groups: those that contain oncogenes, and those that do not (for reviews, see refs. 1-5). Members of the first group (acute, or rapidly transforming retroviruses) induce neoplastic disease in infected animals within a few weeks after infection, and cause rapid transformation of target cells in tissue culture. Viruses of the second group (slowly transforming retroviruses), which lack oncogenes, induce neoplastic disease in animals only after a long latent period (4-12 months), and do not cause transformation of tissue culture cells at detectable frequency.

The oncogenes of acute retroviruses ("v-onc" genes) were derived from normal cellular genes ("proto-onc" or "c-onc" genes) by recombination (2,3). More than fifteen different v-onc genes (and corresponding c-onc genes) have been identified thus far (6). As a consequence of the recombination event that generates an acute retrovirus, the onc gene is placed under the control of viral regulatory sequences. The v-onc gene is thus expressed at constitutively high levels in the infected cell, and does not respond to the regulatory signals that modulate expression of the corresponding c-onc gene. Changes within coding sequences may also accompany the conversion of a c-onc gene to a v-onc gene, but the possible contribution of these changes to the oncogenic potential of the v-onc genes has not been evaluated.

Recent studies have implicated the host c-onc genes more directly in a variety of neoplasms that are not induced by viruses that carry oncogenes. The first such system described was the

B-cell lymphoma induced by avian leukosis virus (ALV). ALV, which lacks an oncogene, induces lymphomas by activating the c-*myc* gene (7), the cellular homolog of the oncogene (v-*myc*) of MC29 virus. More recently, several groups (8,9) have shown that "oncogenes" cloned from human carcinoma cell lines are related to the c-*ras* family of c-*onc* genes. The transforming capacity of these genes, assayed by transfection of NIH/3T3 cells, appears to correlate with the presence of a single point mutation within the coding sequences of the human c-*ras* gene (10,11).

The possibility that chromosomal rearrangements might cause activation of c-*onc* genes has also been proposed (4,7,12). Evidence to support this model is now emerging from studies in a number of laboratories (see below). In this review we will discuss the role of c-*onc* genes in neoplastic disease, focussing on the c-*myc* gene. Activation of the c-*myc* gene has been demonstrated in neoplasms induced by a number of different carcinogenic agents.

The c-myc Gene

The c-*myc* gene, the cellular homolog of the v-*myc* gene of MC29 and other acute viruses (13,14), is present in the genomes of all vertebrates (15). Although little is known about its function in the normal cell, the fact that this gene (as well as other c-*onc* genes) is highly conserved throughout evolution suggests that it is essential for normal development of the organism. Several observations point to a possible role in hematopoietic cell growth control or differentiation. (i) Although c-*myc* is expressed at low levels in most tissues, it is expressed at high levels in hematopoietic tissues (bursa, thymus, spleen, bone marrow) of young birds (up to 4 weeks after hatching) (16; C.-K. Shih, unpublished data), and in isolated B and T cells (17). (ii) The c-*myc* gene is expressed at elevated levels in human tissue culture cells representing early stages of B-cell differentiation (18). (iii) Two c-*onc* genes (c--*myc* and c-*myb*) are expressed at elevated levels in a human promyelocytic leukemia cell line, HL-60 (19,20). Treatment of HL-60 cells with retinoic acid or other compounds induces terminal differentiation, and simultaneous shut-off of c-*myc* (and c-*myb*) expression.

The c-*myc* coding sequences are encoded in two exons, interrupted by an intron of approximately 1 kb (Fig. 1) (21-23). This contrasts with the structure of the v-*myc* gene of MC29, which lacks introns (see Fig. 2). In an effort to localize the promoter for c-*myc* in normal cells, we have used a c-*myc* clone as template in a cell-free transcription system (B.G. Neel, unpublished data). Two initiation sites, located approximately 1 and 1.8 kb upstream from the 5' exon of c-*myc*, were utilized *in vitro* (Fig. 1, arrows). The site of initiation *in vivo* has not been determined. A minor c-*myc* transcript, approximately 4.2 kb in length, can be identified in normal cells (unpublished observation). The size of this putative

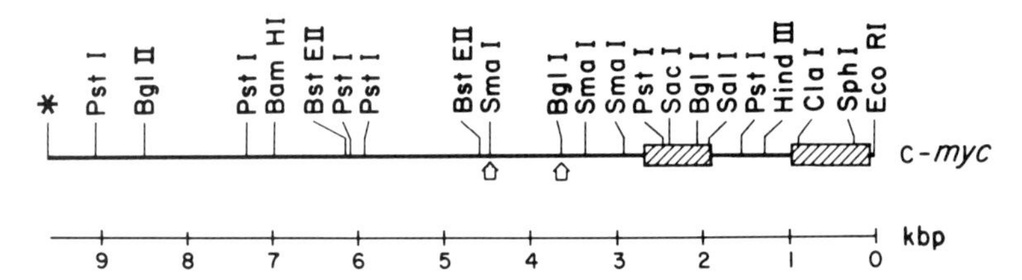

Fig. 1. Structure of the avian c-myc gene. A c-myc clone was selected from a random library of normal chicken DNA, by screening with a myc-specific probe (21). Coding sequences (cross-hatched boxes), identified by comparison with the v-myc gene, are located in two exons (21-23). Two possible promoters (open arrows) were identified by transcription of the c-myc clone in a cell-free system (Neel, unpublished).

primary transcript of c-myc is consistent with initiation from either of these possible promoter sites in vivo, depending on where the poly(A) addition signal is located.

Strategies for myc Gene Activation: MC29 Virus vs. ALV

MC29 is an acute retrovirus that causes a variety of neoplasms, including carcinomas, leukemias, sarcomas, and lymphomas (1,5,16, 24). Neoplasms generally appear between 3 and 6 weeks after infection. MC29 was presumably derived by recombination between an ALV-type virus and the cellular gene, c-myc, which has become inserted in the viral genome, replacing the viral pol gene and portions of gag and env. The myc gene is thus controlled by viral regulatory sequences, and is expressed at high levels in essentially all infected cells.

The v-myc gene product of MC29, a 110K gag-myc fusion protein (13,14,25; see Fig. 2), appears to lack protein kinase activity (25,26). Recent evidence indicates that this protein is a DNA binding protein, and is located primarily in the nucleus (27,28).

In contrast to MC29 virus, the avian leukosis viruses do not cause morphologic transformation of tissue culture cells at detectable frequency, and induce neoplasms in infected birds only after a long latent period (1,4,5). The most frequently observed neoplasm is a B-cell lymphoma, which originates in the bursa of Fabricius.

We have previously presented evidence that ALV induces B-cell lymphomas by activating the host c-myc gene (7). These studies demonstrated that: (i) in nearly all ALV-induced lymphomas a provirus is integrated adjacent to a single cellular gene, c-myc; (ii) levels of c-myc mRNA are elevated 30- to 100-fold in the lymphomas,

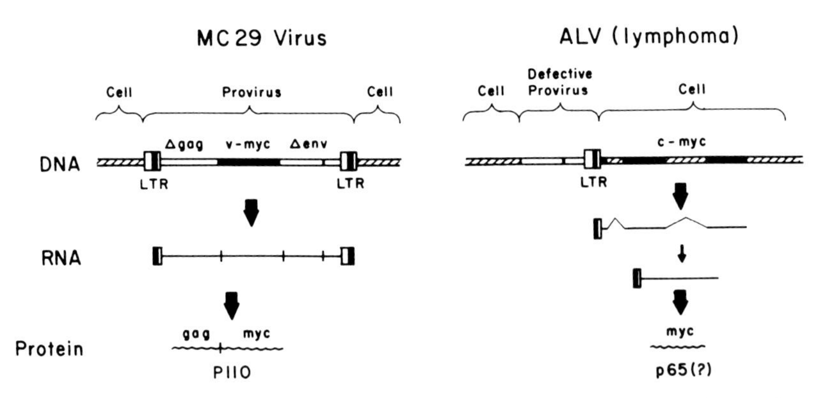

Fig. 2. Strategies for myc gene expression in ALV and MC29 virus-induced tumors. An integrated MC29 provirus in shown on the left. The site of integration does not play a major role in transformation by this virus, since the oncogene (v-myc) is carried in the viral genome, already linked to viral regulatory sequences. Transformation by ALV, however, occurs only when the provirus is inserted adjacent to a c-onc gene. A defective ALV provirus (right) is shown integrated in an orientation such that transcription of c-myc can initiate on the viral promoter, as found in most ALV-induced lymphomas. The myc gene product of MC29 is a gag-myc fusion protein (P110). The putative product of the c-myc gene (p65) in ALV-induced lymphomas has not been identified. DNA and RNA analyses indicate that it does not contain virus-encoded peptides (4,7,21). (From ref. 4.)

compared to equivalent normal tissues; (iii) in most tumors, the myc-specific sequences are present in RNA transcripts that contain approximately 100 nucleotides of viral information derived from the long terminal repeat (LTR) of the integrated provirus. These and other studies (4,7,21,29,30,31) indicate that transcriptional activation of c-myc results from rare integrations of viral regulatory sequences adjacent to the host c-myc gene (Fig. 2).

In the vast majority of B-cell lymphomas the provirus is integrated upstream from the c-myc gene, and in the same transcriptional orientation (7,21,29,31,32) (see Fig. 2). Transcription initiates within the LTR (long terminal repeat) - in most cases the 3' LTR - and reads into the adjacent cellular sequences. Integrations appear to be clustered in several discrete regions located within a 2 kb region immediately upstream from the c-myc coding sequences (21,29, 32). A majority (26/30) of the ALV-induced lymphomas analyzed in our studies (29,32) contained proviruses integrated downstream from both potential c-myc promoters (see Fig. 1), but upstream from the 5' exon of c-myc. In the remaining four tumors, the proviruses were located further upstream, either between the two putative promoters (one tumor) or very close to the more distal promoter (three tumors). All of these proviruses were inserted in the same transcriptional orientation as c-*myc*. Thus, the viral promoter/regulator complex has, in most cases, displaced the cellular promoter and regulatory sequences.

Payne et al. (33) have identified one tumor in which integration was downstream from c-*myc*, and several tumors in which integration was upstream, but in the opposite transcriptional orientation. Since the viral promoter could not be utilized for c-*myc* activation in these tumors, it seems likely that initiation occurs on a cellular promoter. "Enhancer" sequences within the viral LTR may exert some positive regulatory influence over transcription, in a manner analogous to that demonstrated for the 72 bp repeat of SV40 (34,35). This, however, is apparently a less efficient mechanism for activating c-*myc*, since these alternative orientations are found in only a small fraction of ALV-induced lymphomas.

Interestingly, most, and perhaps all, of the proviruses integrated next to c-*myc* are defective (29-33). In most cases, the 5' LTR, plus additional coding information, is deleted. This observation suggests that defectiveness may play an essential role in c-*myc* activation. One explanation is that efficient transcription from the 3' LTR can occur only when active transcription from the 5' LTR is abolished. Since transcription of viral RNA (from the 5' LTR) proceeds into, and beyond, the initiation site within the 3' LTR (36), normal transcription might block efficient utilization of the 3' promoter.

More than fifteen different c-*onc* genes have been identified thus far (6), and it is probable that more will be identified. Thus it was surprising that a single gene, c-*myc*, was involved in such a high proportion of ALV-induced lymphomas (7). Although it is possible that sequences upstream from c-*myc* are preferred sites for ALV integration, this explanation seems unlikely. No specificity for ALV integration has been demonstrated (28). Furthermore, as mentioned above, it is clear that ALV integration near c-*myc* can occur at more than a single site. It seems more likely, therefore, that involvement of c-*myc* in B-cell lymphoma relates to some property of the target cell. Two possibilities can be considered: (i) only certain target cells respond to the c-*myc* product in a way that leads to neoplastic transformation; and (ii) integration may occur preferentially in a transcriptionally active region, due to conformational changes in the chromatin. As mentioned above, c-*myc* is expressed at elevated levels in early bursal cells (16). Relevant to this question is the recent observation that a different c-*onc* gene, c-*erb*, appears to be activated in ALV-induced erythroleukemias (H.-J. Kung, personal communication). Furthermore, neither c-*myc* nor c-*erb* are involved in ALV-induced sarcomas, nephroblastomas, or carcinomas (M.C. Simon; H. Varmus; personal communications).

Integration appears to be specific in these tumors, but the cellular genes involved have not been identified.

Cooper and Neiman (37) have identified a second cellular gene - unrelated to c-*myc* and not linked to proviral sequences - that appears to be involved in lymphomagenesis. These authors have proposed a multi-step process in which the second gene acts at a late stage in tumor development. The mechanism responsible for activation of the second gene is unknown.

Possible Involvement of c-onc Genes in Human Neoplasia

Specific chromosomal abnormalities, especially translocations, are associated with many forms of human cancer (12,38). The classic example is the 9;22 translocation that generates the Philadelphia chromosome associated with chronic myelogenous leukemia (CML) (39). Specific chromosomal abnormalities have also been associated with Burkitt lymphoma (40,41) and acute nonlymphoblastic leukemia (38). One possible explanation for these specific chromosomal abnormalities in human cancers is that rearrangements place a strong transcriptional signal adjacent to the c-*onc* gene (7,12).

To test this hypothesis, we have mapped several human c-*onc* genes, to determine whether they are located at chromosomal sites known to be involved in translocations associated with specific human malignancies (42). Mapping was performed by *in situ* hybridization. With this technique it is possible not only to assign a gene to a specific chromosome, but also to determine the precise map location of the gene on the chromosome.

Two human c-*onc* genes, c-*mos* and c-*myc*, were localized to the long arm of chromosome 8 (42). The human c-*mos* gene was mapped to band position 8q22, and the c-*myc* gene to position 8q24. These sites correspond to the breakpoints involved in specific translocations found in the M-2 subset of acute nonlymphoblastic leukemia (8q22), and Burkitt and other forms of non-Hodgkin's lymphoma (8q24). Direct evidence for translocations of c-*myc* in various mammalian systems has been obtained by restriction analysis of tumor DNAs (M. Cole, C. Croce, P. Leder, personal communications; Hayward et al., unpublished observations).

It seems likely, therefore, that translocations of c-*mos* and/or c-*myc* are causally related to neoplastic disease. It is interesting that, in the Burkitt case, the translocation invariably involves chromosome 8 (position q24, the location of c-*myc*) plus one of three other chromosomes: 2, 14, or 22. These chromosomes carry the Ig κ, heavy, and λ chain Ig loci, respectively (43-45). In two cases, these loci have been mapped precisely to the breakpoints involved in Burkitt lymphoma (44,45). The Ig loci are expressed at high levels in the target cells (early B-cells) involved in Burkitt lymphoma.

Translocation of c-*myc* to an Ig locus might, therefore, place c-*myc* under the control of the transcriptionally active Ig regulatory sequences. Consistent with this interpretation, we have found higher levels of c-*myc* mRNA in cells carrying the 8;14 translocation, compared with other lymphoid cell lines. We have no evidence, however, that this is a direct consequence of the translocation. We could detect no expression of c-*mos* in any cell lines tested.

CONCLUSIONS

Activation of the oncogenic potential of a c-*onc* gene can apparently result from two different types of genetic changes: alterations in regulatory sequences that affect the expression of the gene, and changes within the coding sequences that alter the properties of the gene product.

Activation of c-*myc* by ALV appears to be an example of a regulatory change, although the possibility that point mutations have occurred within the c-*myc* coding sequences cannot be excluded. It seems likely that Burkitt lymphoma, and other human neoplasms in which specific translocations occur, will also fall into this class of neoplasms resulting from regulatory changes in a normal c-*onc* gene. Two groups have demonstrated transformation of NIH/3T3 cells by c-*onc* genes (c-*mos* and c-*ras*) linked to LTR sequences (46,47). Thus, regulatory changes alone are sufficient to activate the oncogenic potential of these c-*onc* genes.

In the acute transforming viruses, a c-*onc* gene has been placed under regulatory control of viral sequences located within the LTR. However, additional changes within the coding sequences (mutations or deletions) are also observed in some cases. The relative importance of these two factors has not yet been evaluated.

An apparent example of a change in coding sequences is the c-*ras* gene, implicated in various human carcinomas (8,9). A single base change within the coding region appears to be responsible for the ability of this gene to transform NIH/3T3 cells (10,11).

The examples cited above include malignancies induced by a wide variety of carcinogenic agents, in hosts ranging from chicken to man. In each case, a genetic change (recombination, proviral integration, translocation, point mutation) affecting either regulatory or coding sequences, has altered the properties of a c-*onc* gene with profound consequences to the organism. Thus, a class of normal cellular genes (c-*onc* genes), first identified as oncogenes in acute retroviruses, appears to play a central role in malignancies of both viral and nonviral origin.

ACKNOWLEDGEMENTS

We thank Nancy Goldberg and Anne Manwell for excellent technical assistance, and Lauren O'Connor for help in preparation of the manuscript. This work was supported by grant CA34502 from the National Institutes of Health, and by the Flora E. Griffin Memorial Fund. B.G.N. is a biomedical fellow in the Medical Scientist Training Program of the National Institutes of Health.

REFERENCES

1. Graf, T., and H. Beug (1978) Avian leukemia viruses. Interaction with their target cells in vivo and in vitro. Biochem. Biophys. Acta Rev. Cancer 516:269.
2. Hanafusa, H. (1981) Cellular origin of transforming genes of RNA tumor viruses. Harvey Lect. 75:255.
3. Bishop, J.M. (1981) Enemies within: The genesis of retrovirus oncogenes. Cell 23:5.
4. Hayward, W.S., B.G. Neel, and S.M. Astrin (1982) Avian leukosis viruses: Activation of cellular "oncogenes." In Advances in Viral Oncology, Vol. 1, George Klein, ed. Raven Press, New York, p. 207.
5. Moscovici, C., and L. Gazzolo (1982) Transformation of hemopoietic cells with avian leukemia viruses. In Advances in Viral Oncology, Vol. 1, G. Klein, ed. Raven Press, New York, p. 83.
6. Coffin, J.M., H.E. Varmus, J.M. Bishop, M. Essex, W.D. Hardy, G.S. Martin, N.E. Rosenberg, E.M. Scolnick, R.A. Weinberg, and P.K. Vogt (1981) A proposal for naming host cell-derived inserts in retrovirus genomes. J. Virol. 49:953.
7. Hayward, W.S., B.G. Neel, and S.M. Astrin (1981) Activation of a cellular onc gene by promoter insertion in ALV-induced lymphoid leukosis. Nature 290:475.
8. Der, C.J., T.G. Krontiris, and G.M. Cooper (1982) Transforming genes of human bladder and lung carcinoma cell lines are homologous to the ras genes of Harvey and Kirsten sarcoma viruses. Proc. Natl. Acad. Sci., USA 79:3637.
9. Parada, L.F., C.J. Tabin, C. Shih, and R.A. Weinberg (1982) Human EJ bladder carcinoma oncogene is homologue of Harvey sarcoma virus ras gene. Nature 297:474.
10. Tabin, C.J., S.M. Bradley, C.I. Bargmann, R.A. Weinberg, A.G. Papageorge, E.M. Scolnick, R. Dhar, D.R. Lowy, and E.H. Change (1982) Mechanism of activation of a human oncogene. Nature 300:143.
11. Reddy, E.P., R.K. Reynolds, E. Santos, and M. Barbacid (1982) A point mutation is responsible for the acquisition of transforming properties by the T24 human bladder carcinoma oncogene. Nature 300:149.
12. Klein, G. (1981) Changes in gene dosage and gene expression: A common denominator in the tumorigenic action of viral oncogenes and non-random chromosomal changes. Nature 294:313.

13. Mellon, P., A. Pawson, K. Bister, G.S. Martin, and P.H. Duesberg (1978) Specific RNA sequences and gene products of MC29 virus. Proc. Natl. Acad. Sci., USA 75:5874.
14. Roussel, M., S. Saule, C. Lagrou, C. Rommens, H. Beug, T. Graf, and D. Stehelin (1979) Three new types of viral oncogene of cellular origin specific for haematopoietic cell transformation. Nature 281:452.
15. Sheiness, D.K., S. Hughes, H.E. Varmus, E. Stubblefield, and J.M. Bishop (1980) Virology 105:415.
16. Hayward, W.S., C.-K. Shih, and C. Moscovici (1982) Induction of bursal lymphoma by myelocytomatosis virus-29 (MC29). In Cetus-UCLA Symposium on Tumor Viruses and Differentiation, Liss, New York.
17. Gonda, T.J., D.K. Sheiness, and J.M. Bishop (1982) Transcripts from the cellular homologs of retroviral oncogenes: Distribution among chicken tissues. Mol. Cell Biol. 2:617.
18. Eva, A., K.C. Robbins, P.R. Andersen, A. Srinivasan, S.R. Tronick, E.P. Reddy, N.W. Ellmore, A.T. Galen, J.A. Lautenberger, T.S. Papas, E.H. Westin, F. Wong-Staal, R.C. Gallo, and S.A. Aaronson (1982) Cellular genes analogous to retroviral onc genes are transcribed in human tumor cells. Nature 295:116.
19. Westin, E.H., R.C. Gallo, S.K. Arya, A. Eva, L.M. Souza, M.A. Baluda, S.A. Aaronson, and F. Wong-Staal (1982) Differential expression of the amv gene in human hematopoietic cells. Proc. Natl. Acad. Sci., USA 79:2194.
20. Westin, E.H., F. Wong-Staal, E.P. Gelman, R.D. Dalla-Favera, T.S. Papas, J.A. Lautenberger, A. Eva, E.P. Reddy, S.R. Tronick, S.A. Aaronson, and R.C. Gallo (1982) Expression of cellular homologues of retroviral onc genes in human hematopoietic cells. Proc. Natl. Acad. Sci., USA 79:2490.
21. Neel, B.G., G.P. Gasic, C.E. Rogler, A.M. Skalka, T. Papas, S.M. Astrin, and W.S. Hayward (1982) Molecular cloning of virus-cell junctions from ALV-induced lymphomas: Comparison with the normal c-myc gene. J. Virol. 44:158.
22. Vennstrom, B., D. Sheiness, J. Zabielski, and J.M. Bishop (1982) Isolation and characterization of c-myc, a cellular homolog of the oncogene (v-myc) of avian myelocytomatosis virus strain 29. J. Virol. 42:773.
23. Robins, T., K. Bister, C. Garon, T. Papas, and P. Duesberg (1982) Structural relationship between a normal chicken DNA locus and the transforming gene of the avian acute leukemia virus MC29. J. Virol. 41:635.
24. Hayman, M.J. (1981) Transforming proteins of avian retroviruses. J. Gen. Virol. 52:1.
25. Bister, K., W.-H. Lee, and P.H. Duesberg (1980) Phosphorylation of the nonstructural proteins encoded by avian acute leukemia viruses and by avian Fujinami sarcoma virus. J. Virol. 26:617.
26. Sefton, B.M., T. Hunter, K. Beemon, and W. Eckhart (1980) Evidence that phosphorylation of tyrosine is essential for cellular transformation by Rous sarcoma virus. Cell 20:807.

27. Abrams, H.D., L.R. Rohrschneider, and R.N. Eisenman (1982) Nuclear location of the putative transforming protein of avian myelocytomatosis virus. Cell 29:427.
28. Donner, P., I. Greiser-Wilke, and K. Moelling (1982) Nuclear localization and DNA binding of the transforming gene product of avian myelocytomatosis virus. Nature 296:262.
29. Neel, B.G., W.S. Hayward, H.L. Robinson, J. Fang, and S.M. Astrin (1981) Avian leukosis virus-induced tumors have common proviral integration sites and synthesize discrete new RNAs: Oncogenesis by promoter insertion. Cell 23:323.
30. Payne, G.S., S.A. Courtneidge, L.B. Crittenden, A.M. Fadly, J.M. Bishop, and H.E. Varmus (1981) Analyses of avian leukosis virus DNA and RNA in bursal tumors: Viral gene expression is not required for maintenance of the tumor state. Cell 23:311.
31. Fung, Y.-K., A.M. Fadly, L.B. Crittenden, and H.-J. Kung (1981) On the mechanism of retrovirus-induced avian lymphoid leukosis: Deletion and integration of the proviruses. Proc. Natl. Acad. Sci., USA 78:418.
32. Rovigatti, U.G., C.E. Rogler, B.G. Neel, W.S. Hayward, and S.M. Astrin (1982) Expression of endogenous oncogenes in tumor cells. In Fourth Annual Bristol-Myers Symposium on Cancer Research, Academic Press, New York, p. 319.
33. Payne, G.S., J.M. Bishop, and H.E. Varmus (1982) Multiple arrangements of viral DNA and an activated host oncogene (c-myc) in bursal lymphomas. Nature 295:209.
34. Moreau, P., R. Hen, B. Wasylyk, R. Everett, M.P. Gaub, and P. Chambon (1981) The SV40 72 base pair repeat has a striking effect on gene expression both in SV40 and other chimeric recombinants. Nucleic Acids Res. 9:6047.
35. Gruss, P., R. Dhar, and G. Khoury (1981) Simian virus 40 tandem repeated sequences as an element of the early promoter. Proc. Natl. Acad. Sci., USA 78:943.
36. Hayward, W.S., and B.G. Neel (1981) Retroviral gene expression. Curr. Top. Microbiol. Immunol. 91:217.
37. Cooper, G.M., and P.E. Neiman (1981) Two distinct candidate transforming genes of lymphoid leukosis virus induced neoplasms. Nature 292:857.
38. Rowley, J.D. (1980) Chromosome abnormalities in human leukemia. Annual Rev. Genet. 14:17.
39. Nowell, P.C., and D.A. Hungerford (1960) A minute chromosome in human chronic granulocytic leukemia. Science 132:1497.
40. Manalov, G., and Y. Manalova (1972) Marker band in one chromosome 14 from Burkitt lymphomas. Nature 237:33.
41. Mitelman, F. (1981) Marker chromosome 14q^{+} in human cancer and leukemia. Adv. in Cancer Res. 34:141.
42. Neel, B.G., S.C. Jhanwar, R.S.K. Chaganti, and W.S. Hayward (1982) Two human c-onc genes are located on the long arm of chromosome 8. Proc. Natl. Acad. Sci., USA 79:7842.
43. Erikson, J., J. Martinis, and C.M. Croce (1981) Assignment of the genes for human immunoglobulin chains to chromosome 22. Nature 294:173.

44. Kirsch, I.R., C. Morton, K. Nakahara, and P. Leder (1982) Human immunoglobulin heavy chain genes map to a region of translocations in malignant B lymphocytes. Science 216:301.
45. Malcolm, S., P. Barton, C. Murphy, M.A. Ferguson-Smith, D.C. Bently, and T.H. Rabbits (1982) Localization of human immunoglobulin light chain variable region genes to the short arm of chromosome 2 by in situ hybridization. Proc. Natl. Acad. Sci., USA 79:4957.
46. Blair, D.G., M. Oskarsson, T.G. Wood, W.L. McClements, P.J. Fischinger, and G.F. Vande Woude (1981) Activation of the transforming potential of a normal cell sequence: A molecular model for oncogenesis. Science 212:941.
47. DeFeo, D., M.A. Gonda, H.A. Young, E.H. Chang, D.R. Lowy, E.M. Scolnick, and R.W. Ellis (1981) Analysis of two divergent rat genomic clones homologous to the transforming gene of Harvey murine sarcoma virus. Proc. Natl. Acad. Sci., USA 78:3328.

HEPATITIS B VIRUS AS AN ENVIRONMENTAL CARCINOGEN

William S. Robinson, Roger H. Miller,
and Patricia L. Marion

Stanford University School of Medicine
Stanford, California 94305

INTRODUCTION

Hepatitis B virus (HBV) is one of the most common and interesting human viruses. Investigation of HBV during the 1970s revealed unique antigenic, ultrastructural, molecular, and biological features which distinguished it from members of all of the recognized virus groups. One of the most notable features of this virus was its DNA structure. Virions contain small circular DNA molecules (1) that are partly single-stranded (2-4) and a DNA polymerase in the virion (5,6) can repair the DNA to make it fully double-stranded (2,3). Among the unique biological features are its liver tropism (7) and the common occurrence of persistent infection with viral antigen and infectious virus in high concentrations in the blood (8-11) continuously for years. This pattern of infection accounts for the transmission of HBV by percutaneous transfer of serum and serum-containing material. Several disease syndromes in addition to acute hepatitis may occur during acute and chronic HBV infection. Included are a serum sickness-like syndrome, membranous glomerulonephritis, and necrotizing vasculitis (polyarteritis), all probably related to viral antigen-antibody complex mediated tissue injury (12). In addition, persistent infection can be associated with a normal liver or with chronic hepatitis. The latter may be severe and progressive, and lead to cirrhosis and in some cases hepatocellular carcinoma (13). Its narrow host range (confined to man and a few higher primates) and failure so far to infect tissue culture cells have made it difficult to investigate some questions about HBV.

Although for several years HBV was considered to be a unique virus, recently similar viruses have been found in three different animal species. The first, woodchuck hepatitis virus (WHV), discovered in 1978 in sera of eastern woodchucks (Marmota monax), a member of the Sciuridae or squirrel family, by Summers, Smolec, and Snyder (14) after Snyder had observed that the most frequent cause of death in captive woodchucks in the Philadelphia Zoo was hepatocellular carcinoma accompanied by chronic hepatitis. The third member of this virus family, ground squirrel hepatitis virus (GSHV), was found in the Beechey ground squirrel (Spermophilus beecheyi), another genus of the Sciuridae family, by Marion, et al. (15) in 1980. The observation of frequent hepatomas in a species of domestic ducks in the People's Republic of China led to the discovery of the fourth member of this virus family, now called duck hepatitis B virus (DHBV), in sera from ducks in China by Summers, London, Sun, Blumberg, et al. (unpublished data). A similar virus has been found in Pekin ducks (Anas domesticus) in the United States (16). Hepatitis B virus and the three related viruses of lower animals share unique ultrastructural, molecular, antigenic, and biologic features and this virus family has been called the hepadna virus group (17). The evidence for a relationship between infection with these viruses and hepatocellular carcinoma (HCC) will be reviewed here. The antigenic, molecular, and ultrastructural features of these viruses and their biological properties have recently been reviewed in detail (18).

EPIDEMIOLOGIC ASSOCIATION OF HEPADNA VIRUS INFECTION AND HEPATOCELLULAR CARCINOMA (HCC)

Hepatocellular carcinoma in man has a worldwide distribution and numerically is one of the major cancers in the world today. Although HCC is rare in many parts of the world, it occurs commonly in sub-Saharan Africa, eastern Asia, Japan, Oceania, Greece, and Italy. In certain areas of Asia and Africa, it is the most common cancer. Geographic areas with the highest incidence of HCC are also areas where persistent HBV infections occur at the highest known frequencies. Within the limits of the data available, there appears to be a good correlation between geographic distribution of HCC and active HBV infection with the highest frequency of both being sub-Saharan Africa and eastern Asia (13). In addition, hepatitis B surface antigen (HBsAg), a marker of active HBV infection, has been found four to six times more frequently in sera of patients with HCC than in tumor negative controls in both high-HCC incidence and low-HCC incidence geographic areas (13). An ongoing prospective study of 22,707 male government workers in Taiwan, 15% of which were HBsAg positive, revealed the incidence of HCC to be more than 300 times higher in HBsAg positive than HBsAg negative individuals (19). Three percent of HBsAg positive patients 50 years or older developed HCC per year, and 43% of all deaths in HBsAg carriers 40 years or older were due to HCC.

The high incidence of persistent HBV infection in mothers of HCC patients, in contrast to that in fathers (20), suggests that transmission from mothers to newborn or infant children may be a frequent mode and the time of HBV infection in HCC patients. The finding of low HBsAg titers, together with the rare occurrence of HB core antigen (HBcAg) and HB e antigen (HBeAg) in most patients (13,21) also suggests that the persistent infections in HCC patients are of long duration. If HBV infection does occur frequently at very early ages in HCC patients, the age distribution of patients when the tumors were clinically recognized in high-incidence areas (22) would suggest that tumors appear after a mean duration of approximately 35 to 40 years of HBV infection. Very few cases of HCC occur in children (22). Between 60% and 90% of HCC patients have coexisting cirrhosis (19,22-24), suggesting that this lesion in association with persistent HBV infection may predispose to HCC, although clearly the presence of cirrhosis is not an absolute requirement. Epidemiological data defining the association of HCC and persistent HBV infection have been reviewed in detail (13,25) and these represent strong evidence for an important role of HBV in HCC formation in man.

Hepatocellular carcinoma appears to occur even more frequently in WHV-infected captive woodchucks (14). In two colonies of woodchucks in this country, approximately one-third of animals persistently infected with WHV have been observed to develop HCC per year. All such animals have histological findings of moderately severe active hepatitis and moderately high levels of DNA containing virions and WHsAg (woodchuck hepatitis surface antigen) in their sera. No hepatomas have developed in uninfected animals.

Hepatitis virus infected ground squirrels have a dramatically different disease response. Although the rate of persistent infection is very high in endemic areas (up to 52%) and levels of DNA containing virions and GSHsAg (ground squirrel hepatitis surface antigen) in sera are always unusually high (15), little or no hepatitis and no HCC has been observed in infected animals (26).

THE OCCURRENCE AND STATE OF HBV IN HCC TISSUE

Immunofluorescent and immunoperoxidase staining of tumor tissue have been used to demonstrate HBsAg and HBcAg in tissue. Tumor cells of patients with HBsAg in the blood and nontumorous liver cells containing HBsAg and/or HBcAg appear most often to be antigen negative. Small numbers of positive cells have been seen in some tumors (reviewed in ref. 25). HBsAg has also been reported to have been detected infrequently by immunoperoxidase staining in tumors of patients without HBsAg in the blood (e.g., ref. 27).

Viral DNA has also been found in HCC tissue. Lutwick and Robinson (28) found viral DNA base sequences in DNA extracted from tumor tissue of a single patient in the United States with HBsAg in the blood (although significantly less than in nontumorous tissue of infected patients) and no viral DNA base sequences in the DNAs of two HCCs from HBsAg negative patients. Summers, et al., (29) found HBV DNA base sequences in the DNAs extracted from nine to ten African HCC patients with HBsAg in the blood and in one of four tumors of patients who were HBsAg negative.

With the development of methods to cleave DNAs at specific sites with restriction endonucleases, separate the resulting DNA fragments by gel electrophoresis, and transfer them to nitrocellulose or other solid media for hybridization with radiolabeled DNA probes (Southern blotting, ref. 30), and to clone specific DNA fragments in bacterial cells, more definitive analysis of the state of viral DNA in HCC tissue has been possible. In addition to free or episomal unit length HBV DNA in HCC tissue from some HBsAg positive patients, evidence for viral DNA sequences integrated in tumor cell DNA has also been found by several investigators (31-38). Evidence for this is the finding of one or more DNA fragments containing viral DNA sequences which are larger than unit length viral DNA (3,200 bp) after but not before digestion of cell DNA with a restriction enzyme (e.g., HindIII) for which no recognition sites exist in the viral DNA. Similar findings of viral sequences have also been reported for some tumors in cirrhotic patients who were HBsAg negative (39).

The sizes of the large DNA fragments containing viral sequences are different in different tumors indicating the integration sites are different and no single integration site common to all tumors is apparent. Such evidence for integration of hepatitis B virus DNA is not unique to hepatocellular carcinomas since similar high molecular weight HindIII fragments with viral sequences have been detected in DNA of chronically infected nontumorous liver (39-41). Again, the specific high molecular weight HindIII DNA fragments containing viral sequences were different in different patients. The ability to detect such a DNA fragment by Southern blotting has been interpreted to mean that viral DNA is integrated in the same site in many different cells of the liver of each patient but the site is different in different patients.

The complete meaning of these findings is not clear, however, since integration of other viral DNAs (e.g., retroviruses) appears to be random so that specific integration sites are not detected in tissue DNA by the experimental strategy described above unless the cells are of clonal origin (e.g., as are cells in most viral-induced tumors). More direct evidence that these high molecular weight DNA fragments from infected liver and HCC actually represent cellular DNA covalently attached to viral DNA is needed. For example, isolation of these fragments by molecular cloning in bacterial cells would permit more definitive characterization of them. High molecular weight DNA forms containing viral sequences without cellular sequences have been detected in infected woodchuck (Rogler and Summers, in press) and ground squirrel (44) liver. These appear not to represent integrated viral DNA but could be confused with integrated sequences when analyzed only by Southern blotting.

Similar findings suggesting viral DNA integrations have been made in HCC from woodchucks infected with WHV. Not only has Southern blot analysis of woodchuck HCC DNA restricted with HindIII revealed high molecular weight DNA fragments containing WHV sequences (42), but such fragments have been cloned in bacterial cells using lambdoid vectors and shown to contain viral DNA sequences covalently attached to cellular DNA (43). The viral DNA showed gross rearrangements and deletions. Thus, many HCC in man and woodchucks contain viral DNA sequences and these appear to be integrated in cellular DNA.

THE STATE OF HBV IN TISSUE CULTURE CELL LINES ISOLATED FROM HUMAN HCC

A tissue culture cell line (PLC/PRF/5 or PLC cells) isolated in 1975 from HCC tissue of an African man with persistent HBV infection has been shown to continuously produce HBsAg in culture (45). The cells produce no detectable HBcAg, whole virions or infectious HBV (45,46). Hepatitis B virus DNA sequences appear to be present exclusively in an integrated form and no unit length free or episomal vital DNA has been detected (31-34). The amount of viral DNA corresponds to only a few (∿10) copies of viral DNA per cell. Sequences in all regions of the viral DNA are represented suggesting that all of the viral DNA is present (31). Restriction endonuclease HindIII (an enzyme which does not cleave HBV DNA) digestion of PLC/PRF/5 cell DNA produces at least eight specific DNA fragments containing integrated HBV DNA and flanking cell DNA sequences (47). The HindIII fragments containing HBV DNA sequences have electrophoretic mobilities of linear DNA fragments of molecular weight 40, 38, 24, 17.2, 11.7, 6.6, 4.8, and 1.9 kbp (47). These results suggest that there are at least eight sites in PLC cell DNA in which HBV DNA is integrated. Viral specific RNAs corresponding to extensive regions of the viral DNA have been demonstrated in the cell (31,34).

The observation that only HBsAg and not HBcAg is expressed in these cells, although the DNA sequences within both genes are present, raises the question of mechanisms controlling viral gene expression. Because methylation of some herpes simplex virus and adenovirus genes appears to turn off their expression in infected cells, methylation of the HBsAg and HBcAg genes in PLC cells has been studied in this lab (47). The methylation of various hepatitis B virus DNA sequences was examined using the restriction endonucleases HpaII and MspI. Although both enzymes recognize the DNA sequences 5'-CCGG-3', HpaII will not cleave the DNA if the internal cytosine is methylated. Hepatitis B virus DNA from Dane particles and virus-infected liver tissue (nontumor) was digested with HpaII or MspI, fractionated by electrophoresis in agarose gels, and the restriction enzyme cleavage pattern examined by Southern blot analysis. No methylation of the 5'-CCGG-3' recognition sequence was detected in either virion DNA or HBV DNA from infected liver tissue. On the other hand, digestion of PLC/PRF/5 DNA with HpaII and MspI showed that the integrated HBV DNA sequences were methylated. Further analysis using probes specific for various regions of the HBV genome demonstrated that some of the hepatitis B viral DNA sequences including those specifying the major surface antigen polypeptide were methylated infrequently or not at all. In contrast, the viral DNA sequences coding for the major core polypeptide were extensively methylated. Since surface antigen is expressed in these cells, while the core antigen is not, the results indicate that DNA methylation could account for the selective expression of HBV genes in this hepatoma cell line.

CONCLUSIONS

Epidemiologic studies clearly show that persistent HBV infection is strongly associated with the occurrence of HCC in man in both high and low prevalence geographic areas of the world. The incidence of HCC in WHV-infected woodchucks is even higher than is HCC in HBV-infected humans.

Viral DNA is frequently found in an integrated state in HCC tissue of infected patients and woodchucks. The presence of viral genes in tumors is a necessary condition for tumor induction by the mechanisms known for the recognized tumor viruses which have been studied extensively in animal systems. Hepatitis B virus DNA has also been found to be present exclusively in an integrated state and in multiple sites in a cell line isolated from human HCC tissue. Although all of viral DNA sequences are apparently present in these cells, only the gene for the HBsAg polypeptide (the viral envelope protein) and not that for the internal HBcAg polypeptide is expressed. The regulation of viral gene expression in these cells may involve selective methylation of viral genes.

The exact role of HBV and WHV in HCC formation is not understood at the molecular level at this time but it is of interest that at least superficially these viruses appear to share some features with retroviruses. Among these features are similarities in genome structure, although the nucleic acid type in virions is different for the two viruses, i.e., DNA in the case of hepadna viruses and RNA in retroviruses. Separation and repair of the cohesive ends of hepadna virus DNAs results in linear molecules with inverted, repeat terminal sequences of approximately 300 bp (48), similar to those of retroviruses. In both viruses all of the viral messenger RNAs appear to be transcribed from the same DNA strand and, thus, in the same direction. In addition, recent evidence (49) suggests that at least the duck virus DNA may replicate through an RNA intermediate utilizing a reverse transcriptase, a mechanism with some analogy to retrovirus replication. A third similarity is that viruses of both groups appear to integrate readily in cellular DNA.

However, it has yet to be shown that hepadna virus DNA integration is a regular and integral event in virus replication, that the viral genome in the integrated state retains its organization, that integration occurs at a specific site in the viral DNA, or that integration is essential to the mechanism of cell transformation; all of which appear to be features of retroviruses. A fourth similarity is that when exclusively integrated in the DNA of infected cells, both hepadna virus (31,46) and retroviruses (50) may express only the gene for their envelope protein. A fifth similarity is tumor formation during infection by at least some members of each virus group. The clear association between HBV and WHV infections, and hepatocellular carcinoma is among the more intriguing features of these viruses and it will be of great interest to investigate in more detail their role in formation of these tumors. It will be important to determine if they may integrate in sites adjacent to oncogenes and function as retroviruses are thought to function in cell transformation and tumor induction (51).

Although the different hepadna viruses all share many unique features and at least HBV and WHV infections are associated with HCC formation, there are interesting differences. The highest rates of HCC formation in man appear to occur only after infection for prolonged periods of time (e.g., 30 years or more) at a time when the virus infection has largely subsided to a low level of virus replication so that frequently only very low titers of incomplete HBsAg forms and no complete viruses remain in the blood. Cirrhosis is a common but not an indispensible accompanying factor. Woodchucks, on the other hand, develop HCC after a shorter time of infection. They regularly have high concentrations of both complete virus and incomplete surface antigen forms in the blood, and have coexisting active hepatitis but never cirrhosis.

Finally, chronically infected ground squirrels have little or no hepatitis or other liver disease, and they have the highest concentrations of complete virus and incomplete surface antigen forms in the blood. Yet infected animals develop HCC, if at all, at a much lower frequency than do infected woodchucks. The relationship of duck hepatitis B virus (DHBV) to HCC in ducks is not yet clear. The differences in liver disease including HCC associated with infection with these very similar viruses is very intriguing and would appear to offer an opportunity to identify factors which are important in pathogenesis of HCC. Further investigations must consider differences in pathogenicity of the viruses, genetic or other differences in their respective hosts, and environmental factors.

REFERENCES

1. Robinson, W.S., D.N. Clayton, and R.L. Greenman (1974) J. Virol. 14:384.
2. Summers, J., A. O'Connell, and I. Millman (1975) Proc. Natl. Acad. Sci., USA 72:4597.
3. Landers, T., H.B. Greenberg, and W.S. Robinson (1977) J. Virol. 23:368.
4. Hruska, J.F., D.A. Clayton, J.L.R. Rubenstein, and W.S. Robinson (1977) J. Virol. 21:666.
5. Kaplan, P.M., R.L. Greenman, J.L. Gerin, et al. (1973) J. Virol. 11:995.
6. Robinson, W.S., and R.L. Greenman (1974) J. Virol. 13:1231.
7. Murphy, B.L., J.M. Petersen, and J.W. Ebert (1975) Intervirology 6:207.
8. Barker, L.F., and R. Murray (1972) Am. J. Med. Sci. 263:27.
9. Barker, L.F., J.E. Maynard, and R.H. Purcell (1975) J. Infect. Dis. 132:451.
10. Shikata, T., K. Karasawa, T. Abe, et al. (1977) J. Infect. Dis. 136:571.
11. Scullard, G., H.B. Greenberg, J.L. Smith, et al. (1982) Hepatology 2:39.
12. Gocke, D.J. (1975) J. Virol. 36:787.
13. Szmuness, W. (1978) Prog. Med. Virol. 24:40.
14. Summers, J., J.M. Smolec, and R. Snyder (1978) Proc. Natl. Acad. Sci., USA 75:4533.
15. Marion, P.L., L.S. Oshiro, D.C. Regnery, et al. (1980-a) Proc. Natl. Acad. Sci., USA 77:2941.
16. Mason, W.S., G. Seal, and J. Summers (1980) J. Virol. 36:829.
17. Robinson, W.S. (1980) Ann. New York Acad. Sci. 354:371.
18. Marion, P.L., and W.S. Robinson (in press).
19. Beasley, R.P., C.C. Kiu, L.Y. Hwang, et al. (1981) Lancet 2:1129-1133.
20. Larouze, B., W.T. London, B.G. Saimot, et al. (1976) Lancet II:534.
21. Nishioka, K., T. Hirayama, T. Sekine, et al. (1973) Gann. Monogr. Can. Res. 14:167.
22. Steiner, P.E. (1960) Cancer 13:1085.
23. Trichopoulos, D., M. Violaki, L. Sparros, et al. (1975) Lancet II:1038.
24. Peters, R.L. (1976) In Hepatocellular Carcinoma, Wiley, New York, p. 107.
25. Kew, D.M. (1978) In Viral Hepatitis: A Contemporary Assessment of Etiology, Epidemiology, Pathogenesis and Prevention, G.N.

Vyas, et al., eds., Franklin Institute Press, Philadelphia, Pennsylvania, p. 439.
26. Marion, P.L., and W.S. Robinson (submitted for publication).
27. Peters, R.L., A.P. Afroudakis, and D. Tatter (1977) Am. J. Clin. Pathol. 68:1.
28. Lutwick, L.I., and W.S. Robinson (1977) J. Virol. 21:96.
29. Summers, J., A. O'Connell, P. Maupas, et al. (1978) J. Med. Virol. 2:207.
30. Southern, E.M. (1975) J. Mol. Biol. 98:503.
31. Marion, P.L., F.H. Salazar, J.J. Alexander et al. (1980) J. Virol. 32:796.
32. Chakraborty, P.R., N. Rviz-Opazo, D. Shouval, et al. (1980) Nature, London 286:531.
33. Brechot, C., C. Pourcel, A. Louise, et al. (1980) Nature, London 286:533.
34. Edman, J., P. Gray, P. Valenzuela, et al. (1980) Nature, London 286:535.
35. Twist, E.M., H.F. Clar, D.P. Aden, et al. (1980) J. Virol. 32:796.
36. Brechot, C., M. Hadchouel, J. Scotto, et al. (1981) Proc. Natl. Acad. Sci., USA 78:3906.
37. Koshy, R., P. Maupas, R. Muller, et al. (1981) J. Gen. Virol. 57:95.
38. Shafritz, D., D. Shouval, H. Sherman, et al. (1981) N. Engl. J. Med. 305:1067.
39. Brechot, C., C. Pourcel, M. Hadchouel, et al. (1982) Hepatology 2:27s.
40. Shafritz, D. (1981) Hepatology 2:365s.
41. Chen, D., B. Hoyer, J. Nelson, et al. (1981) Hepatology 2:42s.
42. Mitamura, K., B. Hoyer, A. Ponzetto, et al. (1982) Hepatology 2:47s.
43. Ogston, C.W., G.J. Jonak, G.B. Tyler, et al. (1981) In Proceedings of the 1981 Symposium on Viral Hepatitis, H. Alter, J. Maynard, and W. Szmuness, eds., pp. 809-810.
44. Marion, P.L., W.S. Robinson, C.E. Rogler, et al. (1982) J. Cell. Biochem., Suppl. 6, p. 203.
45. Macnab, G.M., J.J. Alexander, G. Lecatsas, et al. (1976) Br. J. Cancer 34:509.
46. Marion, P.L., F.H. Salazar, J.J. Alexander, et al. (1979) J. Virol. 32:796.
47. Miller, R., and W.S. Robinson (1983) Proc. Natl. Acad. Sci., USA 80:2534.
48. Sattler, F., and W.S. Robinson (1979) J. Virol. 32:226.
49. Summers, J., W.S. Mason (1982) Cell 29:403.
50. Robinson, H.L., S.M. Astrin, A.M. Senior, et al. (1981) J. Virol. 40:745.
51. Hayward, W., B.G. Neel, and S. Astrin (1981) Nature 290:475.

ROUNDTABLE DISCUSSION

RESEARCH NEEDS ON THE USE OF TUMOR MARKERS IN THE IDENTIFICATION OF CHEMICAL CARCINOGENS

Stewart Sell, Chairman
University of Texas Medical School, Houston
Discussants: Hisashi Shinozuka, University of Pittsburgh School of Medicine; Van R. Potter, McArdle Laboratory - University of Wisconsin Medical School; Edward A. Smuckler, University of California, San Francisco; Arthur C. Upton, New York University Medical Center; Emmanuel E. Farber, Banting Institute - University of Toronoto; Gary M. Williams, American Health Foundation; and Joseph DiPaolo, National Cancer Institute

STEWART SELL: I can't resist starting out with a "Murphy's Law," and this one is: If a problem causes many meetings, the meetings become more important than the problem. The success of a committee is really measured by the reuslt of its effort. The moose, for instance, is an animal designed by a committee; it's too big, it has to kneel down to eat, it spends a lot of time up to its ears in deep water, it has a number of nonfunctional appendages, and it requires government protection for survival.

There are a number of very fundamental questions that have come up during the meeting: How do chemical carcinogens work? What is the importance of binding to DNA, and of DNA repair? What is the role of altering the environment? Is cancer a property of the cell or of the environment, as in the teratocarcinoma model of Pierce, Stevens, Mintz, and Illmintz, et al. when the expression of the "cancer phenotype" depends upon the environment of "normal" cells? What is the nature of preneoplastic lesions?

Although these questions are important, that's really not the major topic of this meeting, which is rather to find out how to determine what compounds are carcinogenic and how we can identify them so that, presumably, we can control the environmental exposure.

Thinking about a simple way to do this, our distinguished colleague, Dr. George Teebor, remarked that the first thing to do is to look at the structure of the compound to see if it is carcinogenic because it probably has certain structural features which carcinogenic compounds have in common. Therefore, one of the ways we can determine whether a substance or agent is carcinogenic is by what it looks like.

Another very important test, which was discussed during the first day, is mutagenicity. There are various assays that measure mutagenicity but some of us don't seem to be very happy with mutagenicity alone. One must conclude, I think, that all mutagens are bad. Whether they are carcinogenic or not may in fact be irrelevant to the problem. If a substance is positive in an assay for mutagenicity it should be looked at very seriously as something that should be controlled. But is this enough? Are all chemical carcinogenic agents mutagenic? Are there some that aren't that we should worry about? Do we need to go to animal assays or are _in vitro_ assays satisfactory? And, if we do need animal assays, how should they be done? At the present state of the art, animal bioassays are dependent on morphologic analysis. There is considerable disagreement on the significance of the changes that are seen. How do we determine by morphologic analysis whether something is carcinogenic or not, particularly if we evaluate the morphologic changes before actual cancer appears?

Are there markers that can be used instead of morphologic analysis? Will, for instance, immunodetection of altered DNA or adducts be accepted in the future? Certainly a very interesting aspect of the beginning of this kind of approach was presented today.

Should we take a stand on promoters in the environment? How do we detect promoters? Do we have to go to complicated systems using animals for detecting promoters?

This discussion is open to the audience and anyone who has anything to say should feel free to do so. Who would like to comment on the first question: How much can we rely on the basis of the chemical structure of a carcinogen. Can we eliminate some compounds from any additional testing simply on the basis of chemical structure?

ELIZABETH K. WEISBURGER: There are inherent pitfalls in examining a chemical structure to determine whether a compound is a carcinogen. One can always be fooled. As time goes on and we learn more and more, new structures pop up and we find carcinogens that we didn't know about before. I can give you a very specific example: ethylene dibromide. It is a simple aliphatic compound and it shouldn't be doing anything, but yet, it is also a potent carcinogen, by inhalation at levels below the threshold limit value. Another compound, the structure of which suggests that it shouldn't be a carcinogen, is diethylhexyl phthalate which turned out to be carcinogenic. On the basis of all the structural and activity groups that we knew before, it should not have been so.

SELL: I think it's quite true that you can't rule all chemicals out by structure, but if a chemical has a certain configuration in its structure, would that be enough?

LIONEL POIRIER: I would like to be a little bit more optimistic than Dr. Weisburger. I think a chemical carcinogen can fall into certain specific classes and we can get a certain measure of predictability, especially, for example, with direct acting alkylating agents. I think they would have a very high probability of being mutagenic. There are exceptions, of course. The number of chemicals that have been shown to be carcinogens over the years has been increasing in part because the chemicals have been selected on the basis of their structure, trying to select those that were more likely to be dangerous.

GARY M. WILLIAMS: I would agree with both of the previous two discussants. Concerning the structure-activity relations to which Dr. Poirier was calling attention to, it seems that within a structural class of compounds that contains a carcinogen that operates as an electrophilic reactant, it is possible now to recognize in the structure of those compounds the site that gives rise to the electrophilicity of the agent if it is direct acting, as in the case of alkylating agents or following metabolism. But the kind of exceptions that Dr. Elizabeth Weisburger pointed out, such as the diethyl hexyl phthalate, possibly represent what John Weisburger and I have suggested are carcinogens that don't operate through the formation of an electrophilic reactant. For such chemicals, the predictivity is almost zero.

As Dr. Weisburger pointed out, there have been more and more surprises in carcinogenicity tests, now that chemicals are tested, not mainly on the basis of their structural analogies to known carcinogens, but on the basis of environmental importance. I think we will continue to be confronted with new chemicals that are tumorigenic through such indirect mechanisms.

EMMANUEL FARBER: I don't understand; what do you mean by "indirect" mechanisms? We have a thing called cancer; what's an "indirect" as compared to a "direct" mechanism?

WILLIAMS: I was referring to the fact that certain carcinogens clearly operate as electrophilic reactants. It is still not certain whether this is in fact the basis for their carcinogenicity, but within those classes of chemicals one can predict with reasonable accuracy whether structurally related compounds will also be carcinogenic. For some carcinogenic agents, however, there is at present no evidence that they form an electrophilic reactant; this is the type that I referred to as being indirectly carcinogenic, possibly operating through a promoting mechanism or other biological effect.

FARBER: If you don't know the mechanism of how a carcinogen acts, how can you know if it is direct or indirect?

WILLIAMS: For chemicals that operate as electrophilic reactants, we recognize an important biological effect: their ability to interact with DNA. This effect has not been established for those other chemicals that nevertheless are carcinogenic in animal studies. But what has been shown for many of these other chemicals, such as saccharin, is that they are capable of exerting a tumor promoting effect. Certain hypolipedimic agents are peroxisome proliferating agents that as a result of chronic toxicity can lead to genetic changes through indirect mechanisms such as aberrant methylation. There are others that generate active oxygen species as a result of chronic toxicity. And I think there is now emerging quite a spectrum of other biological effects that can explain the carcinogenicity of chemicals without a direct attack on genetic material by the agent itself. These types of effects are what I was referring to as "indirect."

SELL: To get back to the question at hand, it seems that there are some chemicals that one can be pretty sure are carcinogenic from their structure, and there are others that aren't. So, let us move on to mutagenicity. We have heard that mutagenicity is a fairly good predictor, but there is disagreement on this. Is mutagenicity alone enough to demand that a chemical be controlled?

EDWARD A. SMUCKLER: I think that perhaps with one exception I am the only one here who has been in a regulatory agency that attempts to determine whether or not you will register any one of several chemicals that we have discussed. It is nice to feel assured that a material may indeed not cause a neoplastic growth. I don't mean to take umbrage with my two colleagues, but "neoplasm" is probably a better term than "tumorigenic" or "cancer" in this case because for a large part we don't know whether some of these compounds are selected because they are biologically active molecules. You wouldn't use them if they didn't do anything. So, if you use a biologically active molecule, it, by its very nature, can be expected to produce a response in a variety of life forms, including in the lower life forms.

The question then is how do you determine whether or not there is a safety margin? I need only remind you that almost all of the chemotherapeutic agents used for treating cancer are carcinogens. And, in fact, the incidence of second tumors in the young people who are successfully cured the first time is not insignificant. This being the case, I would ask the question in a different way: how do you evaluate the relevant level of safety?

SELL: We have heard statements at this meeting that any level, any dose, that is positive in a given test is significant. Would anybody like to discuss this further? For instance, should we do away with fluorescent lights, since fluorescent light is mutagenic?

GEORGE W. TEEBOR: I don't think you have to do away with fluorescent lighting and I have no experience with it as a mutagen.

SMUCKLER: But if an agent is mutagenic, then it should be suspect. The English did a very interesting experiment not so long ago. They found that the Celts were a rowdy lot, and they decided it was better not to have them on the island, so they shipped them off to Australia. As I am sure you are aware, there is a virtual epidemic of melanoma in Australia among those fair-haired people. Now, should we be so rigorous to say that anything that produces a mutational event or causes a cancer should be registered as such? Those of us from California would feel very ill at ease if you attempted to register sunlight!

TEEBOR: I don't think that the issue is so much a question of registration, but I point out that if some substance is entering the marketplace and is shown to be mutagenic, that should be sufficient to warrant concern. I am not recommending universal elimination of all chemicals.

JOSEPH DI PAOLO: There is a tendency to fall back on the DNA model for carcinogenesis because it is the only established molecularly defined mechanism for determining heredity. We do not know the precise event responsible for cellular instability leading to cell autonomy and cancer.

There is a real possibility that there is more than one pathway to cancer such that there is no obligatory step in carcinogenesis. The relationship between the gene and the rest of the cell is not unidirectional. Just as important to cancer is how the cell responds to a carcinogen. The mammalian cell represents the minimal target for the carcinogen because of its importance as a reproductive and functional unit of life. Therefore, it is important to think of the cell's response as results in cancer rather than just being a carcinogen, being mutagenic, producing sister chromatid exchange or causing genotoxicity, whatever that word means. I think this is where we are losing ground just because it is easier to do a microbial or mammalian mutation test.

The question is, are these assays teaching us anything about the process of carcinogenesis? Are these relevant to what is happening in humans? This is where we should be trying to find the biological markers.

In terms of this particular question, some results may be no more relevant to carcinogenesis than the correlation between lung cancer and the number of car accidents that people have at 42nd and Broadway. I ask you people to think more about the biology of the cell as a whole.

SELL: Did you mean to imply we should do carcinogenic testing on humans?

DI PAOLO: Yes. I think we should realize that we are finding more and more potential carcinogens. More time and money should be spent thinking about ways of testing these in human cells. A dilemma exists in which the number of known potential human carcinogens that should be tested increases while the utilization of human diploid cells to dissect the process remains undeveloped. The question is, what are the differences between human and rodent cells that account for the greater facility in transforming rodent cells? After all, extrapolation from experimental carcinogenesis to humans would be more logical if human rather than diploid/aneuploid animal cells were utilized.

SELL: If a carcinogen in an animal or in an animal cell, other than humans, produced cancer, but did not cause transformation of human cells, would you feel that it would not be worth worrying about?

DI PAOLO: I think it is worth worrying about, but I think what we should be doing is asking the questions: Why do human cells not transform? What is the difference between the control mechanisms, the feedback mechanisms in the human cell, and the animal cells?

Look at repair, for example, in the hamster and the human cell. We can ultraviolet irradiate hamster cells and get the same survival as with human cells, yet we find that the human cell will repair pyrimidine dimers much more efficiently than hamster cells. Pyrimidine dimers last three to four times longer in hamster cells than they do in the human cells. There are beautiful transformations in the hamster cells and we get perhaps a transformation, and I say perhaps a transformation in human cells because of a difference in definition. Why? We know that ultraviolet is a human carcinogen. I think we should determine what the differences are in feedback mechanism and the heterozygosity of the human. We should put our money into learning why we are so inept in culturing human cells before we jump to broader subjects and go on screening.

VAN R. POTTER: As to other kinds of attempts, I think it is important to bring up the general subject of intercellular communication because in a higher organism, the whole hierarchy is knitted together by means of negative and positive feedbacks between the cells of different stages. I particularly call attention to the work of Dr. Jim Trosko at Michigan State who has shown that various compounds known as promoters have been effective in blocking intercellular communication, and I understand that he has now done this with human cells. So these compounds that block intercellular communication may include the compounds that are not mutagens. We might be rather cautious in calling them carcinogens if their sole action is to modify the action of other compounds that are carcinogens.

SELL: Alright, we have one vote for additional tests for suspected agents. The task is to test mutagens as possible interrupters of cell communication. Is that easy to do?

POTTER: Yes.

HISASHI SHINOZUKA: I would like to make one comment about testing compounds for metabolic cooperation. A number of compounds have been tested and, as usual, as the number of compounds tested increased there was always some exception. So, there are now many compounds which do not show a positive effect in the test; nevertheless, biologically they are promoters. So, one has to exercise a great deal of caution in applying this test for screening tumor promoters.

The other question is whether two-stage carcinogenesis is applicable to mesenchymal tumor induction. We lack data on two-stage mechanisms in fibrosarcoma, lymphoid, or hematopoietic neoplasms. I don't think fibroblasts or lymphoid cells have the gap junctions for intercellular communication.

FARBER: I must say that I agree very much with Dr. Shinozuka. At least the Ames test is based upon solid scientific data on the metabolism of carcinogens. However, there is no logical bridge between cellular communication and promotion. No one really understands the mechanism and, therefore, it is purely a blind assay - it so happens that a few chemicals happen to fit. To me, it is not a very rational approach to use Chinese hamster ovary cells, a system which has no obvious relevance to the promotion phenomenon as one sees *in vivo*, and then to conclude that these different few chemicals happen to fit, therefore this is a good test for promotion.

It may be a good test for a few promoters, but it certainly has no relevance as far as one can tell now to the phenomenon of promotion. I do not think that one, therefore, should use this until there is a lot more testing done with many hundreds of compounds which, as far as I can see, has not been done.

WILLIAMS: We now have two issues circulating here. As regards the first one of mutagenicity, I think it has to be recognized that there are, in the application of a specific test such as the Ames test, definite pitfalls. Exceptions are now recognized to the correlation with carcinogenicity, such as the naturally occurring flavanoids, Kenbrough's quercetin, and kaempferal. Because these were mutagenic in the Ames test, a rather great deal of concern arose about these natural constituents of foods, and Sugimura's group tested them for carcinogenicity in rats, mice, and hamsters at doses up to 10% in the diet, but did not obtain a clear oncogenic effect. So such tests have the problem of false positives.

I think an important piece of research that needs to be done in this test system is to study some select compounds that have shown a positive response but have not been active _in vivo_, to show that such compounds actually formed covalent adducts in DNA. The molecular basis of mutagenesis is a change in DNA, and if the test is really measuring that effect, it should be possible to measure the formation of covalent adducts in the DNA of these cells. To my knowledge this has not been done with any of these exceptions to the correlation between mutagenicity and carcinogenicity.

The other problem arises with false negatives where, in our institute, there has been a great deal of experience with these because the Ames test was used to try to screen for carcinogenic components in cigarette smoke. After several years of experience by Dr. Hoffman's group doing parallel skin painting along with the mutagenicity tests, a large number of false negatives were encountered. These agents were not mutagenic but were post-initiating agents.

Regarding metabolic cooperation, I think there is a sound rationale for that approach. The presumed relationship between the _in vitro_ effect and promotion _in vivo_, is that the initiated cells are in the latent or dormant stage, held in control by surrounding cells through feedback mechanisms that Dr. Potter referred to, and what the promoting agents that operate at this stage do is to release these dormant cells from the control of surrounding cells by inhibition of intercellular communication.

POTTER: I would like to extend my remarks in that connection. The studies on intercellular communication are still in a very early phase. What's needed are cell systems in which the more mature cells in a lineage that produces the negative feedback are co-cultivated with the progenitor cells which have the receptors for control factors, and then going into that kind of system and testing various compounds that may or may not be promoters. Let's remember that work on intercellular communication is still in its early phase. But, under the general umbrella of intercellular communication is where some very important problems lie.

FARBER: I would like to explore further the point brought up earlier by Dr. Williams, that it is established that there is this latent cell that suddenly becomes malignant. That is one hypothesis for which there are no data, at least as far as I know. There is no evidence that an initiated cell, even if you perturb it, take it out of its context, or put it in tissue culture, develops suddenly into a cancer cell. It takes a long, long time in skin, or liver, or many other organs after initiation in order to have cells grow. So, this is pure figment of the imagination to say that, post-initiation, somehow these cells are kept under some control and therefore intercellular communication is responsible for this. Granted, I agree with Van Potter that this is an exciting area for study. But I don't think it has application to any regulatory agency at this stage and I think that if EPA or anybody else goes this route, they're making a very big mistake, because they are miles ahead of the science.

DI PAOLO: I don't think one has to wait to do the experiments you're suggesting. I think a good cell biologist could clone the target cells and determine directly on them whether or not promotion takes place and this is exactly what's been done. It's possible to seed cells for cloning experiments, add an initiating dose of a carcinogen or a chemical such as perylene - which is only an initiator - wait several days, add the promoting agent and determine the enhanced transformation due to the promoting agent. It has nothing to do with the surrounding cells or releasing other types of cells. In considering how you can reduce cancer frequencies, you can either do it by eliminating all initiating agents or promoting agents, but I think at the beginning you have to decide what you want to work on: is it promotion or is it initiation? In the final analysis, all ways of causing cancer are equal.

ERLANGER: I'm not an expert in carcinogenesis, but I would like to pose a question. Dr. Williams said that tumorigenic agents produce adducts with DNA. Isn't it also possible that tumorigenic agents can react with regulatory proteins which, when modified, cause aberrant regulation of DNA synthesis leading to irreversible, inheritable changes in a cell? In other words, couldn't a compound be a carcinogen without reacting with DNA?

SMUCKLER: Yes, of course. What is the hard data to show that mutational events in Salmonella are related to adduction to the DNA of bacteria?

SELL: Let's explore this further. What is the evidence that the mutagenic effect is is in fact related to binding of the carcinogen to DNA?

MC CORMICK: It seems to me that there are not a lot of data that show covalent binding of the carcinogen to DNA with tritiated compounds bound to DNA. But that's implicit in all of the studies, and certainly in the mutagenicity studies in mammalian cells, where it really has been shown that you have an adduct and the adduct frequency relates to mutagenicity frequency. So, I don't think that there's really much question about that. But Dr. Williams' statement was that there was something to the fact that you had to show the covalent binding of compounds like the quercetins to DNA.

WILLIAMS: I said that it would be interesting to follow-up some of these exceptions.

MC CORMICK: The implication was that all mutagens work through covalent binding and I don't believe that's true. Some of the ICR compounds are very strong frame shift mutagens in mammalian cells and in Salmonella, but show covalent binding to DNA. Most mutagens covalently bind to DNA, but there are certainly some very clear exceptions to that rule and I don't think we can overlook that.

WILLIAMS: Do the ICR compounds bind covalently and intercollate?

MC CORMICK: The ICR compounds intercollate into DNA and then covalently bind. However, the mutation frequencies seen do not correlate with the covalent binding, but rather with the degree of intercollation for the agents that are strong frame-shift mutagens.

ERLANGER: What I had in mind was something, for example, that subverts the editing mechanism or causes inaccurate repair. There are all kinds of regulatory proteins that are involved in the regulation of replication. The binding of a tumorigenic compound to one of the proteins could cause changes in the replication of DNA, leading to cell transformation. This would be a circumstance in which DNA is not the direct target.

FARBER: How do you go from there to the problem of inheritance of the property of cancer in cancer cells?

ERLANGER: The same way that you would if you produced mispairing as a result of a direct reaction with a base in DNA. You produce mispairing by some other type of mechanism - poor editing could be one, but I don't have any specific ideas. For example, a long time ago, it was shown that UV irradiation can affect enzyme activity. There are many studies on the effects of UV irradiation on the active site of trypsin, chymotrypsin, all kinds of enzymes. Nobody thought at all about looking at the DNA in those days. Now we seem to ignore the effect of environmental factors on proteins.

SMUCKLER: I'd like to second what Dr. Erlanger said. There is indeed a fair amount of evidence for nongenetical change. Changing phenotypes of cells during development and differentiation are classical examples. There are gathering data for deletion of genetic information in dermal differentiation, but there's not an overwhelming case for it. To assume that regulation in mammalian systems occurs only because of an immutable code calls into question the identification of transposable elements. As soon as you have that kind of a situation, you've got another regulatory system entirely. As far as the mutagenesis system is concerned, I would agree with Justin McCormick, that it's true, with frameshift mutants, there is no need for direct reaction. In fact, there's a gathering data base by Bruce Ames to show that a large number of these compounds have no interaction with the DNA itself and still produce not only frameshift mutants but even more significant ones. Our ignorance is what is hurting us more than anything else.

SELL: I think we'd better move on. I don't think there is any doubt that the mutagenicity tests are extremely useful. The subject of this discussion is really supposed to be markers. I would like to include not only markers for animal assays but also *in vitro* markers as well as the question of animal bioassays and how they should be done. Should they continue to be done by morphologic analysis, or are there other things that can be done? Everyone thinks of their own system, but I would say that except for one pertinent study on WY14643, every heptocarcinogen that we have looked at has produced a sustained elevation of alphafetoprotein. Is such a marker better than morphology? Are there other markers? Does anybody have a comment about the way that animal bioassays are done? What can we use beside morphology?

ISAAC WITZ: I think it is very appropriate that you pose this question about new ways of assaying markers, and I would like to kick off my comments by mentioning something that Dr. Farber said today, namely that there is a big difference between exposure and risk.

One of the things that we haven't heard at all, or haven't alluded to whatsoever, are markers for risk. Not working on carcinogenesis, I obtained results which seemed very bizarre to me. If you expose animals to a subthreshold dose of dimethylbenzanthracene, some animals develop tumors within a given time, some don't. Among those that develop tumors some develop tumors relatively early, some develop tumors relatively late.

Reading that a meeting on markers, was going to take place in Bethesda, I immediately bought a ticket to the United States in order to get some idea on biological markers that will distinguish these different biological responses to the same subthreshold dose of a carcinogen. To my amazement and disappointment, I didn't find the answer. In view of the fact that the roundtable discussion tonight is on suggestions of what to do in the future, I wish to make some.

Being an immunologist I would like to suggest that immunity plays a role in determining the response of these animals to the subthreshold doses of carcinogens. I would like to propose that we get into immune-toxicology in relation to carcinogenesis.

Studies of this kind have been done in the past, but the major data in this area were generated in the early '60s. The conclusion was that certain carcinogens are immunosuppressive, mainly in suppressing primary adaptive immune responses. Nothing much more is known about mechanisms of the immunosuppressive activity of carcinogens or materials that are suspected of being carcinogens. We know so much about cellular sets and subsets of the immune system, about interrelations, about control systems, about regulatory molecules, and responses to regulatory signals in the immune system, but nothing has been done so far to find out if and what carcinogens do to this complex network of circuits. I would urge very much moving in this direction.

SELL: One of the reasons you probably haven't seen much on carcinogenesis and immune suppression is that most carcinogens have little or no effect on the immune system. On the other hand, many of the chemotherapeutic agents which are carcinogenic reduce the immune response. These agents also interfere with proliferation of normal cells, and their effect on the immune system is almost certainly through that mechanism. They wipe out not only lymphoid cells but also the bone marrow, gastrointestinal lining, etc.

As far as a specific effect of carcinogens, as such, on the immune system, there is very little data that carcinogens cause any kind of specific effect on cell-cell interactions or regulatory mechanisms. On the reference to new systems for studying the immune response, one could go through all of this again, but in my opinion it's unlikely that this will be useful. There are selected carcinogens that, given at high doses, will make an animal so sick that they won't respond to an antigen. That's what, unfortunately, clinical immunosuppression really does. It's not specific. The animal is loaded up with so much of a drug that the lymphoid cells can't proliferate and then it's immunosuppressed.

WITZ: I'd like to make a small comment to answer that. Most of the data were generated on adaptive immune response. In fact I know that essentially nothing has been done on what we call natural immunity or natural defense. Natural antibodies, macrophages, natural killer cells, are all considered today (this may change in the future) to constitute the first line of defense against cells transformed to malignancy. From my very limited experience on the effects of certain carcinogens on such systems, there is a profound effect on certain expressions of natural immunity. For instance, with genetically uniform mice that are subjected to a subthreshold dose of urethan, those animals that maintain a relatively high titer of natural antibodies are significantly more prone to develop lung adenomas than animals with low titers. Another carcinogen, DMBA, seems to knock out the bone-marrow residing stem cells of NK cells. The conclusion from these results is very clear. Various carcinogens exert an effect on natural defense mechanisms. We have to start systematic studies to define what materials affect what functions.

SELL: Unfortunately, I am advised that it is time to terminate this roundtable discussion. We have covered a lot of ground from chemical structure through mechanisms of carcinogenesis and mutagenesis to evaluation of risk. In the discussions on mechanisms of action of carcinogens (DNA binding, DNA repair, nongenetic effects of carcinogens, promotion, role of cell-cell interactions, *in vitro* transformation, etc.) we did not really come up with solid practical suggestions for the EPA on how to test for carcinogens. I am sorry we did not have time to discuss the value of the assay of products of transformed or tumor cells, so called "tumor markers," in evaluation of carcinogen exposure. Perhaps as we have time to think of this more pragmatic aspect of our meeting, we will individually be able to help guide the EPA in its search for more efficacious assays.

Finally, the idea of evaluation of risk in contrast to exposure, as emphasized by Dr. Witz, certainly deserves more consideration.

On behalf of Dr. Milman and the Organizing Committee, I would like to thank each of you for your participation in this meeting, and particularly those of you who have contributed to our roundtable discussion. In closing, let me return to Murphey's Meeting Law and say that the problem has certainly not been solved so that we have a good excuse for more meetings.

THE POLYPEPTIDES OF THE CYTOSOL OF HEPATOCYTE NODULES MAY EXHIBIT A UNIQUE ELECTROPHORETIC PATTERN

L.C. Eriksson, R.K. Ho, M.W. Roomi, R.N. Sharma, R.K. Murray, and E. Farber

Departments of Pathology and Biochemistry
University of Toronto
Toronto, Ontario, Canada M5S 1A8

The polypeptides of the cytosols of hepatocyte nodules produced according to the resistant hepatocyte model (RH) were analysed using SDS-PAGE. Approximately 40 polypeptides were separated in the gel system used. The pattern of the nodules (N = 8) showed a series of differences from that of normal rat liver: a marked decrease of the amounts of polypeptides of 12, 26, and 44 Kd; an increase of polypeptides of 22, 27, and 90 Kd; and a major increase of a polypeptide of 21 Kd.

Cytosols from nodules generated by the Peraino model [2-AAF and phenobarbital (PB) (n = 2)], the Pitot model [diethylnitrosamine (DEN) and PB] (n = 3), and the Lombardi-Shinozuka model [DEN and choline-methionine deficient (CD) diet] (n = 5) showed the same pattern of alterations. Liver cytosols from adult rats treated with DEN plus partial hepatectomy (PH) without 2-AAF (n = 2); 2-AAF, PB, and CD (n = 2); PB alone or 3-methylcholanthrene (n = 4); from fetal and neonatal rats (n = 16); and from rats subjected to PH (n = 10) were quite different.

We conclude that nodular cells, independent of their mode of production, may show a pattern of cytosolic polypeptides useful as an early indicator of liver carcinogenesis. Identification of the major cytosolic polypeptides characteristic of hepatocyte nodules is in progress.

RAT ALLOANTIGENS AS CELLULAR MARKERS FOR HEPATOCYTES IN GENOTYPIC MOSAIC LIVERS DURING CHEMICALLY INDUCED HEPATOCARCINOGENESIS

John M. Hunt, Mark T. Buckley, and Brian A. Laishes

Department of Oncology
University of Wisconsin
Madison, Wisconsin 53706

Genotypic mosaic rat livers were constructed by intravenous transplantation of carcinogen-altered F344 donor liver cells into livers of (WF x F344)F_1 host rats. Donor rats were treated with a carcinogenic regimen consisting of diethylnitrosamine (200 mg/kg i.p.), followed by an experimental regimen of dietary 0.02% 2-acetylaminofluorene (AAF) and two-thirds partial hepatectomy (PH) (AAF/PH regimen). Host rats received the AAF/PH regimen in addition to transplanted donor liver cells. Utilizing alloantiserum specific for the WF major histocompatibility complex haplotype, $RT1^u$, 97% of the γ-glutamyltranspeptidase-positive (GT^+) liver colonies detected in cryostat sections of host rat livers 10-13 days after transplantation were shown to be of F344 donor origin (Hunt et al., Cancer Research, 42:227-236, 1982). Hepatocytes isolated from such genotypic mosaic livers were stained in suspension histochemically and with alloantisera by indirect immunofluorescence to localize GT^+ phenotype and fluorescence in individual hepatocytes: 97% of GT^+ hepatocytes were of F344 origin, consistent with the cryostat section results. Hepatocellular carcinomas arising in genotypic mosaic host rat livers 17 months after donor liver cell transplantation are presently being typed with alloantisera to establish the donor or host origin of the tumor cells.

Rat alloantigens have been detected on isolated hepatocytes by Protein A-SRBC rosette formation, complement-mediated cytolysis, affinity chromatography, and by immunoprecipitation following radioiodination. These alloantigens expressed on hepatocytes are being exploited as "immunological handles" to purify donor- and host-origin hepatocytes isolated from genotypic mosaic rat livers. Purification technology is being refined to yield highly purified premalignant hepatocyte populations for biochemical and immunochemical characterization at sequential stages of hepatocarcinogenesis.

METABOLIC CONSIDERATIONS IN THE TOXICITY

AND MUTAGENICITY OF THE NITRONAPHTHALENES

Dale E. Johnson,[1] Carol A. Benkendorf,[2]
and Herbert H. Cornish[3]

[1]International Research and Development Corporation
Mattawan, Michigan 49071

1-Nitronaphthalene (1-NN) is metabolized to the mutagenic and carcinogenic N-hydroxy-1-naphthylamine (N-OH-1-NA) via the microsomal nitroreductase system. Under "normal" metabolic conditions in rodents, 1-NN appears to be detoxified either by conjugation of N-OH-1-NA or via ring hydroxylation. This apparently accounts for its noncarcinogenicity in bioassay studies in rats and mice. When "normal" conditions are exceeded in the rat (single intraperitoneal injections at doses around the LD_{50} value) target organ tissue toxicity occurs in the liver and lungs (centrilobular necrosis and respiratory distress).

The extent of toxicity and the specific organ affected depend on the metabolic state of the animal. Pretreatment with phenobarbital protects against lung toxicity but increases the severity of the hepatotoxicity. Pretreatment with SKF-525A protects against liver toxicity but respiratory distress still occurs. In *in vitro* assays in *S. typhimurium* (TA100-FR50) where nitroreduction is favored, 1-NN is mutagenic. When ring hydroxylation is favored a decrease in mutagenicity is seen.

In addition, cyclohexene oxide inhibition of epoxide hydrase does not significantly increase the mutagenic response. These findings, when taken together, indicate that the major (if not sole) mutagenic pathway in the metabolism of 1-NN is nitroreduction, while ring hydroxylation, presumably through an epoxide intermediate, is responsible for the acute target organ toxicity.

[2]Ford Motor Company, Dearborn, Michigan; [3]School of Public Health, University of Michigan, Ann Arbor, Michigan

In addition, recent evidence suggests that the pulmonary effects are due to metabolism directly in the lung. The close structural analogue, 2-nitronaphthalene (2-NN), which is also metabolized to a carcinogenic, mutagenic N-hydroxy intermediate, does not show a similar picture of toxicity. High doses with or without altered metabolic states did not produce the same target organ effects. This lack of acute target organ toxicity and the reported increased potential of 2-NN for carcinogenic activity suggest different major "detoxification" pathways.

These studies highlight the importance of route of administration, preferred metabolic pathways, the metabolic state of the animal, and metabolic "overload" when correlating the toxicity and mutagenicity of nitroarenes.

DETECTION AND IDENTIFICATION OF COMMON ONCOFETAL ANTIGENS ON *IN VITRO* TRANSFORMED MOUSE FIBROBLASTS

Fook Hai Lee and Charles Heidelberger

University of Southern California
Comprehensive Cancer Center
Los Angeles, California 90033

Syngeneic (C3H/HeN) antisera to two *in vitro* chemically transformed cell lines (DMBA Cl 2 and MCA Cl 15), and to early embryo cells, as well as sera from multiparous C3H mice, have been found to react with a variety of *in vitro* transformed cell lines derived from C3H/10T1/2Cl 8 mouse embryo fibroblasts, using an ^{125}I-protein A binding assay or immunofluorescence. The degree of this antibody reactivity is correlated positively with the ability of the cells to exhibit anchorage-independent growth. Thus it seems that those cells which grow poorly in agarose will have low binding activities with syngeneic antisera or multiparous serum.

Immunoprecipitation analysis of the ^{125}I-labeled membrane extracts of two transformed clones (DMBA Cl 2 and MCA Cl 15) on sodium dodecyl sulfate (SDS)-polyacrylamide gel electrophoresis, using rat antisera against DMBA Cl 2 or MCA Cl 15, and rabbit anti-rat Igs, showed that at least 3 antigens with apparent molecular weights of 87,000, 85,000 and 26,000 daltons are expressed on the two transformed cell lines. A prominent band of 73,000 daltons was detected from immmunoprecipitates of all transformed cells although a very weak line at the same position was also visible from immunoprecipitates of nontransformed 10T1/2 cells. It appears that the amount of this antigen was greatly increased in transformed cells.

These data indicate that common oncofetal antigens are expressed on *in vitro* transformed cells derived from a cloned parental mouse embryo fibroblast line by treatment with physical or chemical carcinogens.

(Supported by a grant from U.S. EPA, No. R808309-01)

INHIBITION OF METABOLIC COOPERATION BETWEEN

CULTURED MAMMALIAN CELLS BY SELECTED TUMOR PROMOTERS,

CHEMICAL CARCINOGENS, AND OTHER COMPOUNDS

A.R. Malcolm,[a] L.J. Mills,[a] and E.J. McKenna [b]

[a] U.S. Environmental Protection Agency
Narragansett, Rhode Island 02882

Inhibition of metabolic cooperation between cells may result in tumor promotion or teratogenesis. Using a cell culture assay designed to measure such inhibition as a function of mutant cell recovery in reconstruction experiments with cocultivated mutant ($HGPRT^-$) and wild-type ($HGPRT^+$) cells, we tested several chemicals for their ability to inhibit metabolic cooperation.

The potent promoter, phorbol myristate acetate, induced almost complete mutant recovery at 1 ng/ml, whereas the weaker promoter, butylated hydroxytoluene, induced as much as a two-fold recovery at 20-25 μg/ml. The suspected promoter, sodium cyclamate, inhibited metabolic cooperation in a dose-dependent fashion at high (1-4 mg/ml) concentrations. Phenol, a promoter for mouse skin, was inactive at 5-50 μg/ml. Three minor metabolites of phenol (catechol, quinol and hydroxyquinol) were active at 1-2 μg/ml. Two of these metabolites (catechol and quinol) are, however, reported to be inactive as promoters for mouse skin.

The complete carcinogen, 3,4-benzo(a)pyrene, was inactive at noncytotoxic concentrations when tested directly, whereas the apparently nonmutagenic but carcinogenic di-(2-ethylhexyl) phthalate and nitrilotriacetic acid showed dose-response activity. Ethanol was inactive at concentrations as high as 5 mg/ml. Dimethylsulfoxide

[b] Department of Pharmacology and Toxicology, College of Pharmacy, University of Rhode Island, Kingston, Rhode Island 02881

inhibited metabolic cooperation at 5 mg/ml, but was inactive at concentrations of 2.5 mg/ml or less. The monochlorinated biphenyl, 4-chlorobiphenyl, also inhibited metabolic cooperation at noncytotoxic concentrations (10-15 μg/ml).

The assay appears promising as a screening test for some promoters and teratogens, but additional studies are required to better define assay specificity for these toxicants.

A RAPID *IN VITRO* ASSAY FOR DETECTING PHORBOL ESTER-LIKE TUMOR PROMOTERS

Patrick Moore and John F. Ash

Anatomy Department
University of Utah
Salt Lake City, Utah 84132

We are developing a rapid *in vitro* assay for detecting compounds with phorbol ester-like tumor promotion activity. This assay is based on the potentiation of the cytotoxic effect of ouabain by phorbol esters such as Tetra-O-decanoyl, 13-phorbol acetate (TPA) and 3-O-tetradecanoyl ingenol (3TI). Ouabain is a specific inhibitor of the Na^+, K^+ -ATPase, a plasma membrane enzyme responsible for the active maintenance of the sodium and potassium gradients across the cell membrane.

We were investigating the use of TPA to increase the frequency of HeLa cells surviving ouabain selection through amplification of the gene coding for the Na^+, K^+ -ATPase (Varshavsky, 1981). Rather than increasing the frequency of survivors, we found that TPA increased the lethal effect of ouabain when both agents were present in culture medium at low concentrations (ca. 10^{-8} M ouabain, ca. 10 ng/ml TPA or 3TI). TPA induces a rapid increase in $^{86}Rb^+$ influx into cells which is ouabain inhibitable (Moroney, et al., 1978), indicating an interaction of the tumor promoter with the Na^+, K^+ -ATPase. This interaction could simply be due to the production of a sodium leak into cells, similar to the action of vasopressin (Dicker and Rozengurt, 1979), or due indirectly to one of the many alterations of membrane activities initiated by phorbol esters (Weinstein, et al., 1982). In either case, this phenomenon could form the basis for the detection of compounds which, like the phorbol esters, potentiate ouabain killing of cultured cells.

Compounds are screened for their ability to reduce the concentration of ouabain required to inhibit the growth of a culture of HeLa cells by 50%. Tumor promotors such as TPA or 3TI are active in reducing the required ouabain concentration, while nonpromoting phorbol esters have little or no effect. We are presently optimizing this assay and expanding the range of compounds tested. This assay has the potential to complement the assays based on "metabolic cooperation" for the detection of tumor promoters (Trosko, et al., 1982; Murray, Fitzgerald and Guy, 1982) by detecting a different *in vitro* activity of the promoters. In addition, it may prove to be a more rapid and convenient initial assay for mass screening.

Varshavsky, A. (1982) *Cell* 25:561.

Moroney, J., A. Smith, L.D. Tomai, and C.E. Wenner (1978) *J. Cell Physiol.* 95:287.

Dicker, P., and E. Rozengurt (1979) In *Tumor Cell Surfaces and Malignancy*, R.O. Hynes and C.R. Fox, eds. A.R. Liss, New York, p. 527.

Weinstein, I.B., A.D. Horowitz, R.A. Mufson, P.B. Fisher, V. Ivanoic, and E. Greenbaum (1982) In *Carcinogenesis* Vol. 7, E. Hecker, et al., eds. Raven Press, New York, p. 599.

Trosko, J.E., L.P. Yotti, S.T. Warren, G. Tsushimoto, and C.-C. Chang (1982) IBID, p. 565.

Murray, A.W., D.J. Fitzgerald, and G.R. Guy (1982) IBID, p. 587.

INDUCTION OF ENDOGENOUS MURINE RETROVIRUS BY CHEMICAL CARCINOGENS

Ralph J. Rascati

Department of Biological Sciences
Illinois State University
Normal, Illinois 61761

A variety of assay systems exist to test chemical compounds for carcinogenic potential. Many of these assays use mutation as the observational endpoint. However, the patterns of many cancers, leukemias in particular, suggest that the disease may involve stem cell perturbations which reflect altered gene expression in the absence of point mutation. Few, if any, of the assay systems currently in use assay for alterations in the regulation of gene expression. One possibility for such an assay, however, is to measure the ability of chemical compounds to induce the expression of endogenous murine retrovirus in suitable target cells containing the integrated viral genome in an unexpressed but inducible form.

The target cell chosen for preliminary investigation is the AKR mouse embryo fibroblast, a highly inducible continuous cell line. A variety of known or suspected carcinogens were tested for their ability to induce retrovirus expression. Carcinogens that produce bulky adducts to cellular DNA, such as benzo(a)pyrene, methylcholanthrene, aflatoxin B_1, and acetylaminofluorene, were able to induce virus expression. On the other hand, alkylating agents such as methylmethane sulfonate, ethylmethane sulfonate and ethylnitrosourea were unable to induce virus expression under the conditions used. The relationship between the ability to induce virus expression and the carcinogenic potential of certain chemicals, however, is strengthened by the observation that the 7,8-diol- and the 7,8-diol-9,10-epoxy forms of benzo(a)pyrene are more potent inducers of virus expression than the parent compound. This is predicted from the known carcinogenic reactivities of these three compounds.

It would appear, therefore, that the induction of retrovirus expression can be a useful assay system for potential carcinogens. However, certain limitations exist on the type of compounds that can be identified in this assay system.

IN UTERO INDUCTION OF SISTER CHROMATID EXCHANGES TO DETECT TRANSPLACENTAL CARCINOGENS

R.K. Sharma,[1] M. Lemmon,[1] J. Bakke,[1] I. Galperin,[1] and D. Kram[2]

[1]SRI International
333 Ravenswood Avenue
Menlo Park, California 94025

Analysis of sister chromatid exchanges (SCEs) is proving to be a sensitive technique for detecting mutagens and carcinogens. To assess the transplacental carcinogenic potential of chemicals, we are evaluating SCE induction in developing fetuses exposed *in utero* to known carcinogens.

Pregnant ICR mice were each implanted with a 5-bromodeoxyuridine (BrdU) tablet on Day 13 of gestation and injected i.p. (intraperitoneally) with chemicals 30 to 60 min later. Metaphase chromosomes were prepared from maternal bone marrow and fetal liver cells approximately 23 hr after BrdU implantation. Slides were stained using fluorescence-plus-Giemsa procedures, and SCEs were enumerated in maternal and fetal cells.

The dose-response relationships of three carcinogens -- cyclophosphamide (CP, 5-20 mg/kg), dimethylnitrosamine (DMN, 10-100 mg/kg) and ethylnitrosourea (ENU, 40-80 mg/kg) -- were studied. For each chemical, the number of SCEs in both maternal and fetal cells generally increased in relationship to dose. CP induced 16.88-41.68 SCEs/cell in maternal cells and 15.56-32.90 SCEs/cell in fetal cells, compared with control values of 4.95 and 5.22 SCEs-/cell, respectively. DMN induced 6.27-15.07 SCEs/cell in maternal cells and 5.89-15.10 SCEs/cell in fetal cells, compared with respective control values of 7.44 and 6.30 SCEs/cell. ENU induced 12.92-17.47 SCEs/cell in maternal cells and 11.85-14.60 SCEs/cell in fetal cells, compared with control values of 5.70 and 6.65 SCEs/cell.

---

[2]U.S. Environmental Protection Agency, Washington, DC 20460

At some dose levels of CP and ENU, SCE frequencies in maternal cells were significantly different from those in fetal cells. These findings suggest that these chemicals are capable of inducing SCEs in transplacentally exposed developing fetuses as well as in treated mothers. Thus, SCE analysis in fetuses may be a useful approach for detecting transplacental carcinogens.

EVALUATION OF KNOWN GENOTOXIC AGENTS AND Δ^9-TETRAHYDROCANNABINOL FOR SCE INDUCTION BY THE INTRAPERITONEAL 5-BROMO-2'-DEOXYURIDINE INFUSION TECHNIQUE FOR SISTER CHROMATID EXCHANGE VISUALIZATION

N.P. Singh,[1] A. Turturro,[1] and R.W. Hart[2]

National Center for Toxicological Research
Jefferson, Arkansas 72079

A new technique for measurement of sister chromatid exchange (SCE) induction which utilizes intraperitoneal infusion of 5-bromo-2'-deoxyuridine (BrdU) for SCE visualization was evaluated using 7,12-dimethylbenz(a)anthracene (DMBA), a metabolically activated polyaromatic hydrocarbon with multiple development stage-specific biological effects, and methyl methane sulfonate (MMS), a direct-acting alkylating agent, in a transplacental system. Maternal bone marrow and whole fetus were examined. A recently developed procedure for isolating fetal cells was used which permitted, for the first time, the evaluation of SCE induction in fetal brain as well as other fetal organs.

A dose of 15 mg/kg body weight DMBA, or 20 mg/kg body weight MMS, was delivered intraperitoneally to pregnant CD derived Sprague-Dawley rats at different days of gestation. Both agents resulted in increased SCE levels in mother and fetus and SCE induction in maternal bone marrow was equal throughout pregnancy, while the whole fetus at Day 9, and the Day 13 fetal brain and liver showed elevated levels of SCE which declined in the latter stages of gestation. The DMBA-induced SCE levels in early brain and liver were higher than those caused by MMS, probably because of the activation of DMBA by that organ. Also, MMS-induced SCE levels were higher than those induced by DMBA both in bone marrow and later in fetal development in whole fetus.

The technique was also used to evaluate Δ^9-tetrahydrocannabinol (THC), the major psychoactive ingredient of marihauna. THC was

---

[1]Office of the Director; [2]Director

shown to induce SCE in Day 9 fetus and in Day 13 fetal liver, which is the first demonstration of the ability of this compound to induce SCE.

The results presented here, coupled with previous work showing the stability of BrdU blood levels during SCE induction using this paradigm, indicate that the presented technique is able to reliably detect genotoxic agents *in utero*, especially if the early stages in development, which seems to be especially sensitive to genotoxic agents, are investigated and has utility in distinguishing effects of various compounds on different fetal organs.

EFFECT OF 2,3-OXIDE-3,3,3-TRICHLOROPROPANE ON BaP METABOLISM IN MULLET

M. Srivastava[1] and W.P. Schoor[2]

[1]Chemistry Department
Auburn University
Auburn, Alabama 36849

The effect of 2,3-oxide-3,3,3-trichloropropane (TCPO), a potent inhibitor of epoxide hydrase, on liver microsomes prepared from control, 3MC and PB induced mullet has been investigated. HPLC coupled with fluorescence and UV detectors was used to separate and identify the various metabolites of BaP. TCPO inhibited the formation of 9,10-dihydrodiol and 7,8-dihydrodiol by almost 100%. The 4,5-dihydrodiol was inhibited by 60% for control and 80% for both 3MC and PB induced mullet.

In the presence of TCPO, BaP is primarily metabolized to phenols and quinones. Although the profiles of BaP metabolism were similar in all tested cases, their proportions were substantially altered. The total oxidation of BaP is inhibited by 25% in control mullet and 60% in 3MC and PB induced mullet. While the formation of metabolites oxidized at the 4,5 and 2,3 positions increased by a factor of 2 in the TCPO treated control mullet, the oxidation at the 7,8 and 9,10 positions was increased by 5-fold in the 3MC and PB induced TCPO treated mullet. These results suggest that as the metabolism at the K-region increases in TCPO treated control mullet, the induction of enzymes by 3MC and PB shifts the metabolic route towards the bay region in the presence of TCPO.

---

[2]U.S. Environmental Protection Agency, Gulf Breeze Environmental Research Laboratory, Gulf Breeze, Florida 32561

THE EFFECT OF DIMETHYLBENZANTHRACENE ON THE NK ACTIVITY OF MICE

Isaac P. Witz, Margalit Efrati, Elinor Malatzky,
Lea Shochat, and Rachel Ehrlich

The Department of Microbiology
The George S. Wise Faculty of Life Sciences
Tel Aviv University
Tel Aviv, 69978, Israel

The NK activity of BALB/c mice bearing carcinomas induced by intragastric administration of dimethylbenzanthracene (DMBA) is severely depressed. The aim of the present study was to determine when, in relation to tumor appearance or to DMBA administration, was this depression first demonstrable.

The results showed that three weekly administrations of 1 mg DMBA/week caused a transient but significant decrease in the cellularity of the spleen and in the NK activity mediated by splenocytes. The decrease in the NK activity was selective in the sense that while the number of spleen cells decreased 2-3 fold, the number of lytic units per spleen decreased 6-8 fold. A partial recovery in the cellularity of the spleen and in the NK activity was seen 60-70 days after the last DMBA feeding. In mice treated with 6 mg DMBA there was no spontaneous recovery of the NK activity and the decreased activity persisted until the time that these mice developed tumors.

If NK cells normally restrain the emergence of neoplasms then their suppression by carcinogens may be a contributing factor to the progression of transformed cells. The restoration of carcinogen-induced suppressed NK activity may thus lead towards augmentation of the resistance to tumor development. In order to restore DMBA-induced suppressed NK activity we applied interferon and interferon inducers such as poly I:C or double stranded RNA from Ustilago virus to DMBA-treated mice. The materials augmented the DMBA-induced suppressed NK activity. Repeated administration of these biological response modifiers to DMBA-treated animals maintained a normal level of NK activity in these mice for several weeks. However, after about twenty injections the mice became refractory to further treatments and their NK activity could no longer be augmented by these modulators. Lymphoid cell populations adoptively transferred to DMBA treated animals could not restore the depressed NK activity in these mice.

INDEX